International Series in Operations Research & Management Science

Volume 282

More information about this series at http://www.springer.com/series/6161

H. A. Eiselt • Carl-Louis Sandblom

Nonlinear Optimization

Methods and Applications

 Springer

H. A. Eiselt
Faculty of Management
University of New Brunswick
Fredericton, NB, Canada

Carl-Louis Sandblom
Department of Industrial Engineering
Dalhousie University
Halifax, NS, Canada

ISSN 0884-8289   ISSN 2214-7934   (electronic)
International Series in Operations Research & Management Science
ISBN 978-3-030-19464-2   ISBN 978-3-030-19462-8   (eBook)
https://doi.org/10.1007/978-3-030-19462-8

This Springer imprint is published by the registered company Springer Nature Switzerland AG.
The registered company address is: Gewerbestrasse 11, 6330 Cham, Switzerland

Preface

The theory and methods for solving nonlinear programming problems, in which the variables appear nonlinearly in the objective function and/or the constraints, started making rapid progress around the mid-1900s. This more or less coincided with the emergence of linear programming, i.e., formulations, in which the variables appear in a linear fashion in objective and constraints. Since the fields of linear and nonlinear programming (or optimization, to use a more appropriate, but less used, expression) were developed side by side, it is interesting to discuss the features that set them apart. First, maximizing or minimizing a linear function will only make sense in a constrained setting, since an unconstrained linear function, which is not constant, will have no upper or lower bound, so it will approach infinity. This is in contrast to a nonlinear function, which may have any number of finite maxima and/or minima. Accordingly, nonlinear optimization can be divided into the two categories of unconstrained and constrained optimization, i.e., in the absence or the presence of side constraints, which restrict the values that the variables can take.

Another important difference between linear and nonlinear optimization problems is how the level of difficulty can vary for nonlinear problems, based on the type of nonlinearities involved. Broadly speaking, smooth convex functions are easier to deal with than multimodal (i.e., functions with multiple extrema) or nonsmooth functions. This is reflected in the variety of solution methods we will present in later chapters of this book, arranged in order of increasing levels of difficulty. Another difference between linear and nonlinear optimization is worth pointing out. For the purpose of solving linear programming methods, the simplex method has been the dominant technique since its inception in the late 1940s by G.B. Dantzig, and it has stayed that way to this day. From time to time, rival methods have appeared, most notably the so-called interior point methods developed in the 1980s, but none have seriously threatened the hegemony of the simplex method. In nonlinear programming, the situation is quite different. For unconstrained optimization, the *BFGS* method may be the dominant algorithm for numerical solution, but for constrained optimization there are several methods in use, tailored to the properties of the particular problem at hand. In other words, there is no technique in nonlinear programming that can be used to solve just any arbitrary nonlinear problem as is the case in linear programming.

After an introductory chapter covering some basic concepts, we consider the unconstrained optimization of an objective function with one or several variables, followed by a chapter with several different applications. We then continue with the theory of constrained nonlinear optimization, introducing optimality conditions and duality theory. Next are methods for constrained optimization; a number of different algorithms are described in separate chapters, dealing with problems in increasing order of difficulty. As a rule, each algorithm covered in this book is presented in step form and followed by a numerical example.

As a general remark, we should mention that many theoretical results will not be stated with the weakest possible assumptions. This has the advantage that the exposition can be simplified, allowing us to emphasize the main points and issues with a reasonably uncluttered presentation. Our teaching experience has convinced us that this is a pedagogically much superior approach.

Comparing this book with others in the same field, we have put an emphasis on practical applications of nonlinear programming from a variety of areas, hoping that this will illustrate the versatility and wide applicability of nonlinear programming. In an attempt to provide broad coverage of the field, we have also given more attention to nonconvex optimization, particularly geometric programming. Prerequisites for an understanding of our treatment of the material are a rudimentary knowledge of differential calculus and linear algebra and, though only in a few cases, some probability and statistics.

Finally, this volume marks the completion of a five-volume unified treatment of the field of operations research/management science/mathematical optimization. This work has extended over almost two decades and consists of the following volumes, all published by Springer-Verlag:

- Nonlinear optimization: methods and applications (this volume), 2019
- Operations research: a model-based approach (2nd ed.), 2012
- Linear programming and its applications (2007)
- Decision analysis, location models, and scheduling problems (2004)
- Integer programming and network models (2000)

It remains for us to express our sincere thanks to all those who assisted in the creation of this volume. A big thank you is due to Mr. Rauscher of Springer-Verlag for his patience and Mr. Amboy for his help during the production process. Many thanks are also due to Mr. Simranjeet Singh Chadha, who provided us with many figures of the book. Thanks are also due to our dear friend Vladimir Marianov for his advice regarding technical matters. Additional thanks go out to many colleagues and students who helped with different aspects of this book.

Charters Settlement, Canada   H. A. Eiselt
Mosstorp, Sweden   Carl-Louis Sandblom
February 2019

Contents

Symbols

$\mathbb{N} = \{1, 2, \ldots\}$:	Set of natural numbers
$\mathbb{N}_0 = \{0, 1, 2, \ldots\}$:	Set of natural numbers including zero
$\mathbb{R}$:	Set of real numbers
$\mathbb{R}_+$:	Set of nonnegative real numbers
$\mathbb{R}_>$:	Set of positive real numbers
$\mathbb{R}^n$:	n-dimensional real space
$\in$:	Element of
$\subseteq$:	Subset
$\subset$:	Proper subset
$\cup$:	Union of sets
$\cap$:	Intersection of sets
$\varnothing$:	Empty set
$\rightarrow$:	Implies
$\exists$:	There exists at least one
$\forall$:	For all
$|S|$:	Cardinality of the set S
inf:	Infimum
sup:	Supremum
$x \in [a, b]$:	$a \le x \le b$
$x \in [a, b[$:	$a \le x < b$
$x \in]a, b]$:	$a < x \le b$
$x \in]a, b[$:	$a < x < b$
$|x|$:	Absolute value of x
$\|\bullet\|_p$:	Norm
$a := a + b$:	Valuation, a is replaced by $a + b$
$f'(x)$:	First derivative of the function $f(x)$
$\frac{\partial f(x)}{\partial x}$:	Partial derivative of the function $f(x)$ with respect to x
$\nabla f(x)$:	Gradient of the function $f(x)$

$\mathbf{H}_f$:   Hessian of the function f

$\ln x = \log_e x$:   Natural (or Napierian) logarithm (to the base of e) of x

$\log x = \log_{10} x$:   Common (or Briggsian) logarithm (to the base 10) of x

Chapter 1
Introduction

We begin this chapter with a brief section defining the nonlinear programming problem, introducing some basic terminology, and making some general comments. The second section of this chapter is devoted to a discussion of necessary and sufficient conditions for the existence of local optima of nonlinear functions; first in the one-dimensional case, and then for functions of several variables. The third and final section of this chapter introduces the concept of convexity and relates it to the study of optimization.

1.1 The Nonlinear Programming Problem

Using standard notation, we let $\mathbf{x} = [x_1, x_2, \ldots, x_n]^n \in \mathbb{R}^n$ be an n-dimensional vector of real-valued variables. Furthermore, assume that $f(\mathbf{x})$, $g_1(\mathbf{x})$, $g_2(\mathbf{x})$, $\ldots$, $g_m(\mathbf{x})$ are given real-valued functions, defined on $\mathbb{R}^n$. We can then state the following.

Definition 1.1 The problem

P: Min $z = f(\mathbf{x})$
$g_i(\mathbf{x}) \leq 0$, $i = 1, \ldots, m$

is called a *nonlinear programming problem* with *objective function* $f(\mathbf{x})$, *constraints* $g_i(\mathbf{x}) \leq 0$, $i = 1, \ldots, m$, and *feasible region* $\{\mathbf{x}: g_i(\mathbf{x}) \leq 0, i = 1, \ldots, m\}$.

Note that the above formulation is very general: if a problem is given in a slightly different form, one or more of the following transformations may be applied to establish the desired format.

- Any nonnegativity conditions on the variables such as $x_j \geq 0$, rewritten as $-x_j \leq 0$, can be included in the constraints $g_i(\mathbf{x}) \leq 0$, $i = 1, \ldots, m$.

© Springer Nature Switzerland AG 2019
H. A. Eiselt, C.-L. Sandblom, *Nonlinear Optimization*, International Series in Operations Research & Management Science 282,
https://doi.org/10.1007/978-3-030-19462-8_1

- Any maximization problem can be transformed into an equivalent minimization problem (and vice versa) by using the fact that $\{\text{Max } z = f(\mathbf{x})\} \equiv \{\text{Min } -z = -f(\mathbf{x})\}$.
- A constraint of the type $h(\mathbf{x}) \leq b$ can be alternatively be stated as $g(\mathbf{x}) \leq 0$ where $g(\mathbf{x}) = h(\mathbf{x}) - b$.
- A restriction such as $g(\mathbf{x}) = 0$ can be replaced by the two opposing inequalities $g(\mathbf{x}) \leq 0$ and $- g(\mathbf{x}) \leq 0$. For numerical purposes, we will instead sometimes use the approximation $g(\mathbf{x}) - \varepsilon \leq 0$ and $-g(\mathbf{x}) - \varepsilon \leq 0$ for some small $\varepsilon \leq 0$.
- A *minimax problem* is of the type $\text{Min } z = \max_{k}\{f_k(\mathbf{x})\}$, i.e., z is the point-wise maximum (also called the *upper envelope*) of the functions f_k, $k = 1, \ldots, p$. It can be transformed into a standard minimization formulation by introducing a new variable y and writing the problem as $\text{Min } z = y$, subject to the new constraints $y \geq f_k(\mathbf{x}) \; \forall \; k$ in addition to the constraints that were included in the formulation. The *maximin problem* $\text{Max } z = \min_{k} \{f_k(\mathbf{x})\}$ can be dealt with in analogous fashion. The *minimin problem* $\text{Min } z = \min_{k}\{f_k(\mathbf{x})\}$, where z is the point-wise minimum (the *lower envelope*) of the functions $f_k(\mathbf{x})$ $k = 1, \ldots, p$, can be solved by considering the p separate problems $\text{Min } z_k = \min f_k(\mathbf{x})$, $k = 1, \ldots, p$ and then choosing the solution that results in the smallest value of z_k. The *maximax problem* $\text{Max } z = \max_{k} \{f_k(\mathbf{x})\}$ is again solved in similar fashion.
- Absolute values of the objective function lead to minimax or maximax problems as $|f(\mathbf{x})| = \max \{f(\mathbf{x}), -f(\mathbf{x})\}$. The cases of absolute values in the constraints and/or the variables have been dealt with extensively by Eiselt and Sandblom (2007).

Definition 1.2 If the feasible region of a nonlinear programming problem P is the entire real space, i.e., if $\{\mathbf{x}: g_i (\mathbf{x}) \leq 0, i = 1, \ldots, m\} = \mathbb{R}^n$, then P is said to be an *unconstrained* minimization problem. Otherwise, the minimization problem is *constrained*.

Unconstrained problems represent an important class of problems in spite of their seemingly simpler structure. Optimization techniques for unconstrained nonlinear problems will be presented in the next chapter.

By convention, nonlinear programming problems do not include integrality conditions on the variables. We will follow suit, although the reader should be aware of the possibility of including among the constraints integrality requirements as well. For example, if the variable x is only permitted to assume the value of zero or one, then this can be expressed by including the constraints $-x \leq 0, x - 1 \leq 0$, and $(1 - x) x \leq 0$. Every integer programming problem in which upper bounds u_j for all variables x_j are specified, may in principle be transformed into a nonlinear programming problem by replacing the integrality conditions for all variables by the constraints

$$x_j(x_j - 1)\,(x_j - 2)\ldots(x_j - u_j) = \prod_{k=0}^{u_j}(x_j - k) = 0$$
$$\text{for } j = 1, 2, \ldots, n.$$

This relation makes sure that every feasible solution satisfies the integrality constraints. Some upper bounds may be known since if there are any constraints

$$\sum_{j=1}^{n} a_{ij} x_j \le b_i \text{ with } a_{ij} \ge 0 \,\forall\, j,$$

it follows that $x_j \le \min\limits_{i:a_{ij}>0}\left\{\dfrac{b_i}{a_{ij}}\right\} = u_j$; in other cases we may let $x_j \le M_j$, where M_j is sufficiently large.

Linear programming problems are special cases of nonlinear programming problems, in which all functions $f, g_1, \ldots, g_m$ are linear.

1.2   Necessary and Sufficient Conditions for Local Minima

We begin by reviewing some basic results from classical optimization theory, in particular those which characterize the conditions that must hold at a solution point of an unconstrained minimization problem. Similar treatments can be found in Luenberger and Ye (2008), Bazaraa et al. (2013), Bertsekas (2016), and Cottle and Thapa (2017).

Definition 1.3 The *domain* D_f of a function f is the set of points in $\mathbb{R}^n$ for which f is defined.

To investigate the general unconstrained optimization problem it is necessary to distinguish between two types of solution points: the local and the global.

Definition 1.4 A function f has a *local minimum* at a point $\bar{\mathbf{x}} \in D_f$ if there exists an $\varepsilon > 0$ such that $f(\bar{\mathbf{x}}) \le f(\mathbf{x})$ for all $\mathbf{x} \in D_f$ within a distance ε of $\bar{\mathbf{x}}$ (i.e., $\|\mathbf{x}-\bar{\mathbf{x}}\| < \varepsilon$). If $f(\bar{\mathbf{x}}) < f(\mathbf{x})$ for all $\mathbf{x}$ such that $0 < \|\mathbf{x}-\bar{\mathbf{x}}\| < \varepsilon$, we say that f has a *strict local minimum* at $\bar{\mathbf{x}} \in D_f$.

A strict local minimal point will also be a local minimal point, but the converse is not true. This will be clarified in the example below. Analogous definitions can easily be stated for *local maximal* and *strict local maximal* points.

Example Consider the function f whose graphical representation is shown in Fig. 1.1.

While x_1 and x_5 are strict local minimal points; x_1, x_3, and x_5 are local minimal points, f has strict local maxima at x_2, x_4, and x_6, and local maxima at x_2, x_3, x_4, and x_6.

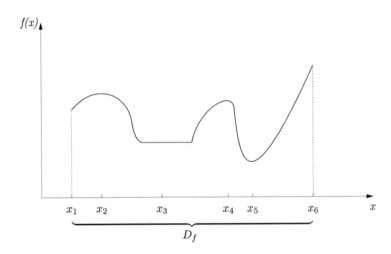

$f(x)$

x_1 x_2 x_3 x_4 x_5 x_6 x

D_f

Fig. 1.1 Local and global maxima and minima of a function

Definition 1.5 A function f has a *global minimum* at a point $\bar{\mathbf{x}} \in D_f$ if $f(\bar{\mathbf{x}}) \leq f(\mathbf{x})$ for all $\bar{\mathbf{x}} \in D_f$. If $f(\bar{\mathbf{x}}) < f(\mathbf{x})$ for all $\mathbf{x} \in D_f$, $\mathbf{x} \neq \bar{\mathbf{x}}$, then $\bar{\mathbf{x}}$ is said to be a *strict global minimum*.

Clearly, a strict global minimal point will also be a global minimal point, but the converse is not true. Again, similar definitions exist for *global maximal* and *strict global maximal* points as well. In the above example, f has a (strict) global minimum at x_5, while x_6 is a strict global maximal point.

We now turn to the linearization of nonlinear functions. For that purpose, we first consider the one-dimensional case and suppose that the function f of one real variable is twice continuously differentiable in the vicinity of a point $\bar{x}$ in the interior of D_f. We can then approximate the function f in the vicinity of $\tilde{x}$, using $f(\tilde{x}), f'(\tilde{x})$, and $f''(\tilde{x})$.

Theorem 1.6 (Linearization) Assume that the function f is continuously differentiable in a vicinity of a point $\tilde{x}$. Then f can be approximated there by the linear function $f_1(x) = f(\tilde{x}) + (x - \tilde{x}) f'(\tilde{x})$. Specifically, $f(x) = f(\tilde{x}) + (x - \tilde{x}) f'(\tilde{x}) + R_1(x) = f_1(x) + R_1(x)$, where the error term $R_1(x)$ is continuous and such that $\lim_{x \to \tilde{x}} \dfrac{R_1(x)}{|x - \tilde{x}|} = 0$.

Another formulation of the approximation that does not involve an error term (but then it is not linear) is $f(x) = f(\tilde{x}) + (x - \tilde{x}) f'(\lambda x + (1 - \lambda)\tilde{x})$, where $0 \leq \lambda \leq 1$, so that $\lambda x + (1 - \lambda)\tilde{x}$ is some linear convex combination of the points x and $\tilde{x}$. Here, the value of λ depends on x and $\tilde{x}$, but we do not specify this linear convex combination any further, all we need to know is that it exists.

Definition 1.7 The expression $f(\tilde{x}) + (x - \tilde{x})f'(\tilde{x})$ in Theorem 1.6 is called the *linearization, linear Taylor formula,* or *first order Taylor formula* for f at the point $\tilde{x}$.

Remark If the function f happens to be linear, then the linearization of f at any point is identical to f itself.

We now take the above development one step further.

Theorem 1.8 (The quadratic Taylor formula) Assume that the function f is twice continuously differentiable in a vicinity of the point $\tilde{x}$. Then f can be approximated there by the quadratic function $f_2(x) = f(\tilde{x}) + (x - \tilde{x})f'(\tilde{x}) + \frac{1}{2}(x - \tilde{x})^2 f''(\tilde{x})$.

Specifically, $f(x) = f(\tilde{x}) + (x - \tilde{x})f'(\tilde{x}) + \frac{1}{2}(x - \tilde{x})^2 f''(\tilde{x}) + R_2(x) = f_2(x) + R_2(x)$

where the error term $R_2(x)$ is continuous and such that $\lim\limits_{x \to \tilde{x}} \dfrac{R_2(x)}{(x - \tilde{x})^2} = 0$.

Another formulation of the approximation that does not involve an error term (but then it is no longer quadratic) is $f(x) = f(\tilde{x}) + (x - \tilde{x})f'(\tilde{x}) + \frac{1}{2}(x - \tilde{x})^2 f''(\lambda x + (1 - \lambda)\tilde{x})$, where $0 \le \lambda \le 1$, so that $\lambda x + (1 - \lambda)\tilde{x}$ is some linear convex combination of the points x and $\tilde{x}$. Here, the value of λ depends on x and $\tilde{x}$, but we do not specify this linear convex combination any further, all we need to know is that it exists.

Definition 1.9 The expression $f(\tilde{x}) + (x - \tilde{x})f'(\tilde{x}) + \frac{1}{2}(x - \tilde{x})^2 f''(\tilde{x})$ in Theorem 1.8 is called the *quadratic Taylor formula* for f at the point $\tilde{x}$.

Remark If the function f happens to be quadratic, then the quadratic Taylor formula for f at any point is identical to f itself.

The following conditions are important for theoretical reasons and in view of the analysis to be carried out in subsequent sections.

Theorem 1.10 (First-order necessary optimality condition) If f has a minimal or a maximal point at $\bar{x}$, then $f'(\bar{x}) = 0$.

Theorem 1.11 (Second-order sufficient optimality conditions) If the point $\bar{x}$ is such that $f'(\bar{x}) = 0$ and $f''(\bar{x}) > 0$ (<0), then f has a strict local minimum (maximum) at $\bar{x}$.

Example Let the function be $f(x) = x^2$. Then $f'(x) = 2x$, $f''(x) = 2$ so that $f'(0) = 0$, $f''(0) = 2$ and therefore f has a local minimum at $x = 0$. The local minimum happens to be global as well in this case.

Secondly, consider the function $f(x) = x^3$. With $f'(x) = 3x^2$ and $f''(x) = 6x$, we find that $f'(0) = f''(0) = 0$, so that the first-order necessary condition for a minimal point is satisfied, but the second-order sufficient condition is not, and the function f does indeed not have a local minimum at $x = 0$. Thirdly, consider the function $f(x) = x^4$ for which $f'(0) = f''(0)$, so that the first-order necessary condition is satisfied, while the second-order sufficient condition is not. Yet, this function has a local (in fact, global) minimum at $x = 0$.

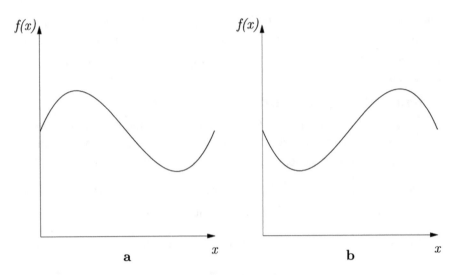

Fig. 1.2 Necessary conditions for sign-restricted optimization

So far we have only considered unconstrained situations. We will now briefly discuss a slight modification and study the problem

P: Min $z = f(x)$
s.t. $x \geq 0$,

i.e., consider sign-restricted optimization. For $x > 0$, the necessary condition for optimality, $f'(x) = 0$, remains, whereas for $x = 0$, $f'(x)$ does not need to vanish at a minimal point. As an example, consider the functions in Fig. 1.2.

In Fig. 1.2a, the function has a local minimal point at $x = 0$ (while $f'(0) > 0$), whereas this is not the case in the function in Fig. 1.2b, where $f'(0) < 0$. We realize that the conditions $x \geq 0$, $f'(x) \geq 0$, and $xf'(x) = 0$ are necessary for a minimal point. Therefore we have

Theorem 1.12 Consider the sign-constrained problem

P: Min $z = f(x)$
s.t. $x \geq 0$.

If problem P has an optimal solution at the point $\bar{x}$, then $\bar{x} \geq 0$, $f'(\bar{x}) \geq 0$, and $\bar{x}f'(\bar{x}) = 0$.

Example Consider the problem

P: Min $z = f(x) = x^3 - 3x^2 + 2x + 1$
s.t. $x \geq 0$.

The graph of f is displayed in Fig. 1.3.

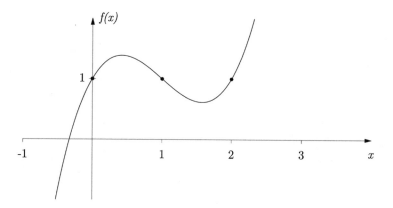

Fig. 1.3 Sign-restricted minimization

With $f'(x) = 3x^2 - 6x + 2$, the necessary optimality conditions for P are $x \geq 0$, $3x^2 - 6x + 2 \geq 0$, and $x(3x^2 - 6x + 2) = 0$. Since $x(3x^2 - 6x + 2) = 3x\left(x - 1 + \frac{1}{\sqrt{3}}\right)\left(x - 1 - \frac{1}{\sqrt{3}}\right)$ we find that the optimality conditions are satisfied for $x_1 = 0$, $x_2 = 1 - \frac{1}{\sqrt{3}} \approx 0.4226$, and $x_3 = 1 + \frac{1}{\sqrt{3}} \approx 1.5774$. Evaluating $f(x)$ at each of these three possible candidates for optimality, we obtain $f(x_1) = 1$, $f(x_2) \approx 1.3849$, and $f(x_3) \approx 0.6151$, so that $\bar{x} = x_3 = 1 + \frac{1}{\sqrt{3}} \approx 1.5774$.

Returning to the general discussion, we point out that the optimality conditions in Theorem 1.12 bear some resemblance to the Karush-Kuhn-Tucker optimality conditions for constrained optimization, which we will discuss in Chap. 4. Actually, the sign-restricted problem P in Theorem 1.12 is a special case of a more general problem to be discussed in Chap. 4. A formal proof of Theorem 1.12 is therefore omitted here.

Let us now consider maximizing the function $g(x) = f(x)/x$ for $x > 0$, where $f(x)$ is a given function. We assume that $f(0) \leq 0$, since otherwise $f(x)/x$ would become unbounded for small positive x. If $f(x)$ is differentiable, then so is $g(x)$ for $x \neq 0$, and we obtain

$$g'(x) = \frac{x f'(x) - f(x)}{x^2},$$

so that, at optimum, $xf'(x) = f(x)$ must hold. Let us now denote by $\hat{x}$ any global maximal point of $f(x)$ and by x^* any global maximal point of $g(x)$. Furthermore, we assume that $\hat{x}$, x^*, $f(\hat{x})$ and $g(x^*)$ are all strictly positive. Therefore we have $f(\hat{x}) \geq f(x)$ for all $x > 0$, implying that $f(\hat{x}) \geq f(x^*)$, as well as $g(x^*) \geq g(x)$, or $f(x^*)/x^* \geq f(x)/x \ \forall \ x > 0$, implying that $f(x^*)/x^* \geq f(\hat{x})/\hat{x}$. It follows that $f(\hat{x})/x^* \geq f(x^*)/x^* \geq f(\hat{x})/\hat{x}$, so that $f(\hat{x})/x^* \geq f(\hat{x})/\hat{x}$, from which we conclude that $x^* \leq \hat{x}$; a result that holds even without requiring $f(x)$ to be differentiable. In

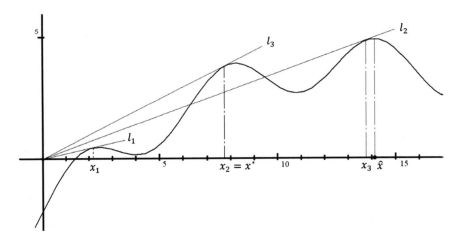

Fig. 1.4 Tangents passing through the origin

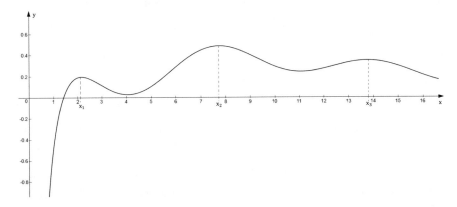

Fig. 1.5 Optima of the sample function

other words, $f(x)/x$ reaches its maximum before $f(x)$ does. Finding x^*, i.e., maximizing $f(x)/x$ also has an interesting geometric interpretation, which is easiest explained by using a numerical example. For that purpose, consider the function $f(x) = -\frac{1}{30}(x-6)^2 + \frac{1}{2}x + \sin x - 1$, which is shown graphically in Fig. 1.4.

Consider also the function $g(x) = f(x)/x$ displayed in Fig. 1.5. In Fig. 1.4, the lines ℓ_1, ℓ_2, and ℓ_3 are tangents to $f(x)$, which pass through the origin, and therefore we must have $x_j f'(x_j) = f(x_j), j = 1, 2, 3$. Now this is precisely the necessary optimality condition for $g(x) = f(x)/x$ to have a maximal point. Since $f'(x_j) = f(x_j)/x_j$, we simply pick the point, for which the slope is highest, i.e., x_2, which corresponds to the tangent with the steepest slope that leads through the origin.

Indeed, in our example $g(x) = f(x)/x$ turns out to have its local maxima at $x_1 \approx 2.1666$ with $g(x_1) \approx 0.1944$, at $x_2 \approx 7.7509$ with $g(x_2) \approx 0.4861$, and at $x_3 \approx 13.7619$ with $g(x_3) \approx 0.3490$, so that $\arg \max (g(x)) = x^* = x_2 \approx 7.7509$ (see

Fig. 1.5). This corresponds to the steepest tangent ℓ_2 in accordance with our discussion above. Furthermore, arg max $(f(x)) = \hat{x} \approx 14.0973$ (see Fig. 1.4), so that we also have $x^* \approx 7.7509 \le 14.0973 \approx \hat{x}$ in agreement with our previously shown inequality $x^* \le \hat{x}$. We should also point out that the "steepest tangent through the origin" principle works without the assumption of differentiability, since the straight lines between the origin and the points $(x, f(x))$ will be at the steepest, when $f(x)/x$ is maximized. All we have to assume is that $f(0) \le 0$, or that we restrict x to be in an interval with a strictly positive left endpoint. Section 3.3 will show an interesting application involving this type of optimization.

We will now see how the results for the one-dimensional case can be generalized to encompass functions of several variables. To do so, consider a real-valued function defined on $\mathbb{R}^n$ and assume that it is twice continuously differentiable.

Definition 1.13 The *gradient* of $f(\mathbf{x})$ denoted by grad $f(\mathbf{x})$ or by $\nabla f(\mathbf{x})$ is the n-dimensional column vector of partial derivatives:

$$\operatorname{grad} f(\mathbf{x}) = \nabla f(\mathbf{x}) = \left[\frac{\partial f}{\partial x_1} \quad \frac{\partial f}{\partial x_2} \quad \cdots \quad \frac{\partial f}{\partial x_n} \right]^T.$$

The gradient $\nabla f(\mathbf{x})$ may exist even if $f(\mathbf{x})$ is not twice continuously differentiable. If $\nabla f(\mathbf{x})$ exists and is a continuous function of $\mathbf{x}$, we say that f is *smooth*.

The gradient (usually meaning incline or slope) of a function is important for several reasons. For instance, if f is a differentiable function in an open set of $\mathbb{R}^n$, then at each point $\mathbf{x}$ of this set for which $\nabla f(\mathbf{x}) \ne \mathbf{0}$, the vector $\nabla f(\mathbf{x})$ points in the direction of maximum increase for f. The number $\|\nabla f(\mathbf{x})\|$ represents the maximum rate of increase. In Sect. 2.2, when discussing the Steepest Descent method, we will see that the directional derivative of a real-valued function f with respect to a unit vector can be written quite easily in terms of the gradient of f.

Example Let $f(x_1, x_2) = e^{x_1 x_2}$. Then $\nabla f(x_1, x_2) = \begin{bmatrix} x_2 \, e^{x_1 x_2} \\ x_1 e^{x_1 x_2} \end{bmatrix}$. It follows that at the point $\mathbf{x}^T = [1, 2]$ the function increases most quickly in the direction $\nabla f(1, 2) = \begin{bmatrix} 2e^2 \\ e^2 \end{bmatrix}$ which has the same direction as the vector $\begin{bmatrix} \dfrac{2}{\sqrt{5}} \\ \dfrac{1}{\sqrt{5}} \end{bmatrix}$ of unit length. The rate of increase in that direction is $\|\nabla f(1, 2)\| = \sqrt{5}e^2$. In the same vein, $\nabla f(-1, 2) = \begin{bmatrix} 2e^{-2} \\ -e^{-2} \end{bmatrix}$ and has direction $\begin{bmatrix} \dfrac{2}{\sqrt{5}} \\ -\dfrac{1}{\sqrt{5}} \end{bmatrix}$ with maximum rate of increase at $(-1, 2)$ equal to $\sqrt{5}e^{-2}$. The maximum rate of decrease occurs in the opposite direction, namely $\begin{bmatrix} -\dfrac{2}{\sqrt{5}} \\ \dfrac{1}{\sqrt{5}} \end{bmatrix}$.

Definition 1.14 The *Hessian matrix* of f, denoted by $\mathbf{H}_f$, is the $[n \times n]$-dimensional matrix of second-order partial derivatives:

$$
\mathbf{H}_f = (h_{ij}) = \left(\frac{\partial^2 f}{\partial x_i \partial x_j} \right) =
\begin{bmatrix}
\dfrac{\partial^2 f}{\partial x_1^2} & \dfrac{\partial^2 f}{\partial x_1 \partial x_2} & \cdots & \dfrac{\partial^2 f}{\partial x_1 \partial x_n} \\[2mm]
\dfrac{\partial^2 f}{\partial x_2 \partial x_1} & \dfrac{\partial^2 f}{\partial x_2^2} & \cdots & \dfrac{\partial^2 f}{\partial x_2 \partial x_n} \\[2mm]
\vdots & \vdots & \ddots & \vdots \\[2mm]
\dfrac{\partial^2 f}{\partial x_n \partial x_1} & \dfrac{\partial^2 f}{\partial x_n \partial x_2} & \cdots & \dfrac{\partial^2 f}{\partial x_n^2}
\end{bmatrix}.
$$

Since $\dfrac{\partial^2 f}{\partial x_i \partial x_j} = \dfrac{\partial^2 f}{\partial x_j \partial x_i}$ it is easily seen that the Hessian matrix is symmetric.

Example Consider the function of three real variables $f(x_1, x_2, x_3) = \dfrac{4}{x_1 x_2 x_3} + 4x_2 x_3 + x_1 x_2 + 2x_1 x_3$. The Hessian matrix is given by

$$
\mathbf{H}_f =
\begin{bmatrix}
\dfrac{8}{x_1^3 x_2 x_3} & \dfrac{4}{x_1^2 x_2^2 x_3} + 1 & \dfrac{4}{x_1^2 x_2 x_3^2} + 2 \\[3mm]
\dfrac{4}{x_1^2 x_2^2 x_3} + 1 & \dfrac{8}{x_1 x_2^3 x_3} & \dfrac{4}{x_1 x_2^2 x_3^2} + 4 \\[3mm]
\dfrac{4}{x_1^2 x_2 x_3^2} + 2 & \dfrac{4}{x_1 x_2^2 x_3^2} + 4 & \dfrac{8}{x_1 x_2 x_3^3}
\end{bmatrix}.
$$

With the aid of the gradient we can approximate f by a linear expression in the vicinity of a given point $\tilde{\mathbf{x}}$.

Theorem 1.15 (Linearization) Assume that the function f is continuously differentiable in a vicinity of the point $\tilde{\mathbf{x}}$. Then f can be approximated there by the linear function $f_1(\mathbf{x}) = f(\tilde{\mathbf{x}}) + (\mathbf{x} - \tilde{\mathbf{x}})^T \nabla f(\tilde{\mathbf{x}})$. Specifically,

$$
f(\mathbf{x}) = f(\tilde{\mathbf{x}}) + (\mathbf{x} - \tilde{\mathbf{x}})^T \nabla f(\tilde{\mathbf{x}}) + R_1(\mathbf{x}),
$$

where the error term $R_1(\mathbf{x})$ is continuous and such that

$$
\lim_{\mathbf{x} \to \tilde{\mathbf{x}}} \frac{R_1(\mathbf{x})}{\|\mathbf{x} - \tilde{\mathbf{x}}\|} = 0.
$$

Another formulation of the approximation which does not involve an error term (but which is then not linear) is

$$f(\mathbf{x}) = f(\tilde{\mathbf{x}}) + (\mathbf{x} - \tilde{\mathbf{x}})\nabla f(\lambda \mathbf{x} + (1 - \lambda)\tilde{\mathbf{x}}),$$

where $0 \le \lambda \le 1$, so that $\lambda \mathbf{x} + (1 - \lambda)\tilde{\mathbf{x}}$ is some linear convex combination of the points $\mathbf{x}$ and $\tilde{\mathbf{x}}$. Here, the value of λ depends on $\mathbf{x}$ and $\tilde{\mathbf{x}}$, but we do not specify this linear convex combination any further; all we need to know is that it exists.

Definition 1.16 The expression $f(\tilde{\mathbf{x}}) + (\mathbf{x} - \tilde{\mathbf{x}})^T \nabla f(\tilde{\mathbf{x}})$ in the formula of Theorem 1.15 is called the *linearization* of f at the point $\tilde{\mathbf{x}}$.

Remark If the function f happens to be linear, then the linearization of f at any point is identical to f itself.

With the aid of the gradient and the Hessian matrix, we can approximate the function f by a quadratic expression in the vicinity of a given point $\tilde{\mathbf{x}}$.

Theorem 1.17 (The quadratic Taylor formula) Assume that the function f is twice continuously differentiable in a vicinity of a point $\tilde{\mathbf{x}}$. Then f can be approximated there by the quadratic function

$$f_2(\mathbf{x}) = f(\tilde{\mathbf{x}}) + (\mathbf{x} - \tilde{\mathbf{x}})^T \nabla f(\tilde{\mathbf{x}}) + \tfrac{1}{2}(\mathbf{x} - \tilde{\mathbf{x}})^T \mathbf{H}_f(\tilde{\mathbf{x}})(\mathbf{x} - \tilde{\mathbf{x}}).$$

Specifically,

$$f(\mathbf{x}) = f(\tilde{\mathbf{x}}) + (\mathbf{x} - \tilde{\mathbf{x}})^T \nabla f(\tilde{\mathbf{x}}) + \tfrac{1}{2}(\mathbf{x} - \tilde{\mathbf{x}})^T \mathbf{H}_f(\tilde{\mathbf{x}})(\mathbf{x} - \tilde{\mathbf{x}})^T$$
$$+ R_2(\mathbf{x}),$$

where the error term $R_2(\mathbf{x})$ is continuous and such that

$$\lim_{\mathbf{x} \to \tilde{\mathbf{x}}} \frac{R_2(\mathbf{x})}{\|\mathbf{x} - \tilde{\mathbf{x}}\|^2} = 0.$$

Another formulation of the approximation, which does not involve an error term is

$$f(\mathbf{x}) = f(\tilde{\mathbf{x}}) + (\mathbf{x} - \tilde{\mathbf{x}})^T \nabla f(\tilde{\mathbf{x}}) + \frac{1}{2}(\mathbf{x} - \tilde{\mathbf{x}})^T \mathbf{H}_f(\lambda \mathbf{x} + (1 - \lambda)\tilde{\mathbf{x}})(\mathbf{x} - \tilde{\mathbf{x}}),$$

where $0 \le \lambda \le 1$, so that $\lambda \mathbf{x} + (1 - \lambda)\tilde{\mathbf{x}}$ is some linear convex combination of the points $\mathbf{x}$ and $\tilde{\mathbf{x}}$. Here, the value of λ depends on $\mathbf{x}$ and $\tilde{\mathbf{x}}$, but we do not specify this linear convex combination any further; all we need to know is that it exists.

Definition 1.18 The expression $f(\tilde{\mathbf{x}}) + (\mathbf{x} - \tilde{\mathbf{x}})^T \nabla f(\tilde{\mathbf{x}}) + \frac{1}{2}(\mathbf{x} - \tilde{\mathbf{x}})^T \mathbf{H}_f(\tilde{\mathbf{x}})(\mathbf{x} - \tilde{\mathbf{x}})$ in Theorem 1.17 is called the *quadratic Taylor formula* for f or the *quadratic approximation* of f at the point $\tilde{\mathbf{x}}$.

Remark If the function f happens to be quadratic, then the quadratic Taylor formula for f at any point is identical to f itself.

Example Let $f(x_1, x_2) = x_1 e^{x_2} - 2x_1 + 3x_2^2$. Then $\nabla f(x_1, x_2) = \begin{bmatrix} e^{x_2} - 2 \\ x_1 e^{x_2} + 6x_2 \end{bmatrix}$ and

$\mathbf{H}_f(x_1, x_2) = \begin{bmatrix} 0 & e^{x_2} \\ e^{x_2} & x_1 e^{x_2} + 6 \end{bmatrix}$. To compute the linearization and the quadratic

Taylor formula for f at the point $\tilde{\mathbf{x}} = [0, 0]^T$, we determine $f(\tilde{\mathbf{x}}) = f(0, 0) = 0$,

$\nabla f(\tilde{\mathbf{x}}) = \nabla f(0, 0) = \begin{bmatrix} -1 \\ 0 \end{bmatrix}$ and $\mathbf{H}_f(\tilde{\mathbf{x}}) = \mathbf{H}_f(0, 0) = \begin{bmatrix} 0 & 1 \\ 1 & 6 \end{bmatrix}$. The linearization

of f at the point $\tilde{\mathbf{x}} = (0, 0)$ is $f(\tilde{\mathbf{x}}) + (\mathbf{x} - \tilde{\mathbf{x}})^T \nabla f(\tilde{\mathbf{x}}) = \mathbf{x}^T \begin{bmatrix} -1 \\ 0 \end{bmatrix} = -x_1$.

The quadratic approximation is

$$f(\tilde{\mathbf{x}}) + (\mathbf{x} - \tilde{\mathbf{x}})^T \nabla f(\tilde{\mathbf{x}}) + \tfrac{1}{2}(\mathbf{x} - \tilde{\mathbf{x}})^T \mathbf{H}_f(\tilde{\mathbf{x}})(\mathbf{x} - \tilde{\mathbf{x}})^T$$
$$= \mathbf{x}^T \begin{bmatrix} -1 \\ 0 \end{bmatrix} + \tfrac{1}{2}\mathbf{x}^T \begin{bmatrix} 0 & 1 \\ 1 & 6 \end{bmatrix} \mathbf{x} = -x_1 + x_1 x_2 + 3x_2^2.$$

Many of the algorithms to be covered in this part of the book will use linear or quadratic approximations of functions.

We will now extend the previous development of necessary and sufficient optimality conditions for functions of one variable to the multidimensional case. However, caution is advised: The behavior of functions of several variables is profoundly different from that of single-variable functions. One should be careful not to accept propositions regarding optimality conditions or other aspects of multivariable functions which may appear intuitively obvious but which are not properly proved. As an example, consider the function $f(x_1 x_2) = (2x_1^2 - x_2)(x_1^2 - 2x_2)$. It can be shown that on each line through the origin, this function (i.e., the single-variable function defined on such a line) will have a local minimum at the point $(x_1, x_2) = (0, 0)$. This would lead us to believe that the function $f(x_1, x_2)$, regarded as a function of two variables, would also have a local minimum at $(x_1, x_2) = (0, 0)$; however, one can show that this is not true.

Theorem 1.19 (First-order necessary optimality condition) If a function f has a minimum or a maximum at a point $\bar{\mathbf{x}}$, then $\nabla f(\bar{\mathbf{x}}) = \mathbf{0}$.

This result is the n-dimensional counterpart of that contained in Theorem 1.10 with the gradient replacing the first-order derivative. In order to formulate second-order optimality conditions, we need to recall a few notions from matrix algebra.

Definition 1.20 A symmetric matrix $\mathbf{A}$ is said to be *positive definite* (*negative definite*) if $\mathbf{x}^T \mathbf{A} \mathbf{x} > 0$ ($\mathbf{x}^T \mathbf{A} \mathbf{x} < 0$) $\forall \mathbf{x} \neq \mathbf{0}$; it is called *positive semidefinite* (*negative semidefinite*) if $\mathbf{x}^T \mathbf{A} \mathbf{x} \geq 0$ ($\mathbf{x}^T \mathbf{A} \mathbf{x} \leq 0$) $\forall \mathbf{x}$. If none of these cases applies, the matrix $\mathbf{A}$ is said to be *indefinite*.

Clearly a positive definite (negative definite) matrix will also be positive semidefinite (negative semidefinite), but the converse is not true.

Before returning to the mainstream of our discussion we will review briefly some useful conditions by which a matrix can be tested for definiteness or semidefiniteness.

Definition 1.21 For a square matrix $\mathbf{A} = (a_{ij})$ of order n, its *leading principal minors* Δ_j are the determinants:

$$\Delta_1 = \det[a_{11}], \Delta_2 = \det\begin{bmatrix} a_{11} & a_{12} \\ a_{21} & a_{22} \end{bmatrix}, \Delta_3 = \det\begin{bmatrix} a_{11} & a_{12} & a_{13} \\ a_{21} & a_{22} & a_{23} \\ a_{31} & a_{32} & a_{33} \end{bmatrix}, \ldots, \Delta_n = \det \mathbf{A}.$$

By deleting from $\mathbf{A}$ rows and columns in matching pairs (i.e., with identical row and column indices), we obtain k-th order determinants D_k, $1 \leq k \leq n$, which are called the *principal minors* of $\mathbf{A}$. For $k < n$, D_k is not unique.

Example Consider the matrix $\mathbf{A} = \begin{bmatrix} 1 & 2 & 3 \\ 4 & 5 & 6 \\ 7 & 8 & 9 \end{bmatrix}$. Its leading principal minors are Δ_1

$= 1, \Delta_2 = \det\begin{bmatrix} 1 & 2 \\ 4 & 5 \end{bmatrix} = -3$ and $\Delta_3 = \det \mathbf{A} = 0$. Its principal minors are:

D_3: $\det \mathbf{A} = 0$,

D_2: $\det\begin{bmatrix} 1 & 2 \\ 4 & 5 \end{bmatrix} = -3$, $\det\begin{bmatrix} 5 & 6 \\ 8 & 9 \end{bmatrix} = -3$, $\det\begin{bmatrix} 1 & 3 \\ 7 & 9 \end{bmatrix} = -12$, and

D_1: $\det [1] = 1$, $\det [5] = 5$, and $\det [9] = 9$.

By using the above definition, one can prove the following result.

Theorem 1.22 The square matrix $\mathbf{A}$ is positive definite if and only if $\Delta_1 > 0$, $\Delta_2 > 0$, $\ldots$, $\Delta_n > 0$. $\mathbf{A}$ is positive semidefinite if and only if $D_1 \geq 0$, $D_2 \geq 0$, $\ldots$, $D_n \geq 0$, for all principal minors of $\mathbf{A}$. $\mathbf{A}$ is negative definite if and only if $\Delta_1 < 0$, $\Delta_2 > 0$, $\ldots$, $(-1)^n \Delta_n > 0$. $\mathbf{A}$ is negative semidefinite if and only if $D_1 \leq 0$, $D_2 \geq 0$, $\ldots$, $(-1)^n D_n \geq 0$ for all principal minors of $\mathbf{A}$.

Remark The conditions $\Delta_1 \geq 0$, $\Delta_2 \geq 0$, $\ldots$, $\Delta_n \geq 0$ are not sufficient to ensure positive semidefiniteness. This can be seen immediately from the counterexample $\mathbf{A} = \begin{bmatrix} 0 & 1 \\ 0 & -1 \end{bmatrix}$ for which $\Delta_1 = \Delta_2 = 0$; when $\mathbf{x} = [0, 1]^T$ we obtain $\mathbf{x}^T \mathbf{A} \mathbf{x} = $

$[0, 1]\begin{bmatrix} 0 & 0 \\ 0 & -1 \end{bmatrix}\begin{bmatrix} 0 \\ 1 \end{bmatrix} = -1$. This shows that $\mathbf{A}$ is not positive semidefinite.

Another convenient way to investigate if matrices are positive definite or otherwise is based upon the eigenvalue concept.

Definition 1.23 The function $f(\lambda) = \det(\mathbf{A} - \lambda \mathbf{I})$, where λ is a scalar variable and $\mathbf{I}$ is the identity matrix, is called the *characteristic polynomial* of the given matrix $\mathbf{A}$. All solutions λ of the polynomial equation $f(\lambda) = 0$ are termed *characteristic values* or *eigenvalues* of $\mathbf{A}$.

Theorem 1.24 The eigenvalues of any symmetric matrix $\mathbf{A}$ are all real-valued. Furthermore, $\mathbf{A}$ is positive definite (positive semidefinite) if and only if all eigenvalues λ satisfy $\lambda > 0$ ($\lambda \geq 0$), $\mathbf{A}$ is negative definite (negative semidefinite) if and only if all eigenvalues λ satisfy $\lambda < 0$ ($\lambda \leq 0$); otherwise $\mathbf{A}$ is indefinite. Finally, if $\lambda_{\min}$ and $\lambda_{\max}$ denote the smallest and largest eigenvalues of $\mathbf{A}$, respectively, then we have $\lambda_{\min}\|\mathbf{x}\|^2 \leq \mathbf{x}^T \mathbf{A} \mathbf{x} \leq \lambda_{\max}\|\mathbf{x}\|^2$ for all $\mathbf{x} \in \mathbb{R}^n$.

Examples As a first illustration, consider the matrix $\mathbf{A} = \begin{bmatrix} 5 & -3 \\ -3 & 3 \end{bmatrix}$. Then $\Delta_1 = 5$ and $\Delta_2 = 6$ so that $\mathbf{A}$ is positive definite. Alternatively the characteristic polynomial is $f(\lambda) = \det \begin{bmatrix} 5-\lambda & -3 \\ -3 & 3-\lambda \end{bmatrix} = (5-\lambda)(3-\lambda) - 9$, so that the roots of $f(\lambda) = 0$ are $\lambda = 4 \pm \sqrt{10} > 0$ which leads to the same conclusion.

Consider now another matrix $\mathbf{A} = \begin{bmatrix} -3 & 2 \\ 2 & -5 \end{bmatrix}$. The leading principal minors are $\Delta_1 = -3$, $\Delta_2 = 11$, which indicates negative definiteness. Since the characteristic polynomial is $f(\lambda) = \det \begin{bmatrix} -3-\lambda & 2 \\ 2 & -5-\lambda \end{bmatrix} = (-3-\lambda)(-5-\lambda) - 4$, the solutions of the equation $f(\lambda) = 0$ are $-4 \pm \sqrt{5} < 0$ confirming the result.

As a third example let the matrix $\mathbf{A} = \begin{bmatrix} 1 & 1 \\ 1 & 1 \end{bmatrix}$ be given. All principal minors are $D_1 = 1$ and $D_2 = 0$, respectively. It follows that the matrix $\mathbf{A}$ is positive semidefinite. Since the characteristic polynomial is given by $f(\lambda) = \det \begin{bmatrix} 1-\lambda & 1 \\ 1 & 1-\lambda \end{bmatrix} = (1-\lambda)^2 - 1 = \lambda^2 - 2\lambda$, the eigenvalues are $\lambda = 0$ and $\lambda = 2$; it follows that positive semidefiniteness is the case.

Finally, take the matrix $\mathbf{A} = \begin{bmatrix} 2 & 3 \\ 3 & 1 \end{bmatrix}$. Since $\Delta_1 = 2$, $\Delta_2 = -7$, $\mathbf{A}$ is indefinite. The same conclusion can be reached by determining the eigenvalues of the characteristic polynomial $f(\lambda) = \det \begin{bmatrix} 2-\lambda & 3 \\ 3 & 1-\lambda \end{bmatrix} = \lambda^2 - 3\lambda - 7$, resulting in $\lambda = \frac{3}{2} \pm \sqrt{\frac{37}{4}}$, i.e., one positive and one negative eigenvalue.

After this brief review of some results from matrix analysis, we return to our discussion about necessary and sufficient optimality conditions. Recall that f is assumed to be twice continuously differentiable.

Theorem 1.25 (Second-order necessary optimality conditions) If a function f has a minimum at a point $\bar{\mathbf{x}}$, then $\nabla f(\bar{\mathbf{x}}) = \mathbf{0}$ and $\mathbf{H}_f(\bar{\mathbf{x}})$ is positive semidefinite.

Examples Consider the function $f(x_1, x_2) = x_1^2(x_2 - 1)^2$. We see that any point on the x_2-axis and any point on the line $x_2 = 1$ is a minimal point. As $\dfrac{\partial f}{\partial x_1} = 2x_1(x_2 - 1)^2$

and $\dfrac{\partial f}{\partial x_2} = 2x_1^2(x_2 - 1)$ we find that $\nabla f(\mathbf{x}) = \mathbf{0}$ at any such point. Furthermore,

$\mathbf{H}_f = \begin{bmatrix} 2(x_2 - 1)^2 & 4x_1(x_2 - 1) \\ 4x_1(x_2 - 1) & 2x_1^2 \end{bmatrix}$. Here, $(x_1, x_2) = (0, 7)$ is a minimal point, and

$\mathbf{H}_f(0,7) = \begin{bmatrix} 72 & 0 \\ 0 & 0 \end{bmatrix}$, which is positive semidefinite. Also, $(x_1, x_2) = (2, 1)$ is another

minimal point, and $\mathbf{H}_f(2,1) = \begin{bmatrix} 0 & 0 \\ 0 & 8 \end{bmatrix}$, which is again positive semidefinite. On the

other hand, consider the function $f(x_1, x_2) = x_1 x_2$ which has no finite minimal point.

Still, we find $\nabla f(0,0) = \begin{bmatrix} 0 \\ 0 \end{bmatrix}$ and $\mathbf{H}_f(0,0) = \begin{bmatrix} 0 & 0 \\ 0 & 0 \end{bmatrix}$. This shows that the

necessary conditions of Theorem 1.25 are not sufficient for a minimal point.

Theorem 1.26 (Second-order sufficient optimality conditions) If a point $\bar{\mathbf{x}}$ is such that $\nabla f(\bar{\mathbf{x}}) = \mathbf{0}$ and $\mathbf{H}_f(\bar{\mathbf{x}})$ is positive (negative) definite, then f has a strict local minimum (maximum) at $\bar{\mathbf{x}}$.

This result is the n-dimensional counterpart of that contained in Theorem 1.11 for the one-dimensional case.

Example Consider the function $f(x_1, x_2) = x_1^2 + 3x_2^2$. We have $\nabla f = \begin{bmatrix} 2x_1 \\ 6x_2 \end{bmatrix}$ so that

$\nabla f(0) = \begin{bmatrix} 0 \\ 0 \end{bmatrix}$. Furthermore $\mathbf{H}_f = \begin{bmatrix} 2 & 0 \\ 0 & 6 \end{bmatrix}$, which gives $\Delta_1 = 2$ and $\Delta_2 = 12$. It

follows that the Hessian matrix $\mathbf{H}_f$ is constant and from Theorem 1.22 that it is positive definite. Consequently, f has a strict local minimum at $(x_1, x_2) = (0, 0)$.

In analogy with our earlier discussion of the one-dimensional case, let us briefly consider problems of the type

P: Min $z = f(\mathbf{x})$
s.t. $\mathbf{x} \geq \mathbf{0}$,

i.e., sign-restricted optimization. For $\mathbf{x} > \mathbf{0}$, i.e., all components of $\mathbf{x}$ are strictly positive, the necessary condition for optimality remains $\nabla f(\mathbf{x}) = \mathbf{0}$, whereas for $\mathbf{x} \geq \mathbf{0}$, in which case some component(s) of $\mathbf{x}$ may be zero, $\nabla f(\mathbf{x})$ need not vanish at the optimal point. However, it can be shown (as we will do in Chap. 4 when discussing the so-called Karush-Kuhn-Tucker conditions) that for an optimal point $\bar{\mathbf{x}}$, either $\bar{x}_i = 0$ or $(\nabla f(\bar{\mathbf{x}}))_i = 0$ or both, for $i = 1, \ldots, m$. This leads to

Theorem 1.27   Consider the sign-constrained problem P: Min $z = f(\mathbf{x})$, s.t. $\mathbf{x} \geq \mathbf{0}$. If P has a solution at the point $\bar{\mathbf{x}}$, then $\bar{\mathbf{x}} \geq \mathbf{0}$, $\nabla f(\bar{\mathbf{x}}) \geq \mathbf{0}$, and $\bar{\mathbf{x}}^T \nabla f(\bar{\mathbf{x}}) = 0$.

Example Let the following problem P be given:

P: Min $z = f(\mathbf{x}) = x_1^2 + 4x_2^2 - 3x_1 x_2 - 2x_1 + 8x_2 + 5$
s.t. $x_1, x_2 \geq 0$.

Computing the gradient of the objective function we find

$$\nabla f(\mathbf{x}) = \left[\begin{array}{c} 2x_1 - 3x_2 - 2 \\ -3x_1 + 8x_2 + 8 \end{array} \right],$$

so that the unique stationary point for f is $(x_1, x_2) = \left(-\frac{8}{7}, -\frac{10}{7}\right)$, which violates the nonnegativity constraints. Therefore, problem P will have no interior solution point, i.e., in the strictly positive orthant. We will therefore search for a point on any of the two coordinate axes that might possibly satisfy the necessary optimality conditions of Theorem 1.27. For $x_1 = 0$, i.e., on the x_2-axis, $(\nabla f(0, x_2))_2 = 8x_2 + 8 > 0$, so that no point on the positive x_2-axis would satisfy the condition $\mathbf{x}^T \nabla f(\mathbf{x}) = 0$; the origin $(0, 0)$ violates the required $\nabla f(\mathbf{x}) \geq 0$. Consequently, only points on the positive x_1-axis might satisfy the optimality condition. With $x_1 > 0$, we must have $(\nabla f(x_1, 0))_1 = 0$, i.e., $2x_1 - 3(0) - 2 = 0$, or $x_1 = 1$. In conclusion, we have found the unique solution $(x_1, x_2) = (1, 0)$ to the necessary optimality conditions of Theorem 1.27. In Chap. 4 we will demonstrate that this point is indeed optimal.

We will now introduce the concept of a *saddle point*, which satisfies the first order optimality condition of having a zero gradient, and yet is neither a minimal nor a maximal point.

Definition 1.28 Let $f(\mathbf{x}, \mathbf{y})$ be a function of the variables $\mathbf{x}$ and $\mathbf{y}$, which may be scalars or vectors. We say that f has a *saddle point* at $(\hat{\mathbf{x}}, \hat{\mathbf{y}})$ if $f(\hat{\mathbf{x}}, \mathbf{y}) \leq f(\hat{\mathbf{x}}, \hat{\mathbf{y}}) \leq f(\mathbf{x}, \hat{\mathbf{y}})$ for all $\mathbf{x}$ and $\mathbf{y}$.

In other words, $\hat{\mathbf{x}}$ is a global minimum point for the function $f(\mathbf{x}, \hat{\mathbf{y}})$ of the variable $\mathbf{x}$ and $\hat{\mathbf{y}}$ is a global maximal point for the function $f(\hat{\mathbf{x}}, \mathbf{y})$ of the variable $\mathbf{y}$. If f has a saddle point at $(\hat{\mathbf{x}}, \hat{\mathbf{y}})$, then we also say that $-f$ has a saddle point there. Furthermore, points having a zero gradient are said to be *stationary*. A minimal or a maximal point is stationary, but the converse is not true, as exemplified by saddle points.

Example Consider the function $f(x_1, x_2) = x_1^2 - x_2^2$ which is graphically shown in Fig. 1.6. It is clear that f has a saddle point at $(x_1, x_2) = (0, 0)$ since $x_1 = 0$ is a global minimal point for the function $f(x_1, 0)$, whereas $x_2 = 0$ is a global maximal point for the function $f(0, x_2)$. We see that $\nabla f(0, 0) = \left[\begin{array}{c} 0 \\ 0 \end{array} \right]$.

1.3   Convex Functions and Quadratic Forms

So far we have considered necessary and sufficient conditions for local minimal or maximal points. Global conditions and global solutions can, in general, only be obtained when the problem is characterized by some convexity properties. In such cases we can claim that any local minimum is also a global minimum. Furthermore,

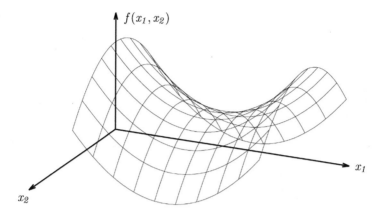

Fig. 1.6 Function with a saddle point

weaker sufficient conditions for a local minimum can be derived when the function is known to be strictly convex.

Definition 1.29 Let f be a given function and let $\mathbf{x}_1 \neq \mathbf{x}_2$ be any two distinct points in the domain of f. If $f(\lambda\mathbf{x}_1 + (1 - \lambda)\mathbf{x}_2) \leq \lambda f(\mathbf{x}_1) + (1 - \lambda) f(\mathbf{x}_2)$ for all $\lambda \in \,]0, 1[$, then the function f is said to be *convex*. Moreover, if the above relation is satisfied as strict inequality, then f is said to be *strictly convex*.

This definition essentially says that a convex function f is never underestimated by a linear interpolation between any two points; i.e., a convex function never lies above its chords. For an illustration of a convex function, see Fig. 1.7.

A *concave* (*strictly concave*) function f is one whose negative $-f$ is convex (strictly convex); it is obviously never overestimated by linear interpolation. A function which is both convex and concave must be linear. More about convexity and related concepts can be found in Rockafellar (1970); see also Luenberger and Ye (2008), Bazaraa et al. (2013), and Cottle and Thapa (2017). In Fig. 1.8 some graphs of convex and concave functions of one variable are shown to illustrate the above definitions.

Examples To check that the function $f(x) = x^2$ in Fig. 1.8 is strictly convex, we proceed in the following manner. The relation

$$[\lambda x_1 + (1 - \lambda)x_2]^2 < \lambda x_1^2 + (1 - \lambda) x_2^2,$$

which we wish to prove, can be written as

$$\lambda^2 x_1^2 + 2\lambda x_1 x_2 - 2\lambda^2 x_1 x_2 - \lambda x_2^2 + \lambda^2 x_2^2 - \lambda x_1^2 < 0,$$

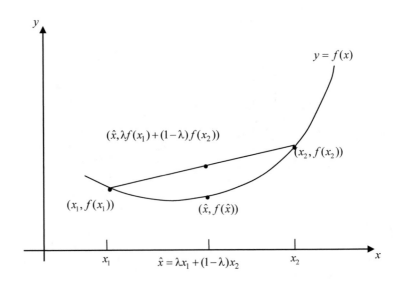

Fig. 1.7 A convex function

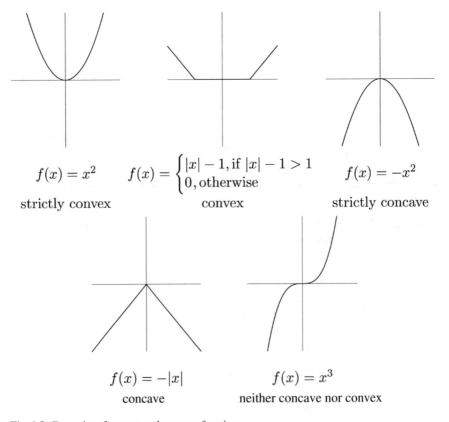

Fig. 1.8 Examples of convex and concave functions

$$\text{i.e., } \lambda(\lambda - 1)(x_1 - x_2)^2 < 0.$$

Since by assumption, $\lambda \in \,]0, 1[$ and $x_1 \neq x_2$, the left-hand side in the above relation is always strictly less than zero and the desired result is obtained. Similarly we can show that $f(x) = x^3$ in Fig. 1.8 is neither convex nor concave. As another example we mention $f(x_1, x_2) = x_1^2 + x_2^2$ which can be shown to be convex. The function $f(x_1, x_2) = x_1^2 - x_2^2$ which we considered in connection with saddle points, is neither convex nor concave. However, one can show that for any fixed $x_1 = \hat{x}_1, f(\hat{x}_1, x_2)$ is a concave function of x_2, and for any fixed $x_2 = \hat{x}_2, f(x_1, \hat{x}_2)$ is a convex function of x_1.

Note also that $f(x_1, x_2) = x_1 x_2$, which is convex (actually: linear) in x_1 for fixed x_2, and convex (actually: linear) in x_2 for fixed x_1, is itself not convex as a function of (x_1, x_2). This can be seen by considering $f(0, 2) = 0 = f(2, 0)$, whereas $f(1, 1) = 1$, i.e., f is at this point above its chord between the points $(0, 2)$ and $(2, 0)$.

Any nonnegative linear combination of convex functions is convex. Any composite function $g(f(x))$ where f is convex and g is a convex and nondecreasing function, is also convex. However, the product of convex functions need not be convex. For example, $f_1(x) = x^2$ and $f_2(x) = x$ are both convex, yet their product $g(x) = f_1(x) f_2(x) = x^3$ is not convex.

Definition 1.30   The *hypograph* of an arbitrary function $y = f(\mathbf{x})$ is the set of points $(\mathbf{x}, y)$ which are on or below the graph of the function, i.e., the set $\{(\mathbf{x}, y): y \leq f(\mathbf{x})\}$. The *epigraph* of any function $y = f(\mathbf{x})$ is the set of points $(\mathbf{x}, y)$ which are on or above the graph of the function, i.e., the set $\{(\mathbf{x}, y): y \geq f(\mathbf{x})\}$.

The sets are depicted in Fig. 1.9 for the function $y = x^2$.

Some interesting and useful relations between the convexity of a function and the convexity of sets are contained in the following theorems.

Theorem 1.31   The epigraph of any convex function as well as the hypograph of any concave function are convex sets. The converse is also true.

Theorem 1.32   If the function f is convex over a convex set C, then the set $\{\mathbf{x} \in C: f(\mathbf{x}) \leq \alpha\}$ is convex for any real number α.

Furthermore, since any intersection of convex sets is also convex, the set of points which simultaneously satisfy the inequalities $g_1(\mathbf{x}) \leq 0, g_2(\mathbf{x}) \leq 0, \ldots, g_m(\mathbf{x}) \leq 0$ where each function g_i is convex, constitutes a convex set.

Remark   The property described in Theorem 1.32 is actually used to define the concept of *quasiconvex functions*, which is more general than the concept of convex functions. Similarly, a *quasiconcave function* is the negative of a quasiconvex function.

Theorem 1.33   If a function f is convex on an open set S, then f is continuous on S.

When a function f is differentiable, convexity can be characterized alternatively as indicated by the following

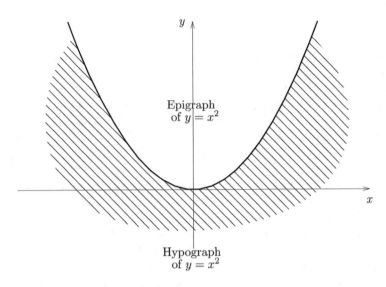

Fig. 1.9 Hypograph and epigraph of a function

Theorem 1.34 Suppose that the function f is differentiable. Then a necessary and sufficient condition for the convexity of f over a convex set C is that $f(\mathbf{y}) \geq f(\mathbf{x}) + (\mathbf{y} - \mathbf{x})^T \nabla f(\mathbf{x})$ for all $\mathbf{x}, \mathbf{y} \in C$.

Example Consider the convex function $f(x_1, x_2) = x_1^2 + x_2^2$. We will now show that the inequality in Theorem 1.34 always holds for this function. With $\mathbf{x} = [x_1, x_2]^T$ and $\mathbf{y} = [y_1, y_2]^T$, we find $\nabla f(\mathbf{x}) = [2x_1, 2x_2]^T$, so that $f(\mathbf{x}) + (\mathbf{y} - \mathbf{x})^T \nabla f(\mathbf{x}) = x_1^2 + x_2^2$

$+ [y_1 - x_1, y_2 - x_2] \begin{bmatrix} 2x_1 \\ 2x_2 \end{bmatrix} = x_1^2 + x_2^2 + 2x_1y_1 - 2x_1^2 + 2x_2y_2 - 2x_2^2 = y_1^2 + y_2^2 -$

$(x_1 - y_1)^2 - (x_2 - y_2)^2 \leq y_1^2 + y_2^2 = f(\mathbf{y})$ for all x_1, x_2, y_1, y_2, thus establishing the desired result.

Should the function f be twice continuously differentiable, then there exists yet another characterization of convexity.

Theorem 1.35 Suppose that f is twice continuously differentiable. Then f is convex over a convex set C if and only if the Hessian matrix of f is positive semidefinite throughout C.

Example Consider the function $f(x_1, x_2) = x_1^2 + 3x_2^2$ from the example after Theorem 1.26. There we found that its Hessian matrix $\mathbf{H}_f = \begin{bmatrix} 2 & 0 \\ 0 & 6 \end{bmatrix}$ was positive definite. Therefore f is convex on $\mathbb{R}^2$.

Theorem 1.35 is in essence the multidimensional analog of the inequality $\dfrac{d^2 f(x)}{dx^2} \geq 0$ which, when satisfied for all x, is a necessary and sufficient condition for the convexity of a single-variable function f with a continuous second derivative.

The following results concerning the optimization of convex functions are of fundamental importance for the connection between local and global analysis.

Theorem 1.36 If f is a convex function defined on a convex set C, then any local minimum of f is also a global minimum.

Proof This theorem will be proved by contradiction. Suppose that f has a local minimum at $\bar{\mathbf{x}}$ and that $\tilde{\mathbf{x}} \in C$ exists with $f(\tilde{\mathbf{x}}) < f(\bar{\mathbf{x}})$. By the convexity of f we obtain

$$f(\lambda \tilde{\mathbf{x}} + (1-\lambda)\bar{\mathbf{x}}) \leq \lambda f(\tilde{\mathbf{x}}) + (1-\lambda)f(\bar{\mathbf{x}}) < \lambda f(\bar{\mathbf{x}}) + (1-\lambda)f(\bar{\mathbf{x}}) = f(\bar{\mathbf{x}})$$

for all $\lambda \in {]}0, 1{[}$. This means that on the entire relative interior of the line segment between $\tilde{\mathbf{x}}$ and $\bar{\mathbf{x}}$, f takes values which are strictly less than $f(\bar{\mathbf{x}})$. This contradicts the local minimal property of $\bar{\mathbf{x}}$. □

Theorem 1.37 Let f be a differentiable convex function on a convex set C. A point $\bar{\mathbf{x}} \in C$ is a global minimum of f if and only if

$$(\mathbf{x} - \bar{\mathbf{x}})^T \nabla f(\bar{\mathbf{x}}) \geq 0 \text{ for all points } \mathbf{x} \in C.$$

Example Consider the function $f(x_1, x_2) = x_1^2 + 3x_2^2$ which was shown to be convex in the example after Theorem 1.35. Let us show that $\bar{\mathbf{x}} = [2, 1]^T$ is a global minimum for f on $C = \{(x_1, x_2): x_1 \geq 2, x_2 \geq 1\}$.

$$\nabla f(x_1, x_2) = \begin{bmatrix} 2x_1 \\ 6x_2 \end{bmatrix}, \text{ so that } \nabla f(2, 1) = [4, 6]^T \text{ and } (\mathbf{x} - \bar{\mathbf{x}})^T \nabla f(\bar{\mathbf{x}})$$

$$= [x_1 - 2, x_2 - 1] \begin{bmatrix} 4 \\ 6 \end{bmatrix} = 4(x_1 - 2) + 6(x_2 - 1) \geq 0 \text{ for all } x_1 \geq 2, x_2 \geq 1,$$

which is the desired result.

We end this section by considering a certain class of functions called *quadratic forms*. These will be of particular importance for quadratic programming problems.

Definition 1.38 A function f which can be written as $f(\mathbf{x}) = \mathbf{x}^T \mathbf{Q} \mathbf{x}$ where $\mathbf{Q}$ is a square matrix, is said to be a *quadratic form*. The matrix $\mathbf{Q}$ is referred to as the *matrix of the form f*.

Examples $\mathbf{A} = \begin{bmatrix} 2 & 3 \\ 3 & 4 \end{bmatrix}$ is the matrix of the quadratic form $f(x_1, x_2)$

$= 2x_1^2 + 6x_1 x_2 + 4x_2^2$ and $\mathbf{B} = \begin{bmatrix} 1 & 0 & 3 \\ 0 & 7 & -2 \\ 3 & -2 & 11 \end{bmatrix}$ is the matrix of the quadratic form

$$f(x_1, x_2, x_3) = x_1^2 + 6x_1x_3 + 7x_2^2 - 4x_2x_3 + 11x_3^2.$$

Remark The matrices $\mathbf{A} = \begin{bmatrix} 1 & 2 \\ 4 & 0 \end{bmatrix}$ and $\mathbf{B} = \begin{bmatrix} 1 & 3 \\ 3 & 0 \end{bmatrix}$ both generate the same quadratic form $f(x_1, x_2) = x_1^2 + 6x_1x_2$. If we only allow symmetric square matrices, however, we find that there is a one-to-one correspondence between all quadratic forms from $\mathbb{R}^n$ to $\mathbb{R}$ and all symmetric square n-th order matrices. When considering quadratic forms, we will therefore restrict ourselves to symmetric matrices. This is not as restrictive as it might appear at first glance. Specifically, for any quadratic form $\mathbf{x}^T\mathbf{A}\mathbf{x}$ where $\mathbf{A}$ is nonsymmetric, we could replace $\mathbf{A}$ by the symmetric matrix $\frac{1}{2}\mathbf{A} + \frac{1}{2}\mathbf{A}^T$, which generates the same quadratic form. This is easily demonstrated, since $\mathbf{x}^T(\frac{1}{2}\mathbf{A} + \frac{1}{2}\mathbf{A}^T)\,\mathbf{x} = \frac{1}{2}\mathbf{x}^T\mathbf{A}\mathbf{x} + \frac{1}{2}\mathbf{x}^T\mathbf{A}^T\mathbf{x} = \frac{1}{2}\mathbf{x}^T\mathbf{A}\mathbf{x} + \frac{1}{2}(\mathbf{x}^T\mathbf{A}\mathbf{x})^T = \mathbf{x}^T\mathbf{A}\mathbf{x}$ for all $\mathbf{x} \in \mathbb{R}^n$.

Definition 1.39 A quadratic form is said to be *positive definite* (*positive semidefinite, indefinite, negative semidefinite, negative definite*), if and only if the matrix of the form is positive definite (positive semidefinite, indefinite, negative semidefinite, negative definite).

Examples The quadratic form $f(x_1, x_2) = 5x_1^2 - 6x_1x_2 + 3x_2^2$ is positive definite, as its matrix $\mathbf{A} = \begin{bmatrix} 5 & -3 \\ -3 & 3 \end{bmatrix}$ was shown to be positive definite in the example following Theorem 1.10. Similarly, $f(x_1, x_2) = -3x_1^2 + 4x_1x_2 - 5x_2^2$ is negative definite, as its matrix $\mathbf{A} = \begin{bmatrix} -3 & 2 \\ 2 & -5 \end{bmatrix}$ was shown to be negative definite. Similarly, we see that $f(x_1, x_2) = x_1^2 + 2x_1x_2 + x_2^2$ with matrix $\mathbf{A} = \begin{bmatrix} 1 & 1 \\ 1 & 1 \end{bmatrix}$ is positive semidefinite, and that $f(x_1, x_2) = 2x_1^2 + 6x_1x_2 + x_2^2$ with matrix $\mathbf{A} = \begin{bmatrix} 2 & 3 \\ 3 & 1 \end{bmatrix}$ is indefinite.

By computing the gradient and the Hessian matrix of a quadratic form we immediately obtain the following.

Theorem 1.40 If f is a quadratic form with matrix $\mathbf{Q}$ and Hessian $\mathbf{H}_f$, then

$$\nabla f(\mathbf{x}) = 2\mathbf{Q}\mathbf{x} \text{ and } \mathbf{H}_f = 2\mathbf{Q}.$$

The significance of this theorem is that all previous results concerning optimality conditions and convexity, which involve the gradient and the Hessian matrix can now immediately be applied to quadratic forms using the above expressions for the gradient and the Hessian. As an example we conclude by stating the following

Theorem 1.41 If a quadratic form is indefinite, then it has a saddle point at the origin.

We will return to the concept of saddle points in Chap. 4, when we deal with the Lagrangean function associated with a constrained nonlinear programming problem.

References

Bazaraa MS, Sherali HD, Shetty CM (2013) *Nonlinear programming: theory and algorithms* (3rd ed.). Wiley, New York

Bertsekas DP (2016) *Nonlinear programming* (3rd ed.). Athena Scientific, Belmont, MA

Cottle RW, Thapa MN (2017) *Linear and nonlinear optimization.* Springer-Verlag, Berlin-Heidelberg-New York

Eiselt HA Sandblom, C-L (2007) *Linear programming and its applications.* Springer-Verlag, Berlin-Heidelberg

Luenberger DL, Ye Y (2008) *Linear and nonlinear programming* (3rd ed.). Springer-Verlag, Berlin-Heidelberg-New York

Rockafellar RT (1970) *Convex analysis.* University Press, Princeton, NJ

Chapter 2
Unconstrained Optimization

This chapter considers a fundamental problem of general optimization theory, namely that of finding the maximal and/or minimal points of a nonlinear function.

If the nonlinear function f is differentiable, then one might use Theorem 1.10 (Theorem 1.19) of the previous chapter and solve the possibly nonlinear equation (system of equations), which result(s) from equating the first derivative (gradient) of f with zero. As this equation (system of equations) may be quite difficult to solve, the approaches in this chapter for finding the maximal and/or minimal points of the function f will often be computationally preferable.

In Chap. 1 we pointed out that the difference in nature between single-variable and multiple variable functions is more profound than it may appear at first sight. Furthermore, few practical problems involve only one decision variable and no constraints. Nonetheless, since many procedures for solving multidimensional optimization problems actually employ one-dimensional techniques as subroutines in some of their iterations, it is natural to start with methods for the unconstrained optimization of functions of one real variable. In the second section of this chapter, functions of several variables are considered.

2.1 Single-Variable Unconstrained Optimization

This section is devoted to a description of some basic techniques for iteratively solving one-dimensional unconstrained optimization problems. These techniques are often referred to as *line searches*.

Problems P for which an optimal solution is sought will be of the type

P: Min $z = f(x)$
s.t. $x \in [a, b]$,

© Springer Nature Switzerland AG 2019
H. A. Eiselt, C.-L. Sandblom, *Nonlinear Optimization*, International Series in
Operations Research & Management Science 282,
https://doi.org/10.1007/978-3-030-19462-8_2

where a and b may be finite or infinite. If a maximization problem Max $z = f(x)$ were to be given instead, we replace it by Min $-z = -f(x)$.

2.1.1   Grid Search

The simplest method for finding the minimal point of a function within a given interval $[a, b]$ is the *grid* or *uniform search*, whereby the interval is divided into subintervals of equal length and the function is evaluated at the resulting grid points. The grid is then refined around the best point obtained and the procedure repeated as many times as required. The procedure is best explained by a numerical example.

Example Consider the problem of minimizing the function

$$f(x) = 2x^2 - 3x + 5e^{-x},$$

pictured in Fig. 2.1. With $f'(x) = 4x - 3 - 5e^{-x}$ and $f''(x) = 4 + 5e^{-x}$, we find that $f''(x) > 0\ \forall\ x$, so that f is convex; furthermore we obtain $f(0) = 5 > f(1) = 0.83940 < f(2) = 2.67668$, so that the global minimum $\bar{x}$ of f must be in the open interval $]0, 2[$.

Evaluating the function f at the points $0, 0.1, 0.2, \ldots, 2.0$, we obtain the function values $f(0) = 5$, $f(0.1) = 4.244$, $f(0.2) = 3.574$, $\ldots$, $f(0.9) = 0.953$, $f(1.0) = 0.839$,

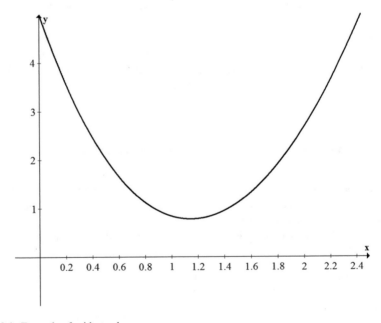

Fig. 2.1 Example of grid search

$f(1.1) = 0.784, f(1.2) = 0.786, f(1.3) = 0.843, f(1.4) = 0.953, \ldots, f(1.9) = 2.268$, and $f(2.0) = 2.677$.

The smallest function value among those obtained in this set is $f(1.1) = 0.784$, so that the new interval is [1.0, 1.2]. Refining the grid, we evaluate the function f at the points 1.00, 1.01, 1.02, ..., 1.19, 1.20, resulting in the functional values $f(1.00) = 0.8394$, $f(1.01) = 0.8313$, $f(1.02) = 0.8238$, ..., $f(1.12) = 0.7802$, $f(1.13) = 0.7790$, $f(1.14) = 0.7783$, $f(1.15) = 0.7782$, $f(1.16) = 0.7786$, ..., $f(1.19) = 0.7833$, and $f(1.20) = 0.7860$. The new interval is then [1.14, 1.16] and the new grid points are 1.140, 1.141, ..., 1.159, and 1.160 that have functional values $f(1.140) = 0.778295$, $f(1.141) = 0.778259$, ..., $f(1.145) = 0.778170$, $f(1.146) = 0.778161$, $f(1.147) = 0.778159$, $f(1.148) = 1.778161$, $f(1.149) = 0.778170, \ldots, f(1.159) = 0.778561$, and $f(1.160) = 0.778631$. We stop here with 1.147 being the current estimate of the minimal point $\bar{x}$ (The true value of $\bar{x}$ to seven decimal points is $\bar{x} = 1.1469894$ with $f(\bar{x}) = f(1.1469894) = 0.77815853$). If more precision is required, we would continue by searching with a finer grid within the interval [1.146, 1.148], and so on.

2.1.2 *Dichotomous Search*

Related to the grid search method is *dichotomous search*, where each function evaluation allows us to cut by about half the interval within which the minimum exists. The method is also referred to as an *elimination technique*, since it successively excludes from consideration subintervals, in which the minimum is assured not to be located. Assuming convexity of the function $f(x)$ and that evaluations cannot be closer to each other than some preset value δ, we proceed as follows. Given an initial interval [a, b], we evaluate f at two points, centered around the midpoint of the interval at a distance of δ apart, i.e. we determine $f(\frac{1}{2}(a + b - \delta))$ and $f(\frac{1}{2}(a + b + \delta))$. Comparing these two function values with each other, we may then exclude from further consideration either the interval [a, $\frac{1}{2}(a + b - \delta)$] (Case (ii) in Fig. 2.2) or the interval [$\frac{1}{2}(a + b + \delta)$, b] (Case (i) in Fig. 2.2). The interval that contains the minimal point will then be either [a, $\frac{1}{2}(a + b + \delta)$] (case (i)) or [$\frac{1}{2}(a + b - \delta)$, b] (case (ii)).

Example Consider again the function $f(x) = 2x^2 - 3x + 5e^{-x}$, the interval [0, 2], and assume that evaluations have to be at least 0.01 units apart. Then we compute $f(\frac{1}{2}(a + b - \delta)) = f(\frac{1}{2}(0 + 2 - 0.01)) = f(0.995) = 0.8437$ and $f(\frac{1}{2}(a + b + \delta)) = f(\frac{1}{2}(0 + 2 + 0.01)) = f(1.005) = 0.8353$. With $f(0.995) > f(1.005)$, i.e., case (ii), we discard the interval [0, 0.995] and consider only [0.995, 2]. Next, we evaluate $f(\frac{1}{2}(0.995 + 2 - 0.01)) = f(1.4925) = 1.1017$ and $f(\frac{1}{2}(0.995 + 2 + 0.01)) = f(1.5025) = 1.1204$. Now, the interval [1.5025, 2] can be discarded and we consider [0.995, 1.5025]. The procedure continues in the same fashion.

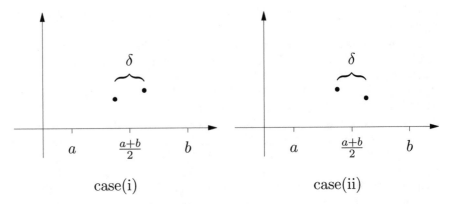

Fig. 2.2 Example of dichotomous search

2.1.3 Bisection Search

For differentiable functions, the *bisection minimization search* method can be used as an alternative to dichotomous minimization search, needing only one function evaluation per iteration. Instead of evaluating $f(x)$ at two points symmetrically located in the interval $[a, b]$, we will now evaluate $f'(\frac{1}{2}(a + b))$. Using the same philosophy as in dichotomous search, we will exclude from further consideration the interval $[a, \frac{1}{2}(a + b)]$ if $f'(\frac{1}{2}(a + b)) < 0$, and the interval $[\frac{1}{2}(a + b), b]$ if $f'(\frac{1}{2}(a + b)) > 0$; if $f'(\frac{1}{2}(a + b))$ happens to be exactly zero (or at least very close to it), we would conclude that $\frac{1}{2}(a + b)$ is the minimal point of the function f in the interval $[a, b]$.

Example Consider again minimizing the function $f(x) = 2x^2 - 3x + 5e^{-x}$. With $f'(x) = 4x - 3 - 5e^{-x}$, we find that $f'(0) = -8 < 0, f'(2) = 4.3233 > 0$, and since $f''(x) = 4 + 5e^{-x} > 0 \; \forall \; x$, the function f is convex. We conclude that the global minimal point of f must be somewhere in the open interval $]0, 2[$ and we proceed with the bisection search with $a := 0$ and $b := 2$. We find $f'(\frac{1}{2}(a + b)) = f'(1) = -0.8394 < 0$, so that $[\frac{1}{2}(a + b), b] = [1, 2]$ is the new interval of interest. We now evaluate f' at the midpoint of this new interval and obtain $f'(\frac{1}{2}(1 + 2)) = f'(1.5) = 1.8843 > 0$, so that the interval $[1.5, 2]$ can be excluded from our search. The next interval is $[1, 1.5]$ with the midpoint 1.25. We find $f'(1.25) = 0.5675 > 0$, so that the new interval is $[1, 1.25]$ with midpoint 1.125. Since $f'(1.125) = -0.1233 < 0$, the new interval is $[1.125, 1.25]$, and with $f'(1.1875) = 0.2251 > 0$, the next interval is $[1.125, 1.1875]$, and the process continues.

We will now take our discussion of elimination techniques further. Given a function $f(x)$ defined on a closed interval $[a, b] = I \subset \mathbb{R}$, our aim is to obtain a subinterval L of I with length $|L|$ containing a point $\bar{x}$, such that $f(\bar{x}) = \min\limits_{x \in I} f(x)$, i.e., the subinterval L should contain the point at which f attains its minimum over the set I. This is done by determining the value of $f(x)$ at one point after another and then using this information to decide at which point $f(x)$ should be evaluated next. The evaluation of such a point is called an *experiment*. First we need the following

Definition 2.1 Assume that the function f has a global minimum point $\bar{x}$ in the interval $I = [a, b]$. If for any x_1 and x_2.

(i) $a \leq x_1 < x_2 \leq \bar{x} \rightarrow f(x_1) > f(x_2)$,
(ii) $\bar{x} \leq x_1 < x_2 \leq b \rightarrow f(x_1) < f(x_2)$,

then we say that $f(x)$ is *strongly unimodal (downwards)*.

The above definition implies that $f(x)$ has a unique minimum over I at $\bar{x}$, that $f(x)$ is strictly decreasing in $[a, \bar{x}]$ and strictly increasing in $[\bar{x}, b]$. It also implies that any function that is strictly convex over I is also strongly unimodal over I. However, the converse is not true as Fig. 2.3-(2) illustrates. Actually, strong unimodality is equivalent to strong quasiconvexity, see Bazaraa et al. (2013).

In the following discussion, the function $f(x)$ is assumed to be strongly unimodal (downwards). Suppose now that we conduct two experiments x_1 and x_2 in I.

If $f(x_1) < f(x_2)$, we know that $\bar{x} \notin [x_2, b]$, as otherwise relation (i) would be violated. But we do know that $\bar{x}$ exists in I. Thus $\bar{x} \in [a, x_2]$ and we can discard the subinterval $[x_2, b]$ in our further search for the minimum. We say that we have reduced the *interval of uncertainty*, i.e., the subinterval which we know contains the point of minimum, from $[a, b]$ to $[a, x_2]$.

If, on the other hand, $f(x_1) > f(x_2)$, we know that $\bar{x} \notin [a, x_1]$, as otherwise relation (ii) would be violated; consequently, $\bar{x} \in [x_1, b]$. If $f(x_1) = f(x_2)$, we see that we can discard both, $[a, x_1]$ and $[x_2, b]$, so that $\bar{x} \in [x_1, x_2]$. However, if in this last case we only discard one of the subintervals $[a, x_1]$ or $[x_2, b]$, we find that, whatever the outcome after we have reduced the interval of uncertainty by aid of the experiments x_1 and x_2, one of these experiments will remain in the new interval of uncertainty. Therefore, the next time we attempt to reduce this interval, we only need one new experiment to match against the old remaining experiment. This way, we can continue to carry out experiments and shorten the interval of uncertainty in successive iterations until some stop criterion is satisfied. Observe that apart from the first round when two experiments are needed to initiate the algorithm, only one experiment is required in each iteration.

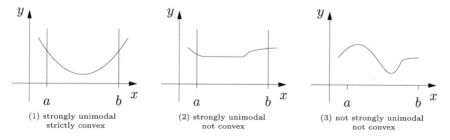

(1) strongly unimodal   (2) strongly unimodal   (3) not strongly unimodal
     strictly convex   not convex   not convex

Fig. 2.3 Strongly unimodal functions

Let us now assume that we have started with an interval of uncertainty L_0 of length $|L_0|$ and that after n experiments, we obtain an interval L_n of length $|L_n|$. Clearly, L_n does not only depend on n, but also on the position of the experiments x_k, $k = 1, 2, \ldots, n$, and on the outcomes $f(x_k)$, $k = 1, 2, \ldots, n$ (i.e., on the function f). By a *search plan* we will understand a strategy of how to position the experiments.

2.1.4   Fibonacci Search

The so-called *Fibonacci search* technique (due to Kiefer 1953) uses a famous numerical sequence called the Fibonacci numbers. They can be defined recursively as shown in

Definition 2.2 The *Fibonacci numbers* F_k are defined by

$$F_0 = F_1 = 1$$
$$F_{k+2} = F_{k+1} + F_k, \ k = 0, 1, 2, \ldots$$

The first ten Fibonacci numbers are $F_0 = F_1 = 1$, $F_2 = 2$, $F_3 = 3$, $F_4 = 5$, $F_5 = 8$, $F_6 = 13$, $F_7 = 21$, $F_8 = 34$, $F_9 = 55$. The Fibonacci sequence occurs quite naturally in a diversity of areas such as genetics, botany, and biology, and its usefulness is far from being restricted to pure mathematics. The son of Bonaccio, Leonardo da Pisa, obtained this sequence around the year 1202 while considering the genetics of rabbit breeding. (Incidentally, "Fibonacci" actually means "the son of Bonaccio.")

We first explain what is meant by an optimal way of positioning our experiments, i.e., the search plan, and inquire about the existence of an optimal search plan. More specifically, we will use a minimax criterion to minimize the maximal length of the interval $|L_n|$ as the function f and the outcomes $f(x_k)$ vary. Furthermore, no experiments are permitted to be closer to each other than a given value ε, and if a corresponding optimal search plan exists, it will be called ε-*minimax*. If no confusion can arise, we will denote by L_k the interval L_k as well as its length $|L_k|$. Also, for the purpose of our illustration, we set the initial interval of uncertainty to $I = [0, 1]$.

Starting with the interval $I = [0, 1]$ and performing two experiments at x_1 and x_2 (with $x_1 < x_2$), we now determine what L_2, the resulting interval of uncertainty, will be. As discussed above, there are two cases to be considered: (i): $f(x_1) < f(x_2)$, and (ii): $f(x_1) \geq f(x_2)$. This results in one of the situations shown in Fig. 2.4.

In case (i), we can discard $[x_2, 1]$, so that $L_2' = x_2 - 0 = x_2$, while in case (ii), we can discard $[0, x_1]$, so that $L_2'' = 1 - x_1$. We see that $\max\{L_2', L_2''\} = \max\{x_2, 1 - x_1\}$. Our goal is to minimize this expression, bearing in mind that we must also have $x_1 - x_2 \geq \varepsilon$. Using the fact that $\max\{\alpha, \beta\} = \frac{1}{2}(\alpha + \beta) + \frac{1}{2}|\beta - \alpha|$ for any α and β, we find

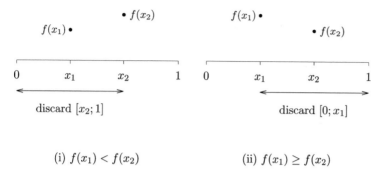

Fig. 2.4 Interval reduction in Fibonacci search

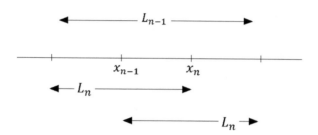

Fig. 2.5 Final step in a symmetric n-experiment plan

$$\max\{L_2', L_2''\} = \max\{x_2, 1 - x_1\} = \tfrac{1}{2}(1 - x_1 + x_2) + \tfrac{1}{2}|1 - x_1 - x_2| \ge \tfrac{1}{2}(1 + \varepsilon),$$

since $x_2 - x_1 \ge \varepsilon$ and $\tfrac{1}{2}|1 - x_1 - x_2| \ge 0$. It follows that the minimal value of max $\{L_2', L_2''\}$ is attained for $x_1 = \tfrac{1}{2} - \tfrac{1}{2}\varepsilon, x_2 = \tfrac{1}{2} + \tfrac{1}{2}\varepsilon$, in which case $L_2' = \tfrac{1}{2} + \tfrac{1}{2}\varepsilon = L_2''$.

Our result shows that the experiments x_1 and x_2 for an ε-minimax search plan with two experiments should be positioned symmetrically around the middle of the initial interval of uncertainty at a distance of ε apart. If L_0 and L_2 were the lengths of our initial and final intervals of uncertainty, respectively, we see that we have (for obvious reasons, we set $L_1 = L_0$):

$$L_1 = L_0 = 2L_2 - \varepsilon.$$

Arguing in the same fashion, we can show that a symmetric n-experiment plan (i.e., with all n experiments symmetrically positioned) must have an end as shown in Fig. 2.5.

Here, $L_{n-1} = L_n - |x_n - x_{n-1}| + L_n = 2L_n - |x_n - x_{n-1}| \le 2L_n - \varepsilon$, i.e., $L_{n-1} \le 2L_n - \varepsilon$ with equality if and only if $|x_n - x_{n-1}| = \varepsilon$. One can show (see, e.g., Avriel and Wilde 1966) that for every nonsymmetric search plan, there is a corresponding symmetric plan that is at least as good, if not better. We can therefore restrict ourselves to considering only symmetric plans. It follows that for an n-

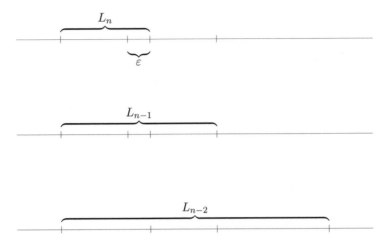

Fig. 2.6 Plan of experiments in a Fibonacci search

experiment symmetric search plan, the position of the first experiment x_1 will completely define the entire plan, because of the symmetry requirement.

Suppose now that we want to generate an n-experiment symmetric search plan by using what we presently know. We can recursively construct a plan as shown in Fig. 2.6.

In general, we have $L_k = L_{k+1} + L_{k+2}, k = 1, 2, \ldots, n - 2$. We then find

$$L_{n-1} = 2L_n - \varepsilon,$$

$$L_{n-2} = L_{n-1} + L_n = 2L_n - \varepsilon + L_n = 3L_n - \varepsilon,$$

$$L_{n-3} = L_{n-2} + L_{n-1} = 5L_n - 2\varepsilon,$$

$$L_{n-4} = L_{n-3} + L_{n-2} = 8L_n - 3\varepsilon,$$

and so on, where the coefficients are precisely the Fibonacci numbers. Therefore we can write

$$L_{n-k} = F_{k+1}L_n - F_{k-1}\varepsilon, \quad k = 1, 2, \ldots, n - 1,$$

where $k = n - 1$ specifies the starting interval $L_1 = F_n L_n - F_{n-2}\varepsilon$. As we know L_1, n, and ε, we can then solve for L_n and obtain

$$L_n = \frac{1}{F_n}(L_1 + \varepsilon F_{n-2}),$$

and, according to our method, it follows that $\varepsilon \le \frac{1}{2}L_n$.

Although the above considerations define the Fibonacci search technique, we still have to compute L_2, the distance from the end points to our initial interval, in which the first two experiments are located at the outset of the search. Setting $k := n - 2$ in the above formula $L_{n-k} = F_{k+1}L_n - F_{k-1}\varepsilon$, we obtain

$$L_2 = F_{n-1}L_n - \varepsilon F_{n-3},$$

but since $L_n = \dfrac{1}{F_n}(L_1 + \varepsilon F_{n-2})$, it follows that

$$L_2 = \frac{F_{n-1}}{F_n}(L_1 + \varepsilon F_{n-2}) - \varepsilon F_{n-3}, \text{i.e.,}$$

$$L_2 = \frac{F_{n-1}}{F_n}L_1 + \frac{F_{n-1}F_{n-2} - F_n F_{n-3}}{F_n}\varepsilon,$$

or simply

$$L_2 = \frac{F_{n-1}}{F_n}L_1 + \frac{(-1)^n}{F_n}\varepsilon.$$

The last formula follows from the fact that $F_{n-1}F_{n-2} - F_n F_{n-3} = (-1)^n$, which can be shown by solving the difference equation that generates the Fibonacci numbers, i.e., $F_{k+2} = F_{k+1} + F_k$, $k = 1, 2, \ldots$ With the initial conditions $F_0 = F_1 = 1$, we obtain

$$F_k = \frac{1}{\sqrt{5}}\left(\frac{1+\sqrt{5}}{2}\right)^{k+1} - \frac{1}{\sqrt{5}}\left(\frac{1-\sqrt{5}}{2}\right)^{k+1}, \quad k = 0, 1, 2, \ldots$$

Once we have computed L_2 and started the search, we simply continue, finding smaller and smaller intervals of uncertainty, and repeatedly locating the new experiment symmetrically to the remaining experiment in the present interval. The procedure is summarized in the following algorithm.

Given that the problem is to find the minimum of a function $f(x)$ on an interval $L_1 = [a, b]$, so that no two evaluation points are closer than a prespecified distance ε, a user-determined number of $n \geq 3$ evaluations are to be placed optimally (i.e., so as to minimize the length of the resulting interval of uncertainty).

The Fibonacci Minimization Search

Step 1: Determine $L_2 = \dfrac{F_{n-1}}{F_n}L_1 + \dfrac{(-1)^n}{F_n}\varepsilon$ and set $y := x_1 := b - L_2$, $z := x_2 := a + L_2$, $\bar{a} := a, \bar{b} := b$, and $k := 2$.

Step 2: Is $f(y) \leq f(z)$?

If yes: Set $L_k := [\bar{a}, z]$ and discard $[z, \bar{b}]$.

If no: Set $L_k := [y, \bar{b}]$ and discard $[\bar{a}, y]$.

Step 3: Is $k = n$?

If yes: Stop, the interval $L_n = [\bar{a}, \bar{b}]$ contains the minimum point of the function $f(x)$.
If no: Determine a new evaluation point x_{k+1} symmetrically positioned in L_k with regard to the existing interior point L_k. Refer to the interior points in L_k, from left to right, as y and z. Let $k := k + 1$, and go to Step 2.

Example Conduct a 4-experiment Fibonacci search on the function used in our previous examples, i.e., $f(x) = 2x^2 - 3x + 5e^{-x}$ in the initial interval of uncertainty $[0, 2]$ with $\varepsilon = 0.1$.

We initialize the search by first computing the distance L_2 from the endpoints 0 and 2 of the given interval. We obtain

$$L_2 = \frac{F_3}{F_4} L_1 + \frac{(-1)^4}{F_4}(0.1) = \tfrac{3}{5}2 + \tfrac{1}{5}(0.1) = 1.22.$$

Starting with $x_1 := 2 - 1.22 = 0.78$ and $x_2 := 1.22$, we find that $f(x_1) = f(0.78) = 1.169 > 0.793 = f(1.22) = f(x_2)$, so that the interval $[0, x_1] = [0, 0.78]$ can be discarded, and $x_3 := 2 - (x_2 - x_1) = 2 - (1.22 - 0.78) = 1.56$. Now $f(x_3) = f(1.56) = 1.238 > 0.793 = f(1.22) = f(x_2)$, so that the interval $[x_3, 2] = [1.56, 2]$ will be discarded, and the new evaluation point is set at $x_4 := x_1 + (x_3 - x_2) = 1.12$. With $f(x_4) = f(1.12) = 0.780 < 0.793 = f(1.22) = f(x_2)$, we finally discard the subinterval $[x_2, x_3] = [1.22, 1.56]$, so that the final interval of uncertainty is $L_4 = [0.78, 1.22]$. This may appear to be rather unimpressive since $L_4/L_1 = 0.44/2 = 0.22$ (i.e., the original interval of uncertainty has only been reduced to about one fourth of its original length); however, the best solution found so far is x_4, which is quite close to the true optimal point at $\bar{x} = 1.147$. Figure 2.7 shows the progression of the method.

Remark If we wish to perform an n-experiment Fibonacci search that ends with a final interval of uncertainty not greater than, say, one tenth of the original interval, we must determine the number of experiments required. In the formula $L_n = \dfrac{1}{F_n}(L_1 + \varepsilon F_{n-2})$, we set $\varepsilon = 0$ and derive $L_n = L_1/F_n$ or $L_n/L_1 = 1/F_n$. Since

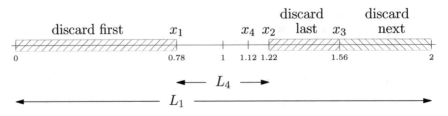

Fig. 2.7 Fibonacci search in the numerical example

we require that $L_n/L_1 = 1/10$, this implies that $F_n \geq 10$, we see that $n = 6$ will suffice (as $F_6 = 13$ is the first Fibonacci number that exceeds the value of 10).

2.1.5 Golden Section Search

When applying the Fibonacci search technique, users have to specify the resolution factor ε, and the number of experiments to be performed must also be given in advance, before the position of each evaluation can be computed. Since a variety of situations exist in which we do not know beforehand how many experiments are needed until a stop criterion is satisfied, the Fibonacci search technique cannot be directly used. However, there exists a closely related technique, called the *Golden Section* search, which, although almost as efficient as the Fibonacci technique, is independent of the number n of experiments to be performed.

In order to devise an experimental plan, let us have a closer look at the formula for starting the Fibonacci search, but ignore ε. Since for $\varepsilon = 0$, the formula $L_2 = \dfrac{F_{n-1}}{F_n} L_1 + \dfrac{(-1)^n}{F_n} \varepsilon$ reduces to $L_2 = \dfrac{F_{n-1}}{F_n} L_1$, we guess that perhaps $\dfrac{F_{n-1}}{F_n}$ has a limit for $n \to \infty$, in which case, for large values of n, the positioning of the experiments at the beginning of the search hardly depends on the specific value of n.

It is not difficult to demonstrate that $\lim\limits_{n \to \infty} \dfrac{F_{n-1}}{F_n} = \dfrac{\sqrt{5} - 1}{2}$. Thus we can see that if starting an infinitely long Fibonacci search, we would take $L_2 = \dfrac{\sqrt{5} - 1}{2} L_1 \approx 0.618 L_1$. Then, as each new step just represents a Fibonacci search with one experiment less, this would imply that $\dfrac{L_n}{L_{n-1}} = \dfrac{\sqrt{5} - 1}{2}$ for all n.

The procedure described by this formula is the *Golden Section* search and can be regarded as the beginning of an infinitely long Fibonacci search. Even though the Golden Section search is not ε-minimax, it is still fairly efficient. It turns out that the final interval of uncertainty after an n-experiment search is only about 17% wider by the Golden Section search than that obtained by the Fibonacci search, when n is large. In fact, reducing an interval of uncertainty to less than 1% of its original length requires only 11 experiments by both methods, while reducing it to less than 0.1% requires 16 experiments by both methods.

The Golden Section search technique is described in algorithmic form below. To restate the purpose of the search, the function $f(x)$ of one real variable is to be minimized over the interval $L_1 = [a, b]$. The procedure terminates when the interval of uncertainty is shorter than a user predetermined value ε.

The Golden Section Minimization Search

Step 1: Determine $L_2 := \dfrac{\sqrt{5}-1}{2} L_1 \approx 0.618 L_1$ and set $y := x_1 := b - L_2, z := x_2 := a + L_2$, $\bar{a} := a, \bar{b} := b$, and the iteration counter $k := 2$.

Step 2: Is $f(y) \leq f(z)$?

If yes: Set $L_k := [\bar{a}, z]$ and discard the subinterval $[z, \bar{b}]$.
If no: Set $L_k := [y, \bar{b}]$ and discard the subinterval $[\bar{a}, y]$.

Step 3: Is $L_k < \varepsilon$?

If yes: Stop, the interval L_k contains the minimum point of the function $f(x)$.
If no: Determine a new search point x_{k+1}, symmetrically positioned in L_k with regard to the existing interior point in L_k. Refer to these interior points, from left to right, as y and z. Set $k := k + 1$ and go to Step 2.

Remark The name "Golden Section" stems from a problem studied by the Greek mathematician Euclid (around the year 300 BC) that involves the division of a line segment into two parts, such that the ratio of the larger part to the whole equals the ratio of the smaller to the larger part. In other words, find a (the larger part) and b (the shorter part), such that $\dfrac{a}{a+b} = \dfrac{b}{a}$. It turns out that $\dfrac{b}{a} = \dfrac{\sqrt{5}-1}{2}$, which is precisely $\dfrac{L_n}{L_{n-1}}$ in the Golden Section search. Dividing a line segment according to these proportions was said to produce the Golden Section, and a rectangle with its sides in this proportion was considered to be especially esthetically pleasing.

Example Consider again the function $f(x) = 2x^2 - 3x + 5e^{-x}$ and locate its minimum point in the interval $[0, 2]$, terminating when seven points have been evaluated. First, we obtain $L_2 = 0.62$, $L_1 = 1.24$, $x_1 = 2 - 1.24 = 0.76$, and $x_2 = 0 + 1.24 = 1.24$. The progress of the technique is shown in Table 2.1.

The conclusion of the process is that the function $f(x)$ attains its minimum value of approximately 0.7786 for $x = 1.16$ with an interval of uncertainty of $[1.12, 1.24]$, a result that was obtained after seven experiments. The progression of the experiments is shown in Fig. 2.8.

Table 2.1 Progression of the Golden Section Minimization Search

k	x_k	$f(x_k)$	Conclusion
1	0.76	1.214	
2	1.24	0.802	discard [0, 0.76]
3	1.52	1.154	discard [1.52, 2]
4	1.04	0.810	discard [0.76, 1.04]
5	1.32	0.860	discard [1.32, 1.52]
6	1.12	0.7802	discard [1.24, 1.32]
7	1.16	0.7786	discard [1.04, 1.12]

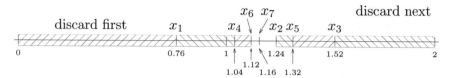

Fig. 2.8 Golden Section search in the numerical example

2.1.6 Quadratic Interpolation

In the two interval elimination techniques just described it was only assumed that the functions being searched were strongly unimodal. If the functions under investigation are also smooth, alternative search procedures have been devised to take advantage of this additional feature. A class of methods of this type is referred to as *approximation* or *curve fitting* techniques, because they consist of evaluating the objective function at different points and then approximating it by means of smooth curves that are passed through the given points. An estimate of the optimal point can then be obtained by differential calculus. In order to proceed, we first need a classical lemma.

Lemma 2.3 (Lagrange interpolation formula) Given n points (a_i, b_i), $i = 1, \ldots, n$ in the plane, such that a_i are all distinct. There is a unique polynomial $y = p(x)$ of degree $n - 1$ which leads through all the n points. This polynomial is given by the expression

$$p(x) = \sum_{i=1}^{n} \frac{\prod\limits_{j:j\neq i} (x - a_j)}{\prod\limits_{j:j\neq i} (a_i - a_j)} \, b_i.$$

It can easily be verified that $p(a_i) = b_i$, $i = 1, \ldots, n$.

The idea of the *quadratic interpolation* method is now the following. Given three x-values a_1, a_2, and a_3, such that $a_1 < a_2 < a_3$ and with function values $f(a_1) > f(a_2) < f(a_3)$ we wish to find the minimal value of the function f in the interval $[a_1, a_3]$. We accomplish this by approximating f by the unique second-degree polynomial that leads through the three points $(a_1, f(a_1))$, $(a_2, f(a_2))$, and $(a_3, f(a_3))$. This polynomial can be expressed using the Lagrange interpolation formula of Lemma 2.3, by which we obtain

$$p(x) = \frac{(x-a_2)(x-a_3)}{(a_1-a_2)(a_1-a_3)}f(a_1) + \frac{(x-a_1)(x-a_3)}{(a_2-a_1)(a_2-a_3)}f(a_2) + \frac{(x-a_1)(x-a_2)}{(a_3-a_1)(a_3-a_2)}f(a_3).$$

One can easily verify that $p(x)$ is a second degree polynomial that leads through the three given points. Differentiating we obtain

$$p'(x) = \frac{2x - a_2 - a_3}{(a_1 - a_2)(a_1 - a_3)} f(a_1) + \frac{2x - a_1 - a_3}{(a_2 - a_1)(a_2 - a_3)} f(a_2) + \frac{2x - a_1 - a_2}{(a_3 - a_1)(a_3 - a_2)} f(a_3).$$

Since $p(x)$ and $f(x)$ coincide at $x = a_1, a_2,$ and a_3, it follows that $f(a_1) = p(a_1) > f(a_2) = p(a_2) < f(a_3) = p(a_3)$, so that p is a convex function. It is now clear that there must exist a unique minimal point $\bar{x} \in]a_1, a_3[$ for which $f'(\bar{x}) = 0$. Approximating $\bar{x}$ by $\hat{x}$ which is obtained by solving the equation $p'(\hat{x}) = 0$ we find after some algebraic manipulation:

$$\hat{x} = \frac{1}{2} \frac{(a_2^2 - a_3^2)f(a_1) + (a_3^2 - a_1^2)f(a_2) + (a_1^2 - a_2^2)f(a_3)}{(a_2 - a_3)f(a_1) + (a_3 - a_1)f(a_2) + (a_1 - a_2)f(a_3)}.$$

Example Consider again mimimizing the function $f(x) = 2x^2 - 3x + 5e^{-x}$, where we are restricting our search to the interval $[0, 2]$, within which we know the global minimum is situated. Using quadratic interpolation based on the three points $(0, 5)$, $(1, 0.83940)$, and $(2, 2.67668)$, we find $\hat{x} = 1.19368$, which we can compare with the true minimal point $\bar{x} = 1.146989$.

We can now repeat the procedure in a neighborhood of the current estimate $\hat{x} = 1.19368$. Using step sizes of length 0.1, we find $f(1.0) = 0.839397 > f(1.1) = 0.784355 < f(1.2) = 0.785971$. We could therefore use $a_1 = 1.0$, $a_2 = 1.1$, and $a_3 = 1.2$; the result of quadratic interpolation is $\hat{x} = 1.147148$, now quite close to the true minimum value of $\bar{x} = 1.146989$. In an application of curve fitting, we will return in Sect. 3.1 to the problem of how to fit a quadratic function to a given set of points.

2.1.7   Cubic Interpolation

Assume that for a differentiable function $f(x)$ we know the function values at two points a and b, with $a < b$, and that we also know that $f'(a) < 0$ and $f'(b) > 0$. This means that the function f must have a minimum in the interval $[a, b]$. As an approximation of this minimal point $\bar{x}$, we will now fit a cubic polynomial and seek its minimal point $\hat{x}$, following a procedure due to Davidon (1959), see also Dixon (1972) and Avriel (1976). In other words, we are given the four values $f(a), f'(a) < 0, f(b),$ and $f'(b) > 0$, see Fig. 2.9. With the *cubic interpolation* method, we now wish to fit a third-degree polynomial $p(x) = c_3 x^3 + c_2 x^2 + c_1 x + c_0$ in such a way that $p(a) = f(a)$, $p'(a) = f'(a)$, $p(b) = f(b)$, and $p'(b) = f'(b)$. By differentiation

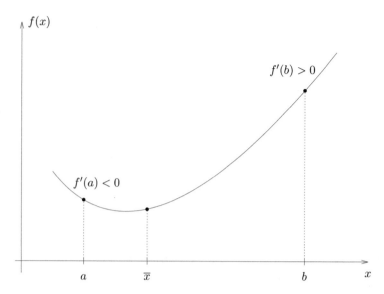

Fig. 2.9 Cubic interpolation

we obtain $p'(x) = 3c_3x^2 + 2c_2x + c_1$, so that we are given $f(a), f'(a), f(b)$, and $f'(b)$, such that

$$c_3a^3 + c_2a^2 + c_1a + c_0 = f(a),$$

$$3c_3a^2 + 2c_2a + c_1 = f'(a),$$

$$c_3b^3 + c_2b^2 + c_1b + c_0 = f(b), \quad \text{and}$$

$$3c_3b^2 + 2c_2b + c_1 = f'(b).$$

Since we are now going to minimize the third-degree polynomial $p(x)$ we wish to express $p'(x)$ with its coefficients given in terms of the known values $f(a), f'(a), f(b)$, and $f'(b)$. This will be facilitated by defining a new variable t by the equation $x :=$ $a + t$. It then follows that $p'(x) = p'(a + t) = 3c_3(a + t)^2 + 2c_2(a + t) +$ $c_1 = 3c_3a^2 + 2c_2a + c_1 + (3c_3a + c_2)2t + 3c_3t^2 = f'(a) + (3c_3a + c_2)2t + 3c_3t^2$. We proceed by attempting to express the coefficient $3c_3a + c_2$ in terms of known quantities. Since $\dfrac{f(b) - f(a)}{b - a} = \dfrac{c_3(b^3 - a^3) + c_2(b^2 - a^2) + c_1(b - a)}{b - a} = c_3(a^2 +$ $ab + b^2) + c_2(a + b) + c_1$, and with $f'(a) + f'(b) = 3c_3(a^2 + b^2) + 2c_2(a + b) + 2c_1$, we obtain, after some calculations, $\dfrac{3((f a) - f(b))}{b - a} + f'(a) + f'(b) = -3c_3ab -$ $c_2(a + b) - c_1$, which we set equal to the new constant α. We then find that $f'(a) +$ $\alpha = 3c_3a^2 + 2c_2a + c_1 - 3c_3ab - c_2(a + b) - c_1 = (3c_3a + c_2)(a - b)$. Therefore we have $p'(x) = f'(a) + \dfrac{f'(a) + \alpha}{a - b}2t + 3c_3t^2$, and we finally wish to express c_3 in terms of given quantities. We just demonstrated that $f'(a) + \alpha = (3c_3a + c_2)(a - b)$, and similarly one can show that $f'(b) + \alpha = (3c_3b + c_2)(b - a)$, from which we obtain f' $(a) + f'(b) + 2\alpha = 3c_3(a - b)^2$, from which c_3 can be obtained.

Summarizing, with $\alpha = \dfrac{3(f(a) - f(b))}{b - a} + f'(a) + f'(b)$, we have $p'(x) = f'(a) - \dfrac{f'(a) + \alpha}{b - a} 2(x - a) + \dfrac{f'(a) + f'(b) + 2\alpha}{(b - a)^2}(x - a)^2$. Once this expression for $p'(x)$ has been obtained, we solve for $p'(x) = 0$, which is a second-degree equation in $x - a = t$, so that we obtain $t = \dfrac{b - a}{f'(a) + f'(b) + 2\alpha} \left[f'(a) + \alpha \underset{(-)}{+} \sqrt{\alpha^2 - f'(a)f'(b)} \right]$ (where the "+" sign must be chosen for minimality), and finally $\hat{x} = a + \dfrac{b - a}{f'(a) + f'(b) + 2\alpha} \left[f'(a) + \alpha + \sqrt{\alpha^2 - f'(a)f'(b)} \right]$, where $\alpha = \dfrac{3(f(a) - f(b))}{b - a} + f'(a) + f'(b)$.

To avoid the issue of the denominator $f'(a) + f'(b) + 2\alpha$ being zero, this formula can, after quite considerable algebraic manipulation, be rewritten as

$$\hat{x} = b - \frac{f'(b) - \alpha + \sqrt{\alpha^2 - f'(a)f'(b)}}{f'(b) - f'(a) + 2\sqrt{\alpha^2 - f'(a)f'(b)}} (b - a).$$

Now, since $f'(a) < 0$ and $f'(b) > 0$, $f'(a)f'(b) < 0$, so that the expression under the square root must be positive; furthermore $f'(b) - f'(a)$ must be positive, so that the denominator $f'(b) - f'(a) + 2\sqrt{\alpha^2 - f'(a)f'(b)}$ must be strictly positive as well.

Example Consider again the function $f(x) = 2x^2 - 3x + 5e^{-x}$, where, as above, we restrict our attention to the interval $[a, b] = [0, 2]$. With $f'(x) = 4x - 3 - 5e^{-x}$, $f'(0) = -8 < 0$ and $f'(2) = 4.3233 > 0$, so that cubic interpolation in the given interval is appropriate. We have $f(a) = f(0) = 5$, $f(b) = f(2) = 2.6767$, and $f'(a) = f'(0) = -8$, $f'(b) = f'(2) = 4.3233$, so that $\alpha = \dfrac{3(f(a) - f(b))}{b - a} + f'(a) + f'(b) = \dfrac{3(5 - 2.6767)}{2 - 0} - 8 + 4.3233 = -0.19175$ and $\sqrt{\alpha^2 - f'(a)f'(b)} = \sqrt{(-0.19175)^2 - (-8)(4.3233)} = \sqrt{34.62317} = 5.8841$, so that $\hat{x} = b - \dfrac{f'(b) - \alpha + \sqrt{\alpha^2 - f'(a)f'(b)}}{f'(b) - f'(a) + 2\sqrt{\alpha^2 - f'(a)f'(b)}} (b - a) = 2 - \dfrac{4.3233 + 0.19175 + 5.8841}{4.3233 + 8 + 2(5.8841)} (2 - 0) = 2 - 0.8633 = 1.1367.$

If we need more accuracy, we need to evaluate the function in the vicinity of $\hat{x} = 1.1367$. Then we find that $f'(1.0) = -0.8394, f'(1.2) = 0.2940$, and we wish to repeat the cubic interpolation procedure for the interval $[a_1, b_1] = [1.0, 1.2]$. We obtain $f(a_1) = f(1.0) = 0.8394, f(b_1) = f(1.2) = 0.7860$, so that $a_1 = \dfrac{3(f(a_1) - f(b_1))}{b_1 - a_1} + f'(a_1) + f'(b_1) = \dfrac{3(0.8394 - 0.7860)}{1.2 - 1.0} - 0.8394 + 0.2940 = 0.25560$, and

$$\sqrt{\alpha^2 - f'(a_1)f'(b_1)} = \sqrt{(0.2556)^2 - (-0.8394)(0.2940)} = \sqrt{0.31211} =$$

0.55867, so that the new point is $\hat{x} = b - \dfrac{f'(b_1) - \alpha_1 + \sqrt{\alpha_1^2 - f'(a_1)f'(b_1)}}{f'(b_1) - f'(a_1) + 2\sqrt{\alpha_1^2 - f'(a_1)f'(b_1)}}$

$$(b_1 - a_1) = 1.2 - \frac{0.2940 - 0.2556 + 0.55867}{0.2940 + 0.8394 + 2(0.55867)}(1.2 - 1.0) =$$

$1.2 - 0.0531 = 1.1469$, which now is quite close to the true minimum $\bar{x} = 1.1470$.

2.1.8 Newton Minimization Search

Perhaps the most important curve fitting method is the *Newton Minimization* search. The idea behind this search is quite simple. Imagine that a function $f(x)$ of one real variable has to be minimized. It is assumed that the second derivative $\dfrac{df(x)}{dx^2} = f''(x)$ exists. Furthermore, at a point x_k where an experiment is placed, the numbers $f(x_k), f'(x_k)$, and $f''(x_k)$ can all be evaluated. Then we approximate the function f in the vicinity of x_k by the following quadratic function $\tilde{f}$ (see the quadratic Taylor formula in Theorem 1.8), where x is the independent variable and $x_k, f(x_k), f'(x_k)$, and $f''(x_k)$ are the coefficients:

$$\tilde{f}(x) = f(x_k) + (x - x_k)f'(x_k) + \tfrac{1}{2}(x - x_k)^2 f''(x_k).$$

The next point x_{k+1} will be the point at which the quadratic function $\tilde{f}(x)$ is minimized, i.e., the point at which $\tilde{f}'(x) = 0$. As $\tilde{f}'(x) = f'(x_k) + (x - x_k)f''(x_k)$, then $\tilde{f}'(x_{k+1}) = 0$ gives $f'(x_k) + (x_{k+1} - x_k)f''(x_k) = 0$, so that we obtain the iteration formula

$$x_{k+1} = x_k - \frac{f'(x_k)}{f''(x_k)}.$$

The solution procedure is initialized with some estimate of $\bar{x}$, the true minimal point of the function f, say x_0. Assume now that the estimate of x_k of the minimal point of f has been found. Then the next point x_{k+1} is found by the above iteration formula, and the procedure is be repeated at the point x_{k+1}. Implicitly, it has been assumed that $f''(x_k) \neq 0 \,\forall\, k$. To help to appreciate the idea, Fig. 2.10 illustrates the geometric aspect of the process. We are searching for a solution $\bar{x}$ to the equation $f'(x) = 0$ in order to find the minimal point of the function f.

The process starts with x_0 and evaluates $f'(x_0)$ as well as $f''(x_0)$. The tangent ℓ_1 to the curve $y = f'(x)$ at the point $(x_0, f'(x_0))$ has the equation $y - f'(x_0) = f''(x_0)$, so that $-f'(x_0) = f''(x_0)(x_1 - x_0)$ and at the point $(x_1, 0)$, where the tangent intersects the x-axis, we obtain $x_1 = x_0 - \dfrac{f'(x_0)}{f''(x_0)}$. Similarly, the tangent ℓ_2 has the equation $y - f'(x_1) = f''(x_1)(x - x_1)$, from which we conclude that $x_2 = x_1 - \dfrac{f'(x_1)}{f''(x_1)}$, and so forth.

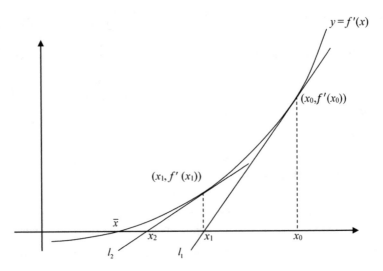

Fig. 2.10 The general principle of Newton minimization search

A different geometric interpretation of the Newton minimization method is obtained by considering $f(x)$ instead of $f'(x)$. From our analysis above, it can then be seen as a succession of quadratic approximations of the function f, each time selecting the minimum point of the quadratic approximation function as the next iteration point. In Sect. 3.1, we will return to the curve-fitting problem as an application of nonlinear optimization. Another observation worth mentioning is that the Newton minimization method aims at finding a point x for which $f'(x) = 0$. If we introduce the function $g(x) = f'(x)$, the iteration formula is then $x_{k+1} = x_k - \dfrac{g(x_k)}{g'(x_k)}$, which we recognize as the famous Newton-Raphson iterate for solving the equation $g(x) = 0$.

We can now summarize the Newton minimization search method in algorithmic form. Recall that the problem is to find a minimal point $\bar{x}$ of a given function $f(x)$. Assume that some stop criterion to be discussed below has been chosen. Furthermore, we start with an initial guess x_0 and set the iteration counter $k := 1$.

The Newton Minimization Search

Step 1: Compute $f'(x_k)$ and $f''(x_k)$ and let $x_{k+1} = x_k - \dfrac{f'(x_k)}{f''(x_k)}$.

Step 2: Is the stop criterion satisfied?

If yes: Stop, x_{k+1} is a solution to the problem.
If no: Set $k := k + 1$ and go to Step 1.

The stop criterion in Step 2 might be the optimality conditions $f'(x_k) = 0$ and $f''(x_k) > 0$. As this criterion is highly unlikely ever to be satisfied in practice, it should be replaced by criteria such as $\left| \dfrac{f(x_{k+1}) - f(x_k)}{\max\{\varepsilon_1, |f(x_{k+1})|, |f(x_k)|\}} \right| \leq \varepsilon_2$ or $\left| \dfrac{x_{k+1} - x_k}{\max\{\varepsilon_3, |x_{k+1}|, |x_k|\}} \right| \leq \varepsilon_4$, where ε_1, ε_2, ε_3, and ε_4 are predetermined positive numbers. These criteria measure the relative changes in $f(x)$ and x, respectively. The two criteria may also be combined; a predetermined number of iterations could also be specified.

Example Consider yet again the function $f(x) = 2x^2 - 3x + 5e^{-x}$ and perform a Newton minimization search, starting with $x_0 = 1$. With $f'(x) = 4x - 3 - 5e^{-x}$ and $f''(x) = 4 + 5e^{-x}$, we obtain the Newton minimization search iteration formula

$$x_{k+1} = x_k - \frac{f'(x_k)}{f''(x_k)} = x_k - \frac{4x_k - 3 - 5e^{-x_k}}{4 + 5e^{-x_k}},$$

from which we obtain $x_0 = 1$, $x_1 = 1.143747$, $x_2 = 1.146988$, $x_3 = 1.146989 = x_4$. We conclude that for six-decimal accuracy, the method converges after no more than three iterations.

Unfortunately, the Newton method does not necessarily converge if the initial point is sufficiently far from the true optimum. Even worse, the function $f(x) = x^{\frac{4}{3}}$ (with a unique minimum of $\bar{x} = 0$) demonstrates that the Newton minimization method may actually fail, regardless of how close to the optimum the initial guess x_0 may be. This is apparent from Fig. 2.11, displaying the first derivative $f'(x) = \dfrac{4}{3}x^{\frac{1}{3}}$.

However, by requiring certain conditions to be satisfied regarding the function and/or the placement of the initial guess x_0, convergence of the Newton minimization method can actually be guaranteed. As an example, we may state the following theorem (see, e.g., Fröberg 1969).

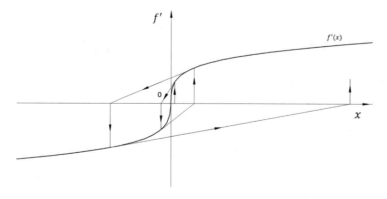

Fig. 2.11 An example of nonconvergence of Newton minimization search

Theorem 2.4 Consider any given interval $[a, b]$, which contains a stationary point of the function $f(x)$. If

(i) $f'''(x)$ exists and does not change sign in $[a, b]$,
(ii) $f'(a)f'(b) < 0$,
(iii) $f''(x) \neq 0$ in $[a, b]$, and
(iv) $\max\left\{ \left| \frac{f'(a)}{f''(a)} \right|, \left| \frac{f'(b)}{f''(b)} \right| \right\} \leq b - a$,

then the Newton minimization method will converge to a point $\bar{x}$ in $[a, b]$ where $f'(\bar{x}) = 0$, no matter where in $[a, b]$ the method was started.

Another sufficient condition for the convergence of the Newton minimization method is simply that $f'(x)$ be convex. We see that the function $f(x) = x^{\frac{4}{3}}$, for which the method failed to converge, does not satisfy this condition. (The conditions of Theorem 2.4 are violated as well, as $f'''(x)$ does not exist at the origin, which happens to be the optimal point).

Let us now briefly explore the nature of how the sequence x_k converges to a minimal point $\bar{x}$ in the Newton minimization method. For that purpose, let ε_k denote the error in estimating $\bar{x}$, in the sense that $\varepsilon_k := |x_k - \bar{x}|$. Then we have the following

Theorem 2.5 Let the function f have a continuous third derivative $f'''(x)$ and let $f''(\bar{x}) \neq 0$ at a minimal point $\bar{x}$ of f. Then for some x_0 sufficiently close to $\bar{x}$, there exists a constant α, such that

$$|x_{k+1} - \bar{x}| \leq \alpha |x_k - \bar{x}|^2, k = 1, 2, \ldots$$

We then say that x_k converges to $\bar{x}$ with at least a *quadratic order of convergence*.

Proof There exist positive constants α_1 and α_2, such that $|f'''(x)| < \alpha_1$ and $|f''(x)| > \alpha_2$ in the vicinity of $\bar{x}$. Now, with $\bar{x}$ being a minimal point of f, we have $f'(\bar{x}) = 0$, so that

$$x_{k+1} - \bar{x} = x_k - \frac{f'(x_k) - f'(\bar{x})}{f''(x_k)} - \bar{x} = -\frac{f'(x_k) - f'(\bar{x}) - (x_k - \bar{x})f''(\bar{x})}{f''(x_k)}.$$

By virtue of the quadratic Taylor formula (Theorem 1.8) for f' at $\bar{x}$, we can write $f'(x_k) = f'(\bar{x}) + (x_k - \bar{x})f''(\bar{x}) + (x_k - \bar{x})^2 f'''(\tilde{x})$, where $\tilde{x}$ is some point in the interval $[x_k, \ \bar{x}]$. Therefore, $x_{k+1} - \bar{x} = -(x_k - \bar{x})^2 \frac{f'''(\tilde{x})}{f''(x_k)}$ and it follows that $|x_{k+1} - \bar{x}| < \frac{\alpha_1}{\alpha_2} |x_k - \bar{x}|^2$, so that by setting $\alpha := \frac{\alpha_1}{\alpha_2}$, the theorem is proved. $\square$

The significance of this result is that the error ε_k in the k-th iteration shrinks in quadratic fashion. Loosely speaking, the number of correct decimals in the current estimate of the optimal value $\bar{x}$ doubles in each iteration.

Let us also point out that the Newton Search Method can be applied to functions given in implicit form. For instance, consider the function $g(x, y)$ of two variables. If g is twice continuously differentiable, then—with certain assumptions—the equation $g(x, y) = 0$ will define y as a function of x, i.e., $y = f(x)$, where f is said to be given in *implicit form*. (For a discussion of functions in implicit form, see, e.g., Bertsekas (2016) and Cottle and Thapa (2017)). The Newton minimization method for finding minimal points of f can now be used, employing the formula

$$x_{k+1} = x_k - \frac{\frac{\partial g}{\partial x}\left(\frac{\partial g}{\partial y}\right)^2}{\frac{\partial^2 g}{\partial x^2}\left(\frac{\partial g}{\partial y}\right)^2 - 2\frac{\partial^2 g}{\partial x \partial y}\frac{\partial g}{\partial x}\frac{\partial g}{\partial y} + \frac{\partial^2 g}{\partial y^2}\left(\frac{\partial g}{\partial x}\right)^2}.$$

All the partial derivatives are evaluated at the point $(x, y) = (x_k, y_k)$, where y_k is defined by $g(x_k, y_k) = 0$, i.e., $y_k = f(x_k)$.

As an illustration, let us minimize y as an implicitly given function of x, where $g(x, y) = \sin x + \cos y - 1 = 0$. Then for $x \in [0, \pi]$, the equation $g(x, y) = 0$ will define y as a function of x. We obtain $\frac{\partial g}{\partial x} = \cos x$, $\frac{\partial g}{\partial y} = -\sin y$, $\frac{\partial^2 g}{\partial x^2} = -\sin x$, $\frac{\partial^2 g}{\partial x \partial y} = 0$,

and $\frac{\partial^2 g}{\partial y^2} = -\cos y$. Therefore the iteration formula can be written as

$$x_{k+1} = x_k - \frac{\frac{\partial g}{\partial x}\left(\frac{\partial g}{\partial y}\right)^2}{\frac{\partial^2 g}{\partial x^2}\left(\frac{\partial g}{\partial y}\right)^2 - 2\frac{\partial^2 g}{\partial x \partial y}\left(\frac{\partial g}{\partial x}\right)\frac{\partial g}{\partial y} + \frac{\partial^2 g}{\partial y^2}\left(\frac{\partial g}{\partial x}\right)^2}$$

$$= x_k - \frac{\cos x_k \sin^2 y_k}{-\sin x_k \sin^2 y_k - \cos y_k \cos^2 x_k}.$$

Instead of computing the y_k values by $g(x_k, y_k) = 0$, we use the fact that $\cos y_k = 1 - \sin x_k$, and $1 - \sin^2 y_k = \cos^2 y_k = (1 - \sin x_k)^2$, therefore $\sin^2 y_k = 1 - (1 - \sin x_k)^2$. Then

$$x_{k+1} = x_k + \frac{\cos x_k \left[1 - (1 - \sin x_k)^2\right]}{\sin x_k \left[1 - (1 - \sin x_k)^2\right] + (1 - \sin x_k) \cos^2 x_k}.$$

Starting with $x_0 = 0.5$, we obtain $x_1 = 1.3525$, $x_2 = 1.574077$, and $x_3 = 1.570796 \approx \frac{1}{2}\pi$. The function $y = f(x)$ is actually $y = \arccos(1 - \sin x)$, whose exact minimum is at $\bar{x} = \frac{1}{2}\pi$.

Other methods similar to the Newton minimization method have been developed, for instance the *Method of False Positions* (*Regula Falsi*), in which the existence of a second derivative is not required. The iteration formula is then

$$x_{k+1} = x_k - (x_k - x_{k-1}) \frac{f'(x_k)}{f'(x_k) - f'(x_{k-1})},$$

see, e.g., Luenberger and Ye (2008) or Cottle and Thapa (2017).

Having considered the minimization problem for various types of functions (differentiable, convex, strictly unimodal), we now turn to a very general type of functions.

2.1.9 Lipschitz Optimization

Definition 2.6 A real-valued function f defined on a set $D_f \in \mathbb{R}$ is called a *Lipschitz continuous function* on D_f, if there is a constant L, such that

$$|f(x_1) - f(x_2)| \le L |x_1 - x_2| \ \forall x_1, x_2 \in D_f.$$

From the definition, it follows that Lipschitz functions are continuous. Linear combinations and products of Lipschitz functions are also Lipschitzian, as are the maxima and minima of such functions. The constant L provides an upper bound on the absolute value of the slope of any segments joining two points on the graph of f. In this sense, Lipschitz functions are said to be of *bounded variation*. On the other hand, the function $f(x) = \sqrt{x}$ is not Lipschitzian on the interval [0, 1], having a vertical slope at the origin.

Applications of Lipschitz functions include solving systems of nonlinear equations and/or inequalities, location problems, and a variety of models for waste water and water pollution models, see, e.g., Pintér (1996). The Lipschitz condition in Definition 2.6 was originally developed as a condition to ensure the existence of unique solutions to differential equations.

Here, we will describe a simple method for minimizing a Lipschitzian function f, due to Danilin and Piyavskii (1967), Piyavskii (1972) and independently developed by Shubert (1972). The idea of the method is to build up a piecewise linear underestimate F of the function f, and then to use the minimal point of F as an estimate of the true minimal point of f.

Consider the problem

P: Min $z = f(x)$
s.t. $x \in I = [a, b] \subseteq \mathbb{R}$

and assume that $f(x)$ is Lipschitzian over the interval I with Lipschitz constant L, so that a minimal point $\bar{x}$ in I exists. Now since for some arbitrary point $x_1 \in I$, we have

$$f(x) - f(x_1) \leq L \mid x - x_1 \mid \forall x \in I,$$

it follows that

$$f(x) \geq f(x_1) - L \mid x - x_1 \mid =: F_1(x),$$

so that $F_1(x)$ underestimates f on the interval I. We then minimize the function $F_1(x)$ over the interval I and thus obtain the point $x_2 \in I$. In other words, $F_1(x_2) = \min_{x \in I} F_1(x)$. Since $f(x) \geq f(x_2) - L \mid x - x_2 \mid$, we have found another underestimate of f on I. Combining the two underestimates, we conclude that

$$F_2(x) := \max_{i=1,2} \{f(x_i) - L|x - x_i|\}$$

is also an underestimate of f on I, see Fig. 2.12, where F_2 is the function that consists of the three line segments making up the upper envelope of $f(x_1) - L \mid x - x_1 \mid$ and $f(x_2) - L \mid x - x_2 \mid$. Minimizing $F_2(x)$ on I yields the solution $x_3 = a$, after which we form $F_3(x) := \max_{i=1,2,3} \{f(x_i) - L|x - x_i|\}$, and so forth. The functions $F_k(x), k = 1, 2, \ldots$ are called *saw-tooth covers* of f on I and yield ever-tighter underestimates of f on I.

We are now ready to describe the Piyavskii method in algorithmic form. Recall that the problem is

P: Min $f(x)$
s.t. $x \in I = [a, b],$

where the function f is Lipschitzian with the constant L. The algorithm is initialized with some arbitrary point x_1 in I and with the iteration counter $k := 1$.

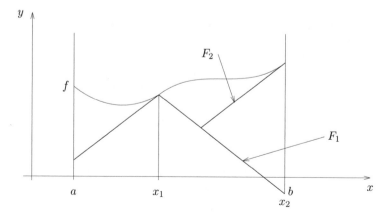

Fig. 2.12 Lipschitz optimization

The Piyavskii Method

Step 1: Solve the piecewise linear optimization problem
$P_k : \underset{x \in I}{\text{Min}}\ z_k = F_k(x) = \underset{i=1,\ldots,k}{\max}\ \{f(x_i) - L|x - x_i|\}$. Denote a solution to P_k by x_{k+1}.

Step 2: Is some stop criterion satisfied?

If yes: Stop with x_{k+1} as the current estimate of $\bar{x}$.
If no: Set $k := k + 1$ and go to Step 1.

By virtue of the construction of the Piyavskii method, we can see that the following lemma must hold:

Lemma 2.7 In the Piyavskii method, where $F_k(x) = \underset{i=1,\ldots,k}{\max}\ \{f(x_i) - L|x - x_i|\}$ and

$x_{k+1} = \arg\ \min F_k(x)$, we have

(i) $f(x_i) = F_k(x_i)\ \forall\ i < k;\ i, k = 1, 2, \ldots$
(ii) $f(x) \geq F_{k+1}(x) \geq F_k(x);\ k = 1, 2, \ldots$
(iii) $f(x_k) \geq f(\bar{x}) \geq F_{k+1}(x_{k+2}) \geq F_k(x_{k+1}),\ k = 1, 2, \ldots$

In the lemma, (i) states that the sawtooth covers F_k coincide with f at the specified points; (ii) indicates that the covers F_k yield ever tighter (in the wide sense) underestimates of f, and (iii) asserts that the points x_k are nondeteriorating estimates of an optimal point $\bar{x}$ for problem P. We will now proceed with a numerical example as a simple illustration of the technique.

Example Consider the function $f(x) = x^2 - 3x + 5\sin x$ over the interval $I = [-1, 4]$, where it is Lipschitz continuous with a constant $L = 2.5$. The search for a minimal point of f is performed over the interval I using four iterations of the Piyavskii method. Arbitrarily, we start with $x_1 = 1$ and obtain the underestimate $F_1(x) = f(1) - 2.5|x - 1| = 2.207 - 2.5|x - 1|$. Now $F_1(x)$ is minimized over I, which results in $x_2 = 4$, so that $F_1(4) = -5.293$. With $f(4) = 0.216$, we obtain $F_2(x) = \max\{2.207 - 2.5|x - 1|, 0.216 - 2.5|x - 4|\}$. The function $F_2(x)$ is the broken line from $(-1, -2.793)$ to $(1, 2.207)$, then to $(2.898, -2.539)$, and on to $(4, 0.216)$. Its minimum is obtained at $x_3 = -1$, at which point we have $F_2(-1) = -2.793$. With $f_3(x) = f(-1) = -0.207$, we set up the function $F_3(x) = \max\{2.207 - 2.5|x - 1|, 0.216 - 2.5|x - 4|, -0.207 - 2.5|x + 1|\}$, which is minimized for $x_4 = 2.898$ with $F_3(x_4) = F_3(2.898) = -2.539$. Since $x_4 = 2.898$, we evaluate $f(x_4) = f(2.898) = 0.910$, and the last sawtooth cover becomes $F_4(x) = \max\{2.207 - 2.5|x - 1|, 0.216 - 2.5|x - 4|, -0.207 - 2.5|x + 1|, 0.910 - 2.5|x - 2.898|\}$, which has its unique minimal point at $x_5 = -0.4828$, and with $f(x_5) = -0.640$, we stop the process, having carried out four iterations. The process is illustrated in Fig. 2.13, where the function values for $x_1, \ldots, x_5$ are shown.

Our estimate of an optimal point $\bar{x}$ for problem P is $x_5 = -0.4828$ with $f(x_5) = -0.640$. The function $f(x) = x^2 - 3x + 5\sin x$ is actually bimodal with its

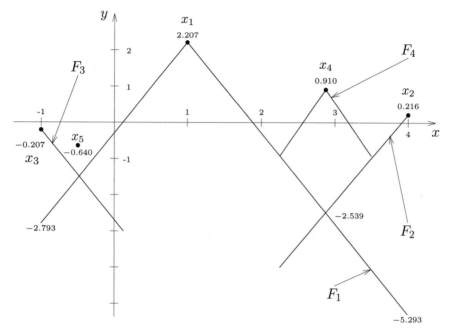

Fig. 2.13 Example of the Piyavskii method

global optimum at $\bar{x} = -0.5847$ with $f(\bar{x}) = -0.6638$, but it also has a local minimum at $\tilde{x} = 3.6651$ with $f(\tilde{x}) = -0.06195$. Our estimate $x_5 = -0.4828$ can therefore be considered reasonable.

In closing, we should note that in practical situations, the interval I will be divided into subintervals and the Piyavskii method applied to each such subinterval, possibly with different Lipschitz constants. The best subinterval is then selected for finer subdivision, whereas dominated subintervals can be dropped from further consideration. It is also clear that the Lipschitz constant L should always be selected as small as possible in order to improve the performance of the Piyavskii method. For a full account of Lipschitz optimization, readers are referred to Hansen and Jaumard (1995) and Horst et al. (2000).

2.2 Multiple-Variable Unconstrained Optimization

One is often tempted to believe that the difference between multivariable and single-variable search problems, which we considered in the previous section, is only one of degree. To see why this is not so, let us assume that in a multivariable minimization problem a point $\mathbf{x}^k$ has been found in the k-th iteration by using a certain method. In order to find the next point $\mathbf{x}^{k+1}$, most methods would proceed in two steps. First a decision would be made in which direction to move from $\mathbf{x}^k$; and secondly, one has to decide on how far to move in the chosen direction, i.e., on the step length. Actually, in so-called trust region methods, to be discussed at the end of this section, the step length is determined before the search direction. Whereas the issue of the step length is the same for one-dimensional and multidimensional problems, the issue of direction is profoundly different. In the one-dimensional case, we can only move in two directions: to the right (which increases the value of $\mathbf{x}$) or to the left (which decreases the value of $\mathbf{x}$). The choice between these two is usually easy, if not trivial. In the multidimensional case, however, we have infinitely many directions to choose from, so that the selection of search direction becomes at least as (if not more) important as the selection of the step length.

2.2.1 Cyclic Coordinate Descent and Simplex Search

A conceptually very simple and appealing technique is the *cyclic coordinate descent* method, in which a minimal point of a function $f(\mathbf{x})$ is sought by sequential one-dimensional searches along the coordinate axes. Given that a starting point $\mathbf{x}^0$, a line search method and a stop criterion are prespecified, the coordinate descent method can formally be described as follows.

Cyclic Coordinate Descent

Step 1: Using the line search method specified and starting from $\mathbf{x}^k$, minimize $f(\mathbf{x})$ as a function of only the k-th component of $\mathbf{x}$. Denote the resulting point by $\mathbf{x}^{k+1}$.
Step 2: Is $k < n$?

If yes: Let $k := k + 1$ and go to Step 1.
If no: Stop with $\hat{\mathbf{x}} = \mathbf{x}^{k+1}$ as the desired point.

Remark The algorithm will finish with $\hat{\mathbf{x}}$ as the result of n one-dimensional minimizations, carried out sequentially along each of the coordinate axes. If the point $\hat{\mathbf{x}}$ is sufficiently good by some measure, the procedure terminates, otherwise the process, starting from $\hat{\mathbf{x}}$, is repeated.

Example Use the cyclic coordinate descent method to minimize the function $f(\mathbf{x}) = 2x_1^2 + x_2^2 + x_1 x_2 - x_1 - 3x_2$, starting from $\mathbf{x}^0 = [1,\ 1]^T$. In Step 1, we find that $f(x_1, 1) = 2x_1^2 + 1 + x_1 - x_1 - 3 = 2x_1^2 - 2$, which is minimized for $x_1 = 0$, therefore $\mathbf{x}^1 = [0,\ 1]^T$. Next, we minimize along the x_2-axis, starting from $\mathbf{x}^1 = [0,\ 1]^T$, so that $f(0, x_2) = x_2^2 - 3x_2$, which is minimized for $x_2 = \tfrac{3}{2}$, and we stop with $\mathbf{x}^2 := \hat{\mathbf{x}} = \left[0, \tfrac{3}{2}\right]^T$. Repeating the procedure and starting with $\hat{\mathbf{x}} = \left[0, \tfrac{3}{2}\right]^T$, we find $f\left(x_1, \tfrac{3}{2}\right) = x_1^2 + \tfrac{1}{2}x_1 - \tfrac{9}{4}$, so that $\mathbf{x}^1 = [-1/4, 3/2]^T$. Next, $f(-\tfrac{1}{4}, x_2) = x_2^2 - \left(\tfrac{13}{4}\right)x_2 + \tfrac{3}{8}$, so that $\mathbf{x}^2 = \left[-\tfrac{1}{4}, \tfrac{13}{8}\right]^T = [-0.25,\ 1.625]^T$, which is now our new $\hat{\mathbf{x}}$. Since the true optimal point is $\bar{\mathbf{x}} = \left[-\frac{1}{7}, \frac{11}{7}\right]^T \approx [-0.143, 1.571]^T$ (obtained by equating the gradients with zero, and checking that the Hessian is positive definite), our solution point $\hat{\mathbf{x}}$, obtained by two sweeps of the cyclical coordinate descent method, is reasonably close.

Although the cyclic coordinate method performed quite well in our example, its convergence properties are in general rather poor; the appeal of the method lies in its simplicity. A slightly more sophisticated method is the *Aitken double sweep* technique, in which, as the name indicates, the search is done along the $x_1, x_2, \ldots, x_n$ axes, and then back again, along the $x_{n-1}, x_{n-2}, \ldots, x_1$ axes. Note that derivatives are not used. In case derivatives are available, we may instead use the *Gauss-Southwell* method, in which the search in each stage is done along the coordinate axis with the steepest descent. Applying the Aitken double sweep or the Gauss-Southwell methods to our example above, we realize that neither method will improve on the performance of basic cyclic coordinate descent. In two dimensions, all we do is keep switching between the two dimensions. Only with three or more dimensions will the Aitken and Gauss-Southwell methods be able to outperform basic cyclic coordinate descent.

Instead of searching along the coordinate axes, the idea underlying the so-called *simplex search method* is quite appealing in its simplicity (pun intended), and it proceeds as follows. Assume that a simplex with corner points $\mathbf{x}^1, \mathbf{x}^2, \ldots, \mathbf{x}^{n+1}$ is given in the n-dimensional space $\mathbb{R}^n$ and that we seek an unconstrained minimal point of a given function $f(\mathbf{x})$. Let $\mathbf{x}^s$ denote the worst of the corners of the given simplex, i.e., $f(\mathbf{x}^s) = \max\limits_{i=1,\ldots n+1} f(\mathbf{x}^i)$. The idea is now to construct a new simplex with the same corners as the previous one, except that $\mathbf{x}^s$ is replaced by its reflection

$\mathbf{x}^{refl}$, defined as $\mathbf{x}^{refl} := \left(\dfrac{2}{n} \sum\limits_{\substack{i=1 \\ i \neq s}}^{n+1} \mathbf{x}^i\right) - \mathbf{x}^s$. Note that $\mathbf{x}^{centr} := \left(\dfrac{1}{n} \sum\limits_{\substack{i=1 \\ i \neq s}}^{n+1} \mathbf{x}^i\right)$ is the center of

the n-dimensional face of the old simplex that does not include $\mathbf{x}^s$, so that $\mathbf{x}^{refl} = \mathbf{x}^s + 2(\mathbf{x}^{centr} - \mathbf{x}^s)$. Figure 2.14 provides an illustration in the two-dimensional space $\mathbb{R}^2$.

Fig. 2.14 The simplex search method

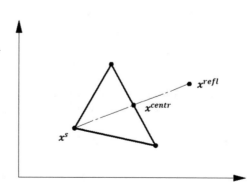

The new simplex now has the corners $\mathbf{x}^i$, $i = 1, \ldots, n + 1$, $i \neq s$, and $\mathbf{x}^{refl}$. At this point, the procedure is repeated. The heuristic algorithm built on this idea needs to be supplemented by a contraction step that reduces the size of the simplex in case cycling occurs, and an expansion step to speed up the search. An initial simplex can be constructed as follows. Given some initial point $\hat{\mathbf{x}}$, add the best of the points $\hat{\mathbf{x}} + \mathbf{e}_j$ and $\hat{\mathbf{x}} - \mathbf{e}_j$, where $\mathbf{e}_j$ denotes the j-th unit vector, and repeat this for all $j = 1, \ldots, n$, the result will be a simplex in $\mathbb{R}^n$. This simplex procedure, not to be confused with the simplex method for linear programming, is originally due to Spendley et al. (1962) and Nelder and Mead (1965). For a detailed description, see Bazaraa et al. (2013) and Bertsekas (2016). It should be noted that by using simplices instead of, say, hypercubes, we avoid the "curse of dimensionality" and the rapidly increasing computational effort as the number n of variables increases. As an example, whereas the simplex in $\mathbb{R}^n$ has only 11 corners to evaluate for $n = 10$, the corresponding hypercube has $2^{10} = 1024$ corners. Without specifying an algorithm for the simplex method, we will use a numerical example to illustrate how to construct an initial simplex and perform a first reflection. For this purpose, we consider the same function and starting point that was used in the numerical example for cyclical coordinate ascent, i.e., $f(\mathbf{x}) = 2x_1^2 + x_2^2 + x_1 x_2 - x_1 - 3x_2$ and the initial point $\hat{\mathbf{x}} = [1, 1]^T$. The initial simplex is then determined by evaluating $f(\mathbf{x})$ at $\hat{\mathbf{x}}$, at $\hat{\mathbf{x}} \pm \mathbf{e}_1$ and $\hat{\mathbf{x}} \pm \mathbf{e}_2$. We thus calculate $f(1, 1) = 0$, $f(2, 1) = 6$, $f(0, 1) = -2$, while $f(1, 2) = 1$ and $f(1, 0) = 1$. The worst, i.e., largest, objective function value is 6, which is obtained for $\mathbf{x} = [2, 1]^T$, so that the initial simplex will have the corners $[1, 1]^T$, $[0, 1]^T$, and $[1, 0]^T$ with associated function values of 0, -2, and 1, respectively. With $\mathbf{x}^s = [1, 0]^T$ being the worst corner of the initial simplex, we obtain $\mathbf{x}^{centr} = [\frac{1}{2}, 1]^T$ and $\mathbf{x}^{refl} = [0, 2]^T$, so that the next simplex will have the corners $[0, 1]^T$, $[1, 1]^T$, and $[0, 2]^T$. From there, the procedure continues.

Returning to the Gauss-Southwell method, a natural extension of the idea of searching along that coordinate axis, which has the steepest descent of the function being minimized is provided by the next method considered here.

2.2.2 The Gradient (Steepest Descent) Method

The *Gradient, Cauchy,* or *Steepest Descent Method* was originally suggested by Cauchy (1847) in the context of solving systems of simultaneous equations, and many other techniques are built upon it. Since the direction of the minimization is based on the gradient only, and not on any higher-order derivatives, it is said to be a *first-order method*.

To clarify what is meant by "steepest descent," assume that from a given point $\mathbf{x}^k \in \mathbb{R}^n$, we move to a new point $\mathbf{x}^{k+1} = \mathbf{x}^k + \alpha\mathbf{d}$, where $\mathbf{d} \in \mathbb{R}^n$ and $\alpha \in \mathbb{R}$. In other words, we move into the direction $\mathbf{d}$ and the step length of this move is expressed by α. If the function f is continuously differentiable, we know from Theorem 1.15 that

$$f\left(\mathbf{x}^k + \alpha\mathbf{d}\right) = f\left(\mathbf{x}^k\right) + \alpha\mathbf{d}^T\nabla f\left(\mathbf{x}^k\right) + R\left(\mathbf{x}^k + \alpha\mathbf{d}\right),$$

where

$$\frac{R\left(\mathbf{x}^k + \alpha\mathbf{d}\right)}{\|\alpha\mathbf{d}\|} \to 0 \text{ as } \alpha \to 0.$$

Hence

$$\frac{f\left(\mathbf{x}^k + \alpha\mathbf{d}\right) - f\left(\mathbf{x}^k\right)}{\alpha} = \mathbf{d}^T\nabla f\left(\mathbf{x}^k\right) + \frac{R\left(\mathbf{x}^k + \alpha\mathbf{d}\right)}{\alpha},$$

so that

$$\lim_{\alpha \to 0} \frac{f\left(\mathbf{x}^k + \alpha\mathbf{d}\right) - f\left(\mathbf{x}^k\right)}{\alpha} = \mathbf{d}^T\nabla f\left(\mathbf{x}^k\right).$$

This limit is called the *directional derivative* of f at $\mathbf{x}^k$ in the direction $\mathbf{d}$. The direction of steepest descent is now defined as the vector $\bar{\mathbf{d}}$ of unit length, which solves the problem $\min_{\|\mathbf{d}\|=1} \mathbf{d}^T\nabla f\left(\mathbf{x}^k\right)$. It is not difficult to show that this problem has the unique solution $\bar{\mathbf{d}} = \dfrac{-\nabla f\left(\mathbf{x}^k\right)}{\|f(\mathbf{x}^k)\|}$ as long as $\nabla f(\mathbf{x}^k) \neq \mathbf{0}$.

As we now know that the vector $-\nabla f(\mathbf{x}^k)$ points into the direction of steepest descent from $\mathbf{x}^k$, it may be sensible to start with a good guess $\mathbf{x}^0$, take a step in the direction $-\nabla f(\mathbf{x}^0)$ to find $\mathbf{x}^1$, and continue this process until some stop criterion is satisfied. We could describe this by writing $\mathbf{x}^{k+1} = \mathbf{x}^k - \alpha_k \nabla f(\mathbf{x}^k), k = 0, 1, 2, \ldots,$ where $\alpha_k > 0$. Of course, we want to choose only improving values of $\mathbf{x}^k$, so that $f(\mathbf{x}^{k+1}) < f(\mathbf{x}^k)$. If $\nabla(\mathbf{x}^k) \neq \mathbf{0}$, it is evident from our discussion above that such an $\alpha_k > 0$ always exists. One can prove that under certain assumptions, this procedure will converge towards a minimal point (albeit possibly only local) of $f(\mathbf{x})$; see, e.g., Luenberger and Ye (2008) for details. The Gradient Method finds values of α_k by minimizing $f(\mathbf{x}^k) - \alpha\nabla f(\mathbf{x}^k)$ for $\alpha > 0$. As $\mathbf{x}^k$ and $\nabla f(\mathbf{x}^k)$ are here known and constant, this is but a real-valued function of one real variable and one of the one-dimensional line searches of the previous section can be applied. Let us now summarize the procedure in algorithmic form. We recall that the problem is to minimize the function $f(\mathbf{x}), \mathbf{x} \in \mathbb{R}^n$. Denote $-\nabla f(\mathbf{x}^k)$ by $\mathbf{s}^k$.

The Steepest Descent Method

Step 1: Start with an initial point $\mathbf{x}^0$. Compute $\mathbf{s}^0 = -\nabla f(\mathbf{x}^0)$, and let $k := 0$. Is $\mathbf{s}^0 = \mathbf{0}$?

If yes: Stop, $\mathbf{x}^0$ is the desired point.

If no: Go to Step 2.

Step 2: Find α_k, the value of $\alpha > 0$ that minimizes $f(\mathbf{x}^k + \alpha \mathbf{s}^k)$.

Step 3: Let $\mathbf{x}^{k+1} := \mathbf{x}^k + \alpha_k \mathbf{s}^k$, and compute $\mathbf{s}^{k+1} := -\nabla f(\mathbf{x}^{k+1})$. Is $\mathbf{s}^{k+1} = \mathbf{0}$?

If yes: Stop, $\mathbf{x}^{k+1}$ is the desired point.

If no: Set $k := k + 1$ and go to Step 2.

Remark 1 We observe that $\dfrac{d}{d\alpha} f(\mathbf{x}^k + \alpha \mathbf{s}^k) = (\mathbf{s}^k)^T \nabla f(\mathbf{x}^k + \alpha \mathbf{s}^k)$ and since this is equal to zero for $\alpha = \alpha_k$, it follows that $0 = (\mathbf{s}^k)^T \nabla f(\mathbf{x}^k + \alpha_k \mathbf{s}^k) = (\mathbf{s}^k)^T \nabla f(\mathbf{x}^{k+1}) = -(\mathbf{s}^k)^T \mathbf{s}^{k+1}$, indicating that successive directions are orthogonal to each other.

Remark 2 As the Steepest Descent method will usually not converge in a finite number of steps, a stop criterion such as the one for the one-dimensional Newton method would be more useful than requiring the gradient to be zero (i.e., $\mathbf{s}^{k+1} = \mathbf{0}$) in Step 3. The simplest such criterion would call for the procedure to terminate after a predetermined number of iterations. Alternatively, one might wish to stop when successive function values do not change very much, i.e., when $\left| \dfrac{f(\mathbf{x}^{k+1}) - f(\mathbf{x}^k)}{\max\{\varepsilon_1\}, |f(\mathbf{x}^{k+1})|, |f(\mathbf{x}^k)|} \right| \leq \varepsilon_2$, or when successive points do not change significantly, i.e., when $\left| \dfrac{\mathbf{x}^{k+1} - \mathbf{x}^k}{\max\{\varepsilon_3\}, |\mathbf{x}^{k+1}|, |\mathbf{x}^k|} \right| \leq \varepsilon_4$, where ε_1, ε_2, ε_3, and ε_4 are preselected positive numbers.

Example The function of two variables $f(x) = f(x_1, x_2) = 2x_1^2 + x_2^2 + x_1 x_2 - x_1 - 3x_2$ is to be minimized using the Steepest Descent method, starting at the point $\mathbf{x}^0 = [1, 1]^T$ and working through three iterations. With $-\nabla f(\mathbf{x}) = \begin{bmatrix} -4x_1 - x_2 + 1 \\ -x_1 - 2x_2 + 3 \end{bmatrix}$ and the initial point $\mathbf{x}^0 = [1, 1]^T$, we obtain $\mathbf{s}^0 = -\nabla f(\mathbf{x}^0) = [-4, 0]^T \neq [0, 0]^T$. We now calculate $f(\mathbf{x}^0 + \alpha \mathbf{s}^0) = f(1 - 4\alpha, 1) = 2(1 - 4\alpha)^2 - 2$, which is minimized for $1 = 4\alpha$, hence $\alpha_0 = \frac{1}{4}$. We now have $\mathbf{x}^1 = \mathbf{x}^0 + \alpha \mathbf{s}^0 = \begin{bmatrix} 1 \\ 1 \end{bmatrix} + \frac{1}{4}\begin{bmatrix} -4 \\ 0 \end{bmatrix} = \begin{bmatrix} 0 \\ 1 \end{bmatrix}$ as well as $\mathbf{s}^1 = -\nabla f(\mathbf{x}^1) = \begin{bmatrix} 0 \\ 1 \end{bmatrix} \neq \begin{bmatrix} 0 \\ 0 \end{bmatrix}$.

The next iteration determines $f(\mathbf{x}^1 + \alpha \mathbf{s}^1) = f(0, 1 + \alpha) = (1 + \alpha)^2 - 3(1 + \alpha) = (\alpha - \frac{1}{2})^2 - \frac{9}{4}$, which is minimized for $\alpha = \frac{1}{2}$, hence $\alpha_1 = \frac{1}{2}$. This leads to $\mathbf{x}^2 = \mathbf{x}^1 + \alpha_1 \mathbf{s}^1 = \begin{bmatrix} 0 \\ 1 \end{bmatrix} + \frac{1}{2}\begin{bmatrix} 0 \\ 1 \end{bmatrix} = \begin{bmatrix} 0 \\ 3/2 \end{bmatrix}$ and $\mathbf{s}^2 = -\nabla f(\mathbf{x}^2) = \begin{bmatrix} -\frac{1}{2} \\ 0 \end{bmatrix} \neq \begin{bmatrix} 0 \\ 0 \end{bmatrix}$. The next iteration has $f(\mathbf{x}^2 + \alpha \mathbf{s}^2) = f(-\frac{1}{2}\alpha, \frac{3}{2}) = \frac{1}{2}\alpha^2 - \frac{1}{4}\alpha - \frac{9}{4} = \frac{1}{2}(\alpha - \frac{1}{4})^2 - \frac{73}{32}$, which is minimized for $\alpha = \frac{1}{4}$, hence $\alpha_2 = \frac{1}{4}$. We now have $\mathbf{x}^3 = \mathbf{x}^2 + \alpha_2 \mathbf{s}^2 = \begin{bmatrix} 0 \\ 3/2 \end{bmatrix} + \frac{1}{4}\begin{bmatrix} -\frac{1}{2} \\ 0 \end{bmatrix} = \begin{bmatrix} -\frac{1}{8} \\ 3/2 \end{bmatrix}$, and $\mathbf{s}^3 = -\nabla f(\mathbf{x}^3) = \begin{bmatrix} 0 \\ \frac{1}{8} \end{bmatrix} \neq \begin{bmatrix} 0 \\ 0 \end{bmatrix}$. After the required three iterations, the procedure terminates. As mentioned above, the exact optimum of the function f occurs at the point $\begin{bmatrix} -1/7 \\ 11/7 \end{bmatrix} \approx \begin{bmatrix} -0.143 \\ 1.571 \end{bmatrix}$, which should be

compared with $\mathbf{x}^3 = \begin{bmatrix} -1/8 \\ 3/2 \end{bmatrix} = \begin{bmatrix} -0.125 \\ 1.500 \end{bmatrix}$, which is in close proximity.

Remark 3 As successive descent directions are orthogonal to each other, the convergence of the steepest descent method may be very slow for some functions. If the level surfaces of the objective function are hyperspheres, convergence will be obtained in one stage. On the other hand, if the level surfaces are elongated hyperellipsoidal, a phenomenon called "zigzagging" occurs, as displayed in Fig. 2.15, where the steepest descent method is applied to the function $f(x_1, x_2) = x_1^2 + 16x_2^2$. The tendency to zigzag may be reduced by modifying the calculation of the next evaluation point as $\mathbf{x}^{k+1} = \mathbf{x}^k + \beta\alpha_k\mathbf{s}^k$, where a "dampening" factor β is used, with a positive value of less than one. We will return to the discussion about zigzagging at the end of this chapter.

Remark 4 The α_k in Step 2 can easily be obtained, if the function f is quadratic and strictly convex. In that case we have $f(\mathbf{x}) = \frac{1}{2}\mathbf{x}^T\mathbf{Q}\mathbf{x} + \mathbf{c}^T\mathbf{x} + c_0$ with some symmetric positive definite matrix $\mathbf{Q}$ (see Theorem 1.35). In our illustration we have $\mathbf{Q} = \begin{bmatrix} 4 & 1 \\ 1 & 2 \end{bmatrix}$, $\mathbf{c} = \begin{bmatrix} -1 \\ -3 \end{bmatrix}$, and $c_0 = 0$. Then $\mathbf{s}^k = -\nabla f(\mathbf{x}^k) = -\mathbf{Q}\mathbf{x}^k - \mathbf{c}$

and $f(\mathbf{x}^k + \alpha\mathbf{s}^k) = \frac{1}{2}\left[(\mathbf{x}^k)^T + \alpha(\mathbf{s}^k)^T\right]\mathbf{Q}(\mathbf{x}^k + \alpha\mathbf{s}^k) + \mathbf{c}^T(\mathbf{x}^k - \alpha\mathbf{s}^k) + c_0$

$= \frac{1}{2}\alpha^2(\mathbf{s}^k)^T\mathbf{Q}\mathbf{s}^k + \alpha\left[(\mathbf{x}^k)^T\mathbf{Q}\mathbf{s}^k + \mathbf{c}^T\mathbf{s}^k\right] + \frac{1}{2}(\mathbf{x}^k)^T\mathbf{Q}\mathbf{x}^k + \mathbf{c}^T\mathbf{x}^k + c_0$. It follows that α minimizes $f(\mathbf{x}^k + \alpha\mathbf{s}^k)$ if and only if $\alpha(\mathbf{s}^k)^T\mathbf{Q}\mathbf{s}^k + (\mathbf{x}^k)^T\mathbf{Q}\mathbf{s}^k + \mathbf{c}^T\mathbf{s}^k = 0$. Hence

we find that $\alpha_k = -\dfrac{(\mathbf{x}^k)^T\mathbf{Q}\mathbf{s}^k + \mathbf{c}^T\mathbf{s}^k}{(\mathbf{s}^k)^T\mathbf{Q}\mathbf{s}^k} = -\dfrac{\left[(\mathbf{x}^k)^T\mathbf{Q} + \mathbf{c}^T\right]\mathbf{s}^k}{(\mathbf{s}^k)^T\mathbf{Q}\mathbf{s}^k} = \dfrac{(\mathbf{s}^k)^T\mathbf{s}^k}{(\mathbf{s}^k)^T\mathbf{Q}\mathbf{s}^k}$, where $\mathbf{s}^k = -(\mathbf{Q}\mathbf{x}^k + \mathbf{c})$.

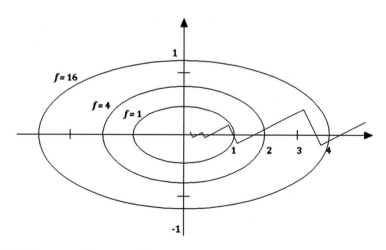

Fig. 2.15 The steepest descent method

2.2.3 Inexact Line Searches

Many methods for minimizing a function of several variables will employ a line search as part of an iteration of the method. By line search we mean the one-dimensional minimization of the function along a given direction. This is the case of the gradient method of the previous section, where in Step 2 we searched for the minimal point of the function in the direction of the negative of the current gradient, or $\mathbf{s}_k = -\nabla f(\mathbf{x}^k)$, in the notation of Sect. 2.2.2. Unless the contour surfaces of the function happen to be close to orthogonal to the true minimal point of the function, which appears unlikely, particularly in the early stages of the procedure, when we are likely far from optimum, it is probably not a useful strategy to spend a lot of computational resources to search along a direction, which may not be leading close to the minimal point. Therefore it makes sense to sacrifice some accuracy in the line searches. At least in the early stages of the computations in order to reduce total computational effort. One way to cut down on the computational burden of a line search is to stop it prematurely, i.e., before the specific line search algorithm has converged. Another way is to use some simple heuristic procedure to find some point on the search line, which is (hopefully) not much poorer than the best. Such a heuristic procedure is provided by the so-called *Armijo's Rule* (from Armijo 1966), which is described below.

In our usual notation, we have a function $f(\mathbf{x})$ to be minimized. Let $\tilde{\mathbf{x}}$ be the current point, for which a line search is to be performed in the given direction $\tilde{\mathbf{s}}$. In the previous section we had $\tilde{\mathbf{s}}_k = -\nabla f(\tilde{\mathbf{x}}^k)$, but in the following sections we will also use other search directions. We assume that $\tilde{\mathbf{s}}^T \nabla f(\tilde{\mathbf{x}}) < 0$, so that $\mathbf{s}$ is an improving direction (remember that we are minimizing), meaning that for small values of α, we have $h(\alpha) := f(\tilde{\mathbf{x}} + \alpha \tilde{\mathbf{s}}) < f(\tilde{\mathbf{x}}) = h(0)$. Let us consider the univariate function $h(\alpha)$, the minimization of which is the purpose of the line search. Since $h'(\alpha) = \dfrac{\partial}{\partial \alpha} f(\tilde{\mathbf{x}} + \alpha \tilde{\mathbf{s}}) = \tilde{\mathbf{s}}^T \nabla f(\tilde{\mathbf{x}} + \alpha \tilde{\mathbf{s}})$, so that $h'(0) = \tilde{\mathbf{s}}^T \nabla f(\tilde{\mathbf{x}}) < 0$. It follows that the tangent to $h(\alpha)$ for $\alpha = 0$ is the line $y = h(0) + \alpha h'(0)$, which has a negative slope, see Fig. 2.16. For a preselected ε with $0 < \varepsilon < 1$, we also consider the line $h(0) + \varepsilon \alpha h'(0)$, which, since $\varepsilon < 1$, has a negative slope that is less steep than that of the tangent (see again Fig. 2.16). Assuming that a finite minimal point exists, these two lines will intersect at one, not necessarily unique, point $\bar{\alpha}$, which, in Armijo's Rule, will be used as an upper bound on the allowable step length α. So as to preclude unacceptably small values of α, another parameter $\delta > 1$ is preselected, and α is required to be sufficiently large to satisfy $h(\delta \alpha) > h(0) + \varepsilon \delta \alpha h'(0)$. This means that α should stay between $\dfrac{\bar{\alpha}}{\delta}$ and $\bar{\alpha}$. The acceptable range for α-values is shown in Fig. 2.16 with $\delta := 2$.

A simple procedure for finding a value α in the acceptable range is to start with some tentative value and test if $h(\alpha) \leq h(0) + \varepsilon \alpha h'(0)$ is satisfied. If so, α is repeatedly increased by the factor δ, until the test is not satisfied. In that case, the last acceptable value of α is chosen.

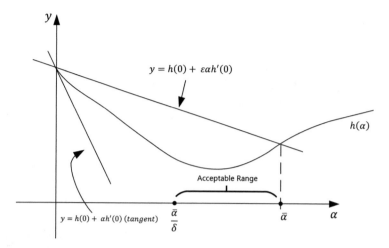

Fig. 2.16 The acceptable range in Armijo's rule

If, on the other hand, the initial value of α is such that $h(\alpha) > h(0) + \varepsilon\alpha h'(0)$, then α is repeatedly reduced by the factor δ until $h(\alpha) \leq h(0) + \varepsilon\alpha h'(0)$ for the first time. We will now formally describe this heuristic procedure. Recall that the problem under consideration is to minimize a differentiable function $f(\mathbf{x})$, starting from a given point $\tilde{\mathbf{x}} \in \mathbb{R}^n$ and moving in a given direction $\tilde{\mathbf{s}} \in \mathbb{R}^n$, which is improving in the sense that $\tilde{\mathbf{s}}^T \nabla f(\tilde{\mathbf{x}}) < 0$. We are given two preselected parameters ε with $0 < \varepsilon < 1$ and $\delta > 1$. Commonly, $\varepsilon = 0.2$ and $\delta = 2$.

Armijo's Rule

Step 1: Define the function $h(\alpha)$ by $h(\alpha) := f\left(\tilde{\mathbf{x}} + \alpha\tilde{\mathbf{s}}\right)$ and pick some α with an arbitrary value.

Step 2: Is the condition $h(\alpha) \leq h(0) + \varepsilon\alpha h'(0)$ satisfied? (Condition *)

If yes: Go to Step 3.
If no: Go to Step 4.

Step 3: Increase the value of α repeatedly by the factor δ, until the condition (*) is violated for the first time. Let $\tilde{\alpha}$ denote the last value of α before the violation occurred and go to Step 5.

Step 4: Reduce the value of α repeatedly by the factor α until the condition (*) is satisfied for the first time. Let $\tilde{\alpha}$ denote this value of α and go to Step 5.

Step 5: Stop the line search with the point $\tilde{\mathbf{x}} + \tilde{\alpha}\tilde{\mathbf{s}}$.

Example Consider the function $f(\mathbf{x}) = f(x_1, x_2) = 2x_1^2 + x_1 x_2 + x_2^2 - x_1 - 3x_2$ and find a minimal point using the Steepest Descent Method with Armijo's Rule line search, starting at the point $\mathbf{x}^0 = [1, 1]^T$ and working through three iterations. This is the same function used in the example in the previous section, we refer to the calculations there. With $\mathbf{x}^0 = [1, 1]^T$, we find $\mathbf{s}^0 = -\nabla f(\mathbf{x}^0) = [-4, 0]^T$, so that

$h(\alpha) = f(\mathbf{x}^0 + \alpha \mathbf{s}^0) = f(1 - 4\alpha, 1) = 2(1 - 4\alpha)^2 - 2$ and $h(0) = 0$. Furthermore, $h'(\alpha) = 64\alpha - 16$, so that condition (*) becomes $2(1 - 4\alpha)^2 - 2 \leq (0.2)\alpha(-16)$, which for $\alpha > 0$ reduces to $\alpha \leq 0.4$. Arbitrarily starting with $\alpha = 1$, we find the condition (*): $\alpha \leq 0.4$ violated, so that α is reduced by the factor $\delta = 2$, which equals $\alpha = \frac{1}{2}$. Condition (*) is still violated, so that $\alpha = \frac{1}{2}$ is once more reduced by a factor $\delta = 2$, resulting in $\alpha = \frac{1}{4}$. At this point, condition (*) is satisfied, so that Armijo's Rule ends with $\tilde{\alpha} = \frac{1}{4}$, which is inserted into the Steepest Descent Method as α_0, the same value that was obtained in Sect. 2.2.2. Now we find that $\mathbf{x}^1 = \mathbf{x}^0 + \alpha \mathbf{s}^0 = [0, 1]^T$ and $\mathbf{s}^1 = [0, 1]^T$, so that $h(\alpha) = f(\mathbf{x}^1 + \alpha \mathbf{s}^1) = f(0, 1 + \alpha) = (1 + \alpha)^2 - 3(1 + \alpha)$, with $h(0) = -2$ and $h'(\alpha) = 2\alpha - 1$. Condition (*) now becomes $(1 + \alpha)^2 - 3(1 + \alpha) \leq -2 + (0.2)\alpha(-1)$, which, for $\alpha > 0$, reduces to $\alpha \leq 0.8$.

Again, arbitrarily starting with $\alpha = 1$, we find condition (*) violated, so that α is reduced by the factor $\delta = 2$, resulting in $\alpha = 0.4$. With this value, condition (*) is satisfied, so that we terminate Armijo's Rule with $\tilde{\alpha} = 0.4$, i.e., $\alpha_1 = 0.4$. Therefore, $\mathbf{x}^2 = \mathbf{x}^1 + \alpha_1 \mathbf{s}^1 = [0, 1]^T + 0.4[0, 1]^T = [0, 1.4]^T$ and $\mathbf{s}^2 = -\nabla f(\mathbf{x}^2)$ $= \begin{bmatrix} -4x_1^2 - x_2^2 + 1 \\ -x_1^2 - 2x_2^2 + 3 \end{bmatrix} = [-0.4, 0.2]^T$, and then we obtain $h(\alpha) = f(\mathbf{x}^2 + \alpha \mathbf{s}^2) = f(-0.4\alpha,$ $1.4 + 0.2\alpha) = 0.28\alpha^2 - 0.2\alpha - 2.24$, so that $h(0) = -2.24$ and $h'(\alpha) = 0.56\alpha - 0.2$. Condition (*) is now $0.28\alpha^2 - 0.2\alpha - 2.24 \leq -2.24 + (0.2)\alpha(-0.2)$, which, for $\alpha > 0$, simplifies to $\alpha \leq \frac{4}{7}$.

As before, we arbitrarily start with $\alpha = 1$, which violates condition (*), and consequently, we reduce α by the factor $\delta = 2$, yielding $\alpha = 0.5$. This value satisfies the condition (*), so that Armijo's Rule terminates with $\tilde{\alpha} = 0.5$, i.e., $\alpha_2 = 0.5$. Now $\mathbf{x}^3 := \mathbf{x}^2 + \alpha \mathbf{s}^2 = [0, 1.4]^T + 0.5[-0.4, 0.2]^T = [-0.2, 1.5]^T$. We also compute $\mathbf{s}^3 = -\nabla f(\mathbf{x}^3)$, which is $\mathbf{s}^3 = [0.3, 0.4]^T$, and with three iterations of the Steepest Descent Method in conjunction with Armijo's Rule line search, we stop. Comparing the solutions obtained with the procedures, we find the exact optimal solution $\bar{\mathbf{x}} = \left[-\frac{1}{7}, \frac{11}{7}\right]^T \approx [0.143, 1.571]^T$, whereas the Steepest Descent Method after three iterations yielded $\mathbf{x}^3 = [-0.125, 1.5]^T$ with exact line search versus $\mathbf{x}^3 = [-0.2, 1.5]^T$ with the inexact Armijo's Rule line search. It appears that the computational relief offered by the inexact line search has only marginally deteriorated the performance of the Steepest Descent Method with exact line search. As a remark, we should also point out that in each iteration of the main algorithm, $h(0)$ corresponds to the current value of the objective function.

There are alternative ways to define the lower end point of the range for the step length α in Armijo's Rule. Specifically, the so-called *Goldstein Rule* (Goldstein 1965) uses the parameter ε, $0 < \varepsilon < \frac{1}{2}$ and requires α to be large enough to satisfy the inequality $h(\alpha) > h(0) + (1 - \varepsilon)\alpha h'(0)$.

Finally, *Wolfe's Rule* (Wolfe 1969) requires α large enough to satisfy $h'(\alpha) > (1 - \varepsilon)h'(0)$. Convergence properties and further discussion of the above and other inexact line search techniques can be found in Cottle and Thapa (2017), Bertsekas (2016), Luenberger and Ye (2008), and Sun and Yuan (2006).

2.2.4 The Davidon–Fletcher–Powell Method

A class of methods known as *conjugate gradient* algorithms will actually produce second derivative information without calculating more than gradients. One of them is the *deflected gradient procedure* developed by Fletcher and Powell (1963) based on earlier work by Davidon (1959). The conjugate gradient algorithms can be regarded as extensions of the steepest descent method, where the search directions are modified. These modified search directions, the conjugate gradients, are obtained by premultiplying the negative of the gradients by square matrices, which turn out to be approximations of the inverse of the Hessian matrix. To be specific, we need to define the concept of conjugate directions.

Definition 2.8 Let $\mathbf{Q}$ be a given symmetric matrix. The directions $\mathbf{s}^k$, $k = 0, \ldots,$ $n - 1$, are said to be $\mathbf{Q}$-*conjugate* if $\left(\mathbf{s}^k\right)^{\mathrm{T}}\mathbf{Q}\mathbf{s}^\ell = 0$ for $k \neq \ell$.

We see that if the $\mathbf{Q}$-matrix is the identity matrix, conjugate directions are simply orthogonal directions, $(\mathbf{s}^k)^T\mathbf{s}^\ell = 0$, $k \neq l$. The following lemma establishes an important property of conjugate directions.

Lemma 2.9 If the symmetric matrix $\mathbf{Q}$ is positive definite, then nonzero $\mathbf{Q}$-conjugate directions are linearly independent.

Proof Suppose that λ_k are any real numbers for $k = 1, \ldots, n - 1$, such that $\sum_k \lambda_k \mathbf{s}^k = \mathbf{0}$, where $\mathbf{s}^k$ are $\mathbf{Q}$-conjugate directions. Then

$$0 = \left(\sum_k \lambda_k \mathbf{s}^k\right)^T \mathbf{Q}\left(\sum_\ell \lambda_\ell \mathbf{s}^\ell\right) = \sum_k \sum_\ell \lambda_k \lambda_\ell \left(\mathbf{s}^k\right)^T \mathbf{Q}\mathbf{s}^\ell = \sum_k (\lambda_k)^2 \left(\mathbf{s}^k\right)^T \mathbf{Q}\mathbf{s}^k.$$

As $\mathbf{Q}$ is positive definite and $\mathbf{s}^k \neq \mathbf{0}$, we have $(\mathbf{s}^k)^T\mathbf{Q}\mathbf{s}^k > 0$ for $k = 0, 1, \ldots, n - 1$, so that the above expression equals zero only if $\lambda_k = 0$ for $k = 0, 1, \ldots, n - 1$. Consequently, the directions $\mathbf{s}^k$ are linearly independent. $\square$

In view of the above discussion, the minimization procedure will now be stated in the following way.

The Davidon–Fletcher–Powell (*DFP*) Method

Step 1: Start with an initial point $\mathbf{x}^k$ and any symmetric positive definite $\mathbf{H}^k$ (e.g., the identity matrix) and let $k := 0$. Is $\nabla f(\mathbf{x}^k) = 0$?

If yes: Stop, $\mathbf{x}^k$ is the desired point.
If no: Go to Step 2.

Step 2: Let $\mathbf{s}^k = -\mathbf{H}^k\nabla f(\mathbf{x}^k)$. Find α_k that minimizes $f(\mathbf{x}^k + \alpha\mathbf{s}^k)$ for $\alpha > 0$, and let $\mathbf{x}^{k+1} = \mathbf{x}^k + \alpha_k\mathbf{s}^k$ (i.e., $= \mathbf{x}^k - \alpha_k\mathbf{H}^k \nabla f(\mathbf{x}^k)$).

Step 3: Is $\nabla f(\mathbf{x}^{k+1}) = \mathbf{0}$?

If yes: Stop, $\mathbf{x}^{k+1}$ is the desired point.
If no: Set $\mathbf{y}^k := \nabla f(\mathbf{x}^{k+1}) - \nabla f(\mathbf{x}^k)$ and go to Step 4.

Step 4: Compute $\mathbf{H}^{k+1} := \mathbf{H}^k + \dfrac{\alpha_k \mathbf{s}^k (\mathbf{s}^k)^T}{(\mathbf{s}^k)^T \mathbf{y}^k} - \dfrac{\mathbf{H}^k \mathbf{y}^k (\mathbf{y}^k)^T \mathbf{H}^k}{(\mathbf{y}^k)^T \mathbf{H}^k \mathbf{y}^k}$, set $k := k + 1$ and go to

Step 2.

Fletcher and Powell have proposed the two following alternative stop criteria (see Step 3 above and also the stop criteria suggested for the steepest descent method): Stop, when $\|\mathbf{s}^k\|$ or the greatest component of $\mathbf{s}^k$ is sufficiently small, or work through at least n iterations and apply the test to $\alpha_k \mathbf{s}^k$ as well as to $\mathbf{s}^k$. To understand why the method converges, we need the following lemmas. For more details and proofs see Fletcher and Powell (1963), Eiselt et al. (1987) or Bazaraa et al. (2013).

Lemma 2.10 If $\mathbf{H}^k$ is positive definite, then so is $\mathbf{H}^{k+1}$.

As $\mathbf{H}^0$ was chosen to be positive definite, the above lemma tells us that all $\mathbf{H}^k$ will be positive definite. From this we can conclude that the single variable function $f(\mathbf{x}^k + \alpha \mathbf{s}^k)$ will be decreased in Step 2 with some $\alpha_k > 0$. The argument is as follows. We have $\nabla f(\mathbf{x}^k) \neq \mathbf{0}$ and

$$\left[\frac{d}{d\alpha} f(\mathbf{x}^k + \alpha \mathbf{s}^k)\right]_{\alpha=0} = (\mathbf{s}^k)^T \nabla f(\mathbf{x}^k) = -\nabla f(\mathbf{x}^k)^T \mathbf{H}^k \nabla f(\mathbf{x}^k) < 0,$$

as $\mathbf{H}^k$ was assumed to be positive definite. Hence $f(\mathbf{x}^k + \alpha \mathbf{s}^k)$ is a strictly decreasing function of α at $\alpha = 0$; it cannot be minimized for $\alpha_k = 0$ and therefore we must have $\alpha_k > 0$.

By inserting $\alpha = \alpha_k$ instead of $\alpha = 0$ in the derivative above, we immediately obtain the following result.

Lemma 2.11

$$\left(\mathbf{s}^k\right)^T \nabla f\left(\mathbf{x}^{k+1}\right) = 0, k = 0, 1, \ldots$$

The next lemma assumes that the objective function is quadratic.

Lemma 2.12 Assume that $f(\mathbf{x}) = \dfrac{1}{2}\mathbf{x}^T \mathbf{Q} \mathbf{x} + \mathbf{c}^T \mathbf{x} + c_0$ with some symmetric positive definite matrix $\mathbf{Q}$. Applying the Davidon–Fletcher–Powell method, we have:

$$\left(\mathbf{s}^k\right)^T \mathbf{Q} \mathbf{s}^\ell = 0, k \neq \ell \text{ and } \mathbf{H}^k \mathbf{Q} \mathbf{s}^\ell = \mathbf{s}^\ell, k > \ell.$$

Lemma 2.12 states that the directions $\mathbf{s}^k$ are $\mathbf{Q}$-conjugate and that for $k > \ell$, $\mathbf{s}^\ell$ is an eigenvector of $\mathbf{H}^k \mathbf{Q}$ with eigenvalue 1. We are now ready to state the main convergence theorem.

Theorem 2.13 When applied to the quadratic function $f(\mathbf{x}) = \frac{1}{2}\mathbf{x}^T\mathbf{Q}\mathbf{x} + \mathbf{c}^T\mathbf{x} + c_0$, where the $[n \times n]$-dimensional matrix $\mathbf{Q}$ is symmetric and positive definite, the Davidon–Fletcher–Powell method converges to the minimal point of f in at most n steps. Furthermore, if n steps are needed, then we have $\mathbf{H}^n = \mathbf{Q}^{-1}$, i.e., the inverse of the Hessian matrix of the function f.

Proof From Lemma 2.12, we know that the directions $\mathbf{s}^k$, $k = 0, 1, \ldots, n-1$ (if convergence has not been obtained earlier), are nonzero and $\mathbf{Q}$-conjugate. By virtue of Lemma 2.9 they are therefore linearly independent, so that they span the entire space $\mathbb{R}^n$. From Lemma 2.11, we know that $\mathbf{H}^n\mathbf{Q}\mathbf{s}^k = \mathbf{s}^k$, $k = 0, 1, \ldots, n-1$. Hence we must have $\mathbf{H}^n\mathbf{Q}\mathbf{x} = \mathbf{x} \; \forall \; \mathbf{x} \in \mathbb{R}^n$. Consequently, $\mathbf{H}^n\mathbf{Q}$ must be the identity matrix, so that $\mathbf{H}^n = \mathbf{Q}^{-1}$. Lemma 2.12 also indicates that $(\mathbf{s}^k)^T\mathbf{Q}\mathbf{s}^n = 0$, $k = 0, 1, \ldots,$ $n-1$, i.e., $(\mathbf{s}^k)^T\mathbf{Q}\mathbf{H}^n \nabla f(\mathbf{x}^n) = 0$ or $\nabla f(\mathbf{x}^n) = \mathbf{0}$. Using again the fact that $\mathbf{s}^k$, $k = 0, 1, \ldots, n-1$ span $\mathbb{R}^n$, we conclude that $\nabla f(\mathbf{x}^n) = \mathbf{0}$, so that $\mathbf{x}^n$ is the minimal point. $\square$

Example To see how the Davidon–Fletcher–Powell method works, we consider the same numerical example as before where the function was defined as

$$f(x_1, x_2) = 2x_1^2 + x_2^2 + x_1x_2 - x_1 - 3x_2.$$

Starting with $\mathbf{x}^0 = \begin{bmatrix} 1 \\ 1 \end{bmatrix}$ and the identity matrix $\mathbf{H}^0 = \begin{bmatrix} 1 & 0 \\ 0 & 1 \end{bmatrix}$, we first obtain

$\mathbf{s}^0 = -\mathbf{H}^0 \nabla f(\mathbf{x}^0) = -\nabla f(\mathbf{x}^0) = \begin{bmatrix} -4 \\ 0 \end{bmatrix}$, and, from the optimum gradient example,

$\alpha_0 = \frac{1}{4}$. Then $\mathbf{x}^1 = \begin{bmatrix} 0 \\ 1 \end{bmatrix}$, $\nabla f(\mathbf{x}^1) = \begin{bmatrix} 0 \\ -1 \end{bmatrix}$, and $\mathbf{y}_0 = \nabla f(\mathbf{x}^1) - \nabla f(\mathbf{x}^0) = \begin{bmatrix} -4 \\ -1 \end{bmatrix}$.

Furthermore, $\dfrac{\alpha_0\mathbf{s}^0(\mathbf{s}^0)^T}{(\mathbf{s}^0)^T\mathbf{y}_0} = \frac{1}{4}\begin{bmatrix} 1 & 0 \\ 0 & 0 \end{bmatrix}$ and $\dfrac{\mathbf{H}^0\mathbf{y}_0(\mathbf{y}^0)^T\mathbf{H}^0}{(\mathbf{y}^0)^T\mathbf{H}^0\mathbf{y}_0} = \frac{1}{17}\begin{bmatrix} 16 & 4 \\ 4 & 1 \end{bmatrix}$. This gives

$$\mathbf{H}^1 = \begin{bmatrix} 1 & 0 \\ 0 & 1 \end{bmatrix} + \frac{1}{4}\begin{bmatrix} 1 & 0 \\ 0 & 0 \end{bmatrix} - \frac{1}{17}\begin{bmatrix} 16 & 4 \\ 4 & 1 \end{bmatrix} = \begin{bmatrix} \dfrac{21}{68} & -\dfrac{4}{17} \\ -\dfrac{4}{17} & \dfrac{16}{17} \end{bmatrix} \text{ and } \mathbf{s}^1 = -\mathbf{H}^1 \nabla f$$

$(\mathbf{x}^1) = \dfrac{4}{17}\begin{bmatrix} -1 \\ 4 \end{bmatrix}$. We then obtain $f(\mathbf{x}^1 + \alpha\mathbf{s}^1) = f\left(-\dfrac{4}{17}\alpha, 1 + \dfrac{16}{17}\alpha\right) = 2\left(\dfrac{4}{17}\alpha\right)^2$

$+ (1 + \dfrac{16}{17}\alpha)^2 + \left(-\dfrac{4}{17}\alpha\right)\left(1 + \dfrac{16}{17}\alpha\right) + \dfrac{4}{17}\alpha - 3\left(1 + \dfrac{16}{17}\alpha\right).$

This function is minimized for $\alpha = \alpha_1 = \dfrac{17}{28}$, so that $\mathbf{x}^2 = \mathbf{x}^1 + \alpha_1\mathbf{s}^1 = \begin{bmatrix} 0 \\ 1 \end{bmatrix} +$

$\dfrac{17}{28}\dfrac{4}{17}\begin{bmatrix} -1 \\ 4 \end{bmatrix} = \begin{bmatrix} -\dfrac{1}{7} \\ \dfrac{11}{7} \end{bmatrix}$, for which the gradient is zero, so that it is optimal. In

agreement with Theorem 2.13, convergence was obtained in two steps for this two-variable problem.

2.2.5 The Fletcher–Reeves Conjugate Gradient Method

Since the development of the Davidon–Fletcher–Powell method, many other conjugate gradient methods have been proposed. They are motivated by the desire to
accelerate the often slow convergence associated with the steepest descent method,
while avoiding the information and computation requirements associated with the
evaluation, storage, and inversion of the Hessian matrix.

We will here consider the method proposed by Fletcher and Reeves (1964). It
implements a conjugate gradient algorithm for convex problems by essentially
requiring only line searches. As for the Davidon–Fletcher–Powell method, it converges in n steps for a quadratic function. The idea behind the method is to generate
conjugate directions in a simple fashion. Once conjugate directions are found, the
rest is straightforward, as specified by the following result, which follows along the
lines of our previous discussion.

Theorem 2.14 Given n nonzero $\mathbf{Q}$-conjugate directions, where $\mathbf{Q}$ is a positive
definite symmetric $[n \times n]$-dimensional matrix, we can minimize the quadratic
function $f(\mathbf{x}) = \frac{1}{2}\mathbf{x}^T\mathbf{Q}\mathbf{x} + \mathbf{c}^T\mathbf{x} + c_0$ in at most n steps.

Proof Starting with $\mathbf{x}^0$, assume that $\mathbf{x}^k$ has been found. Then we determine the
value α_k, which minimizes $f(\mathbf{x}^k + \alpha\mathbf{s}^k)$. This requires that $(\mathbf{s}^k)^T \nabla f(\mathbf{x}^{k+1}) = 0$.
Letting $\mathbf{x}^{k+1} = \mathbf{x}^k + \alpha_k\mathbf{s}^k$, we obtain $(\mathbf{s}^k)^T \nabla f(\mathbf{x}^{k+1}) = 0$. Also,
$\mathbf{x}^\ell = \mathbf{x}^{k+1} + \sum_{v=k+1}^{\ell-1} \alpha_v\mathbf{s}^v, 1 \leq k+1 < \ell \leq n$. However, as $\nabla f(\mathbf{x}) = \mathbf{Q}\mathbf{x} + \mathbf{c}$, it follows
that, by premultiplying the above equation by $\mathbf{Q}$ and adding $\mathbf{c}$, we have $\nabla f(\mathbf{x}^\ell) =$
$\nabla f(\mathbf{x}^{k+1}) + \sum_{v=k+1}^{\ell-1} \alpha_v\mathbf{Q}\mathbf{s}^v$, $1 \leq k+1 < \ell \leq n$, and since $0 = (\mathbf{s}^k)^T \nabla f(\mathbf{x}^{k+1})$, we have
$$(\mathbf{s}^k)^T f(\mathbf{x}^\ell) = (\mathbf{s}^k)^T \nabla f(\mathbf{x}^{k+1}) + \sum_{v=k+1}^{\ell-1} \alpha_v(\mathbf{s}^k)^T\mathbf{Q}\mathbf{s}^v = \sum_{v=k+1}^{\ell-1} \alpha_v(\mathbf{s}^k)^T\mathbf{Q}\mathbf{s}^v, 1 \leq k+1 < \ell$$
$\leq n$. The fact that $\mathbf{s}^k$ are $\mathbf{Q}$-conjugate for $k = 0, 1, \ldots, n-1$ implies that $(\mathbf{s}^k)^T \nabla f$
$(\mathbf{x}^\ell) = 0$ for $0 \leq k < \ell \leq n$, which, when $\ell = n$, turns into $(\mathbf{s}^k)^T \nabla f(\mathbf{x}^n) = 0$, $k = 0$,
$1, \ldots, n-1$. Now from Lemma 2.9 above we already know that the vectors $\mathbf{s}^k$ are
linearly independent, so that $\nabla f(\mathbf{x}^n) = \mathbf{0}$ as it is orthogonal to every vector in $\mathbb{R}^n$.
Hence $\mathbf{x}^n$ is the minimal point of f. $\square$

Our task is now reduced to generating $\mathbf{Q}$-conjugate directions. Starting with $\mathbf{s}^0 =$
$-\nabla f(\mathbf{x}^0)$, we try to find $\mathbf{s}^{k+1}$ as a sum of $-\nabla f(\mathbf{x}^{k+1})$ and a linear combination of $\mathbf{s}^0$,
$\mathbf{s}^1, \ldots, \mathbf{s}^k$, so that the $\mathbf{Q}$-conjugacy is preserved. It can be shown (see, e.g., Bazaraa
et al. 2013) that this is accomplished by Step 4 of the Fletcher–Reeves method
below, which is now described in algorithmic form. Recall that the problem is to
minimize the function $f(\mathbf{x})$.

The Fletcher–Reeves Method

Step 1: Start with an initial point $\mathbf{x}^0$, let $\mathbf{s}^0 = -\nabla f(\mathbf{x}^0)$ and set $k := 0$.

Is $\nabla f(\mathbf{x}^0) = \mathbf{0}$?

If yes: Stop, $\mathbf{x}^0$ is the desired point.
If no: Go to Step 2.

Step 2: Find α_k which minimizes $f(\mathbf{x}^k + \alpha \mathbf{s}^k)$ for $\alpha > 0$.
Step 3: Let $\mathbf{x}^{k+1} := \mathbf{x}^k + \alpha_k \mathbf{s}^k$. Is $\nabla f(\mathbf{x}^{k+1}) = \mathbf{0}$?

If yes: Stop, $\mathbf{x}^{k+1}$ is the desired point.
If no: Go to Step 4.

Step 4: Let $\left(\mathbf{s}^k\right)^T := -\nabla f\left(\mathbf{x}^{k+1}\right) + \dfrac{\left\|\nabla f\left(\mathbf{x}^{k+1}\right)\right\|^2}{\left\|\nabla f(\mathbf{x}^k)\right\|^2} \mathbf{s}^k$, set $k := k + 1$, and go to Step 2.

Our assumption that $\alpha_k > 0$ in Step 2 is clearly legitimate for $k = 0$, and because for $k > 1$, we have

$$\left(\mathbf{s}^k\right)^T \nabla f\left(\mathbf{x}^k\right) = \left[-\nabla f\left(\mathbf{x}^k\right) + \frac{\left\|\nabla f\left(\mathbf{x}^k\right)\right\|^2}{\left\|\nabla f(\mathbf{x}^{k-1})\right\|^2} \mathbf{s}^{k-1}\right]^T \nabla f\left(\mathbf{x}^k\right) = -\left\|\nabla f\left(\mathbf{x}^k\right)\right\|^2,$$

due to Lemma 2.11, so that the function $f(\mathbf{x})$ is strictly decreasing at $\mathbf{x}^k$ in the direction $\mathbf{s}^k$. We also need a stop criterion, which could be any of the stop criteria suggested for the Steepest Descent or Davidon–Fletcher–Powell methods of previous sections.

Example We now apply the Fletcher–Reeves method to minimize the same function considered above, i.e.,

$$f(x_1, x_2) = 2x_1^2 + x_2^2 + x_1 x_2 - x_1 - 3x_2.$$

We start from $\mathbf{x}^0 = \begin{bmatrix} 1 \\ 1 \end{bmatrix}$ and obtain $\mathbf{s}^0 = \begin{bmatrix} -4 \\ 0 \end{bmatrix}$, $\alpha_0 = \frac{1}{4}$, $\mathbf{x}^1 = \begin{bmatrix} 0 \\ 1 \end{bmatrix}$, and $\nabla f(\mathbf{x}^1)$

$= \begin{bmatrix} 0 \\ -1 \end{bmatrix}$ as for the Davidon–Fletcher–Powell method. Next, $\mathbf{s}^1 = -\nabla f(\mathbf{x}^1) +$

$\dfrac{\left\|\nabla f(\mathbf{x}^1)\right\|^2}{\left\|f(\mathbf{x}^0)\right\|^2} \mathbf{s}^0 = -\begin{bmatrix} 0 \\ -1 \end{bmatrix} + \dfrac{1}{16} \begin{bmatrix} -4 \\ 0 \end{bmatrix} = \begin{bmatrix} -\frac{1}{4} \\ 1 \end{bmatrix}$. We then obtain $f(\mathbf{x}^1 + \alpha \mathbf{s}^1) = f$

$(-\frac{1}{4}\alpha, 1 + \alpha) = 2(\frac{1}{4}\alpha)^2 + (1 + \alpha)^2 + (-\frac{1}{4}\alpha)(1 + \alpha) + \frac{1}{4}\alpha - 3(1 + \alpha)$. This function is

minimized for $\alpha = \alpha_1 = \dfrac{4}{7}$, so that $\mathbf{x}^2 = \mathbf{x}^1 + \alpha_1 \mathbf{s}^1 = \begin{bmatrix} 0 \\ 1 \end{bmatrix} + \dfrac{4}{7} \begin{bmatrix} -\frac{1}{4} \\ 1 \end{bmatrix} = \begin{bmatrix} -\frac{1}{7} \\ \frac{11}{7} \end{bmatrix}$, which is optimal. This is the same result as that obtained for the Davidon–Fletcher–Powell method.

Since the development of conjugate gradient methods in the early 1960s, a considerable number of different unconstrained minimization algorithms have been proposed. Most of them fall into the category of co-called *quasi-Newton* or *variable metric methods*. This is a class of techniques that does not require the computationally costly evaluation and inversion of the Hessian matrix, but which nevertheless builds up some information corresponding to the knowledge of the Hessian matrix. Thus the Steepest Descent method does not belong into this class, whereas the Davidon–Fletcher–Powell method and the *BFGS* method (to be described in the next section) do.

The Davidon–Fletcher–Powell method is sometimes referred to as a "rank two method," since the matrix by which $\mathbf{H}^k$ is updated to $\mathbf{H}^{k+1}$ in Step 4 can be shown to be of rank two. Techniques have been constructed which use a slightly different updating formula for $\mathbf{H}^k$, *viz.*,

$$\mathbf{H}^{k+1} = \mathbf{H}^k + \frac{\left(\mathbf{y}^k - \alpha_k \mathbf{H}^k \mathbf{s}^k\right)\left(\mathbf{y}^k - \alpha_k \mathbf{H}^k \mathbf{s}^k\right)^T}{\alpha_k \left(\mathbf{y}^k - \alpha_k \mathbf{H}^k \mathbf{s}^k\right)^T \mathbf{s}^k}.$$

The second term on the right-hand side can be shown to be a matrix of rank one, and consequently such techniques are referred to as "rank one methods," see, e.g., Luenberger and Ye (2008).

2.2.6 The BFGS Method

Although about half a century old, the so-called *BFGS* method is still generally considered the most efficient of the quasi-Newton methods. Its name derives from the fact that it was independently and simultaneously developed by four different individuals: Broyden (1970), Fletcher (1970), Goldfarb (1970), and Shanno (1970). It is quite similar to the Davidon–Fletcher–Powell method described above; we discuss this similarity in more detail below. Rather than provide a formal description of the procedure, we just outline the major differences to the Davidon–Fletcher–Powell method. Recall that the problem at hand is to minimize the function $f(\mathbf{x})$. The method can then be described as follows.

The *BFGS* Method

Steps 1–3 are identical to those of the Davidon–Fletcher–Powell method.

Step 4: Set $\mathbf{H}^{k+1} := \mathbf{H}^k + \left[1 + \dfrac{\left(\left(\mathbf{y}^k\right)^T \mathbf{H}^k \mathbf{y}^k\right)}{\alpha_k \left(\mathbf{s}^k\right)^T \mathbf{y}^k}\right]\left[\dfrac{\alpha_k \mathbf{s}^k \left(\mathbf{s}^k\right)^T}{\left(\mathbf{s}^k\right)^T \mathbf{y}^k}\right] - \dfrac{\mathbf{s}^k \left(\mathbf{y}^k\right)^T \mathbf{H}^k + \mathbf{H}^k \mathbf{y}^k \left(\mathbf{s}^k\right)^T}{\left(\mathbf{s}^k\right)^T \mathbf{y}^k},$

set $k := k + 1$, and go to Step 2.

Considerations regarding stop criteria other than a vanishing gradient (as in Steps 1 and 3) are the same as for the Davidon–Fletcher–Powell method.

Example As a numerical illustration of the *BFGS* method, we use the same function

$f(x_1, x_2) = 2x_1^2 + x_2^2 + x_1 x_2 - x_1 - 3x_2$ and start from $\mathbf{x}^0 = \begin{bmatrix} 1 \\ 1 \end{bmatrix}$ with the identity

matrix $\mathbf{H}^0 = \begin{bmatrix} 1 & 0 \\ 0 & 1 \end{bmatrix}$. As with the Davidon–Fletcher–Powell method, we obtain

$\mathbf{s}^0 = -\mathbf{H}^0 \nabla f(\mathbf{x}^0) = -\nabla f(\mathbf{x}^0) = \begin{bmatrix} -4 \\ 0 \end{bmatrix}$, then $\alpha_0 = \frac{1}{4}$, so that $\mathbf{x}^1 = \begin{bmatrix} 0 \\ 1 \end{bmatrix}$, $\nabla f(\mathbf{x}^1) =$

$\begin{bmatrix} 0 \\ -1 \end{bmatrix} \neq \begin{bmatrix} 0 \\ 0 \end{bmatrix}$, and $\mathbf{y}_0 = \nabla f(\mathbf{x}^1) - \nabla f(\mathbf{x}^0) = \begin{bmatrix} -4 \\ -1 \end{bmatrix}$. Now $\left(\mathbf{s}^0\right)^T \mathbf{y}^0 = [-4 \ \ 0] \begin{bmatrix} -4 \\ -1 \end{bmatrix}$

$= 16$, $\alpha_0 \mathbf{s}^0 (\mathbf{s}^0)^T = \frac{1}{4} \begin{bmatrix} -4 \\ 0 \end{bmatrix} [-4 \ \ 0] = \begin{bmatrix} 4 & 0 \\ 0 & 0 \end{bmatrix}$. Next, $(\mathbf{s}^0)^T \mathbf{y}^0 = [-4 \ \ 0] \begin{bmatrix} -4 \\ -1 \end{bmatrix} =$

16, $\left(\mathbf{y}^0\right)^T \mathbf{H}^0 \mathbf{y}^0 = [-4 \ \ -1] \begin{bmatrix} 1 & 0 \\ 0 & 1 \end{bmatrix} \begin{bmatrix} -4 \\ -1 \end{bmatrix} = 17$, so that $\left[1 + \dfrac{(\mathbf{y}^0)^T \mathbf{H}^0 \mathbf{y}^0}{\alpha_0 (\mathbf{s}^0)^T \mathbf{y}^0}\right]$

$\left[\dfrac{\alpha_0 \mathbf{s}^0 (\mathbf{s}^0)^T}{(\mathbf{s}^0)^T \mathbf{y}^0}\right] = \left[1 + \dfrac{17}{\frac{1}{4}(16)}\right] \dfrac{1}{16} \begin{bmatrix} 4 & 0 \\ 0 & 0 \end{bmatrix} = \begin{bmatrix} \frac{21}{16} & 0 \\ 0 & 0 \end{bmatrix}$. Furthermore, $\mathbf{s}^0 (\mathbf{y}^0)^T \mathbf{H}^0 =$

$\begin{bmatrix} -4 \\ 0 \end{bmatrix} [-4 \ \ -1] \begin{bmatrix} 1 & 0 \\ 0 & 1 \end{bmatrix} = \begin{bmatrix} 16 & 4 \\ 0 & 0 \end{bmatrix}$, so that $\dfrac{\mathbf{s}^0 (\mathbf{y}^0)^T \mathbf{H}^0 + \mathbf{H}^0 \mathbf{y}^0 (\mathbf{s}^0)^T}{(\mathbf{s}^0)^T \mathbf{y}^0} =$

$\dfrac{1}{16} \begin{bmatrix} 32 & 4 \\ 4 & 0 \end{bmatrix} = \begin{bmatrix} 2 & \frac{1}{4} \\ \frac{1}{4} & 0 \end{bmatrix}$. Finally then, $\mathbf{H}^1 = \begin{bmatrix} 1 & 0 \\ 0 & 1 \end{bmatrix} + \begin{bmatrix} \frac{21}{16} & 0 \\ 0 & 0 \end{bmatrix} - \begin{bmatrix} 2 & \frac{1}{4} \\ \frac{1}{4} & 0 \end{bmatrix} =$

$\begin{bmatrix} \frac{5}{16} & \frac{1}{4} \\ -\frac{1}{4} & 1 \end{bmatrix}$. Then $\mathbf{s}^1 = -\mathbf{H}^1 \nabla f(\mathbf{x}^1) = -\begin{bmatrix} \frac{5}{16} & \frac{1}{4} \\ -\frac{1}{4} & 1 \end{bmatrix} \begin{bmatrix} 0 \\ -1 \end{bmatrix} = \begin{bmatrix} -\frac{1}{4} \\ 1 \end{bmatrix}$. Therefore, in

Step 2 of the next iteration, we have $f(\mathbf{x}^1 + \alpha \mathbf{s}^1) = f(-\frac{1}{4}\alpha, 1 + \alpha) = 2(\frac{1}{4}\alpha)^2$

$+ (1 + \alpha)^2 + (-\frac{1}{4}\alpha)(1 + \alpha) + \frac{1}{4}\alpha - 3(1 + \alpha)$, which is minimized for $\alpha^1 = \dfrac{4}{7}$, which

yields $\mathbf{x}^2 = \mathbf{x}^1 + \alpha_1 \mathbf{s}^1 = \begin{bmatrix} 0 \\ 1 \end{bmatrix} + \dfrac{4}{7} \begin{bmatrix} -\frac{1}{4} \\ 1 \end{bmatrix} = \begin{bmatrix} -\frac{1}{7} \\ \frac{11}{7} \end{bmatrix}$ which, as we know from our

previous discussion, is optimal. Again, as with the Davidon–Fletcher–Powell and Fletcher–Reeves methods, we have converged to the true optimal point in two steps.

In comparing the $\mathbf{H}^{k+1}$ updates for the *DFP* (Davidon–Fletcher–Powell) and *BFGS* methods, we find that they look similar. It turns out that if we form a linear convex combination of the two updates, $\mathbf{H}^{DFP}$ for the *DFP* method and $\mathbf{H}^{BFGS}$ for the *BFGS* method, to obtain the "compromise" update $\mathbf{H}^{\lambda} = \lambda\mathbf{H}^{DFP} + (1 - \lambda)\mathbf{H}^{BFGS}$, where $\lambda \in [0, 1]$ is some chosen number, we obtain a whole class of updates called the *Broyden family*, and using these updates, we call the corresponding method a *Broyden method*. One may also let the value of λ vary between iterations of the algorithm. It is also possible to obtain updating formulas for directly computing the Hessian matrix instead of its inverse. For details, see, e.g. Luenberger and Ye (2008).

2.2.7   The Newton Minimization Method

So far we have dealt with multiple-variable minimization methods that require no more than first-order derivatives. If second-order derivatives are available, so that not only the gradient but also the Hessian matrix can be computed, we can apply the classical *Newton Minimization Method* for finding local minima of a given function $f(\mathbf{x})$. This method is the multidimensional analog of the one-dimensional Newton minimization search described in Sect. 2.1.8. The basic idea of the method is to produce a quadratic approximation of the function f in the vicinity of the current estimate of a minimal point. The minimal point of this quadratic approximation is then the next estimate of a minimal point, and the process continues.

To be specific, assume that we seek to minimize a twice continuously differentiable given function $f(\mathbf{x})$, where $\mathbf{x} \in \mathbb{R}^n$, with the gradient $\nabla f(\mathbf{x})$ and symmetric Hessian matrix $\mathbf{H}_f(\mathbf{x})$. Given a current estimate $\mathbf{x}^k$ of a minimal point, we can use Theorem 1.17 of Sect. 1.2 to approximate f in a vicinity of the point $\mathbf{x}^k$ by the quadratic function

$$\tilde{f}(\mathbf{x}) = f(\mathbf{x}^k) + (\mathbf{x} - \mathbf{x}^k)^T \nabla f(\mathbf{x}^k) + \frac{1}{2}(\mathbf{x} - \mathbf{x}^k)^T \mathbf{H}_f(\mathbf{x}^k)(\mathbf{x} - \mathbf{x}^k).$$

Then the next estimate $\mathbf{x}^{k+1}$ will be the point, at which the quadratic function $\tilde{f}(\mathbf{x})$ is minimized, i.e., the point at which $\nabla \tilde{f}(\mathbf{x}) = \nabla f(\mathbf{x}^k) + \mathbf{H}_f(\mathbf{x}^k)(\mathbf{x} - \mathbf{x}^k) = \mathbf{0}$. Therefore, $\nabla f(\mathbf{x}^k) + H_f(\mathbf{x}^k)(\mathbf{x}^{k+1} - \mathbf{x}^k) = \mathbf{0}$, and assuming that for all $\mathbf{x} \neq \bar{\mathbf{x}}$ in a vicinity of an optimal point $\bar{\mathbf{x}}$, the Hessian matrix $\mathbf{H}_f(\mathbf{x})$ is positive definite, its inverse $\mathbf{H}_f(\mathbf{x})^{-1}$ will exist. Then we obtain

$$\mathbf{x}^{k+1} = \mathbf{x}^k - \mathbf{H}_f^{-1}(\mathbf{x}^k)\nabla f(\mathbf{x}^k)$$

as the iteration formula for the Newton method. The resemblance to the one-dimensional Newton method of Sect. 2.1.8 becomes more obvious if we write the iteration formula for that method as

$$x^{k+1} = x^k - \left(f''(x^k)\right)^{-1} f'(x^k).$$

Repeating the analogy with our development in Sect. 2.1.8, in which we briefly referred to the Newton–Raphson method for solving an equation, we would here consider a system of equations written as $\mathbf{g}(\mathbf{x}) = \mathbf{0}$, where $\mathbf{g} : \mathbb{R}^n \to \mathbb{R}^n$ is a differentiable function. With $\mathbf{g}(\mathbf{x})$ replacing $\nabla f(\mathbf{x})$, the Newton–Raphson iteration formula for solving the equations $\mathbf{g}(\mathbf{x}) = \mathbf{0}$ is then

$$\mathbf{x}^{k+1} = \mathbf{x}^k - \left(\frac{\partial \mathbf{g}}{\partial \mathbf{x}}\right)^{-1}_{\mathbf{x}=\mathbf{x}^k} \mathbf{g}(\mathbf{x}^k).$$

One word of caution concerning the multidimensional Newton method is that it requires the evaluation of the inverse $\mathbf{H}_f^{-1}$, which can be quite cumbersome in spaces $\mathbb{R}^n$ of higher dimension. The quasi-Newton methods in the previous sections do not require the inversion of square matrices, while still providing an estimate of the inverse Hessian at optimality. On the other hand, they require line searches, which the Newton method does not. Also, for quadratic functions, we realize that the Newton method will converge in one single step, just as its one-dimensional counterpart.

The algorithm can now formally be described as follows. The problem is that of minimizing a twice continuously differentiable function $f(\mathbf{x})$, where $\mathbf{x} \in \mathbb{R}^n$. We assume that some stop criterion is given, e.g., a criterion similar to those proposed in the previous sections.

The Newton Minimization Method

Step 1: Start with an initial guess $\mathbf{x}^0$ of the minimal point. Let $k := 0$.
Step 2: Compute $\nabla f(\mathbf{x}^k)$ and $\mathbf{H}^{-1}(\mathbf{x}^k)$. Let $\mathbf{x}^{k+1} := \mathbf{x}^k - \mathbf{H}_f^{-1}(\mathbf{x}^k) \nabla f(\mathbf{x}^k)$.
Step 3: Is the stop criterion satisfied?

If yes: Stop, a solution $\mathbf{x}^{k+1}$ has been found.
If no: Let $k := k + 1$ and go to Step 2.

Example 1 To make comparisons with the previous methods possible, the Newton technique will now be applied to the same numerical example as before. Since f is a quadratic function, the minimum will be attained in just one iteration. With $f(x_1, x_2) = 2x_1^2 + x_2^2 + x_1 x_2 - x_1 - 3x_2$, we have $\nabla f(\mathbf{x}) = \begin{bmatrix} 4x_1 + x_2 - 1 \\ x_1 + 2x_2 - 3 \end{bmatrix}$ and $\mathbf{H}_f(\mathbf{x}) = \begin{bmatrix} 4 & 1 \\ 1 & 2 \end{bmatrix}$.

As f is quadratic, all its second order partial derivatives are constant, so that $\mathbf{H}_f$ will be constant. $\mathbf{H}_f$ is positive definite, reflecting the fact that f is strictly convex. Hence

$$\mathbf{H}_f^{-1} = \begin{bmatrix} 2/7 & -1/7 \\ -1/7 & 4/7 \end{bmatrix}.$$

Step 1: We start with $\mathbf{x}^0 = \begin{bmatrix} 1 \\ 1 \end{bmatrix}$ as before.

Step 2: $\nabla f(\mathbf{x}^0) = \begin{bmatrix} 4 \\ 0 \end{bmatrix}$ and $\mathbf{H}_f = \begin{bmatrix} 4 & 1 \\ 1 & 2 \end{bmatrix}$, so that $\mathbf{x}^1 = \mathbf{x}^0 - \mathbf{H}_f^{-1}(\mathbf{x}^0)\nabla f(\mathbf{x}^0)$

$$= \begin{bmatrix} 1 \\ 1 \end{bmatrix} - \begin{bmatrix} 2/7 & -1/7 \\ -1/7 & 4/7 \end{bmatrix}\begin{bmatrix} 4 \\ 0 \end{bmatrix} = \begin{bmatrix} -1/7 \\ 11/7 \end{bmatrix} \approx \begin{bmatrix} -0.143 \\ 1.57 \end{bmatrix}.$$

Step 3: As $\nabla f(\mathbf{x}^1) = \begin{bmatrix} 0 \\ 0 \end{bmatrix}$ and $\mathbf{H}_f$ is positive definite, $\mathbf{x}^1$ is the exact minimal point and the process terminates after just one iteration.

In Example 1 above, the function to be minimized was a quadratic function, the same as that used in previous sections, for comparative purposes. However, since Newton's method applied to quadratic functions will converge in one single step, we will now show how the iterative procedure works, using an example with a nonquadratic function.

Example 2 Consider the function $f(x_1, x_2) = 2e^{-x_1} + x_1 + 3e^{-x_2} + x_2$ and minimize it, using Newton's Method. We will start from the arbitrary initial point $\mathbf{x}^0 = \begin{bmatrix} 0 \\ 0 \end{bmatrix}$ and carry out three iterations. First, we obtain $\nabla f(\mathbf{x}) = \begin{bmatrix} -2e^{-x_1} + 1 \\ -3e^{-x_2} + 1 \end{bmatrix}$ and $\mathbf{H}_f(\mathbf{x}) = \begin{bmatrix} 2e^{-x_1} & 0 \\ 0 & 3e^{-x_2} \end{bmatrix}$, which is positive definite for all $\mathbf{x}$. Since $\mathbf{H}_f^{-1}(\mathbf{x}) = \begin{bmatrix} \frac{1}{2}e^{x_1} & 0 \\ 0 & \frac{1}{3}e^{x_2} \end{bmatrix}$, we find that the Newton iterative formula becomes $\mathbf{x}^{k+1} = \mathbf{x}^k$

$$-\mathbf{H}_f^{-1}(\mathbf{x}^k)\nabla f(\mathbf{x}^k) = \begin{bmatrix} x_1^k \\ x_2^k \end{bmatrix} - \begin{bmatrix} \frac{1}{2}e^{x_1^k} & 0 \\ 0 & \frac{1}{3}e^{x_2^k} \end{bmatrix}\begin{bmatrix} -2e^{-x_1^k} + 1 \\ -3e^{-x_2^k} - 1 \end{bmatrix} = \begin{bmatrix} x_1^k - \frac{1}{2}e^{x_1^k} + 1 \\ x_2^k - \frac{1}{3}e^{x_2^k} + 1 \end{bmatrix}.$$

Starting with $\mathbf{x}^0 = \begin{bmatrix} x_1^0 \\ x_2^0 \end{bmatrix} = \begin{bmatrix} 0 \\ 0 \end{bmatrix}$, we then obtain $\mathbf{x}^1 = \begin{bmatrix} x_1^0 - \frac{1}{2}e^{x_1^0} + 1 \\ x_2^0 - \frac{1}{3}e^{x_1^0} + 1 \end{bmatrix} = \begin{bmatrix} \frac{1}{2} \\ \frac{2}{3} \end{bmatrix}$,

from which we calculate $\mathbf{x}^2 = \begin{bmatrix} x_1^1 - \frac{1}{2}e^{x_1^1} + 1 \\ x_2^1 - \frac{1}{3}e^{x_2^1} + 1 \end{bmatrix} = \begin{bmatrix} 1\frac{1}{2} - \frac{1}{2}e^{\frac{1}{2}} \\ 1\frac{1}{3} - \frac{1}{3}e^{\frac{2}{3}} \end{bmatrix} \approx \begin{bmatrix} 0.67564 \\ 1.01742 \end{bmatrix}$, and

finally $\mathbf{x}^3 = \begin{bmatrix} x_2^1 - \frac{1}{2}e^{x_2^1} + 1 \\ x_2^2 - \frac{1}{3}e^{x_2^2} + 1 \end{bmatrix} \approx \begin{bmatrix} 0.69299 \\ 1.09540 \end{bmatrix}$, and we stop here, having carried out

three complete iterations of Newton's method. The exact optimal solution obtained by solving the equation $\nabla f(\bar{\mathbf{x}}) = \mathbf{0}$ is $\bar{\mathbf{x}} = \begin{bmatrix} \ln 2 \\ \ln 3 \end{bmatrix} \approx \begin{bmatrix} 0.69315 \\ 1.09861 \end{bmatrix}$. Recall that the second-order sufficient condition for a minimal point is satisfied, since the Hessian matrix has been shown to be positive definite for all $\mathbf{x}$. We conclude that after three iterations, the Newton method has come to within a three-digit accuracy of the true optimal solution.

Just as its one-dimensional version, the multidimensional Newton method suffers from convergence problems, if the function to be minimized is not strictly convex. The Hessian matrix might then not be positive definite and several methods have been devised to get around this problem. A drawback of the method is also that the required inversion of the Hessian matrix is a cumbersome operation, which needs judicious consideration.

Addressing the issue of ensuring that we have a positive definite matrix to be inverted when the Newton formula is employed, a common approach is the following. It is clear that one can modify the Hessian matrix $\mathbf{H}_f(\mathbf{x}^k)$ by adding multiples r_k of the identity matrix $\mathbf{I}$, so that the matrix $\mathbf{H}_f(\mathbf{x}^k) + r_k\mathbf{I}$ becomes positive definite. The argument is then as follows. According to Theorem 1.24 in Chap. 1, if $\lambda_{\min}$ denotes the smallest eigenvalue of $\mathbf{H}_f(\mathbf{x}^k)$, then $\mathbf{x}^T\mathbf{H}_f(\mathbf{x}^k)\mathbf{x} \geq \lambda_{\min}\|\mathbf{x}\|^2$ for all $\mathbf{x} \in \mathbb{R}^n$. Therefore, $\mathbf{x}^T(\mathbf{H}_f(\mathbf{x}^k) + r_k\mathbf{I})\mathbf{x} = \mathbf{x}^T\mathbf{H}_f(\mathbf{x}^k)\mathbf{x} + r_k\|\mathbf{x}\|^2 \geq (\lambda_{\min} + r_k)\|\mathbf{x}\|^2$, and, if we selected r_k sufficiently large, so that $\lambda_{\min} + r_k > 0$, then $\mathbf{H}_f(\mathbf{x}^k) + r_k\mathbf{I}$ will be positive definite.

We then define the direction $\mathbf{s}^k := -(\mathbf{H}_f(\mathbf{x}^k) + r_k\mathbf{I})^{-1}\nabla f(\mathbf{x}^k)$ and perform a line search by minimizing $f(\mathbf{x}^k + \alpha\mathbf{s}^k)$, denoting an optimal solution by α^k. We see that this modification is a compromise between the Newton Method (with $r_k = 0$) and the Steepest Descent Method (with $r_k \gg 0$). Concerning the drawback of the Newton Method coming from having to invert a matrix, this can be handled by special techniques. They involve so-called *Cholesky factorization* of the Hessian matrix, which has been extensively studied. For a full description we refer to the works of Cottle and Thapa (2017), Bertsekas (2016), and Sun and Yuan (2006).

Finally, we should mention that Newton-like methods may suffer from convergence problems in the presence of constraints. This is exemplified by the so-called *Maratos effect*, for which we refer to Sun and Yuan (2006), see also Bazaraa et al. (2013) and Bertsekas (2016).

2.2.8 Trust Region Methods

In the previous section on the Newton minimization method we pointed out two concerns. First, the need to ensure that the matrix to be inverted (in the basic case the Hessian matrix) is positive definite. Secondly, the procedure for numerically inverting the matrix, equivalent to solving a system of simultaneous linear equations, needs efficient computational procedures for its practical implementation. A third concern, which we will deal with in this section, is that the basic idea underlying Newton's method rests on being able to approximate the function f to be minimized by its quadratic Taylor formula, at least locally around the current iteration point $\mathbf{x}^k$. Specifically, and referring to Theorem 1.17 and Definition 1.18 from Chap. 1, we know that the quadratic function f_2 given by $f_2(\mathbf{x}) = f(\mathbf{x}^k) + (\mathbf{x} - \mathbf{x}^k)^T \nabla f (\mathbf{x}^k) + \frac{1}{2}(\mathbf{x} - \mathbf{x}^k)^T \mathbf{H}_f(\mathbf{x}^k)(\mathbf{x} - \mathbf{x}^k)$ is a good approximation of the function $f(\mathbf{x})$ in a small vicinity of the current point $\mathbf{x}^k$. Unfortunately, a step with the ordinary Newton method may take us outside this vicinity. One remedy to this situation is to consider a restricted Newton step, which consists of minimizing the quadratic function $f_2(\mathbf{x})$, restricting $\mathbf{x}$ to stay within a hypersphere centered at the current point $\mathbf{x}^k$ and with a preselected radius of Δ_k, which is called the *trust parameter*. The hypersphere is called the *trust region*, indicating that this is the neighborhood of the current point $\mathbf{x}^k$ within which the quadratic function f_2 can be trusted to closely approximate the original function f. The trust region subproblem is therefore

P_k: Min $z_k = f_2(\mathbf{x})$
s.t. $\|\mathbf{x} - \mathbf{x}^k\| \leq \Delta_k$
$\mathbf{x} \in \mathbb{R}^n$,

and, assuming that $\nabla f(\mathbf{x}^k) \neq \mathbf{0}$, the solution $\hat{\mathbf{x}}^k$ to the subproblem P_k will then obviously satisfy $f_2(\hat{\mathbf{x}}^k) < f_2(\mathbf{x}^k)$, but not necessarily $f(\hat{\mathbf{x}}^k) \leq f(\mathbf{x}^k)$. For now, we will postpone a discussion of how to solve the constrained minimization subproblem P_k. The question is then what to do next. There are three alternatives: either accept $\hat{\mathbf{x}}^k$ as the next point from which to continue, i.e., set $\mathbf{x}^{k+1} := \hat{\mathbf{x}}^k$, or change the size of the current hypersphere, to bigger or to smaller, and resolve the trust region subproblem. In order to decide if the current hypersphere is suitably sized, we calculate the ratio R_k defined as $R_k := \dfrac{f(\mathbf{x}^k) - f(\hat{\mathbf{x}}^k)}{f_2(\mathbf{x}^k) - f_2(\hat{\mathbf{x}}^k)}$, which is the ratio of the actual (using f) to the predicted (using f_2) decrease of the objective function. It makes sense to reduce the size of the trust region, if the ratio R_k is small, which indicates a poor quadratic approximation f_2 to the original function f. On the other hand, if R_k is close to one, the approximation is good and the trust region can be expanded. We might even obtain a negative value for R_k, indicating that even though the point $\hat{\mathbf{x}}^k$ is an improvement over the point $\mathbf{x}^k$ as judged by the quadratic f_2, it is in fact a poorer solution in terms of the original objective function f. A sensible strategy is now this: if $R_k < \beta_1$, the latter being some preselected parameter (typically, $\beta_1 = 0.2$ is chosen), then the new trust region radius Δ_{k+1} is reduced to become $\Delta_{k+1} := \gamma_1 \|\hat{\mathbf{x}}^k - \mathbf{x}^k\|$, where γ_1 is some preselected parameter, typically $\gamma_1 = 0.25$. Note that $\|\hat{\mathbf{x}}^k - \mathbf{x}^k\| \leq \Delta_k$, since $\hat{\mathbf{x}}^k$ is a feasible solution to P_k, so that with $\gamma_1 = 0.25$, the radius Δ_{k+1} is one quarter of the radius Δ_k if $\hat{\mathbf{x}}^k$ is located on the boundary of the trust region hypersphere. On the other hand, if $\hat{\mathbf{x}}^k$ is an interior feasible point, it is less than one quarter. If $R_k > \beta_2$, another preselected parameter typically set at 0.75, and also $\|\hat{\mathbf{x}}^k - \mathbf{x}^k\| = \Delta_k$, so that the boundary solution case is at hand and we have an indication that within the trust region, the quadratic approximation is good. Then the trust region radius Δ_{k+1} is expanded to become $\Delta_{k+1} := \gamma_2 \Delta_k$, where γ_2 is again some preselected parameter, typically set to 2. If neither of the above two cases applies, we set $\hat{\mathbf{x}}^k := x^{k+1}$ and proceed to the next iteration with $\Delta_{k+1} = \Delta_k$.

We are now ready to summarize the trust region method in algorithmic form. Note that below, we will discuss the question of how to solve the subproblem P_k. Recall that the function to be minimized is $f(\mathbf{x})$ with a quadratic approximation $f_2(\mathbf{x})$ as given above. Start with some preselected initial point $\mathbf{x}^0$, an initial trust region radius of Δ_0, and the parameters β_1, β_2, γ_1, and γ_2, such that $0 < \beta_1 < \beta_2 < 1$ and $0 < \gamma_1 < 1 < \gamma_2$. Set the iteration counter $k := 0$.

A Trust Region Method

Step 1: Is $\nabla f(\mathbf{x}^k) = \mathbf{0}$?

If yes: Stop, an optimal solution has been found.
If no: Go to Step 2.

Step 2: Using some procedure, solve the subproblem

P_k: Min $z_k = f_2(\mathbf{x})$
s.t. $\|\mathbf{x} - \mathbf{x}^k\| \leq \Delta_k$
$\mathbf{x} \in \mathbb{R}^n$,

where $f_2(\mathbf{x})$ is the quadratic approximation to $f(\mathbf{x})$ at the point $\mathbf{x}^k$, defined as

$$f_2(\mathbf{x}) := f(\mathbf{x}^k) + (\mathbf{x} - \mathbf{x}^k)^T \nabla f(\mathbf{x}^k) + \frac{1}{2}(\mathbf{x} - \mathbf{x}^k)^T \mathbf{H}_f(\mathbf{x}^k)(\mathbf{x} - \mathbf{x}^k).$$

Let $\hat{\mathbf{x}}^k$ denote an optimal solution to P_k.
Step 3: Is $f(\hat{\mathbf{x}}^k) < f(\mathbf{x}^k)$?

If yes: Set $\hat{\mathbf{x}}^k := \mathbf{x}^{k+1}$ and compute the ratio $R_k := \dfrac{f(\mathbf{x}^k) - f(\mathbf{x}^{k+1})}{f_2(\mathbf{x}^k) - f(\mathbf{x}^{k+1})}$, and go to Step 4.
If no: Set $\Delta_{k+1} := \gamma_1 \|\hat{\mathbf{x}}^k - \mathbf{x}^k\|$, let $k := k + 1$, and go to Step 1.

Step 4: If $R_k < \beta_1$, then set $\Delta_{k+1} := \gamma_1 \|\hat{\mathbf{x}}^k - \mathbf{x}^k\|$; and if $R_k > \beta_2$ and $\|\hat{\mathbf{x}}^k - \mathbf{x}^k\| = \Delta_k$, then set $\Delta_{k+1} := \gamma_2 \Delta_k$. Otherwise, set $\Delta_{k+1} := \Delta_k$. Set $k := k + 1$ and go to Step 1.

A few comments are in order. The subproblem P_k in Step 2 is a constrained minimization problem, and we are not dealing with how to solve such problems until Chap. 5. However, there are specialized methods for this particular problem; see the book by Conn et al. (2000). Below we will use an approximate method for solving P_k, the so-called *dogleg method* due to Powell (1970). Another comment is that we are using the Euclidean (or ℓ_2) norm here. If the infinity (or Chebyshev or ℓ_∞) norm is used instead, the trust regional will be hypercubes instead of hyperspheres, and the resulting method will be the so-called *box step method*; see, e.g., Marsten et al. (1975). As starting value for the trust region radius, Sun and Yuan (2006) suggest $\Delta_0 = 1$ or $\Delta_0 = (0.1)\|\nabla f(\mathbf{x}^0)\|$. In general, it appears that the trust region method is rather insensitive to the choice of parameter values.

Fig. 2.17 Cauchy point and
Newton point in the dogleg
method

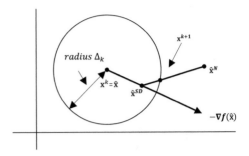

Now to the issue of how to solve the subproblem P_k, i.e.,

P_k: Min $z_k = f_2(\mathbf{x}) := f(\mathbf{x}^k) + (\mathbf{x} - \mathbf{x}^k)^T \nabla f(\mathbf{x}^k) + \frac{1}{2}(\mathbf{x} - \mathbf{x}^k)^T \mathbf{H}_f(\mathbf{x}^k)(\mathbf{x} - \mathbf{x}^k)$
s.t. $\|\mathbf{x} - \mathbf{x}^k\| \leq \Delta_k$
$\mathbf{x} \in \mathbb{R}^n$.

We can see that P_k is a constrained minimization problem with a quadratic
objective function, whose Hessian matrix of second-order partial derivatives we
will assume is positive definite, and a single quadratic inequality constraint. An
approximate and efficient way to solve P_k is the *dogleg method* (Powell 1970), which
can be seen as a compromise between the Steepest Descent and Newton methods.
Consider the current point $\mathbf{x}^k := \hat{\mathbf{x}}$, the center of the trust region with radius Δ_k,
furthermore let $\hat{\mathbf{x}}^{SD}$ denote the result of a Steepest Descent iteration starting at $\hat{\mathbf{x}}$,
and similarly let $\hat{\mathbf{x}}^N$ denote the result of a Newton step from $\hat{\mathbf{x}}$. Some authors refer to
$\hat{\mathbf{x}}^{SD}$ as the *Cauchy point* and $\hat{\mathbf{x}}^N$ as the *Newton point*. Then draw a line from $\hat{\mathbf{x}}$ to $\hat{\mathbf{x}}^{SD}$
and from $\hat{\mathbf{x}}^{SD}$ to $\hat{\mathbf{x}}^N$ as shown in Fig. 2.17.

The point $\hat{\mathbf{x}}^{SD}$ must always be closer to $\hat{\mathbf{x}}$ than $\hat{\mathbf{x}}^N$, unless the negative gradient
$-\nabla f(\hat{\mathbf{x}})$ points straight at $\hat{\mathbf{x}}^N$, in which case $\hat{\mathbf{x}}^{SD}$ and $\hat{\mathbf{x}}^N$ coincide, i.e., $\hat{\mathbf{x}}^{SD} = \hat{\mathbf{x}}^N$. A
formal proof of this fact will be provided below, after the numerical example for the
Dogleg method. For now, we will use an intuitive argument as follows. Assume that
$-\nabla f(\hat{\mathbf{x}})$ does not point at $\hat{\mathbf{x}}^N$. Since $f_2(\mathbf{x})$ is a quadratic function, the Newton method
will converge in a single step, so that $\hat{\mathbf{x}}^N$ is the true minimal point for f_2. The level sets
for f_2 are ellipsoids, centered at $\hat{\mathbf{x}}^N$ and with identical axes, and $\hat{\mathbf{x}}^{SD}$ is on a level set at
the point, at which the negative gradient is its tangent. It is then apparent that the
angle of the dogleg $\hat{\mathbf{x}} - \hat{\mathbf{x}}^{SD} - \hat{\mathbf{x}}^N$ must be obtuse, so that $\hat{\mathbf{x}}^{SD}$ is closer to $\hat{\mathbf{x}}$ than $\hat{\mathbf{x}}^N$;
for a visualization, see again Fig. 2.17. Now, if the point $\hat{\mathbf{x}}^N$ is outside the trust
region, the intersection of the dogleg with the boundary will be the next point $\mathbf{x}^{k+1}$,
otherwise we set $\mathbf{x}^{k+1} := \hat{\mathbf{x}}^N$. As mentioned earlier, the Dogleg method can be seen as a
compromise between the Steepest Descent and Newton methods. If $\hat{\mathbf{x}}^{SD}$ is outside the
trust region, we have taken a pure Steepest Descent step, whereas if $\hat{\mathbf{x}}^N$ is inside the
trust region, we have taken a pure Newton step.

Since the function $f_2(\mathbf{x})$ to be minimized with the Dogleg method is quadratic, we
can use the results of Remark 4 in Sect. 2.2.2, where the Steepest Descent method
was applied to such a function. Accordingly, we would find the optimal step length

factor $\hat{\alpha}$ by the expression $\hat{\alpha} = \dfrac{\hat{\mathbf{s}}^T\hat{\mathbf{s}}}{\hat{\mathbf{s}}^T\mathbf{H}_f(\hat{\mathbf{x}})\hat{\mathbf{s}}}$, where $\hat{\mathbf{s}}$ is the search direction $-\nabla f(\hat{\mathbf{x}})$.

Therefore we find $\hat{\mathbf{x}}^{SD} = \hat{\mathbf{x}} + \hat{\alpha}\hat{\mathbf{s}} = \hat{\mathbf{x}} + \dfrac{\hat{\mathbf{s}}^T\hat{\mathbf{s}}}{\hat{\mathbf{s}}^T\mathbf{H}_f(\hat{\mathbf{x}})\hat{\mathbf{s}}}\hat{\mathbf{s}}$, and if the step length is at least

as large as the trust region radius Δ_k, i.e., $\dfrac{\hat{\mathbf{s}}^T\hat{\mathbf{s}}}{\hat{\mathbf{s}}^T\mathbf{H}_f(\hat{\mathbf{x}})\mathbf{s}}\|\hat{\mathbf{s}}\| \geq \Delta_k$, so that $\hat{\mathbf{x}}^{SD}$ is outside

the trust region, then we set $\mathbf{x}^{k+1} := \mathbf{x}^k + \dfrac{\Delta_k}{\|\hat{\mathbf{s}}\|}\hat{\mathbf{s}} = \mathbf{x}^k - \dfrac{\Delta_k}{\|\nabla f(\mathbf{x}^k)\|}\nabla f(\mathbf{x}^k)$, which is

where the line from $\mathbf{x}^k$ along the negative gradient intersects the trust region boundary. On the other hand, if $\hat{\mathbf{x}}^{SD}$ is inside and $\hat{\mathbf{x}}^N$ is outside the trust region,

i.e., if we have $\dfrac{\hat{\mathbf{s}}^T\hat{\mathbf{s}}}{\hat{\mathbf{s}}^T\mathbf{H}_f(\hat{\mathbf{x}})\hat{\mathbf{s}}}\|\hat{\mathbf{s}}\| < \Delta_k$ and $\left\|\mathbf{H}_f^{-1}(\hat{\mathbf{x}})\hat{\mathbf{s}}\right\| > \Delta_k$, then we set

$\mathbf{x}^{k+1} := \lambda\hat{\mathbf{x}}^{SD} + (1-\lambda)\hat{\mathbf{x}}^N$, where the value of the variable λ, $0 < \lambda < 1$, is determined via $\left\|\lambda\hat{\mathbf{x}}^{SD} + (1-\lambda)\hat{\mathbf{x}}^N\right\| = \Delta_k$, so that $\mathbf{x}^{k+1}$ lies on the boundary of the trust region. Finally, if $\left\|\mathbf{H}_f^{-1}(\hat{\mathbf{x}})\hat{\mathbf{s}}\right\| \leq \Delta_k$, then $\hat{\mathbf{x}}^N$ is on the boundary of or inside the trust region, and we set $\mathbf{x}^{k+1} := \hat{\mathbf{x}}^N$.

We are now ready to summarize the Dogleg method in algorithmic form. The method is initialized with a quadratic function $f_2(\mathbf{x})$ and a trust region with radius Δ_k centered around a given point $\hat{\mathbf{x}} = \mathbf{x}^k$.

The Dogleg Method

Step 1: Set $\hat{\mathbf{s}} := -\nabla f(\hat{\mathbf{x}})$, i.e., the negative gradient at $\hat{\mathbf{x}}$ and compute the point $\hat{\mathbf{x}}^{SD}$, obtained by the Steepest Descent method and the point $\mathbf{x}^N$, obtained by the Newton method, as follows: $\hat{\mathbf{x}}^{SD} := \hat{\mathbf{x}} + \dfrac{\hat{\mathbf{s}}^T\hat{\mathbf{s}}}{\hat{\mathbf{s}}^T\mathbf{H}_f(\hat{\mathbf{x}})\hat{\mathbf{s}}}\hat{\mathbf{s}}$, $\hat{\mathbf{x}}^N := \hat{\mathbf{x}} + \mathbf{H}_f^{-1}(\hat{\mathbf{x}})\hat{\mathbf{s}}$.

Step 2: Is $\dfrac{\hat{\mathbf{s}}^T\hat{\mathbf{s}}}{\hat{\mathbf{s}}^T\mathbf{H}_f(\hat{\mathbf{x}})\hat{\mathbf{s}}}\|\hat{\mathbf{s}}\| \geq \Delta_k$?

If yes: Set $\mathbf{x}^{k+1} := \mathbf{x}^k + \dfrac{\Delta_k}{\|\hat{\mathbf{s}}\|}\hat{\mathbf{s}}$ and go to Step 5.
If no: Go to Step 3.

Step 3: Is $\dfrac{\hat{\mathbf{s}}^T\hat{\mathbf{s}}}{\hat{\mathbf{s}}^T\mathbf{H}_f(\hat{\mathbf{x}})\hat{\mathbf{s}}}\|\hat{\mathbf{s}}\| < \Delta_k$ and $\left\|\mathbf{H}_f^{-1}(\hat{\mathbf{x}})\hat{\mathbf{s}}\right\| > \Delta_k$?

If yes: Set $\mathbf{x}^{k+1} := \lambda\hat{\mathbf{x}}^{SD} + (1-\lambda)\hat{\mathbf{x}}^N$, where λ solves $\left\|\lambda\hat{\mathbf{x}}^{SD} + (1-\lambda)\hat{\mathbf{x}}^N\right\| = \Delta_k$. Go to Step 5.
If no: Go to Step 4.

Step 4: Set $\mathbf{x}^{k+1} := \hat{\mathbf{x}}^N$ and go to Step 5.
Step 5: Stop with $\mathbf{x}^{k+1}$ as the solution.

We will now illustrate the Dogleg algorithm by means of a numerical

Example Consider the quadratic function $f_2(x_1, x_2) = x_1^2 + \dfrac{3}{2}x_2^2 - x_1 - 2x_2 + 5$ and perform one iteration of the dogleg method, starting from the initial point $\hat{\mathbf{x}} = [0, 0]^T$ and using the trust region radius $\Delta_k = 1$.

Step 1 determines that $\hat{\mathbf{s}} = \begin{bmatrix} 1 \\ 2 \end{bmatrix}$ and $\mathbf{H}_f(\mathbf{x}) = \begin{bmatrix} 2 & 0 \\ 0 & 3 \end{bmatrix}$ as well as $\mathbf{H}_f^{-1}(\hat{\mathbf{x}}) =$

$\begin{bmatrix} \frac{1}{2} & 0 \\ 0 & \frac{1}{3} \end{bmatrix}$, so that $\dfrac{\hat{\mathbf{s}}^T \hat{\mathbf{s}}}{\hat{\mathbf{s}}^T \mathbf{H}_f(\hat{\mathbf{x}})\hat{\mathbf{s}}} = \dfrac{\begin{bmatrix} 1 & 2 \end{bmatrix}\begin{bmatrix} 1 \\ 2 \end{bmatrix}}{\begin{bmatrix} 1 & 2 \end{bmatrix}\begin{bmatrix} 2 & 0 \\ 0 & 3 \end{bmatrix}\begin{bmatrix} 1 \\ 2 \end{bmatrix}} = \dfrac{5}{13}$ and $\hat{\mathbf{x}}^{SD} := \hat{\mathbf{x}} + \dfrac{\hat{\mathbf{s}}^T \hat{\mathbf{s}}}{\hat{\mathbf{s}} \mathbf{H}_f(\hat{\mathbf{x}})\hat{\mathbf{s}}} \hat{\mathbf{s}}$

$= \begin{bmatrix} 0 \\ 0 \end{bmatrix} + \dfrac{5}{13}\begin{bmatrix} 1 \\ 2 \end{bmatrix} = \begin{bmatrix} \frac{5}{13} \\ \frac{10}{13} \end{bmatrix}$, $\hat{\mathbf{x}}^N := \hat{\mathbf{x}} + \mathbf{H}_f^{-1}(\hat{\mathbf{x}})\hat{\mathbf{s}} = \begin{bmatrix} 0 \\ 0 \end{bmatrix} + \begin{bmatrix} \frac{1}{2} & 0 \\ 0 & \frac{1}{3} \end{bmatrix}\begin{bmatrix} 1 \\ 2 \end{bmatrix} = \begin{bmatrix} \frac{1}{2} \\ \frac{2}{3} \end{bmatrix}$,

and we go to Step 2, where we compute $\dfrac{\hat{\mathbf{s}}^T \hat{\mathbf{s}}}{\hat{\mathbf{s}}^T \mathbf{H}_f(\hat{\mathbf{x}})\hat{\mathbf{s}}} \|\hat{\mathbf{s}}\| = \dfrac{5}{13}\sqrt{1^2 + 2^2} = \dfrac{5\sqrt{5}}{13}$,

which is smaller than $\Delta_k = 1$. Therefore we set $\mathbf{x}^{k+1} := \mathbf{x}^k + \dfrac{\Delta_k}{\|\hat{\mathbf{s}}\|}\hat{\mathbf{s}} =$

$\begin{bmatrix} 0 \\ 0 \end{bmatrix} + \dfrac{1}{\sqrt{5}}\begin{bmatrix} 1 \\ 2 \end{bmatrix} = \begin{bmatrix} \frac{1}{\sqrt{5}} \\ \frac{2}{\sqrt{5}} \end{bmatrix} \approx \begin{bmatrix} 0.4472 \\ 0.8944 \end{bmatrix}$, and we terminate after one iteration as

specified in the beginning.

As promised above, we will now formally prove that the step length by Steepest Descent is no longer than the step length of the Newton method, i.e., we will show

that $\left\| \dfrac{\hat{\mathbf{s}}^T \hat{\mathbf{s}}}{\hat{\mathbf{s}}^T \mathbf{H}_f(\hat{\mathbf{x}})\hat{\mathbf{s}}} \hat{\mathbf{s}} \right\| \leq \left\| \mathbf{H}_f^{-1}(\hat{\mathbf{x}})\hat{\mathbf{s}} \right\|$. Putting it differently, the Cauchy point $\hat{\mathbf{x}}^{SD}$ is no

further away from $\hat{\mathbf{x}}$ than the Newton point $\hat{\mathbf{x}}^N$. For ease of notation, we set $\mathbf{H}_f(\hat{\mathbf{x}}) := \mathbf{Q}$, so that $\mathbf{H}_f^{-1}(\hat{\mathbf{x}}) = \mathbf{Q}^{-1}$, and set $\hat{\mathbf{s}} := \mathbf{s}$. Recalling that $\mathbf{Q}$, and therefore also $\mathbf{Q}^{-1}$ are positive definite, $\mathbf{s}^T \mathbf{Q}\mathbf{s}$ and $\mathbf{s}^T \mathbf{Q}^{-1}\mathbf{s}$ must be strictly positive, so that $\left\| \dfrac{\mathbf{s}^T \mathbf{s}}{\mathbf{s}^T \mathbf{Q}\mathbf{s}}\mathbf{s} \right\| =$

$\dfrac{\|\mathbf{s}\|^3}{\mathbf{s}^T \mathbf{Q}\mathbf{s}} \leq \dfrac{\|\mathbf{s}\|^3}{\mathbf{s}^T \mathbf{Q}\mathbf{s}} \dfrac{\|\mathbf{s}^T\|\|\mathbf{Q}^{-1}\mathbf{s}\|}{\mathbf{s}^T \mathbf{Q}^{-1}\mathbf{s}} = \dfrac{\|\mathbf{s}\|^4}{(\mathbf{s}^T \mathbf{Q}\mathbf{s})(\mathbf{s}^T \mathbf{Q}^{-1}\mathbf{s})}\|\mathbf{Q}^{-1}\mathbf{s}\| \leq \|\mathbf{Q}^{-1}\mathbf{s}\|$, where the last inequality follows from Proposition A.9 (the Kantorovich matrix inequality). The Newton step length is therefore at least as long as that of Steepest Descent.

Using the Kantorovich inequality, we can also strengthen the argument in Remark 3 of Sect. 2.2.2 about the zigzagging behavior of the Steepest Descent Method. Specifically, we can show that for a quadratic function with positive Hessian matrix $\mathbf{Q}$, for which the smallest eigenvalue is denoted by $\lambda_{\min}$, and the largest eigenvalue is denoted by $\lambda_{\max}$, the following inequality holds:

Table 2.2 Technical requirements of solution methods and their convergence properties

Method	Gradient required?	Hessian required?	Steps to convergence for a quadratic objective function
Cyclic coordinate descent	No	No	∞
Simplex search	No	No	∞
Cauchy Steepest Descent	Yes	No	∞
Davidon–Fletcher– Powell	Yes	No	n
Fletcher–Reeves	Yes	No	n
Rank one	Yes	No	n
BFGS	Yes	No	n
Newton search	Yes	Yes	1
Trust region	Yes	Yes	1

$$\left\| \mathbf{x}^{k+1} - \bar{\mathbf{x}} \right\| \leq \left(\frac{\lambda_{\max} - \lambda_{\min}}{\lambda_{\max} + \lambda_{\min}} \right)^2 \left\| \mathbf{x}^k - \bar{\mathbf{x}} \right\|,$$

see, e.g., Bertsekas (2016). For $\lambda_{\max} \approx \lambda_{\min}$, i.e., nearly hyperspherical level curves of the quadratic function, convergence to the optimal point $\bar{\mathbf{x}}$ will be fast. On the other hand, $\lambda_{\max} > \lambda_{\min}$ indicates elongated hyperellipsoidal level curves and a slow rate of convergence.

To summarize the results of our discussion of multivariable minimization methods, the computational requirements and convergence properties of the algorithms described in this section of the chapter are displayed in Table 2.2.

References

Armijo L (1966) Minimization of functions having Lipschitz continuous first partial derivatives. *Pacific Journal of Mathematics* **16**: 1-3

Avriel M (1976) *Nonlinear programming: analysis and methods.* Prentice-Hall, New Jersey

Avriel M, Wilde DJ (1966) Optimality proof for the symmetric Fibonacci search technique. *Fibonacci Quarterly Journal* **4**: 265-269

Bazaraa MS, Sherali HD, Shetty CM (2013) *Nonlinear programming: theory and algorithms* (3rd ed.) Wiley, New York

Bertsekas DP (2016) *Nonlinear programming* (3rd ed.) Athena Scientific, Belmont, MA

Broyden CG (1970) The convergence of a class of double rank minimization algorithms. Parts I and II of the *Journal for the Institute of Mathematics and its Applications* **6**: 76-90, 222-231

Cauchy A (1847) Méthode générale pour la résolution des systèmes d'équations simultanées. *Comptes rendus de la Academie des Sciences*, Paris **25**: 536-538

Conn AR, Gould NIM, Orban D, Toint PL (2000) A primal-dual trust-region algorithm for non-convex nonlinear programming. *Mathematical Programming* **B87**: 215-249

Cottle RW, Thapa MN (2017) *Linear and nonlinear optimization*. Springer-Verlag, Berlin-Heidelberg-New York

Danilin YM, Piyavskii SA (1967) On an algorithm for finding the absolute minimum. *Theory of Optimal Decisions*, vol. II. Institute of Cybernetics, Kiev, (in Russian).

Davidon WC (1959) Variable metric method for minimization. *Research and Development Report* ANL-5990 (Rev.) Argonne National Laboratory, U. S. Atomic Energy Commission

Dixon LCW (1972) *Nonlinear optimisation*. English Universities Press, London.

Eiselt HA, Pederzoli G, Sandblom C-L (1987) *Continuous optimization models*. W. de Gruyter, Berlin – New York.

Fletcher R (1970) A new approach to variable metric algorithms. *The Computer Journal* **13**: 317-322

Fletcher R, Powell MJD (1963) A rapidly convergent descent method for minimization. *The Computer Journal* **6**: 163-168

Fletcher R, Reeves CM (1964) Function minimization by conjugate gradients. *The Computer Journal* **7**: 149-154

Fröberg C-E (1969) *Introduction to numerical analysis* (2nd ed.) Addison-Wesley, Reading, MA

Goldfarb D (1970) A family of variable metric methods derived by variational means. *Mathematics of Computation* **24**: 23-26

Goldstein AA (1965) On steepest descent. *SIAM Journal on Control* **3**: 147-151.

Hansen P, Jaumard B (1995) Lipschitz optimization. pp 407-493 in Horst R, Pardalos PM (eds.) *Handbook of global optimization*. Kluwer, Boston, MA

Horst R, Pardalos PM, Thoai NV (2000) *Introduction to global optimization*, vol. 2. Kluwer, Dordrecht, The Netherlands

Kiefer J (1953) Sequential minimax search for a maximum. *Proceedings of the American Mathematical Society* **4**: 502-506

Luenberger DL, Ye Y (2008) *Linear and nonlinear programming* (3rd ed.) Springer-Verlag, Berlin-Heidelberg-New York

Marsten RE, Hogan WW, Blankenship JW (1975) The Boxstep Method for large-scale optimization. *Operations Research* 23/3: 389-405

Nelder JA, Mead R (1965) A simplex method for function minimization. *The Computer Journal* **7**: 308-313

Pintér JD (1996) *Global optimization in action*. Kluwer, Boston, MA

Piyavskii SA (1972) An algorithm for finding the absolute extremum of a function. *USSR Computational Mathematics and Mathematical Physics* **12**: 57-67

Powell MJD (1970) A new algorithm for unconstrained optimization. In: Rosen JB, Mangasarian OL, Ritter K (eds.), *Nonlinear programming*. Academic Press, pp. 31-36

Shanno DF (1970) Conditioning of quasi-Newton methods for function minimization. *Mathematics of Computation* **24**: 647-656

Shubert BO (1972) A sequential method seeking the global maximum of a function. *SIAM Journal of Numerical Analysis* **9**: 379-388

Spendley W, Hext GR, Himsworth FR (1962) Sequential application of simplex designs in optimization and evolutionary operation. *Technometrics* **4**: 441-461

Sun W, Yuan Y (2006) *Optimization theory and methods. Nonlinear programming*. Springer-Verlag, Berlin-Heidelberg-New York

Wolfe P (1969) Convergence conditions for ascent methods. *SIAM Review* **11**, 226-235

Chapter 3
Applications of Nonlinear Programming

The first two chapters of this book were devoted to the theory and methods of unconstrained nonlinear optimization. In this chapter, we will switch our attention to nonlinear models, and present a variety of examples, in which a problem can be formulated as a constrained or unconstrained nonlinear optimization model. The process of formulating models—albeit complex and challenging—has been covered elsewhere, see, e.g., Eiselt and Sandblom (2007). Our purpose is to provide examples of problems that naturally lead to formulations in the form of nonlinear optimization models. We are not concerned with how these models will be solved; instead, we describe the formulation and the conclusions that can be drawn from their solutions.

The first two sections of curve fitting and reference point programming discuss tools rather than applications. Next, we consider the optimal time to harvest trees in a forestry operation. It turns out that trees should be cut earlier than one might intuitively believe, as judged by the tree biomass growth curve. Thereafter follows a section about a queuing system, in which the number of service facilities needs to be determined together with the optimal level of training effort spent on the operator of the service facilities.

The composition of financial securities in an investment portfolio is the topic of Sect. 3.5, where we concentrate on a variety of ways to measure market risk of the securities selected, and Sect. 3.6 discusses a model that locates electrical substations in an electrical network so as to minimize the costs of siting the substations and the costs of losses in the network.

The next section considers a taxation problem: what level of taxation of the income earners in a society will maximize the state's income? The discussion is based on the assumption that the higher the rate of taxation, the less amount of time is spent by the tax payers on work that generates taxable income. Following are some topics in location analysis; first, how supermarkets should be located geographically in order to maximize their respective market share. Secondly, we look at the issue of locating undesirable or obnoxious facilities, such as polluting industries or industrial

© Springer Nature Switzerland AG 2019
H. A. Eiselt, C.-L. Sandblom, *Nonlinear Optimization*, International Series in
Operations Research & Management Science 282,
https://doi.org/10.1007/978-3-030-19462-8_3

dump sites, so as to minimize their negative effects on the population. Thirdly, we consider a packing problem, in which a maximal number of objects are to be placed in a geographically restricted area. The final two sections of the chapter consider economic planning problems. Section 3.9 investigates the issue of maximizing profit in a production process, as well as inventory decisions in a situation, in which alternative sizes of an inventory facility are taken into account. Finally, in the last section we look at the macroeconomic planning problem of optimizing fiscal policy so as to achieve acceptable levels of unemployment and inflation in the economy.

We should also mention that a few applications have been placed at the end of Chap. 8, since they require a background knowledge of geometric programming theory and techniques, to which that chapter is devoted, and therefore do not fit into this chapter.

3.1 Curve Fitting

This section is devoted to the so-called curve-fitting problem, which, in a general setting, can be described as follows. Given a set of points $(x_j, y_j) \in \mathbb{R}^2$, $j = 1, \ldots,$ n and a class $\mathscr{F}$ of functions, find a function $f \in \mathscr{F}$, for which $f(x_j)$ in some sense "best" approximates y_j for all $j = 1, \ldots, n$. One way to define what a best approximation is, would be to find a function $f \in \mathscr{F}$, for which the sum of the absolute deviations $|y_j - f(x_j)|$, $j = 1, \ldots, n$ is smallest, i.e., solving the problem

$$P_{abs}: \quad \underset{f \in \mathscr{F}}{\text{Min}} \; z_{abs} = \sum_{j=1}^{n} |y_j - f(x_j)|.$$

The most prevalent measure in curve fitting is to seek to minimize the squared deviations $(y_j - f(x_j))^2$, i.e., to solve the problem

$$P_{squ}: \underset{f \in \mathscr{F}}{\text{Min}} \; z_{squ} = \sum_{j=1}^{n} (y_j - f(x_j))^2.$$

This is the problem we will consider for different classes $\mathscr{F}$ of functions f. Figure 3.1 may illustrate the situation for $n = 6$ and $\mathscr{F}$ the set of third-degree polynomials.

Before proceeding, let us first consider some special cases. If $\mathscr{F}$ is the set of straight lines in $\mathbb{R}^2$, i.e., f can be written as $y = \alpha + \beta x$ with given constants α and β, then problem P_{squ} is that of least squares linear regression, a basic problem in statistics, which will take the special form

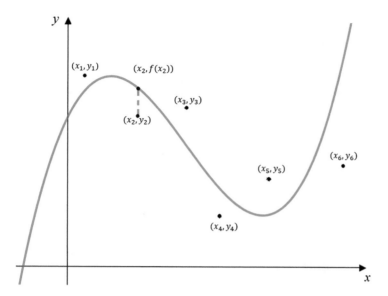

Fig. 3.1 Curve fitting a set of points

$$P_{\alpha\beta} : \underset{\alpha, \beta}{\text{Min}} \ z_{\alpha\beta} = \sum_{j=1}^{n} \left(y_j - (\alpha + \beta x_j) \right)^2.$$

Somewhat surprisingly, for that case a closed-form optimal solution of $P_{\alpha\beta}$, for $j \geq 2$, can be determined. It is

$$\bar{\beta} = \frac{n \sum_{j=1}^{n} x_j y_j - \left(\sum_{j=1}^{n} x_j \right) \left(\sum_{j=1}^{n} y_j \right)}{n \sum_{j=1}^{n} x_j^2 - \left(\sum_{j=1}^{n} x_j \right)^2}, \text{ and}$$

$$\bar{\alpha} = \frac{1}{n} \left(\sum_{j=1}^{n} y_j - \bar{\beta} \sum_{j=1}^{n} x_j \right). \text{ The optimal objective function value}$$

$$\bar{z}_{\alpha\beta} = \sum_{j=1}^{n} y_j^2 - \bar{\alpha} \sum_{j=1}^{n} y_j - \bar{\beta} \sum_{j=1}^{n} x_j y_j$$

is then referred to as the *sum of squared errors*. For details, see, e.g., Kutner et al. (2004). We may additionally require the straight line $y = \alpha + \beta x$ to be horizontal, i.e., $\beta = 0$. The problem is then

$$P_\alpha : \underset{\alpha}{\text{Min}}\ z_\alpha = \sum_{j=1}^{n} \left(y_j - \alpha\right)^2,$$

which has the optimal solution $\overline{\alpha} = \dfrac{1}{n}\sum_{j=1}^{n} y_j$, i.e., the arithmetic mean of the y_j values.

Yet another linear model is obtained if we require the regression line to lead through the origin, i.e., to have the form $y = \beta x$. The problem is then

$$P_\beta : \underset{\beta}{\text{Min}}\ z_\beta = \sum_{j=1}^{n} \left(y_j - \beta x_j\right)^2,$$

which has the optimal solution $\overline{\beta} = \dfrac{\sum_{j=1}^{n} x_j y_j}{\sum_{j=1}^{n} x_j^2}$, see Kutner *op cit.*

It is also possible to consider the class $\mathscr{F}$ of broken linear functions. In Eiselt and Sandblom (2000) we describe how to formally handle such functions using integer programming. The drawback with that approach is that the number of parameters to be estimated increases considerably. The next level of sophistication is achieved by considering $\mathscr{F}$ as the class of second-degree polynomials $y = \alpha + \beta x + \gamma x^2$, for which our curve-fitting problem becomes

$$P_{\alpha\beta\gamma} : \underset{\alpha,\,\beta,\,\gamma}{\text{Min}}\ z_{\alpha\beta\gamma} = \sum_{j=1}^{n} \left(y_j - \left(\alpha + \beta x_j + \gamma x_j^2\right)\right)^2.$$

Since $P_{\alpha\beta\gamma}$ is an unconstrained convex quadratic minimization problem, its solution can easily be obtained by setting its partial derivatives with respect to α, β, and γ to zero and solving the resulting system of three linear equations in the three unknowns α, β, and γ. Quadratic functions are already used in a curve-fitting context in Sects. 2.1.6 (quadratic interpolation) and 2.1.8 (Newton minimization search) for the purpose of finding minimal points of a given function. Furthermore, if we let $\mathscr{F}$ denote the class of degrees of polynomials of degree $n - 1$, then the Lagrangean interpolation formula in Lemma 2.3 will again provide a closed-form solution with $\overline{z} = 0$, i.e., a polynomial f can be found with an exact fit in the sense that $f(x_j) = y_j$, $j = 1, \ldots, n$.

As an example, consider the six points $(x_j, y_j), j = 1, \ldots, 6$ in $\mathbb{R}^2$ given by $(-0.5, 6)$, $(0, 2)$, $(0.5, -1)$, $(1, -2)$, $(2, 0)$, and $(3, 5)$, see Fig. 3.2.

For this set of data, the least-squares regression model $P_{\alpha\beta}$ yields the fitted line $\ell_{\alpha\beta}$ with equation $y = 1.7255 - 0.05882x$, displayed in Fig. 3.2, as are the lines ℓ_α, with equation $y = 1.6667$ and ℓ_β with equation $y = 0.6552x$, obtained by solving the models P_α and P_β, respectively. None of these three lines give a reasonable fit to the data; for model $P_{\alpha\beta}$ the minimal value of the objective function, i.e., the sum of

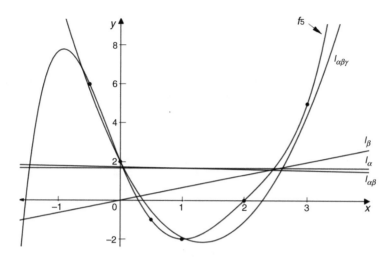

Fig. 3.2 Different regression models to fit a set of points

squared errors, is $\bar{z}_{\alpha\beta} = 53.3039$. We are therefore encouraged to try a quadratic fit by solving the model $P_{\alpha\beta\gamma}$, which turns out to have the optimal solution $y = 2.0021 - 6.2648x + 2.4535x^2$, the optimal objective function value is $\bar{z}_{\alpha\beta\gamma} = 0.926$, indicating quite a good fit, see the curve $\ell_{\alpha\beta\gamma}$ in Fig. 3.2. Finally, using the Lagrangean interpolation formula, we obtain a perfect fit to the six given points with the fifth-degree polynomial function $f_5(x)$ given by $f_5(x) = 0.247619x^5 - 1.54286x^4 + 2.56667x^3 + 2.38571x^2 - 7.65714x + 2$, see the curve f_5 in Fig. 3.2.

Returning to the general discussion, other classes $\mathscr{F}$ of functions are also used. In models with biological, chemical, or economic growth, functions involving exponentials are common, e.g., $y = \alpha + \beta e^{\gamma x}$. In statistical applications bell curve functions of the type $y = a + be^{-\gamma x^2}$ are useful, and in situations, in which asymptotic phenomena occur, fractions functions, such as $y = \alpha + \beta x + \gamma/x$ can be used. So-called *splines*, which are piecewise polynomial functions, have also been used for curve fitting. In this approach, typically cubic functions are connected in such a way that a smooth function results. For details, readers may consult Boor (2001) or other reference texts on splines.

We will now consider another situation that leads to curve fitting considerations. In many growth processes (economic, biological, bacterial), some entity will exhibit growth of the type $\left(1 + \dfrac{i\%}{100}\right)^n$, which expresses the proportional growth of the entity after n periods, each of which has caused an increase of $i\%$. The equation $\left(1 + \dfrac{i\%}{100}\right)^n = 2$ will therefore connect values of i and n, for which the entity has doubled in value. Setting $x := i\%/100$, we obtain the equation $(1 + x)^n = 2$, which can be written as $n \ln(1 + x) = \ln 2$, or $n = \dfrac{\ln 2}{\ln(1 + x)}$, so that $xn = \dfrac{x \ln 2}{\ln(1 + x)}$ expresses growth rate multiplied by doubling time. Defining the function

Table 3.1 Values of the function (ln 2)$f(x)$

x	(ln 2) $f(x)$
0.001	0.6349
0.005	0.6949
0.01	0.6966
0.02	0.7001
0.04	0.7069
0.05	0.7103
0.08	0.7205
0.10	0.7273
0.11	0.7306
0.12	0.7340
0.15	0.7439
0.20	0.7604

$f(x) = \dfrac{x}{\ln(1+x)}$, we wish to study the problem of fitting a straight line $y = \alpha + \beta x$ to the nonlinear function (ln 2)$(f(x))$, hoping that a good fit might be achieved with a value β close to zero. In other words, we will consider the problem

$$P : \underset{\alpha,\,\beta}{\text{Min}} \; z = \underset{\underline{b} \le x \le \bar{b}}{\max} \; \{|\alpha + \beta x - (\ln 2)f(x)|\},$$

where the concern is the behavior of $f(x)$ in the preselected interval $\underline{b} \le x \le \bar{b}$. First we will show that $f(x)$ is an increasing function for $x > 0$. Differentiation yields
$f'(x) = \dfrac{\ln(1+x) - \frac{x}{1+x}}{(\ln(1+x))^2}$ for $x > 0$. Defining $f(0) := 1$, one can show that $f'(0) = 0$.

Furthermore, $\dfrac{d}{dx}\left(\ln(1+x) - \dfrac{x}{1+x}\right) = \dfrac{x}{(1+x)^2} > 0$ for $x > 0$, and equal to zero for $x = 0$. Consequently, $f'(x) > 0$ for all $x > 0$. For different selected values of x, we can compute the values (ln 2) $f(x)$, which are shown in Table 3.1.

From Table 3.1, we see that for $\underline{b} = 0.04$ and $\bar{b} = 0.12$, and with $\beta = 0$, $\alpha = 0.72$ is a good approximate solution to problem P. In other words, for $4 \le i\% \le 12$, $i\%n$ varies between 71 and 73, making $i\%n \approx 72$ a good approximation. In economics, the "rule of 72" states that the interest rate of an asset multiplied by its doubling time equals 72, which is verified by the above discussion. From Table 3.1, we can also see that $(i\%)n$ ranges from 70 for $i = 1\%$ to 74 for $i = 15\%$, thus covering most realistic interest rates in the financial area.

The rule of 72 is most useful as a quick and simple guide in the evaluation of various investment possibilities. For instance, with an interest rate of 6%, an investment will double in value in 12 years, while an investment that pays 8% will double in 9 years, and an investment that pays 12% will double in value in approximately 6 years.

A different, albeit related, question is that of shrinking populations, radioactive decay, or monetary values, whose purchasing power is reduced through inflation. In these circumstances, one may ask how long it takes for an amount to be reduced to half its size. When dealing with radioactive matter, this is referred to as it *half-life*.

We could then consider the equation $\left(1 + \dfrac{i\%}{100}\right)^n = \frac{1}{2}$, where $i\%$ is the percentage, by which the size of a population shrinks per period (e.g., a year), and n is the number of periods, until the size is reduced to half. By copying the development for doubling times, we arrive at what we may call the "rule of 67." More specifically, $(i\%)n$ ranges from 68 for $i\% = 4$ to 66 for $i = 11$, and from 69 for $i\% = 1$ to 64 for $i\% = 15$.

Similar rules have been developed for tripling (quadrupling, etc.) times, as well as for third-life, quarter-life times, and so forth.

3.2 Reference Point Programming with Euclidean Distances

This section describes a technique that allows decision makers to find compromise solutions in the context of multiobjective optimization problems. In general, multiobjective optimization problems (for simplicity, we assume that they have linear constraints), can be described as

$$
\begin{aligned}
\text{P:} \quad & \text{"Max"} \ z_1 = f_1(\mathbf{x}) \\
& \text{"Max"} \ z_2 = f_2(\mathbf{x}) \\
& \cdots \\
& \text{"Max"} \ z_p = f_p(\mathbf{x}) \\
\text{s.t.} \quad & \mathbf{Ax} \le \mathbf{b} \\
& \mathbf{x} \ge \mathbf{0},
\end{aligned}
$$

where $\mathbf{x}$ is an n-dimensional vector of decision variables, $\mathbf{b}$ is an m-dimensional vector of right-hand side values, $\mathbf{A}$ is an $[m \times n]$-dimensional matrix, f_k is a (potentially nonlinear) function, and we follow the custom to put the "Max" in quotes to indicate that we cannot optimize all p functions simultaneously. Rather than applying optimality—a concept that loses its meaning for more than a single objective—we typically employ Pareto-optimality, also referred to as efficiency, noninferiority, or nondominance. Loosely speaking, a feasible point $\mathbf{x}$ is called Pareto-optimal if there exists no other feasible point $\mathbf{x}'$ such that $f_k(\mathbf{x}') \ge f_k(\mathbf{x})$ for $k = 1, \ldots, p$ with the inequality being strict for at least one k. An in-depth discussion of ways to deal with multiobjective optimization problems can be found in Eiselt and Sandblom (2004).

A typical straightforward way to deal with multiobjective optimization problems (other than enumerating all Pareto-optimal points and let the decision maker choose one) is the *weighting method*, suggested by Cohon (1978). The technique simply transforms the given p objective functions into a single composite objective by associating weights $w_1, w_2, \ldots, w_p$ with the objectives and adding up the objectives. These weights are supposed to express the relative importance of the individual objectives. More specifically, the ratio w_k/w_ℓ is designed to express how much more one unit of achievement of the k-th objective is worth than one unit of achievement of the ℓ-th objective. In doing so, the weight ratios are also designed to deal with potential noncommensurabilities of the objectives. The single objective is then Max $z = \sum_{k=1}^{p} w_k f_k(\mathbf{x})$, which is optimized with respect to the given constraints. As long as all functions are linear, the optimal solution to this problem is a Pareto-optimal point.

Another approach is the *reference point method*, pioneered by Wierzbicki (1982); there is also the *Topsis* method by Hwang and Yoon (1981), see also Eiselt and Sandblom (2004). The idea of the reference point method is fairly simple: choose achievement levels of the individual objectives that are higher than anything we can actually reach and, given that reference point, fine a feasible solution that minimizes the distance between itself and the reference point. The result is typically a compromise solution. This solution will be determined by a number of factors that must be chosen by the decision maker:

- the chosen ideal (or reference) point,
- the choice of distance function, and
- the weights used for the individual coordinates.

Formally, let $\mathbf{x}^*$ denote the ideal point, at which the achievements of the individual objectives are $z_k(\mathbf{x}^*)$, $k = 1, \ldots, p$. The only requirement is that $\mathbf{x}^*$ cannot be reached by any linear combination of the objectives. We can then formulate the problem as follows:

$$\mathbf{P'}\text{: Max } z = \left[\sum_{k=1}^{p} |z_k(\mathbf{x}) - z_k(\mathbf{x}^*)|^{\beta} \right]^{\frac{1}{\beta}}$$
$$\text{s.t. } \mathbf{Ax} \leq \mathbf{b}$$
$$\mathbf{x} \geq \mathbf{0}.$$

In other words, we are using Minkowski distances with a parameter β. (For a discussion of Minkowski distances, see, e.g., Sect. 3.8). For the purpose of this example, we will demonstrate the approach by using Euclidean distances, the distance with $\beta = 2$. Consider the bi-objective optimization problem

$$\mathbf{P}\text{: ``Max'' } z_1 = 4x_1 - x_2$$
$$\text{``Max'' } z_2 = -x_1 + 3x_2$$
$$\text{s.t.} 3x_1 + 4x_2 \leq 12 \qquad\qquad \text{(I)}$$
$$3x_1 + x_2 \leq 6 \qquad\qquad\quad \text{(II)}$$
$$x_1, \ x_2 \geq 0.$$

The graphical representation of the problem is shown in Fig. 3.3.

Optimizing the first objective with respect to the constraints results in an optimal solution $\bar{x}_1^1 = 2$, $\bar{x}_2^1 = 0$ with objective values $\bar{z}_1^1 = 8$ and $\bar{z}_2^1 = -2$. Similarly, we can optimize the second objective with respect to the same constraints, which results in the optimal solution $\bar{x}_1^2 = 0$, $\bar{x}_2^2 = 3$ with objective values $\bar{z}_1^2 = -3$ and $\bar{z}_2^2 = 9$. These are the extreme points of what is known as the *nondominated frontier* with the value

Fig. 3.3 Feasible region and nondominated frontier

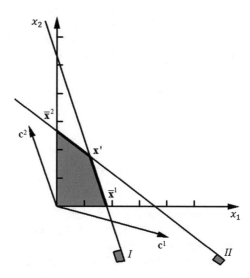

Fig. 3.4 Compromise solutions for two ideal points

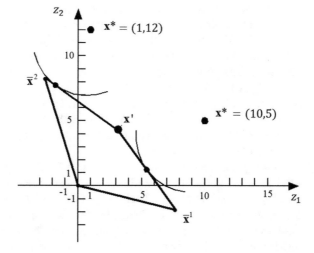

of the first objective function ranging from -3 to 8 and that of the second objective being between -2 and 9. The point $\mathbf{x}'$ is located at an extreme point of the feasible set between the two solutions that optimize the individual objectives z_1 and z_2, respectively. The nondominated frontier is shown as bold line segments in Fig. 3.3 between $\bar{\mathbf{x}}^1$ and $\mathbf{x}'$, and between $\mathbf{x}'$ and $\bar{\mathbf{x}}^2$, respectively.

Another representation of the problem is in the objective space (z_1, z_2). Each extreme point in the (x_1, x_2)-space maps to a corresponding point in (z_1, z_2)-space. In our example, the extreme points $(x_1, x_2) = (0, 0)$, $(2, 0)$, $\left(1\frac{1}{3}, 2\right)$, and $(0, 3)$ are mapped into the objective space as $(z_1, z_2) = (0, 0)$, $(8, -2)$, $\left(3\frac{1}{3}, 4\frac{2}{3}\right)$, and $(-3, 9)$, respectively. The resulting space is shown in Fig. 3.4.

Suppose now that we choose the ideal point $z_1^* = 10$ and $z_2^* = 5$. The unweighted problem with the variables x_1, x_2, z_1, and z_2 (the last two could be replaced by the objectives, but we leave them here for clarity) can then be written as

$$P' : \operatorname{Min} z = \left[(10 - z_1)^2 + (5 - z_2)^2 \right]^{\frac{1}{2}}$$

$$\text{s.t. } z_1 = 4x_1 - x_2$$
$$z_2 = -x_1 + 3x_2$$
$$\text{s.t. } 3x_1 + 4x_2 \leq 12 \tag{I}$$
$$3x_1 + x_2 \leq 6 \tag{II}$$
$$x_1, x_2 \geq 0, \ z_1, z_2 \in \mathbb{R}.$$

The optimal solution to this nonlinear, single-objective optimization problem is $\bar{\mathbf{x}} = (1.6242, 1.1275)$ with objective values $\bar{\mathbf{z}} = (5.3691, 1.7584)$. On the other hand, using the ideal point $\mathbf{x}^* = (1, 12)$, we obtain the optimal solution $\bar{\mathbf{x}} = (0.2792, 2.7906)$ with objective values $\bar{\mathbf{z}} = (-1.6736, 8.0925)$. Both ideal points and their resulting solutions are shown in Fig. 3.4.

The compromise that results from this procedure can be modified by using weights. For instance, suppose that we were to choose weights of $w_1 = 1$ and $w_2 = 3$, meaning that deviating by one unit of distance from the ideal solution of the second objective is three times as a deviation of one distance unit by the first objective. The optimal compromise solution given the ideal point $\mathbf{x}^* = (10, 5)$ is then $\bar{\mathbf{x}} = (2, 0)$ with objective values of $\bar{\mathbf{z}} = (8, -2)$. It is not much of a surprise that this solution, in comparison with the unweighted problem discussed earlier, has the second objective closer to its ideal point, as the penalty for being distant is much larger than before. We wish to point out that this approach is applicable to problems with any number of objectives.

At this point, we have two tools to express the decision maker's preferences: the ideal point and the weights. Results for a number of combinations of ideal points and weights are shown in Table 3.2, where the first two columns show the ideal point and the weights of the objectives chosen by the decision maker, respectively, while the third and fourth column show the solution and the individual respective values, respectively. The first two rows in Table 3.2 deal with the examples shown in Fig. 3.4.

We should note that using squared Euclidean distances $(\ell_2)^2$ instead of (ℓ_2) results in exactly the same results in all instances above.

A possible extension considers the possibility that the decision maker—not unlike target values—has specified not an ideal point, but an ideal or *desirable region*. In our example, suppose that the decision maker has specified that it would be desirable to have objective values of at least 6 and 4 for the two individual objectives, respectively, as well as a combined (unweighted) value of 12 for the two objectives together. The corresponding constraints $z_1 \geq 6$, $z_2 \geq 4$, and $z_1 + z_2 \geq 12$ describe an unbounded polytope in (z_1, z_2)-space. We will now rewrite our problem P in terms of z_1 and z_2 by using the definitions of z_1 and z_2 provided in the objective

Table 3.2 Optimal solutions for various ideal points and weights

Ideal point	Weights	$\bar{\mathbf{x}} = (\bar{x}_1, \bar{x}_2)$	$\bar{\mathbf{z}} = (\bar{z}_1, \bar{z}_2)$
10, 5	1, 1	1.6242, 1.1275	5.3691, 1.7584
1, 12	1, 1	0.2792, 2.7906	-1.6736, 8.0925
5, 10	1, 1	1.0491, 2.2132	1.9830, 5.5906
10, 2	1, 1	1.8255, 0.5235	6.7785, -0.2550
10, 10	1, 1	1.33, 2.00	3.33, 4.67
50, 10	1, 1	2, 0	8, -2
10, 5	2, 1	1.7879, 0.6364	6.5152, 0.1212
10, 5	3, 1	1.8866, 0.3401	7.2065, -0.8664
10, 5	5, 1	2, 0	8, -2
10, 5	1, 2	1.4940, 1.5181	4.4578, 3.0602
10, 5	1, 3		
10, 5	1, 5	1.3880, 1.8361	3.7158, 4.1202
10, 5	1, 10	1.3460, 1.9619	3.4223, 4.5396

functions in problem P. In other words, taking $z_1 = 4x_1 - x_2$ and $z_2 = -x_1 + 3x_2$ and solving the two equations for x_1 and x_2 results in $x_1 = \frac{3}{11}z_1 + \frac{1}{11}z_2$ and $x_2 = \frac{1}{11}z_1 + \frac{4}{11}z_2$. In addition, define additional variables y_1 and y_2, which express a point in the desirable region. The objective is then to find points $z = (z_1, z_2)$ in the feasible set and $y = (y_1, y_2)$ in the desirable set, so as to minimize the distance between them. We can then write the problem as

$$P' : \underset{y,z}{\text{Min}} \left[(z_1 - y_1)^2 + (z_2 - y_2)^2 \right]^{\frac{1}{2}}$$

$$\text{s.t. } \frac{13}{11}z_1 + \frac{19}{11}z_2 \leq 12 \qquad (I')$$

$$\frac{10}{11}z_1 + \frac{7}{11}z_2 \leq 6 \qquad (II')$$

$$\frac{3}{11}z_1 + \frac{1}{11}z_2 \geq 0 \qquad (x_1 \geq 0)$$

$$\frac{1}{11}z_1 + \frac{4}{11}z_2 \geq 0 \qquad (x_2 \geq 0)$$

$$y_1 \geq 6$$

$$y_2 \geq 4$$

$$y_1 + y_2 \geq 12$$

$$z_1, z_2; y_1, y_2 \in \mathbb{R},$$

where constraints I' and II' are the original constraints I and II, written in terms of z_1 and z_2. The result of the problem has $\bar{z}_1 = 3.5838$, $\bar{z}_2 = 4.3086$, $\bar{y}_1 = 6$, and $\bar{y}_2 = 6$. Reverting the z-variables back into x-variables, we obtain $\bar{x}_1 = 1.3690$, and $\bar{x}_2 = 1.8924$. In Fig. 3.3, we can see that the optimal solution is a point on constraint II just below the point $\mathbf{x}'$, so that constraint I is loose, while constraint II is binding at

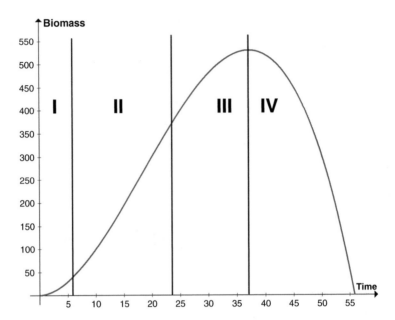

Fig. 3.5 Sigmoidal function showing tree growth

that point. Clearly, the choice of the desirable set will influence the chosen solution in a way similar to that of the choice of ideal point. In addition, it is also possible to use a weighted distance in the objective of P′.

3.3   Optimal Rotation Time in Forestry

One of the major questions in forestry management is to decide when a forest should be harvested. This is known as the optimal rotation time. Clearly, the harvesting schedule will depend on the specific species under consideration. Of particular interest is the growth function of the species. What is of most interest to the harvesting companies is the usable volume of the tree, which is a proportion of the entire biomass of a tree, which, in turn, is a function of time, i.e., the age of the tree. Similar functions have been used in modeling the growth of cells or populations. Often, modelers use a logistic function, i.e., a sigmoidal function with its typical "S" shape. A polynomial function (actually, a cubic) with similar features is shown in Fig. 3.5.

In its juvenile phase "I," sometimes referred to as the "lag phase," the tree prepares its growth. From the outside, nothing much can be seen as far as growth is concerned. Then the tree enters a period of exponential growth, here shown as phase "II," followed by a time of maturity "III," in which growth slows down. Finally, the tree reaches a time of equilibrium "IV," in which the rate of decay equals that of new growth, and then, once the tree dies, the amount of usable biomass decreases to zero, i.e., the functional values drop down gradually to zero.

Cumulative growth functions such as the one shown above were pioneered by von Bertalanffy (1951), and refined by Richards (1959) and Chapman (1961). An alternative model was put forward by Schnute (1981), which was originally devised for fisheries. A generalized logistic function or Richards' curve is of the form

$$f(t) = \left[\frac{\alpha}{1 - e^{\beta(t-\hat{\imath})}} \right]^{-\gamma}, \tag{3.1}$$

where t is the actual age of the tree, $f(t)$ is the biomass, $\hat{\imath}$ is the t-value of the function midpoint, and α, β, and γ are parameters that dictate the shape of the function. They can be finetuned depending on the species under consideration.

Whatever the specific growth function under consideration, the main idea is to determine a time t^*, which maximizes the annual gain of the biomass. In other words, we attempt to find the time t^*, which maximizes $f(t)/t$. In the absence of constraints, this problem is easy to solve by way of differential calculus. In Sect. 1.2 we showed that at optimality, $t^* f'(t^*) = f(t^*)$ must hold. Using the same function as that shown in Fig. 3.5, viz., $f(t) = t + 1.1t^2 - 0.02t^3$, we determine that the function has a maximum at $\hat{\imath} = 37.1157$ years, while the point that maximizes the annual production is at $t^* = 27.5$ years. Incidentally, this is the point at which a tangent to the sigmoidal curve, which leads through the origin, has its maximal slope. In our example, the tangent is $f(t) = 16.125t$ (indicating an annual production of 16.125 units of biomass) which touches the growth function at t^*, where a total of 443.4375 units of biomass have been produced. In contrast, the maximal biomass is reached at 37.1157 years. Figure 3.6 shows the sigmoidal biomass function as well as the tangent that leads through the origin.

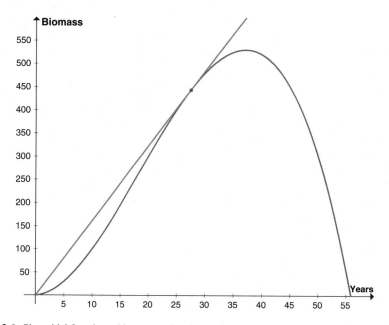

Fig. 3.6 Sigmoidal function with tangent showing optimal rotation time

In other words, we maximize the annual biomass production, if we harvest every 27.5 years. Thus trees should be harvested significantly earlier than when the maximal biomass has been obtained. Intuitively, this is not surprising, considering the fact that the growth rate diminishes as the age of the tree approaches the time of its maximal biomass, thereby reducing the average yield over time.

Interesting other nonlinear optimization problems arise from forestry problems that include different species, multiple lots, and capacities regarding the equipment used for harvesting, see Davis et al. (2001) and Rönnqvist (2003). The result will be a long-term viable harvesting schedule, whose main unknowns are, of course, the prices for the wood of the individual species. This type of optimization also occurs in the farming of fish, poultry, and other livestock and vegetables.

3.4   Queuing Optimization

In a service system, customers arrive at a set of service stations, where they are served by clerks. In order to serve customers well, clerks must constantly be trained. Training has three effects: it costs money, it removes the clerks temporarily from service, and it enables them to increase their service rate. While standard queuing modes at their core have the task to determine characteristics of the system, such as waiting time, the average time customers have to wait and various probabilities, they can very well be embedded in optimization problems. In this section, we will investigate two different problems: in the first, we consider the well-known problem of finding the optimal number of clerks in a system, while in the second, we determine the optimal retraining schedule of a clerk.

First consider the problem of determining the optimal number of identical clerks in a queuing system "supermarket" style, which, in case of s service stations, has s separate waiting lines. In simple words, we look at s separate $M/M/1$ systems. For introductions to queuing, see, e.g., Hillier et al. (2017) or Eiselt and Sandblom (2012). In this context, we have two different cost types: on the one hand, there are the costs of the service station (here referred to as "clerks"), while the other concerns the customers of the service stations. Often in the literature, this type of model is put in the context of a tool crib, in which workers (the customers in our description above), who are in need of a specialized tool, have tick up that tool from a service desk, manned by a clerk. Clearly, the workers and the clerks present a tradeoff: more clerks means more costs for the clerks, but less waiting time for the workers, and thus less lost time associated with them. In other contexts, the costs associated with customers are likely to be measured in loss of goodwill, which is difficult to assess numerically.

Defining c_{serv} and c_{cust} as the unit (e.g., hourly) cost of a clerk (service station) and a worker (customer), respectively, and suppose that L_s denotes the average number of workers that are in a system. With λ denoting the average number of arrivals at the system per hour (the arrival rate), whereas μ denotes the service rate, i.e., the average number of customers a clerk can deal with in 1 h. As usual, we assume that all clerks have the same capabilities and service rates. For s separate service stations, the effective arrival rate at any one of the systems is $\lambda^{eff} = \lambda/s$. The costs per time unit can then be written as $C = c_{serv}s + c_{cust}sL_s$. Given that $L_s = \lambda^{eff}/(\mu - \lambda^{eff})$ for each of the s parallel systems, we can formulate the problem as

$$P_1: \underset{s=1,2,\dots}{\text{Min}} \ C = c_{serv}s + \frac{c_{cust}\lambda s}{\mu s - \lambda},$$

with s being the only variable, and where we must assume that $\mu s > \lambda$, i.e., the combined service rate of the system exceeds the overall arrival rate. Note that the cost function is only defined for integer values of s. Furthermore, note that the cost function, having a strictly positive second derivative for $\mu s > \lambda$, is convex. Ignoring the integrality requirement for a moment, we can determine the derivative of the cost function with respect to s, set it equal to zero, and solve the resulting quadratic equation for s. The result may then be any real number, so, due to the convexity of the function, we only have to explore the two neighboring integers in order to determine the optimal value of s. Standard calculations show that the optimal number of clerks is

$$s^* = \left[1 \pm \sqrt{\frac{c_{cust}}{c_{serv}}}\right]\rho,$$

where $\rho = \lambda/\mu$ denotes the traffic intensity of the system. As a matter of fact, for $c_{cust}/c_{serv} > 1$, only the "+" in the square brackets leads to a feasible solution. In case of $c_{cust}/c_{serv} < 1$, we invoke the feasibility requirement $\mu s > \lambda$ or $\rho = \lambda/\mu < s$, which leads to the necessity that the term in square brackets must be greater than one. Again, only the "+" sign in the brackets applies.

As an example, suppose that the unit costs are $c_{serv} = 20$ and $c_{cust} = 5$, while the overall arrival rate is $\lambda = 100$ customers per hour, and the service rate at each of the service stations is $\mu = 20$. Clearly, since we must have $\mu s > \lambda$, at least 6 service stations are needed. The formula will reveal that $s* = \left[1 + \sqrt{\dfrac{5}{20}}\right]5 = 7.5$. Hence we have to inspect costs for $s = 7$ and $s = 8$, which are $C(s = 7) = 227.5$ and $C(s = 8) = 226.67$, so that the choice of $s = 8$ is optimal.

Consider now the second problem, in which clerks can be trained in order to improve service. Again, we assume that the assumptions of the standard $M/M/1$ model apply.

It is now possible to (re-)train the clerks for better service. If the service they provide deals with technology (e.g., software), constant training is required. This is the case under consideration here. The function that relates the training hours t and the service rate is specified to be

$$\mu = (\alpha - \beta^{-t})\gamma,$$

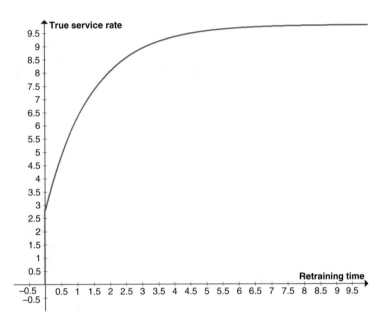

Fig. 3.7 True service rate as a function of retraining time

with constants α, β, and γ. In this example, we are given $\alpha = 1.4$, $\beta = 2$, and $\gamma = 7$, so that $\mu(t) = (1.4 - 2^{-t})(7)$. This function is shown in Fig. 3.7.

Here, zero training time results in a service rate of 2.8 (the clerk will have to look up a lot of manuals before he can actually help a customer), it then ascends quickly, before settling just below 10 (at 2 h of training time, it has already reached a service rate of 8). From these improving service rates, we must, of course, deduct the time for the training, resulting in the function

$$\mu^{\text{true}}(t) = (\alpha - \beta^{-t})\gamma - t,$$

which, in our example, can be visualized in Fig. 3.8.

In this function, zero training time $t = 0$ results in a service rate of $\mu^{\text{true}} = 2.8$, reaches a peak at $t = 2.28$ h of training time and a service rate of $\mu^{\text{true}} = 8.36$, and at $t = 9.8$, it reaches a service rate of zero (at which point the clerk is trained very well, but the training consumes so much of his time, so that there is no more actual service).

In order to now describe the model, assume that the arrival rate is $\lambda = 4$ customers per hour, and the original service rate is 2.8 customers per hour with the improvement function $\mu = (1.4 - 2^{-t})(7)$ as shown above. The costs are $c_{serv} = \$10$ per hour for the clerks, $c_{cust} = \$50$ per hour for the customers, and $c_t = \$5$ per hour for retraining. The total cost function can then be written as $\mathcal{C} = $ (wages for the clerks) $+$ (training costs for clerks) $+$ (opportunity costs for customers), i.e., the problem can be formulated as

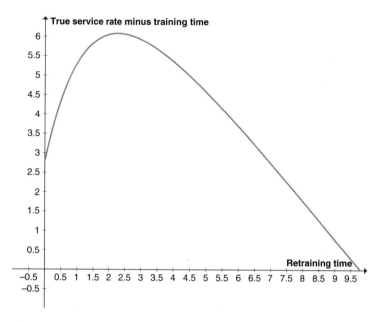

Fig. 3.8 True service rate minus training time as a function of training time

$$P_2 : \operatorname*{Min}_{s,\,t} C = c_{serv}s + c_t ts + c_{cust}sL_s$$

$$= c_{serv}s + c_t ts + c_{cust}s\,\dfrac{\dfrac{\lambda}{s}}{[(\alpha - \beta^{-t})\gamma - t] - \dfrac{\lambda}{s}},$$

which in our example takes the form

$$C = 10s + 5ts + 50s\,\dfrac{\dfrac{4}{s}}{[(1.4 - 2^{-t})7 - t] - \dfrac{4}{s}}.$$

Figure 3.9 shows the cost functions for $s = 2$ clerks (the red curve), $s = 3$ clerks (the blue curve), and $s = 4$ clerks (the green curve).

If we were to adopt the model with two clerks, the optimal retraining time would be $t = 1.489$ h, and the total costs would be $C = 87.2846$. With three clerks, the optimal retraining time is $t = 1.1009$ with total costs of $C = 95.2676$, while with four clerks, the optimal retraining time is $t = 0.8618$ h and the total costs are then $C = 106.1795$. Consequently, the solution with two clerks should be adopted.

As a final remark, a few different queuing models are formulated and solved with geometric programming techniques, see, e.g., Illustration 4 in Sect. 8.5.

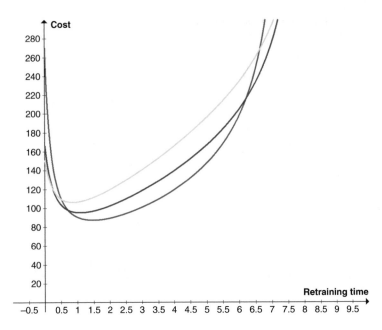

Fig. 3.9 Cost as a function of training time

3.5 Portfolio Selection Analysis

Consider the problem of an investor who is building a portfolio consisting of a variety of financial securities, such as stocks, bonds, cash, treasury bills, commodity certificates, and other types of investments. The investor wants to purchase various amounts of these securities within an overall budget. The resulting portfolio of assets will then achieve a performance over time, which depends on the performance of the individual assets in the portfolio as well as their relative weights in the financial mix. An important question is then by what measure the performance of the portfolio should be judged, and what constraints the investor might put on the selection of securities. One possible way to measure the portfolio concerns its anticipated return (which, clearly, the investor attempts to maximize over some specific period of time), and/or the potential risk or volatility (which is to be minimized). One possibility is to optimize some combination of the aforementioned two concerns.

In the makeup of the portfolio, there are a number of constraints that are typically considered. For instance, it may be required to achieve a lowest allowable level of return and/or a highest acceptable level of risk. There may also be bounds on the allowable proportions of the constituent assets in the portfolio. In all cases, there will be an overall budget constraint that restricts the amount of money invested. In order to introduce the mathematical formulation of the portfolio selection problem, we return to its roots found in the work of Markowitz (1952). Assume that there are

n securities to be considered for the portfolio with random returns $R_j, j = 1, \ldots, n$. Let $\mu_j := E(R_j)$ denote the expected return of security j and $\sigma_j^2 := V(R_j) = E(R_j - \mu_j)^2$ the variance (or risk) of the j-th security, $j = 1, \ldots, n$. Commonly, $\sigma_j = \sqrt{V(R_j)}$ is referred to as the *volatility* of the security. We also have the symmetric covariance matrix $\mathbf{V} = (\sigma_{ij})$, whose elements are the covariances $Cov(R_i, R_j) := E((R_i - \mu_i)(R_j - \mu_j)) := \sigma_{ij}; i, j = 1, \ldots, n$, where $\sigma_{jj} = \sigma_j^2$, as well as the correlation coefficients ρ_{ij} for the returns R_i and R_j, which satisfy the equations $\sigma_{ij} = \rho_{ij}\sigma_i\sigma_j$. Let now the variable x_j denote the weights (i.e., the proportion) of the j-th asset in our portfolio. Then the expected value of the return R of the entire portfolio can be expressed as $E(R) = E\left(\sum_{j=1}^n R_j x_j\right) = \sum_{j=1}^n E(R_j)x_j = \sum_{j=1}^n \mu_j x_j$. For the variance $V(R)$ of the total return R we obtain

$$V(R) = V\left(\sum_{j=1}^n R_j x_j\right) = \sum_{i=1}^n \sum_{j=1}^n x_i \sigma_{ij} x_j.$$ Using matrix notation with $\boldsymbol{\mu} := [\mu_1, \ldots, \mu_n]$ and $\mathbf{x} := [x_1, \ldots, x_n]^T$, we can write $E(R) = \boldsymbol{\mu}\mathbf{x}$ and $V(R) = \mathbf{x}^T\mathbf{V}\mathbf{x}$. The Markowitz model now minimizes the risk z_{risk} of the entire portfolio, as measure by the variance $V(R)$, subject to some constraints. These constraints include a predefined lowest acceptable level of return R_{min}, the requirements that the weights are nonnegative and sum up to one, i.e., $x_j \geq 0, j = 1, \ldots, n$ and $\sum_{j=1}^n x_j = 1$, in addition to any other restrictions on the portfolio, which we will write as $\mathbf{A}\mathbf{x} \leq \mathbf{b}$. In summary, the basic Markowitz problem can be written as

$$\text{BM}: \quad \underset{\mathbf{x}}{\text{Min}}\, z_{risk} = \mathbf{x}^T\mathbf{V}\mathbf{x}$$
$$\text{s.t. } \boldsymbol{\mu}\mathbf{x} \geq R_{min}$$
$$\mathbf{A}\mathbf{x} \leq \mathbf{b}$$
$$\mathbf{e}\mathbf{x} = 1$$
$$\mathbf{x} \geq \mathbf{0},$$

where $\mathbf{e} = [1, 1, \ldots, 1]$ denotes the n-dimensional summation row vector. We see that the problem BM has a quadratic objective function and linear constraints; it is therefore a quadratic programming problem, which can be solved by various techniques, some of which are presented in Chap. 6.

The minimal risk level $\bar{z}_{risk}$ will obviously depend on the required return R_{min} of the optimal portfolio. It is apparent that by raising the requirement level R_{min}, the planner has to accept an increased level of risk in the portfolio, and the feasible region of BM will become smaller, thus leading to a poorer (i.e., larger, since we are minimizing) optimal value of the objective function. The model will therefore display a tradeoff between a desirable level of return and a harmful risk. Alternatively, we could maximize the return subject to an acceptable level of risk $\hat{z}_{risk}$. We will call this the first modification MM(a) of the basic Markowitz model. The problem can then be written as

$$\text{MM(a)}: \quad \underset{\mathbf{x}}{\text{Max}}\; z_{return} = \boldsymbol{\mu}\mathbf{x}$$
$$\text{s.t.}\; \mathbf{x}^T\mathbf{V}\mathbf{x} \le \hat{z}_{risk}$$
$$\mathbf{Ax} \le \mathbf{b}$$
$$\mathbf{ex} = 1$$
$$\mathbf{x} \ge \mathbf{0}.$$

One may argue that an ordinary investor will rather maximize return subject to an acceptable level of risk than minimize risk subject to an acceptable level of return; therefore model MM(a) would be preferable to the basic model BM. However, since we are able to determine tradeoffs between risk and return in both models, we can view them as equivalent. For a further discussion, see Cornuejols and Tütüncü (2007).

A more bothersome aspect of the Markowitz model is the way in which risk is measured. By computing the variance, return values above the mean are counted in exactly the same way as are values below the mean, which is not meaningful: after all, an investor will dislike returns below the mean, while returns above the mean are most welcome and do not constitute risk. In other words, the proxy "variance" does not appear to be a suitable expression for "risk." A more appropriate measure of risk (usually referred to as *downside risk*) is provided by the *semivariance* of returns, a concept actually proposed by Markowitz himself (Markowitz 1959), but which has only lately become more prevalent. The reasons for this delay are theoretical and computational.

More specifically, recall that the variance of the j-th security is given by $V(R_j) := E((R_j - \mu_j)^2)$. Given some benchmark return r_j we then define the semivariance or downside risk as $SV(R_j, r_j) := E((\min\{R_j - r_j, 0\})^2)$. The benchmark value r_j is preselected by the investor and could be some desired value or μ_j, or simply zero. We see that the semivariance only considers instances in which $R_j < r_j$, i.e., those in which the security underperforms relative to the benchmark, will count. The instances for which $R_j \ge r_j$ will not contribute towards the semivariance, as intended. One can easily show that if the benchmark is chosen to be the mean, i.e., $r_j = \mu_j$, then the semivariance can never be larger than the regular variance.

In our discussion above we referred to the expression $\sqrt{V(R_j)}$ as the volatility of the j-th security; in analogy $SV(R_j r_j)$ is called the *downside volatility* of the security. Carrying the discussion further, we define the *semicovariance* between the i-th and the j-th assets as $SCov(R_i, r_i; R_j r_j) := E((\min\{R_i - r_i, 0\})(\min\{R_j - r_j, 0\}))$, where $i, j = 1, \ldots, n$. A semicovariance is always nonnegative, being the sum of products of nonpositive numbers. Furthermore, with means as the chosen benchmark, the semicovariances can never be larger than the absolute value of the corresponding covariances. The semicovariance matrix $\mathbf{SV}(r_1, \ldots, r_n)$ or $\mathbf{SV}$ for short, is then the matrix formed by the semicovariances between the securities. For a portfolio, in which the j-th security enters with weight $x_j, j = 1, \ldots, n$, we then define the portfolio downside risk $\underline{z}_{risk}$ as $\underline{z}_{risk} = \mathbf{x}^T \mathbf{SVx}$, which leads to the second modification MM (b) of the basic Markowitz model as

$$\text{MM(b):} \quad \underset{\mathbf{x}}{\text{Min}}\, z_{risk} = \mathbf{x}^T \mathbf{SVx}$$
$$\text{s.t.} \; \boldsymbol{\mu}\mathbf{x} \geq R_{min}$$
$$\mathbf{Ax} \leq \mathbf{b}$$
$$\mathbf{ex} = 1$$
$$\mathbf{x} \geq \mathbf{0}.$$

If we argue, as before, that an investor would rather maximize return, we can, in analogy to our previous discussion, obtain the third modification MM(c) of the basic Markowitz model as follows:

$$\text{MM(c):} \quad \underset{\mathbf{x}}{\text{Max}}\, z_{return} = \boldsymbol{\mu}\mathbf{x}$$
$$\text{s.t.} \; \mathbf{x}^T \mathbf{SVx} \leq \hat{\underline{z}}_{risk}$$
$$\mathbf{Ax} \leq \mathbf{b}$$
$$\mathbf{ex} = 1$$
$$\mathbf{x} \geq \mathbf{0},$$

where the parameter $\hat{\underline{z}}_{risk}$ is an investor-selected acceptable upper limit on the downside risk. One can claim that this third modification is the most realistic portfolio selection model we have developed so far.

A small numerical example is now presented to illustrate our discussion. An investor wishes to put together a portfolio that consists of three different mutual funds: an equity fund in stocks, a bond fund, and a money market fund. The portfolio is not allowed to include more than 60% of its assets in stocks; furthermore, the amount invested in money market funds cannot exceed more than half of the money invested in bonds. The 5-year returns (net of inflation and taxes) have been recorded at the market closing time at the end of ten consecutive weeks with figures shown in Table 3.3.

These data will yield the covariance matrix $\mathbf{V} = \begin{bmatrix} 7 & -1.2 & 6.3 \\ -1.2 & 1 & -4.5 \\ 6.3 & -4.5 & 40.6 \end{bmatrix}$, so that

the basic Markowitz model for minimizing the total risk using the covariance matrix subject to the aforementioned constraints can then be written as

Table 3.3 Five-year returns

	Week										
Five year return R_i	1	2	3	4	5	6	7	8	9	10	Mean μ_i
$i = 1$: stocks	17	19	18	19	20	21	23	25	22	16	20
$i = 2$: bonds	13	12	12	11	11	11	11	12	13	14	12
$i = 3$: money market	5	10	13	16	20	25	25	14	8	14	15

$$\text{BM}: \quad \underset{\mathbf{x}}{\text{Min}} \; z_{risk} = [x_1, x_2, x_3] \begin{bmatrix} 7 & -1.2 & 6.3 \\ -1.2 & 1 & -4.5 \\ 6.3 & -4.5 & 40.6 \end{bmatrix} \begin{bmatrix} x_1 \\ x_2 \\ x_3 \end{bmatrix}$$

s.t. $[20, 12, 15][x_1, x_2, x_3]^T \geq 14$ (minimally acceptable return)

$\quad\;\; [1, 0, 0][x_1, x_2, x_3]^T \leq 0.6$ (maximal proportion of stocks)

$\quad\;\; [0, -1, 2][x_1, x_2, x_3]^T \leq 0$ (bonds vs money market relationship)

$\quad\;\; [1, 1, 1][x_1, x_2, x_3]^T = 1$

$\quad\;\; [x_1, x_2, x_3]^T \geq \mathbf{0}.$

We can also write this model as

$$\text{BM}: \; \underset{x_1, x_2 x_3}{\text{Min}} \; z_{risk} = 7x_1^2 - 2.4x_1x_2 + 12.6x_1x_3 + x_2^2 - 9x_2x_3 + 40.6x_3^2$$

$$\text{s.t.}\; 20x_1 + 12x_2 + 15x_3 \geq 14$$
$$x_1 \qquad\qquad\qquad \leq 0.6$$
$$-x_2 + 2x_3 \leq 0$$
$$x_1 + x_2 + x_3 = 1$$
$$x_1, \quad x_2, \quad x_3 \geq 0.$$

The model BM is a standard quadratic programming problem, i.e., an optimization problem with a quadratic objective function and linear constraints. Its unique optimal solution is $\bar{x}_1 = 0.2287$, $\bar{x}_2 = 0.7146$, $\bar{x}_3 = 0.0567$, resulting in $\bar{z}_{risk} = 0.4139$. In other words, the optimal portfolio consists of 22.87% stocks, 71.46% bonds, and 5.67% money market funds. The total return of the portfolio is 14% over the 5 year planning period, exactly the minimally required return. From a technical point of view, the first constraint is tight at optimality, whereas the next two constraints are not. Since the minimal risk is $\bar{z}_{risk} = 0.4139$, we obtain the volatility of the optimal portfolio as $\sqrt{0.4139} = 0.6433$. In financial theory, this portfolio would be called efficient, since it has the minimal variance among all portfolios with at least 14% return. If we require an overall return of only 13% instead of 14% and reoptimize, the minimal variance turns out to be better (i.e., smaller) at 0.3146. Requiring a return of 15% leads to a minimal variance of 0.7748. Mapping out the optimal variances as a function of the required overall returns, we obtain the efficient frontier of the portfolio universe.

Note that if a total return of more than 20% were to be required, the problem has no solution. This is evident since none of the assets will return more than 20%. Actually, since the portfolio proportion of the 20%-yielding stocks can be no more than 0.6, the maximal possible return will be less than 20%. Similarly, if the minimal return required is set at less than 12%, the first constraint will be nonbinding for all feasible portfolios. Table 3.4 displays the optimal solutions for various levels of the required return μ.

For increasing values of the minimally required return μ, the proportion x_1 of stocks (the most profitable of the three assets) will increase, until it reaches, and stays at, its maximally allowed level 0.6, which occurs when $\mu = 16.8\%$. Meanwhile, the

Table 3.4 Solutions for various levels of μ

μ (in %)	z_{risk}	x_1 (in %)	x_2 (in %)	x_3 (in %)
≤ 13	0.314	11.1	80.9	8.0
14	0.414	22.9	71.4	5.7
15	0.774	36.4	60.6	3.0
16	1.400	49.9	49.8	0.3
16.2	1.556	52.5	47.5	0.0
16.8	2.104	60.0	40.0	0.0
17	2.636	60.0	33.3	6.7
17.2	3.617	60.0	26.7	13.3
>17.2	Infeasible			

proportions x_2 and x_3 of the other two types of assets keep shrinking; x_2 to reach a value of 0.475 and x_3 to become zero for $\mu = 16.2\%$. Until μ reaches the level of 16.8%, the first constraint $20x_1 + 12x_2 + 15x_3 \geq \mu$ is tight, whereas the second constraint $x_1 \leq 0.6$ as well as the third constraint $-x_2 + 2x_3 \leq 0$ are loose. For $\mu \in [16.8\%, 17.2\%]$, the first and second constraints are both tight and only the third constraint is loose. It is then natural for x_3 to increase in value, since its return is better than that of x_2. Finally, if μ is required to exceed 17.2%, no feasible solution exists.

Consider now the first modification MM(a) of the basic Markowitz model, where the maximization of the expected return is the objective, while the variance appears among the constraints. With a linear objective function and one of the constraints involving a quadratic function, this is not a quadratic programming problem, and it is more difficult to solve. Specifically, our problem can be written as

$$\text{MM(a)}: \underset{x_1, x_2, x_3}{\text{Max}} \ z_{return} = 20x_1 + 12x_2 + 15x_3$$

$$\text{s.t.} \ 7x_1^2 - 2.4x_1x_2 + 12.6x_1x_3 + x_2^2 - 9x_2x_3 + 40.6x_3^2 \leq 0.4139$$

$$x_1 \qquad\qquad\qquad \leq 0.6$$
$$-x_2 + 2x_3 \leq 0$$
$$x_1 + x_2 + x_3 = 1$$
$$x_1, \quad x_2, \quad x_3 \geq 0.$$

Since we have set the maximal risk level at 0.4139, which was the optimal level achieved in the basic model, it should come as no surprise that the optimal solution is obtained as $\bar{x}_1 = 0.2287$, $\bar{x}_2 = 0.7146$, $\bar{x}_3 = 0.0567$ with a return of $\bar{z}_{return} = 14\%$, i.e., the same solution as that of the BM model. In this way, the two models are equivalent, even though the previous model focused on the risk and the present model centers on return. Again, with a maximally acceptable risk level of 0.35, the maximal return is 13.65%, whereas no feasible solution exists if we require a risk level of no more than 0.3. If a risk level of 0.5 were to be deemed acceptable, the optimal return is 14.32%, while a risk level of 0.6 causes the optimal return to be 14.6%, etc.

Table 3.5 Solutions for various levels of z_{risk}

z_{risk}	μ (in %)	x_1 (in %)	x_2 (in %)	x_3 (in %)
≤ 0.3	Infeasible			
0.4	13.94	22.0	72.1	5.9
0.5	14.32	27.2	68.0	4.8
0.6	14.60	31.0	64.9	4.1
0.7	14.84	34.2	62.4	3.4
0.8	15.05	37.1	60.1	2.8
0.9	15.24	39.6	58.1	2.3
1.0	15.41	41.9	56.2	1.9
1.5	16.13	51.6	48.4	0.0
2.0	16.70	58.7	41.3	0.0
2.5	16.96	60.0	34.6	5.4
3.0	17.08	60.0	30.5	9.5
3.5	17.18	60.0	27.3	12.7
≥ 4.0	17.20	60.0	26.7	13.3

Table 3.5 displays the optimal results of MM(a) for various levels of maximally acceptable risk z_{risk}.

For risk levels of $z_{risk} \leq 0.3$, there exists no feasible solution. Actually, as we have seen earlier, 0.314 is the lowest risk level that can be achieved. For increasing risk levels, the total return increases to a maximal level of 17.2%. The proportion of stocks in the optimal portfolio will grow from a minimal level of 11.1% to its maximal level of 60%. Again, for a risk level of 1.485, the proportion of money market funds has decreased to zero, only to increase again as explained above. We should note that Tables 3.4 and 3.5 both contain information about the efficient frontier. In both tables, the frontier is shown by the pairs (z_{risk}, μ). Both tables convey essentially the same information.

We will now use our numerical example to illustrate the effect of using the semicovariance instead of the covariance matrix to express the risk inherent in the portfolio. For our data, we obtain the semicovariance matrix as

$$SV = \begin{bmatrix} 3.1 & 0.1 & 4.3 \\ 0.1 & 0.4 & 0 \\ 4.3 & 0 & 18 \end{bmatrix}$$, where we have used the individual means as benchmarks.

Minimizing the downside risk $\underline{z}_{risk}$, our modified Markowitz model MM(b) can then be written as

$$MM(b): \underset{x_1, x_2, x_3}{Min} \ \underline{z}_{risk} = 3.1x_1^2 + 0.2x_1x_2 + 8.6x_1x_3 + 0.4x_2^2 + 18x_3^2$$

$$\begin{aligned} s.t. \quad & 20x_1 + 12x_2 + 15x_3 \geq 14 \\ & x_1 \qquad\qquad\quad \leq 0.6 \\ & \qquad -x_2 + 2x_3 \leq 0 \\ & x_1 + x_2 + x_3 = 1 \\ & x_1, \quad x_2, \quad x_3 \geq 0. \end{aligned}$$

Table 3.6 Solutions for various levels of μ

μ (in %)	$\underline{z}_{risk}$	x_1 (in %)	x_2 (in %)	x_3 (in %)
≤ 12	0.373	9.1	90.9	0
13	0.376	12.5	87.5	0
14	0.456	25.0	75.0	0
15	0.639	37.5	62.5	0
16	0.925	50.0	50.0	0
16.8	1.228	60.0	40.0	0
17	1.624	60.0	33.3	6.7
17.2	2/184	60.0	26.7	13.3
>17.2	Infeasible			

Again, MM(b) is a standard quadratic programming problem. Its unique optimal solution is $\bar{x}_1 = 0.25$, $\bar{x}_2 = 0.75$, and $\bar{x}_3 = 0$ with a downside risk of $\underline{z}_{risk} = 0.4562$. In other words, the optimal portfolio now comprises 25% stocks, 75% bonds, and no money market funds; the total return of the portfolio over 5 years is 14%. With a minimal downside risk of 0.4562, the downside volatility is $\sqrt{0.4562} = 0.6755$. Table 3.6 displays results for various values of required return μ.

Finally, we consider our model MM(c), which maximizes the overall portfolio return z_{return} subject to a maximally allowable preselected downside risk $\underline{z}_{risk}$ along with the other portfolio constraints. The problem can be written as

$$\text{MM(c)}: \max_{x_1, x_2, x_3} z_{return} = 20x_1 + 12x_2 + 15x_3$$

$$\text{s.t.} \, 3.1x_1^2 + 0.2x_1x_2 + 8.6x_1x_3 + 0.4x_2^2 + 18x_3^2 \leq 0.4562$$

$$
\begin{aligned}
x_1 & & & \leq 0.6 \\
& -x_2 + & 2x_3 & \leq 0 \\
x_1 + & x_2 + & x_3 & = 1 \\
x_1, & x_2, & x_3 & \geq 0.
\end{aligned}
$$

Following the pattern from the solutions to problems BM and MM(a) above, we find a unique optimal solution with $\bar{x}_1 = 0.25$, $\bar{x}_2 = 0.75$, $\bar{x}_3 = 0$ with an objective value of $\bar{z}_{return} = 14\%$. Again, we can regard models MM(b) and MM(c) as equivalent in some sense. For various levels of downside risk $\underline{z}_{risk}$ we obtain the results shown in Table 3.7.

The pattern is similar to that of model MM(a): for increasingly generous levels of acceptable downside risk, the total return rises to a maximal level of 17.2%. The proportion of stocks in the optimal portfolio will increase to its maximally allowable level of 60%, whereas money market funds come into play only after stocks have reached their maximal level.

It is now interesting to look at how using downside risk instead of ordinary risk will affect the composition of optimal portfolios, based on our numerical example. This amounts to studying how the efficient frontiers differ with the two different

Table 3.7 Solutions for various levels of z_{risk}

z_{risk}	μ (in %)	x_1 (in %)	x_2 (in %)	x_3 (in %)
≤ 0.3	Infeasible			
0.4	13.45	18.2	81.8	0.0
0.5	14.30	28.7	71.3	0.0
0.6	14.83	35.3	64.7	0.0
0.7	15.25	40.6	59.4	0.0
0.8	15.61	45.1	54.9	0.0
0.9	15.92	49.1	50.9	0.0
1.0	16.22	52.7	47.3	0.0
1.5	16.94	60.0	35.2	4.8
2.0	17.14	60.0	28.7	11.3
≥ 2.5	17.2	60.0	26.7	13.3

approaches of measuring risk. We compare the optimal portfolios, based on the data in Tables 3.4 and 3.6 and consider values of μ between 13% and 16%. It turns out that whereas with ordinary risk all portfolios contain some money market assets, these are all absent from the downside risk portfolios. For stocks, the downside risk portfolios contain a marginally larger proportion than the ordinary risk portfolios; a tendency that is a bit more pronounced for bonds. We note that based on variances as well as on semivariances, bonds are the least and money market are the riskiest assets in the mix. The portfolios do not drastically differ using the two approaches, but it appears that when using the downside risk the optimal strategy is more risk averse than it is with the ordinary risk measure.

Returning to our model, a few comments are in order. The requirement $\mathbf{ex} = 1$ in all models considered so far could be relaxed to $\mathbf{ex} \leq 1$, allowing some funds in the portfolio not to be invested. Negative values of some of the variables may also be permitted, indicating so-called short-selling, i.e., selling a security that the investor does not own. An alternative way of balancing return and risks in portfolio selection is to include both in the objective function reminiscent of the weighting method for multiobjective optimization. This would be accomplished by maximizing a so-called risk-adjusted return function $z_{adj}(\delta) = \mu\mathbf{x} - \delta\mathbf{x}^T\mathbf{V}\mathbf{x}$, where δ is a preselected parameter that serves as an indicator of the tradeoff between return and risk.

In our description above of the semivariance and semicovariance, we have set the benchmark return r_j equal to the mean μ_j of the j-th security. This avoids the problem of the semicovariance matrix becoming dependent on the weights x_j in the portfolio, as pointed out by Estrada (2008). Finally, the reader may wonder why in our example the downside risk levels achieved were not smaller than the ordinary risk levels, in particular since the elements of the semicovariance matrix $\mathbf{SV}$ were smaller than the absolute values of the corresponding elements of the regular covariance matrix $\mathbf{V}$. A clue to this lies in the fact that some off-diagonal elements in $\mathbf{V}$ were negative, leading to negative terms when computing the risk function. In contrast, the downside risk function contains only terms with positive sign, since all elements of the semicovariance matrix are nonnegative.

Referring to the models discussed above, we may also consider mean absolute deviation $E\left(\left|R_j - \mu_j\right|\right)$ instead of the variance $V(R_j) = E((R_j - \mu_j)^2)$ as a measure of risk. Similarly, downside risk could be measured by $E(|\min\{R_j - r_j, 0\}|) = E(\max\{r_j - R_j, 0\})$. Another possibility is to build models that attempt to minimize the maximal loss that could occur.

3.6 Optimizing an Electrical Distribution Network

Electrical networks have long been the focus of optimization. The positioning of generating stations and substations, the design of the network, the voltage used for transmission and distribution lines, the diameter of the lines, and many other issues. Some examples of optimization in electrical networks is found in Molzahn et al. (2017), Momoh (2017) or Zhu (2009). Since the complexity of models that consider a large variety of aspects of the design make them computationally prohibitive, individual authors focus on specific aspects of the overall problem. An example is Willis et al. (1995), which is a typical reference when it comes to the design of electrical distribution networks. Another example in this context is Lakhera et al. (2011), which provides a formulation, which minimizes the losses in an electrical distribution network. Their model is the basis of the problem described in this section.

Suppose now that there is a set S of points that represent potential sites for substations and a set D of customers. We define x_j as a zero-one variable that assumes a value of one, if we locate a substation at site $j \in S$, and zero otherwise. Substations have a capacity of κ. Using standard terminology and notation, we denote by $\mathbf{i}$ the vector of currents, and $\mathbf{Y}$ is the admittance matrix that is determined by way of Kirchoff's equations. Another variable is v_j, the voltage at point j. The voltages are collected in the voltage vector $\mathbf{v}$. We can then write the equations

$$\mathbf{Yv} = \mathbf{i}. \tag{3.2}$$

The voltages are normalized, so that at any load node j, $\underline{v}_j$ denotes the minimally acceptable voltage at site j, and "1" as its upper bound. In order to formulate demand constraints, we have to ensure that the voltage at any site j assumes a value between the minimally acceptable voltage at that point and 1. This can be written as

$$\left(1 - x_j\right)\underline{v}_j + x_j \le v_j \le 1 \quad \forall j \in S \cup D. \tag{3.3}$$

In case there is no substation at point j (i.e., just a demand point), x_j assumes a value of 0, and the relation reads $\underline{v}_j \le v_j \le 1$, which are the desired bounds. On the other hand, if a substation is located at point j, then $x_j = 1$ and we have $1 \le v_j \le 1$, i.e., $v_j = 1$.

Another set of constraints requires that the current delivered at some demand point equals the demand, i.e., full service, while substations have to respect the capacity constraints. Defining ℓ_j as the load at site j, these constraints can be written as

$$\left(1 - x_j\right)\ell_j \leq i_j \leq \left(1 - 2x_j\right)\ell_j + \kappa x_j \quad \forall j \in S \cup D. \tag{3.4}$$

For all sites j without a substation, $x_j = 0$, and the condition reads $\ell_j \leq i_j \leq \ell_j$, and hence $i_j = \ell_j$; in other words, the demand at site j must be satisfied. For those sites j at which a substation is located, we have $x_j = 1$. Note that a node with a substation may nonethless have a positive load. We then obtain $0 \leq i_j \leq \kappa - \ell_j$. In other words, the current at site j must be somewhere between 0 and the spare capacity at site j.

Before we consider the objective function, we first have to deal with the binary variables x_j. We could either leave them as binary integer variables, which adds another degree of difficulty to the problem. Alternatively, we could define x_j as continuous variables

$$0 \leq x_j \leq 1, \tag{3.5}$$

but penalize all noninteger values. Noting that the quadratic expression $x_j(1 - x_j)$ is represented as an inverted parabola. Hence, multiplying it by an arbitrarily large penalty parameter $M \gg 0$, the term $x_j(1 - x_j)M$ penalizes all noninteger values of x_j, and, for sufficiently large values of M, all noninteger values of x_j are prohibitive in a minimization objective.

The objective function of the model will include costs of losses in the electrical network as well as the cost of establishing substations. These two components are combined in one objective similar to the weighting method. Defining C_L as the cost of loss, since $\mathbf{Yv} = \mathbf{i}$ and electrical power is defined as $\mathbf{p} = \mathbf{vi}$, the first part of the objective, written as $C_L\mathbf{v}^T\mathbf{Yv}$, minimizes the electrical losses in the network. Defining C_S as the cost of siting a facility and avoiding including the zero-one variables explicitly in the model and replacing them by variables x_j that are restricted to the zero-one interval, we can include a smoothened nonlinear version $1 - e^{-\alpha x_j}$ in the objective function. Finally, as alluded to above, the zero-one conditions are replaced by the nonlinear expression $x_j(1 - x_j)M$. Then the objective function can be written as the nonlinear optimization problem

$$P : \min_{\mathbf{v}, \mathbf{x}, \mathbf{i}} z = C_L\mathbf{v}^T\mathbf{Yv} + C_S \sum_{j \in S \cup D} 1 - e^{-\alpha x_j} + \sum_{j \in S \cup D} x_j(1 - x_j)M$$

$$\text{s.t. constraints } (3.2), (3.3), (3.4), \text{ and } (3.5).$$

As a general observation, we would like to point out that the objective function is the sum of a convex function, a concave function, and another concave function, and as such, it is d.c., see Sect. 7.4.1. Potential extensions of the model may include different types of substations with different capacities and costs, different types of transmission cables (viz., different diameters, hollow, stranded, etc.), and others.

3.7 Optimal Taxation

The problem discussed in this section deals with a strongly idealized situation, in which two planners, an individual and the state, both rational and equipped with their own respective objective functions, make decisions that impact each other. First consider the individual planner, who behaves as a "rational man" (or *homo economicus*), see, e.g., Simon (1959). In order to plan his time, the individual can either work or engage his time in the pursuit of leisure (where we take, say, 8 h of sleep as given and thus not available). For each of the two activities, work and leisure, the individual has certain utilities. It appears reasonable to assume that both utility functions are concave, i.e., we face decreasing marginal utilities of work and leisure. The interdependence is clear: if the individual works more, there is more income and thus a higher utility from work (via the pay it generates), while, at the same time, there will be less time for leisure, which results in less utility from leisure. The idea for the individual is then to find an optimal mix between the two.

On the other hand, the second player, viz., the state, is interested not in leisure but only in the individual's work time, as there is no way that leisure time can reasonably be taxed. The state's objective is then to optimally set the level of taxation, so as to maximize tax revenues. This, in turn, will affect the individual, who, with a higher tax rate, will find that the utility that derives from his work, has now decreased, as a larger part of his salary is now taken away. Typically, he will react by working less, which will decrease the state's total tax revenues.

The game described above is a typical leader-follower game first described by the economist Stackelberg (1943). Such a game will have one of the two players (the "leader") move first and make an irreversible decision. The other player (the "follower) will then take the leader's decision as given and optimize on that basis. Clearly, in the context described above, the state is the leader and the individual is the follower. Viewing the situation as an optimization problem, the problem is a bilevel optimization problem, in which the leader's (upper-level) problem has a constraint that requires a solution that is considered, i.e., feasible to be an optimal solution to the follower's problem. This nested structure does, of course, make the leader's problem generally very difficult. The follower's best possible reaction to each decision made by the leader is summarized as the follower's reaction function, which has to be taken into consideration by the leader. For details, see, e.g., Eiselt and Sandblom (2004).

The problem is very well known in marketing, where the question is whether or not the first mover, i.e., the leader does generally have an advantage. In the context of competitive location problems, Ghosh and Buchanan (1988) considered the situations, in which the first mover would not have an advantage, as the "first entry paradox."

In our context, the leader, i.e., the state, will then attempt to learn about individuals' utility functions and set the tax rate t accordingly, so as to maximize tax receipts T. This is usually summarized in what is known as the *Laffer curve*, which plots the tax rates against the tax income by the state. Pertinent references can be found in Wanniski (1978) and Laffer (2004). Without doing any calculations, we know that for $t = 0$, clearly, $T = 0$ as well, while for $t = 1$, $T = 0$ as well as nobody would work just for the state. Somewhere in $t \in]0, 1[$, T will be positive and reach its maximum. Without further information, we do not know exactly where this maximum will be, we can think of the Laffer curve as an inverted parabola with x-intercepts at $t = 0$ (no tax at all) and 1 (100% tax, i.e., total confiscation). The exact shape of the curve will depend on many details: the utility functions by the individuals, the type of taxation used by the state, how the individuals view the future (a high degree of uncertainty and a negative outlook will cause people to forego leisure in order to have some funds to fall back upon in difficult times in the future), etc.

In order to further explain the concept, define x_w as the number of hours per day that an individual is planning to work. Similarly, x_L is the leisure time that the individual has planned. Given 8 h of sleep, we have $x_w + x_L = 16$. Furthermore, assume that the individual's utility for leisure time is $u_L = 0.15x_L^{0.5} = 0.15(16 - x_w)^{0.5}$, while his utility for work is $u_w = 0.2(1 - t)x_w^{0.5}$ with a flat tax rate of t. Given a wage rate w, the individual's disposable income is then $(1 - t)wx_w$. As a result, the individual's overall utility is $u = u_w + u_L = 0.2(1 - t)x_w^{0.5} + 0.15(16 - x_w)^{0.5}$, while the state's tax receipt is $T = tx_w$.

Our first task is now to compute the reaction function for each of the leader's choices. In other words, for each tax rate, we need to determine the optimal choice of the follower's regarding his income and leisure. The leader will then choose among all possible tax rates the one, which is best for him. For simplicity, we have chosen possible tax rates $t = 0, 0.1, 0.2, \ldots, 1$. The individual's utilities given his hours of work is shown in Fig. 3.10 for the 11 tax rates chosen for this example.

Optimizing his utility function, the individual will now determine for each given tax rate t the number of hours of work that will result in his maximal utility. This is done by solving the problem P_t for each given tax rate t. This problem can be formulated as

$$P_t: \operatorname*{Max}_{x_w} \; z_t = 0.2(1 - t)x_w^{0.5} + 0.15(16 - x_w)^{0.5}.$$

The resulting hours of labor and tax receipts are summarized in Table 3.8.

We can now plot these figures into a (t, T) diagram, which, given appropriate interpolation, results in the Laffer curve in Fig. 3.11.

The optimal tax rate in this example is about $t = 0.44$. More realistic examples will be obtained by using empirically verified utility functions, different tax rates (e.g., the usual "progressive" tax), and others.

The answer to the often asked question whether or not tax cuts will actually result in increased tax receipts depends on where we presently are. Assuming a smooth shape of the curve similar to the one shown in Fig. 3.11, if the present level of taxation is to the left of the maximum, then a decrease of the flat tax rate t, i.e., tax cut will result in reduced tax income by the state. If, however, the present tax rate is higher than its maximum, then a tax cut will indeed result in an increase of the total tax collected by the state.

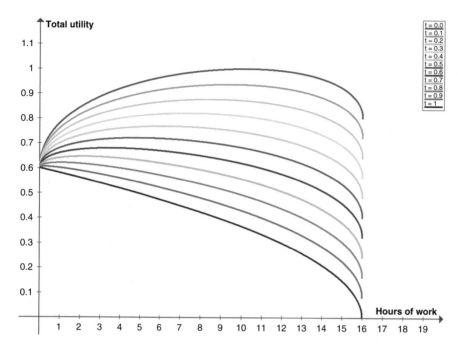

Fig. 3.10 Utility as a function of working time

Table 3.8 Labor hours and tax receipts

Tax rate t	Hours of labor x_w	Total taxes T
0.0	10.24	0
0.1	9.442	0.9442
0.2	8.516	1.7032
0.3	7.449	2.2347
0.4	6.244	2.4976
0.5	4.923	2.4615
0.6	3.543	2.1258
0.7	2.207	1.5448
0.8	1.062	0.8498
0.9	0.279	0.2511
1.0	0	0

3.8 Location Problems

This section will examine some location problems, which result in nonlinear optimization formulations. In order to facilitate the discussion, we need to discuss two concepts before we commence with details of the location problems. In its simplest version, a location problem comprises a number of *customers*, whose numbers and locations are, at least temporarily, fixed and known, a number of *facilities*, which are

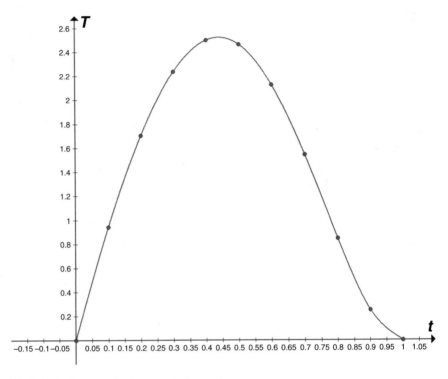

Fig. 3.11 Laffer curve for the numerical example

to be located (their number may or may not be known), and some measure of *distance* (used here as a synonym for disutility) in the *space* that customers and facilities are (to be) located in. Two types of spaces are typically used, d-dimensional real spaces (mostly two-dimensional spaces, even though three-dimensional applications exist), and networks. As far as distances in two dimensions are concerned, we typically use Minkowski distances, so-called ℓ_p distances, with some parameter p, which are defined between points i and j with coordinates (a_i, b_i) and (a_j, b_j), respectively, as

$$d_{ij}^p = \left[\left|a_i - a_j\right|^p + \left|b_i - b_j\right|^p\right]^{1/p}.$$

Specific and well-known versions of this distance function are known for some values of the parameter p. First, given $p = 1$, we obtain $d_{ij}^1 = \mid a_i - a_j \mid + \mid b_i - b_j \mid$, which is the so-called ℓ_1, *rectilinear*, or *Manhattan* distance function. In essence, it assumes that all movements between points are parallel to the axes. City street networks on a grid such as Manhattan (with the exception of Broadway) are typical applications for this distance function. Second, for $p = 2$, we obtain

$$d_{ij}^2 = \left[\left(a_i - a_j\right)^2 + \left(b_i - b_j\right)^2\right]^{1/2},$$ which are the ℓ_2, *Euclidean*, or *straight-line*

distances. They are typically applied in case where straight-line travel is applicable, such as when helicopters or ships are concerned (assuming there are no forbidden regions, such as no-fly zones in case of air travel, or islands or shoals in case of travel on water). Finally, for $p \rightarrow \infty$, we obtain $d_{ij}^{\infty} = \max\{|a_i - a_j|, |b_i - b_j|\}$, the so-called ℓ_{∞} or *Chebyshev* distance. As pointed out by Daskin (2013) and Eiselt and Sandblom (2004), its use can be found in measuring travel times in case of specific technologies.

One of the main tools for location modeling are *Voronoi diagrams*, which are also known as *Thiessen polygons* (often by geographers), *Dirichlet tessellations* (frequently by mathematicians), or *Wigner-Seitz cells* (by chemists). They can easily be defined as follows. For a set of given points (or *seeds*) $P_1, P_2, \ldots, P_n$, we assign a set $S(P_i)$ to each seed, which includes all points in the given space, which are closer to P_i than to any other seed $P_j \neq P_i$. In the case of ℓ_2 distances, the lines that separate points P_i and P_j are their perpendicular bisectors, in the case of ℓ_1 and ℓ_{∞} distance functions, the boundary lines are piecewise linear, while for other Minkowski functions, they are nonlinear.

Okabe et al. (2000) demonstrate many applications of Voronoi diagrams. For efficient techniques that compute Voronoi diagrams in various metrics, see, e.g., also Klein (1989). As a small example, consider a 10×10 mile square with seeds located at $P_1 = (0, 0)$, $P_2 = (8, 5)$, $P_3 = (10, 10)$, $P_4 = (5, 5)$, $P_5 = (2, 6)$, and $P_6 = (4, 9)$. The Voronoi diagram spanned by these points given the Euclidean metric is shown in Fig. 3.12.

The Voronoi diagrams for the ℓ_1 and ℓ_{∞} distance functions have piecewise linear boundaries while Voronoi diagrams for all other Minkowski metrics with finite values of $p \neq 1, 2$, have nonlinear boundaries.

Suppose that in a highly simplified context, the 10×8 rectangle in Fig. 3.13 grid represents an area in which customer demand is uniformly distributed. Furthermore, assume that customers always patronize the retail facility, such as a supermarket, on

Fig. 3.12 Voronoi diagram for the numerical example

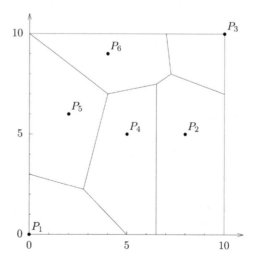

Fig. 3.13 Voronoi diagram
for the optimization problem

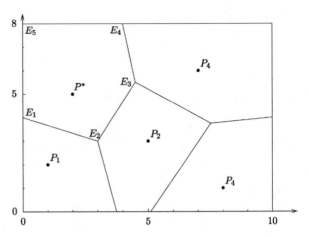

the basis of distance, i.e., they will make their purchases at the supermarket closest to them. In this scenario, the size of a facility's area is deemed to be proportional to the facility's revenue. Maximizing its revenue then reduces to maximizing its Voronoi area, an issue that we will investigate in the following paragraphs.

Consider the case in Fig. 3.13, which shows the Voronoi diagram with five facilities $P^* = (2, 5)$, $P_1 = (1, 2)$, $P_2 = (5, 3)$, $P_3 = (7, 6)$, and $P_4 = (8, 1)$. Assume now that P^* is the facility of interest. We will first determine its revenue/market share by finding the area of its Voronoi set, which is a polyhedron bounded by the extreme points $E_1, E_2, \ldots, E_5$. Then we need to triangulate the set, i.e., determine the areas of the triangle E_1, E_2, E_3, the triangle E_1, E_3, E_4, and finally the triangle E_1, E_4, E_5. We will denote these areas by Δ_{123}, Δ_{134}, and Δ_{145}, respectively. The coordinates of the extreme points can be calculated as $E_1 = (0, 4)$, $E_2 = \left(2\frac{19}{22}, 3\frac{1}{22}\right)$, $E_3 = (4\frac{1}{2}, 5\frac{1}{2})$, $E_4 = (4, 8)$, and $E_5 = (0, 8)$. The areas of the triangles are then determined using the well known relation

$$\Delta_{ijk} = \tfrac{1}{2}\det \begin{vmatrix} E_i & 1 \\ E_j & 1 \\ E_k & 1 \end{vmatrix},$$

resulting in

$\Delta_{123} = 8\frac{13}{22}$, $\Delta_{134} = 6$, and $\Delta_{145} = 8$ for a total area of 22.5909.

Suppose now that it were possible to relocate, i.e., move the point P^*, while the competitors would stay at their respective present locations. This means that rather than having fixed coordinates $P^* = (2, 5)$, we now have $P^* = (x^*, y^*)$. Consider any move of P^* out of its present location graphically. If P^* were to move into a northwesterly direction, the boundary between E_2 and E_3 will tilt clockwise and also move in a northwesterly direction, while the boundary between E_3 and E_4 will move and tilt in a clockwise fashion, while the boundary between E_1 and E_2 will also move northwesterly and tilt counterclockwise. The expression of the extreme points E_i are then ratios of polynomials of the variables x^* and y^*, and so are the expressions for the areas after the triangulation. This allows the optimization of the facility that is presently located at P^*, at least within the neighborhood of its current location. For larger changes, some of the boundary lines may shrink to a point (for instance, if P^* stays put but P_2 were to move towards E_3, the boundary of its polytope that that coincides with the abscissa will shrink to a point).

The second special area in the field of location theory that is discussed here deals with the location of undesirable facilities. Traditionally, but without ever stating it explicitly, location science has been dealing with the location of facilities, which are deemed desirable in the sense that customers like to be as close to them as possible. However, as first pointed out by Goldman and Dearing (1975), there are many facilities, which are undesirable, due to their dangerous and/or polluting nature. Power plants, heavy industries, trucking terminals, landfills, and similar facilities are but a few examples. In all of these cases, customers would like to maximize, rather than minimize, the distance between themselves and the facility. In order to get the main points acrosss, consider the simple case of locating a single undesirable facility among a fixed number of existing customers.

More specifically, let there be a single undesirable facility, whose coordinates are (x, y), and suppose that customers are located at known sites with locations (a_1, b_1), (a_2, b_2), . . ., (a_n, b_n). The population at site i is assumed to be w_i. The idea is to locate the undesirable facility, so as to impact as few people as possible.

Given a distance function $f[(a_i, b_i), (a_j, b_j)]$ that assigns a distance between any two points i and j, one of the main issues is how to measure the effect of the polluting facility on the population. There is a variety of possibilities. One way to measure the impact is to determine the actual pollution at each customer point and multiply it with the population at that point. This total pollution is then to be minimized, resulting in a minisum objective. Another approach is to ensure that the most affected member of the population is affected as little as possible. Such a minimax approach causes the maximally polluted member of the population to be polluted as little as possible. Yet another approach will define legal, i.e., tolerable limits of pollution and then either minimize the number of people who are polluted beyond the legal limit, or minimize the total amount of pollution beyond the legal limit that reaches the population. In addition, we may introduce costs associated with the location (x, y) of the undesirable facility and seek to minimize the costs, subject to satisfying the legal limits for some or all of the given customer sites.

In order to illustrate the point, consider the first option and minimize the total pollution that reaches the population. A customer at site i is then polluted from a site at (x, y) by the amount of pollution π emitted by the facility and some function of the distance between the facility and the customer $f[(a_i, b_i), (x, y)]$, where $f(\bullet)$ is the pollution decay function. The total pollution at customer site i is then πw_i, where w_i denotes the population at site i that is affected by the pollution. We can then minimize the total pollution by formulating the following only lightly constrained problem.

$$P: \text{Min} z = \sum_i w_i \pi f\left[(a_i, b_i), (x, y)\right]$$
$$\text{s.t.} (x, y) \in S,$$

where $S \subseteq \mathbb{R}^2$ is some set, on which the facility location can be chosen. As is the case in all locations of undesirable facilities, we need to require the facility be located within this finite set, as otherwise an optimal location, i.e., one that minimizes effects on people, will be towards infinity.

Similarly, we can formulate the minimax version of the problem as

$$P: \text{Min} z = \max_i \left\{ \pi f\left[(a_i, b_i), (x, y)\right] \right\}$$
$$\text{s.t.} (x, y) \in S.$$

Given Euclidean distances and a polluting factor $\pi = 1$, the formulation becomes

$$P : \text{Min } z = \max_i \left[(a_i - x)^2 + (b_i - y)^2\right]^{\frac{1}{2}} \}$$
$$\text{s.t.} (x, y) \in S,$$

in which the bounded set S ensures that the optimal solution is finite. This version has an interesting geometric interpretation. It is the so-called *largest empty circle* problem, in which a facility is to be located, such that a circle with the facility location at its center that does not include any one of the given customer points, has the largest possible radius. This was first pointed out by Shamos and Hoey (1975), who solve the problem via Voronoi diagrams in polynomial time, specifically in $O(n \log n)$ time; see, e.g., Toussaint (1983).

Note that in general, minimax problems can be transformed into regular minimization problems by using standard transformations, see, e.g., Eiselt and Sandblom (2007).

Example Consider a problem, in which ten customers are located at the points $(0, 0)$, $(0, 4)$, $(1, 2)$, $(1, 6)$, $(2, 0)$, $(3, 4)$, $(4, 1)$, $(4, 5)$, $(6, 3)$, and $(6, 6)$, and the set of feasible locations is bounded by $0 \leq x, y \leq 6$. We first rewrite the maximin problem as the equivalent maximization problem with an additional variable z, a linear objective and quadratic constraints, resulting in the formulation

$$P: \text{Max} z$$

$$\text{s.t. } z \leq \left[(0-x)^2 + (0-y)^2\right]^{\frac{1}{2}} \quad z \leq \left[(3-x)^2 + (4-y)^2\right]^{\frac{1}{2}}$$

$$z \leq \left[(0-x)^2 + (4-y)^2\right]^{\frac{1}{2}} \quad z \leq \left[(4-x)^2 + (1-y)^2\right]^{\frac{1}{2}}$$

$$z \leq \left[(1-x)^2 + (2-y)^2\right]^{\frac{1}{2}} \quad z \leq \left[(4-x)^2 + (5-y)^2\right]^{\frac{1}{2}}$$

$$z \leq \left[(1-x)^2 + (6-y)^2\right]^{\frac{1}{2}} \quad z \leq \left[(6-x)^2 + (3-y)^2\right]^{\frac{1}{2}}$$

$$z \leq \left[(2-x)^2 + (0-y)^2\right]^{\frac{1}{2}} \quad z \leq \left[(6-x)^2 + (6-y)^2\right]^{\frac{1}{2}}$$

$$0 \leq x, y \leq 6.$$

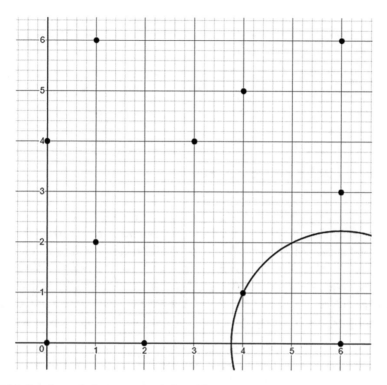

Fig. 3.14 Solution to the largest empty circle problem

This problem has an optimal solution at $(\bar{x}, \bar{y}) = (6, 0)$ with $\bar{z} = 2.236068$. The solution is shown in Fig. 3.14. It is typical for this type of problem to have solutions that are located on the boundary of the feasible set.

The third, and last, class of problems deals with packing. Packing problems were pioneered by Minkowski and their main idea is to find an arrangement of given shapes in a given body. For instance, we could be interested in finding an arrangement that packs the largest number of 3 ft × 5 ft × 2 ft boxes into a standard 20 ft container. In that, the problem is somewhat related to cutting stock problems. Another example in two dimensions is how to maximize the number of small trees in an orchard (disks with the trunk at the center) in a lot of a given size. Figure 3.15a shows a 50 ft × 28 ft rectangular field, in which as many trees as possible are to be planted, given that we expect each tree to have a radius of 5 ft. In the example shown in Fig. 3.15a, the arrangements packs 10 circles into the given space, which utilizes 785.40 sq ft out of 1400 sq ft for a utilization of 56.1%. Considering the fact that the trees use the lot all the way to its boundary on three out of four sides, this is actually not a very dense packing. A different arrangement is shown in Fig. 3.15b. Here, the space accommodates 14 disks (trees), thus using a total space of 1099.56 sq ft, for a usage rate of 78.5%. Interestingly, the minimal size rectangle that could

(a)   (b)

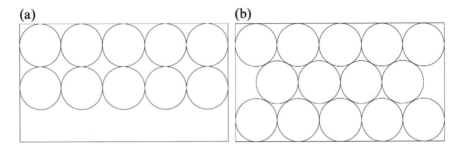

Fig. 3.15 Two solutions to the packing problem

accommodate the packing arrangement in Fig. 3.15a would be of size 50 ft × 20 ft for a packing density of $\pi/4 \approx 78.4\%$, the minimal size rectangle that could accommodate the packing arrangement in Fig. 3.15b could be 50 ft × $10(1 + \sqrt{3})$ ft   or   50   ft   ×   27.32   ft,   for   a   packing   density   of $14 \times 25\pi/500(1 + \sqrt{3}) = 7\pi/10(1 + \sqrt{3}) \approx 80.5\%$. As a matter of fact, if the given space were very large, the Fig. 3.15a pattern would have a usage rate of $\pi/4 = 78.54\%$, while the Fig. 3.15b pattern utilizes $\dfrac{\pi}{2\sqrt{3}} \approx 90.69\%$ of the space.
Clearly, realistic shapes will render the problem significantly more difficult.

It is important to realize that small changes in the lot or the tree can result in dramatically different solutions. The arrangement in Fig. 3.15a requires a minimum lot size of 20 ft × 50 ft, the moment that either width or depth decrease by an arbitrarily small amount the number of trees in the lot will decrease. For instance, if the lot were only 19 ft × 50 ft in size, the second row of trees could no longer be accommodated and this pattern would only be able to pack a single row, thus cutting the number of trees in half. However, two rows of trees of the pattern of Fig. 3.15b would still be possible with the loss of only a single tree.

3.9   Production and Inventory Planning

In this section we will treat a few production and inventory planning models which will be formulated as nonlinear programming problems. Let us first consider a situation where the unit price of a product decreases as its supply increases. To keep the discussion simple, let the relationship between the unit price p and the number of units sold x be linear, i.e. expressed by an equation such as $p = a - xb$ where a and b are nonnegative real constants and $x \in [0, a/b]$ in order to ensure nonnegative x and p. The revenue function is then defined as $R(x) = (a - bx)x$, and if an increasing convex cost function $C(x)$ is given and the purpose is to maximize total profit, then the objective function is $z = R(x) - C(x) = ax - bx^2 - C(x)$. Even with a linear cost function, this expression is obviously nonlinear. We have thus formulated the following nonlinear programming model:

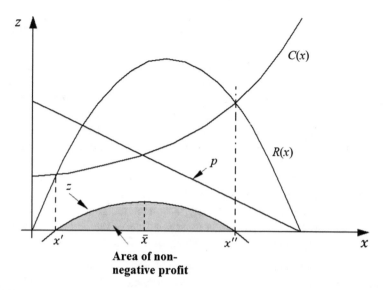

Fig. 3.16 Profit function for a price-quantity relation

$$P : \text{Max } z = ax - bx^2 - C(x)$$
$$\text{s.t.} 0 \le x \le a/b.$$

This problem is illustrated in Fig. 3.16.

x' and x'' represent the breakeven points between which lies the area of nonnegative profit. The shaded profit-area has been obtained by subtracting $C(x)$ from $R(x)$ graphically. The maximal profit occurs when the quantity produced is equal to $\bar{x}$.

To illustrate the above model numerically, consider the price-quantity relation $p(x) = 100 - 2x$, $x \in [0, 50]$ and the cost function $C(x) = x^2 + 10x + 5$. Then the revenue function is $R(x) = -2x^2 + 100x$ and the profit is $z = -3x^2 + 90x - 5$. The breakeven points are calculated by setting z equal to 0, resulting in $x' = 15 - \sqrt{670/3} \approx 0.056$ and $x'' = 15 + \sqrt{670/3} \approx 29.944$. Using differential calculus, it can be shown that the quantity x^* that results in the maximal revenue is $x^* = 25$ with $p(x^*) = 50$, $R(x^*) = 1250$, $z^* = 370$ and $C(x^*) = 880$.

Moreover, the quantity that maximizes the profit is $\bar{x} = 15$ with $p(\bar{x}) = 70$, $R(\bar{x}) = 1050$, $\bar{z} = 670$ and $C(\bar{x}) = 380$. (One can show that in general $\bar{x} \le x^*$, as long as the cost function is nondecreasing). Suppose now that management cannot decide if maximizing profits or revenues is more important; still one can determine all noninferior solutions. (For a detailed discussion of noninferiority, see, e.g., Eiselt and Sandblom 2004). From the above it is obvious that solutions with quantities $x < 15$ are inferior, since revenue as well as profit can be increased by increasing x. On the other hand, solutions with quantities $x > 25$ are also inferior, since by decreasing x, both revenue and profit increase. Formally, we can determine all noninferior solutions by introducing a weight $w \in [0, 1]$ and maximize the objective $z = wR(x) + (1 - w)z$. In our example, differentiation with respect to x results in

$2xw + 10w - 6x + 90 = 0$ or $w = \dfrac{6x - 90}{2x + 10}$. Then $w \geq 0$ implies $x \geq 15$ and $w \leq 1$ implies $x \leq 25$, which is the desired interval of noninferior solutions.

We will now introduce inventory considerations in our planning models and deal with the production and inventory decision of a company. In this context, we will restrict ourselves to operational decisions concerning order and production quantities in the presence of limited storage capabilities. In general, the ordering and production decisions for the products are separate and decomposable, with the common inventory being the strand that connects all of these. In our illustration, we consider three products and assume that the "make or buy" decisions have already been made. All models are based on the economic order quantity (*EOQ*) model from inventory theory. This means that in all cases, we assume that the rate of demand is constant, there are no quantity discounts, the goods are not perishable; moreover, we add all the other standard assumptions of the pertinent models; see, e.g., Eiselt and Sandblom (2012). The details are as follows.

Product 1 We have decided to buy the good and have to determine the order quantity. We assume that stockouts are not permitted (this is the standard *EOQ* model). Given the usual definitions of the parameters c_{o1} (unit ordering costs), c_{h1} (unit holding or unit carrying costs), D_1 (annual demand), and TC_1 (total inventory-related costs), and the single decision variable Q_1 (the order quantity), we can write the total cost function as $TC_1 = c_o \dfrac{D_1}{Q_1} + c_h \tfrac{1}{2} Q_1$, which has the optimal unconstrained solution $Q_1^* = \sqrt{\dfrac{2D_1 c_{o1}}{c_{h1}}}$, thus the maximal inventory level at any point in time is Q_1^*.

Product 2 The second product will also be purchased rather than made. We use the same type of parameters as for product 1, except with the subscript "2." However, for this product we allow shortages and backorders. The additional unit shortage costs are denoted by c_{s2}, and the planned shortages are symbolized by the variable S_2. The total cost function can then be written as

$$TC_2 = c_{o2} \frac{D_2}{Q_2} + c_{h2} \frac{(Q_2 - S_2)^2}{2Q_2} + c_{s2} \frac{S_2^2}{2Q_2}$$

and the optimal (unconstrained) solution amounts are $Q_2^* = \sqrt{\dfrac{2D_2 c_{o2}}{c_{h2}} \dfrac{c_{h2} + c_{s2}}{c_{s2}}}$ and $S_2^* = \dfrac{c_{h2}}{c_{h2} + c_{s2}} Q_2^*$. The maximal inventory level at any time is $Q_2^* - S_2^*$.

Product 3 As far as the third product is concerned, we have decided to make it ourselves. In addition to the parameters defined for product 1 (now, of course, subscripted with "3,") we have the daily demand rate d_3, which is defined as the annual demand D_3 divided by the number of days in a year, which is typically defined as 250 working days, and we have a daily production rate r_3. Clearly, $r_3 \geq d_3$ is required for feasibility. We then obtain the cost function

$$TC_3 = c_{o3} \frac{D_3}{Q_3} + c_{h3}(r_3 - d_3) \tfrac{1}{2} \frac{Q_3}{r_3},$$

whose optimal solution is $Q_3^* = \sqrt{\dfrac{2D_3 c_{o3}}{c_{h3}} \dfrac{r_3}{r_3 - d_3}}$. The highest inventory level throughout the cycles is $(r_3 - d_3)Q_3/r_3$.

In a specific scenario, we assume that the following parameters are given:

Product 1 $c_{o1} = 100$, $D_1 = 1{,}000{,}000$, $c_{h1} = 2$, so that the unconstrained order quantity is $Q_1^* = 10{,}000$, which is also the maximal inventory level throughout the cycle.

Product 2 $c_{o2} = 50$, $D_2 = 50{,}000$, $c_{h2} = 6$, and $c_{s2} = 30$, so that the unconstrained order quantity is $Q_2^* = 1000$, the highest planned shortage level is $S^* = 166\frac{2}{3}$, and the highest inventory level at any point in time is therefore $833\frac{1}{3}$.

Product 3 $c_{o3} = 20$, $D_3 = 3000$ (which, with 250 working days per year, results in $d_3 = 12$), $c_{h3} = 0.6$, and $r_3 = 60$, so that the unconstrained optimal order quantity $Q_3^* = 500$. The highest inventory level during a cycle is $(r_3 - d_3)Q_3^*/r_3 = 400$.

Suppose now that one unit of Product 1 requires 1 sq ft of storage space, one unit of Product 2 requires 8 sq ft, and one unit of Product 3 requires 20 sq ft. If we (somewhat unreasonably) assume that the three maximal inventory levels were to occur at the same time (or, at least safeguard against the possible case in which they are), we need space for $1Q_1^* + 8(Q_2^* - S^*) + 20(r_3 - d_3)Q_3^*/r_3 = 1(10{,}000) + 8(833\frac{1}{3}) + 20(400) = 24{,}666\frac{2}{3}$. If we were to have that much space or more, the space constraint is not an issue and we can use the three unconstrained solutions of the decomposed problem.

The problem with an inventory space restriction can then be formulated as

$$P:\ \text{Min}\, TC = 100{,}000{,}000 Q_1^{-1} + Q_1 + 2{,}500{,}000 Q_2^{-1} + 6\frac{(Q_2 - S_2)^2}{2Q_2}$$
$$+ 30\frac{S_2^2}{2Q_2} + 60{,}000 Q_3^{-1} + 0.24 Q_3,$$
$$\text{s.t.}\ Q_1 + 8(Q_2 - S_2) + 16 Q_3 \le B,$$
$$Q_1, Q_2, Q_3, S_2 \ge 0,$$

where B denotes the amount of space available to the planner. Table 3.9 shows the individual production/order quantities as well as the total inventory-related costs for

Table 3.9 Costs and quantities for various space availabilities

Space availability B	TC-value (costs)	Q_1	Q_2	Q_3
30,000	25,240.00	10,000	1000	500
25,000	25,240.00	10,000	1000	500
20,000	25,314.85	9781.5	945.0	249.6
15,000	26,052.42	8729.5	745.8	107.0
10,000	29,426.89	6745.8	521.0	55.6
5000	40,892.17	4553.2	408.3	31.3

various space availabilities B. Recall that if the available space B is in excess of 24,666⅔ sq ft, the space constraint will be loose, which is borne out by the numbers in Table 3.9.

A few other inventory models are formulated and solved with geometric programming techniques, see Illustration 2 in Sect. 8.5.

3.10   Economic Planning

A classical problem in the field of quantitative macroeconomic policy is the constrained minimization of a welfare cost function. The constraints appear in the form of statistically estimated difference equations, which constitute a macroeconomic model that describes the economic relationships in the economy under consideration. To be specific, we assume that the econometric model under study is linear and written in a so-called reduced form with the model equations formulated as

$$\mathbf{x}^0 = \mathbf{b}^0$$
$$\mathbf{x}^t = \mathbf{A}\mathbf{x}^{t-1} + \mathbf{B}\mathbf{y}^t + \mathbf{b}^t, \quad t = 1, \ldots, T,$$

where $\mathbf{x}^t$, $t = 1, \ldots, T$ are n-dimensional column vectors, i.e., state (endogenous) variables, such as private consumption, private investment, and unemployment rate. The m-component column vector $\mathbf{y}^t$ comprises the control (policy) variables, e.g., government spending and the prime interest rate set by the central bank. The n-component vector $\mathbf{b}^t$ expresses the effects of exogenously given variables, including any random disturbances. Finally, $\mathbf{A}$ and $\mathbf{B}$ are known and constant $[n \times n]$- and $[m \times n]$-dimensional coefficient matrices. The time period t is typically a quarter of a year. Although the model formulated above includes time lags of only one period, this is not a restriction, since time lags of more than one period can be accommodated by introducing auxiliary variables, each lagged by one period. An objective (welfare cost) function is now added. It is

$$z = \sum_{t=1}^{T} f^t(\mathbf{x}^t) + g^t(\mathbf{y}^t),$$

where f^t and g^t are known and given increasing functions: $f^t(\mathbf{x}^t)$ represents welfare costs related to endogenous variables and $g^t(\mathbf{y}^t)$ are costs of economic policy. Now assume that the purpose of the policy is to keep the values of the endogenous variables close to some given ideal time path $\hat{\mathbf{x}}^t$. A natural way to express this is by using a piece-wise quadratic function f^t, which penalizes deviations from the ideal path in some progressive fashion, whereas the policy costs g^t could be linear. To be specific, assume that the "ideal" rate of unemployment is $\hat{x}_1^t$, and the "ideal" rate of inflation $\hat{x}_2^t$, $t = 1, \ldots, T$, whereas the policy cost is $\mathbf{c}\mathbf{y}^t$ for come given cost row vector $\mathbf{c}$. The policy optimization problem can then be formulated as

$$P: \operatorname{Min} z = \sum_{t=1}^{T} \left[\left(\max\{x_1^t, \hat{x}_1^t\} - \hat{x}_1^t \right)^2 + \left(\max\{x_2^t, \hat{x}_2^t\} - \hat{x}_2^t \right)^2 + \mathbf{c}\mathbf{y}^t \right]$$

$$\text{s.t.}\ \ \mathbf{x}^0 = \mathbf{b}^0$$

$$\mathbf{x}^t = \mathbf{A}\mathbf{x}^{t-1} + \mathbf{B}\mathbf{y}^t + \mathbf{b}^t, \ \ t = 1, \ldots, T.$$

The welfare costs associated with unemployment are illustrated in Fig. 3.17 for a case of an ideal unemployment rate of 4% and a quarterly rate of inflation of ½%, i.e., an annual inflation rate of 2%.

The significant advantage of using piecewise instead of ordinary quadratic functions is that we avoid penalizing situations, in which unemployment and/or the rate of inflation are below the ideal targets. In this way, the encouragement of inadvertent "perverse" policy actions is removed. This is the same logic that was used in Sect. 3.5, where minimizing or restricting semicovariance in security port-folios was considered preferable to minimizing or restricting the ordinary covariance.

The policy problem P may also be augmented by adding upper and/or lower bounds on the controls, i.e., using constraints of the form $\boldsymbol{\ell}^t \leq \mathbf{y}^t \leq \mathbf{u}^t, t = 1, \ldots, T$, where $\boldsymbol{\ell}^t$ and $\mathbf{u}^t$ are preselected lower and upper bounds on the acceptable values of the policy controls. To avoid the sometimes occurring "bang bang" controls, where policy variables jump between their lower and upper bounds in adjacent time periods, we may also impose bounds on the change of the values of some policy variables between time periods.

The numerical feasibility of solving problem P was first demonstrated by Fair (1974), after which a number of applications have been reported, see, e.g., Lasdon and Waren (1980). The bounded control approach was used by Sandblom (1985). The size n of the econometric models employed in solving problem P varies from just a few to hundreds of equations, and the number T of time periods over which the optimization is performed, is typically 12–20.

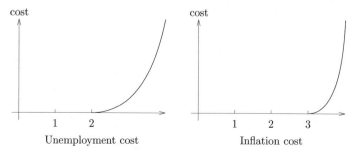

Fig. 3.17 Penalty functions for unemployment and inflation

Another aspect when solving the macroeconometric policy problem P is that of uncertainty. If the model is to be used for practical policy making in the future, the values $\mathbf{b}^t$ of exogenous variables must be forecast. Once a policy has been optimized and the first period has potentially been implemented, revised forecasts will have become available in the next period. Also, the first-period forecast will have been replaced by actually observed values of the exogenous variables. It does, therefore, make sense to reoptimize problem P, incorporating the new information, including the forecast T periods ahead, which has now moved one step further into the future. This has been called the "sliding window" approach with only the first-period policy ever executed. For an illustration of this approach, see, e.g., Sandblom (1984).

For another application of nonlinear optimization in macroeconomic planning, readers are referred to Illustration 3 in Sect. 8.5, where geometric programming techniques are used to optimize a model that uses the Cobb-Douglas production function.

References

Chapman DG (1961) Statistical problems in population dynamics. *Proceedings of the Fourth Berkeley Symposium on Mathematical Statistics and Probability*. University of California Press, Berkeley and Los Angeles, pp. 153-168

Cohon JL (1978) *Multiobjective programming and planning*. Academic Press, New York

Cornuejols G, Tütüncü R (2007) *Optimization methods in finance*. Cambridge University Press, Cambridge, UK

Daskin MS (2013) *Network and discrete location*. Wiley-Interscience, New York, NY

Davis LS, Johnson KN, Bettinger PS, Howard TE (2001) *Forest management: to sustain ecological, economic, and social values*. 4th ed., Mc Graw-Hill

de Boor C (2001) *A practical guide to splines*. (rev. ed.) Springer-Verlag, New York

Eiselt HA, Sandblom C-L (2000) *Integer programming and network models*. Springer-Verlag, Berlin-Heidelberg-New York

Eiselt HA, Sandblom C-L (2004) *Decision analysis, location models, and scheduling problems*. Springer-Verlag, Berlin-Heidelberg-New York

Eiselt HA, Sandblom, C-L (2007) *Linear programming and its applications*. Springer-Verlag, Berlin-Heidelberg

Eiselt HA, Sandblom C-L (2012) *Operations research: a model-based approach*. (2nd ed.) Springer-Verlag, Berlin-Heidelberg-New York

Estrada J (2008) Mean-semivariance optimization: a heuristic approach. *Journal of Applied Finance*, Spring/summer: 57-72

Fair RC (1974) On the solution of optimal control problems as maximization problems. *Annals of Economic and Social Measurement* **3**: 135–153

Ghosh A, Buchanan B (1988) Multiple outlets in a duopoly: a first entry paradox. *Geographical Analysis* **20**: 111–121

Goldman AJ, Dearing PM (1975) Concepts of optimal location for partially noxious facilities. *Bulletin of the Operations Research Society of America* **23**, Supplement 1: 331

Hillier FS, Lieberman GJ, Nag B, Basu P (2017) *Introduction to operations research*. (10th ed.), McGraw Hill India

Hwang CL, Yoon K (1981) *Multiple attribute decision making: methods and applications*. Springer-Verlag, New York

Klein R (1989) *Concrete and abstract Voronoi diagrams.* Lecture Notes in Computer Science 400. Springer-Verlag, Berlin-New York

Kutner M, Nachtsheim C, Neter J, Li W (2004) *Applied linear statistical models.* (5th ed.). McGraw-Hill/Irwin, New York

Laffer A (2004) The Laffer curve: past, present, and future. Available online at https://www.heritage.org/taxes/report/the-laffer-curve-past-present-and-future. Last accessed 2/10/2019

Lakhera S, Shanbhag UV, McInerney MK (2011) Approximating electrical distribution networks via mixed-integer nonlinear programming. *Electrical Power and Energy Systems* **33**: 245–257

Lasdon LS, Waren AD (1980) Survey of nonlinear programming applications. *Operations Research* **28**: 1029-1073

Markowitz HM (1952) Portfolio selection. *Journal of Finance* **7**: 77-91

Markowitz HM (1959) *Portfolio selection: efficient diversification of investments.* Cowles Foundation Monographs #16, Wiley, New York

Molzahn DK, Dörfler F, Sandberg H, Low SH, Chakrabarti S, Baldick R, Lavaei J (2017) A survey of distributed optimization and control algorithms for electric power systems. *IEEE Transactions on Smart Grid* **8/6**: 2941-2962

Momoh JA (2017) *Electric power system applications of optimization.* (2nd ed.) CRC Press, Boca Raton, FL

Okabe A, Boots B, Sugihara K, Chiu S-N (2000) *Spatial tessellations: concepts and applications of Voronoi diagrams* (2nd ed.) Wiley, Chichester

Richards FJ (1959) A flexible growth function for empirical use. *Journal of Experimental Botany* **10/2**: 290–300

Rönnquist M (2003) Optimization in forestry. *Mathematical Programming* **97**: 267-284

Sandblom C-L (1984) Optimizing economic policy with sliding windows. *Applied Economics* **6**: 45-56

Sandblom C-L (1985) Economic policy with bounded controls. *Economic Modelling* **2**: 135-148

Schnute J (1981) A versatile growth model with statistically stable parameters. *Canadian Journal of Fisheries Aquatic Sciences* **38**: 1128-1140

Shamos MI, Hoey D (1975) Closest-point problems. *Proceedings of the 16th Annual IEEE Symposium on Foundations of Computer Science*: 151-162

Simon HA (1959) Theories of decision-making in economics and behavioral science. *The American Economic Review* **49/3**: 253-283

Stackelberg H von (1943) *Grundlagen der theoretischen Volkswirtschaftslehre* (translated as: *The Theory of the Market Economy*). W. Hodge & Co. Ltd., London, 1952

Toussaint GT (1983) Computing largest empty circles with location constraints. *International Journal of Computer and Information Sciences* **12/5**: 347-358

von Bertalanffy L (1951) *Theoretische Biologie.* (Band II). Franke, Bern

Wanniski J (1978) Taxes, revenues, and the "Laffer curve." *National Affairs* **50**: 3-16, Winter 1978. Available online at https://nationalaffairs.com/public_interest/detail/taxes-revenues-and-the-laffer-curve, last accessed on 2/10/2019

Wierzbicki AP (1982) A mathematical basis for satisficing decision making. *Mathematical Modelling* **3**: 391–405

Willis HL, Tram H, Engel MV, Finley L (1995) Optimization applications to power distribution. *IEEE Computer Applications in Power* **10**: 12-17

Zhu JZ (2009) *Optimization of power systems operations.* Wiley-IEEE Press

Chapter 4
Optimality Conditions and Duality Theory

This chapter studies optimization problems with constraints under special consideration of necessary and sufficient conditions for optimality of solution points. Although the theory to be presented is not immediately concerned with computational aspects of solution techniques, it nevertheless represents the foundation for the development of algorithms which will be introduced in the following chapters. Apart from their intrinsic relevance, some of these results also provide useful information about the sensitivity of an optimal solution.

The discussion begins with the definition of the Lagrangean function which plays an essential role in the entire field of mathematical programming and is particularly important for the duality theory associated with nonlinear programming problems. The same remarks can be made about the importance of the so-called Lagrangean multipliers: their interpretation as prices is analogous to that of dual variables in linear programming and they are therefore very valuable in almost all optimization settings.

After considering the relationship between saddle points and stationary points of Lagrangean functions, the Karush-Kuhn-Tucker theory will be discussed. Historically, shortly after the formalization of the linear programming model and the appearance of the first version of the simplex method, John (1948) and Kuhn and Tucker (1951) published their results on necessary and sufficient optimality conditions for solutions of nonlinear programming problems. These conditions helped in establishing solid foundations for optimization with constraints. Actually, Karush (1939) appears to have been first, apart from Lagrange himself, to deal with the study of constrained optimization. The results of Karush did not have any significant impact at the time, though.

The third section in this chapter is devoted to duality and sensitivity analysis. Since nonlinear programming became an active field of study, it took almost

© Springer Nature Switzerland AG 2019
H. A. Eiselt, C.-L. Sandblom, *Nonlinear Optimization*, International Series in
Operations Research & Management Science 282,
https://doi.org/10.1007/978-3-030-19462-8_4

10 years before a nonlinear duality theory started to develop. This is much longer than for the linear case and perhaps a bit surprising, since in nonlinear programming the entities that play the role of dual variables, i.e., the Lagrange multipliers, predated modern nonlinear programming theory by more than a century. Contrary to the situation in linear programming, there are different non-equivalent formulations of the dual of a nonlinear programming problem. We first treat the Wolfe (1961) dual and then general Lagrangean duality. The connections with linear programming duality will be explored and the usefulness of duality theory for sensitivity analysis purposes will be explained. The chapter ends with a discussion on computational aspects.

4.1   The Lagrangean Function

In this and the following section we assume that the general nonlinear programming problem is given by

$$P: \text{Min } z = f(\mathbf{x})$$
$$\text{s.t. } g_i(\mathbf{x}) \leq 0, \; i = 1, \ldots, m$$
$$\mathbf{x} \geq \mathbf{0},$$

where the objective function $f(\mathbf{x})$ as well as all constraints $g_i(\mathbf{x})$ are convex functions in $\mathbb{R}^n$. Occasionally, we will consider problems without the nonnegativity constraints and/or problems with additional equation constraints. Although the resulting formulas may have to be modified, this does not cause any loss of generality in our treatment, due to equivalence of the various formulations, see Definition 1.1 and the following discussion. Here, we choose the formulation with the nonnegativity constraints, so as to facilitate the comparison with similar results in linear programming.

Defining $\mathbf{g}(\mathbf{x}) = [g_1(\mathbf{x}), \; g_2(\mathbf{x}), \; \ldots, \; g_m(\mathbf{x})]^T$, the above formulation can be rewritten as

$$P: \text{Min } z = f(\mathbf{x})$$
$$\text{s.t. } \mathbf{g}(\mathbf{x}) \leq \mathbf{0}$$
$$\mathbf{x} \geq \mathbf{0}.$$

Definition 4.1 Let $\mathbf{u} = [u_1, u_2, \ldots, u_m]$ be a row vector of m real variables. Then the *Lagrangean function* $L(\mathbf{x}, \mathbf{u})$ for the problem P is the sum of the objective function and the weighted constraints, namely

$$L(\mathbf{x}, \mathbf{u}) = f(\mathbf{x}) + \sum_{i=1}^{m} u_i g_i(\mathbf{x}) = f(\mathbf{x}) + \mathbf{u}\mathbf{g}(\mathbf{x}),$$

where the variables u_i are called the *Lagrangean multipliers*.

Definition 4.2 A point $(\bar{\mathbf{x}}, \bar{\mathbf{u}}) \geq (\mathbf{0}, \mathbf{0})$ is said to be a *saddle point* of the Lagrangean function $L(\mathbf{x}, \mathbf{u})$ if

$$L(\bar{\mathbf{x}}, \mathbf{u}) \leq L(\bar{\mathbf{x}}, \bar{\mathbf{u}}) \leq L(\mathbf{x}, \bar{\mathbf{u}}) \quad \forall \ (\mathbf{x}, \mathbf{u}) \geq (\mathbf{0}, \mathbf{0}).$$

The Lagrangean multipliers in Definitions 4.1 and 4.2 correspond to the dual variables in linear programming, as we will demonstrate below. Definition 4.2 should be compared to Definition 1.28 of a saddle point in general. Note that for a fixed $\bar{\mathbf{u}}$, the Lagrangean function is a convex function in $\mathbf{x}$ because it is a nonnegative linear combination of the convex functions $f(\mathbf{x})$ and $g_i(\mathbf{x})$, see the discussion of convexity in Sect. 1.3. On the other hand, for a fixed $\bar{\mathbf{x}}$, the Lagrangean function is simply linear in $\mathbf{u}$ since as long as $\bar{\mathbf{x}}$ does not vary, $f(\bar{\mathbf{x}})$ as well as all $g_i(\bar{\mathbf{x}})$ are constants.

Furthermore, for a fixed $\bar{\mathbf{u}}$, the Langrangean function is minimized at $\bar{\mathbf{x}}$ (due to the second inequality of the relationship in Definition 4.2) whereas for a fixed $\bar{\mathbf{x}}$, the Lagrangean function is maximized at $\bar{\mathbf{u}}$ (which follows from the first inequality of the relationship in Definition 4.2). This allows us to state

Lemma 4.3 If there exists a saddle point $(\bar{\mathbf{x}}, \bar{\mathbf{u}})$ of the Lagrangean function $L(\mathbf{x}, \mathbf{u})$, then $\bar{\mathbf{x}}$ is an optimal solution for problem P.

Proof The first part of the saddle point criterion can be stated as

$$f(\bar{\mathbf{x}}) + \mathbf{u}\mathbf{g}(\bar{\mathbf{x}}) \leq f(\bar{\mathbf{x}}) + \bar{\mathbf{u}}\mathbf{g}(\bar{\mathbf{x}}),$$

or simply as $(\mathbf{u} - \bar{\mathbf{u}})\mathbf{g}(\bar{\mathbf{x}}) \leq 0$. Now $\mathbf{g}(\bar{\mathbf{x}}) \leq \mathbf{0}$, because if there exists some k such that $g_k(\bar{\mathbf{x}}) > 0$, one could set $u_k := \bar{u}_k + 1$, $u_i := \bar{u}_i \ \forall \ i \neq k$, and thus violate the above inequality. Hence $\mathbf{g}(\bar{\mathbf{x}}) \leq \mathbf{0}$, so that $\bar{\mathbf{x}}$ satisfies all the constraints of problem P. Moreover, $\bar{\mathbf{u}}\mathbf{g}(\bar{\mathbf{x}}) = 0$, as otherwise there would exist some k, such that $\bar{u}_k > 0$ and $g_k(\bar{\mathbf{x}}) < 0$. By setting $u_k := 0$ and $u_i := \bar{u}_i \ \forall \ i \neq k$, the above inequality would again be violated. Now the second part of the saddle point criterion reads as

$$f(\bar{\mathbf{x}}) + \bar{\mathbf{u}}\mathbf{g}(\bar{\mathbf{x}}) \leq f(\mathbf{x}) + \bar{\mathbf{u}}\mathbf{g}(\mathbf{x})$$

or, by using the fact that $\bar{\mathbf{u}}\mathbf{g}(\bar{\mathbf{x}}) = 0$, as $f(\bar{\mathbf{x}}) \leq f(\mathbf{x}) + \bar{\mathbf{u}}\mathbf{g}(\mathbf{x})$. Since $\mathbf{u} \geq \mathbf{0}$ and, as long as $\mathbf{x}$ is feasible for problem P (so that $\mathbf{g}(\mathbf{x}) \leq \mathbf{0}$) also $\bar{\mathbf{u}}\mathbf{g}(\mathbf{x}) \leq 0$ applies, it follows that $f(\bar{\mathbf{x}}) \leq f(\mathbf{x})$. This shows that $\bar{\mathbf{x}}$ is indeed an optimal solution for problem P. $\square$

For a geometrical illustration of the above discussion, consider the problem

$$\text{P: } \text{Min} \, z = e^x - 2x$$
$$\text{s.t. } x^2 - 4x + 3 \leq 0$$
$$x \geq 0.$$

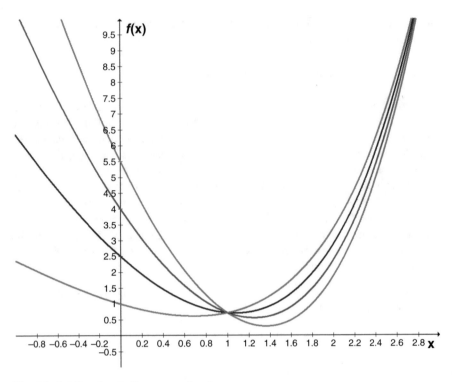

Fig. 4.1 Saddle point of a Lagrangean function

The Lagrangean function associated with P is $L(x,\ u) = e^x - 2x + u(x^2 - 4x + 3)$, which is graphically portrayed in Fig. 4.1 for various values of the Lagrangean multiplier: $u = 0$ (green), $u = 0.5$ (purple), $u = 1$ (red) and $u = 1.5$ (teal). Using partial differentiation, one can easily show that L has a unique saddle point located at $(\bar{x}, \bar{u}) = (1, \tfrac{1}{2}e - 1) \approx (1,\ 0.3591)$. Since the feasible region of P consists of the closed interval [1, 3] and since the objective function $f(x) = e^x - 2x$ is strictly increasing (as $f'(x) = e^x - 2 > 0$) over the interval, the unique optimal solution of P is $\bar{x} = 1$, in agreement with Lemma 4.3.

To see that a saddle point does not necessarily exist, consider the problem

$$P: \text{Min } z = -x_1 - x_2$$
$$\text{s.t. } x_1^2 + x_2 \le 0$$
$$x_1,\ x_2 \ge 0.$$

The hatched area in Fig. 4.2 represents the set of feasible points when only the nonlinear constraint is considered. If all the constraints are included, then the origin is the only feasible point and hence it must be optimal as well, i.e., $\bar{x}_1 = \bar{x}_2 = 0$. From Lemma 4.3 it is known that if the Lagrangean function of problem P has a saddle point $(\hat{x}_1,\ \hat{x}_2,\ \hat{u})$, then $(\hat{x}_1,\ \hat{x}_2)$ is optimal. But $(\bar{x}_1,\ \bar{x}_2)$ is the unique

Fig. 4.2 Lagrangean function without a saddle point

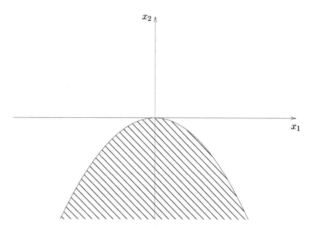

optimal point, so that if a saddle point $(\hat{x}_1, \ \hat{x}_2, \ \hat{u})$ were to exist, it would have to be $(\hat{x}_1, \ \hat{x}_2) = (0, \ 0)$. By using these values in the saddle point criterion we obtain

$$0 \leq 0 \leq -x_1 - x_2 + \bar{u}x_1^2 + \bar{u}x_2 \ \forall \ \mathbf{x} \geq \mathbf{0}.$$

Now if $\bar{u} = 0$, by setting x_1 and x_2 equal to any positive value, one can see that the above relation is violated: if $\bar{u} > 0$, then let $x_1 := \dfrac{1}{\bar{u}+1}$ and $x_2 := 0$, so that the above relation becomes $0 \leq -\dfrac{1}{\bar{u}+1} + \dfrac{\bar{u}}{(\bar{u}+1)^2}$ or simply $0 \leq \dfrac{-1}{(\bar{u}+1)^2}$, which is obviously a contradiction. Thus, problem P has no saddle point for its Lagrangean function. Extending the discussion of the proof of Lemma 4.3, we can establish the following

Theorem 4.4 Assume that $(\bar{\mathbf{x}}, \bar{\mathbf{u}}) \geq (\mathbf{0}, \mathbf{0})$. Then $(\bar{\mathbf{x}}, \bar{\mathbf{u}})$ is a saddle point of $L(\mathbf{x}, \mathbf{u})$ if and only if

(a) $\bar{\mathbf{x}}$ minimizes $L(\mathbf{x}, \bar{\mathbf{u}})$ over $\mathbf{x} \in \mathbb{R}_+^n$,
(b) $\bar{\mathbf{u}}\mathbf{g}(\bar{\mathbf{x}}) = 0$, and
(c) $\mathbf{g}(\bar{\mathbf{x}}) \leq \mathbf{0}$.

Proof If $(\bar{\mathbf{x}}, \ \bar{\mathbf{u}})$ is a saddle point, then (a) follows from the definition of a saddle point and (b) and (c) were shown in the proof of Lemma 4.3. Conversely, assume that (a), (b), and (c) are true. We then only need to prove that $\bar{\mathbf{u}}$ maximizes $L(\bar{\mathbf{x}}, \ \mathbf{u})$ over $\mathbf{x} \in \mathbb{R}_+^n$. For all $\mathbf{u} \geq \mathbf{0}$, we have $\mathbf{u}\mathbf{g}(\bar{\mathbf{x}}) \leq 0$, so that

$$L(\bar{\mathbf{x}}, \ \bar{\mathbf{u}}) = f(\bar{\mathbf{x}}) + \bar{\mathbf{u}}\mathbf{g}(\bar{\mathbf{x}}) = f(\bar{\mathbf{x}}) \geq f(\bar{\mathbf{x}}) + \mathbf{u}\mathbf{g}(\bar{\mathbf{x}}) \quad \forall \ \mathbf{u} \geq \mathbf{0},$$

and the theorem is proved. $\square$

Note that condition (a) in the theorem is concerned with sign-constrained minimization, which was discussed in Sect. 1.2; see Theorem 1.27 with example. To facilitate the discussion we now introduce the following

Definition 4.5 A point $(\bar{\mathbf{x}}, \bar{\mathbf{u}}) \geq (\mathbf{0}, \mathbf{0})$, such that conditions (a), (b), and (c) in Theorem 4.4 are fulfilled is said to satisfy the *optimality conditions* of problem P. Condition (b) is referred to as the *complementarity condition*.

Theorem 4.4 states that $(\bar{\mathbf{x}}, \bar{\mathbf{u}})$ is a saddle point of $L(\mathbf{x}, \mathbf{u})$ if and only if it satisfies the optimality conditions of problem P.

Let us consider the complementarity condition $\mathbf{u}\mathbf{g}(\mathbf{x}) = 0$ for an arbitrary feasible point $\bar{\mathbf{x}}$, i.e., satisfying $\mathbf{x} \geq \mathbf{0}$ and $\mathbf{g}(\mathbf{x}) \leq \mathbf{0}$, and for $\mathbf{u} \geq \mathbf{0}$. With $u_i \geq 0$ and $g_i(\mathbf{x}) \leq 0$, we have $u_i g_i(\mathbf{x}) \leq 0$ for $i = 1, \ldots, m$, so that $\mathbf{u}\mathbf{g}(\mathbf{x}) = \sum_{i=1}^{m} u_i g_i(\mathbf{x})$ is a sum of m nonpositive terms. Therefore, $\mathbf{u}\mathbf{g}(\mathbf{x}) = 0$ is equivalent to the m equations $u_i g_i(\mathbf{x}) = 0$, $i = 1, \ldots, m$. Complementarity then implies the following: For any positive u_i, $g_i(\mathbf{x})$ must be zero, and for any negative $g_i(\mathbf{x})$, u_i must be zero. This consequence of the complementarity condition will from time to time be exploited later on.

Definition 4.6 The vector $\bar{\mathbf{u}}$ is said to be *an optimal Lagrangean multiplier vector* of problem P if there exists an $\bar{\mathbf{x}}$, such that $(\bar{\mathbf{x}}, \bar{\mathbf{u}})$ satisfies the optimality conditions of problem P as outlined in Definition 4.5.

Note that convexity properties have not been used so far in this chapter, nor has the assumption $\mathbf{x} \geq \mathbf{0}$ been used. In order to obtain the converse of Lemma 3.3 we need a so-called *constraint qualification*.

Definition 4.7 *Slater's constraint qualification* is satisfied for problem P, if there exists an $\mathbf{x} \geq \mathbf{0}$ such that $\mathbf{g}(\mathbf{x}) < \mathbf{0}$.

Its purpose is to rule out certain kinds of irregular behavior on the boundary of the set of feasible solutions. Slater's constraint qualification implies that there exists at least one interior point in the given feasible set. Note that this was not the case in one of the above examples, in which the feasible set actually degenerated to just one point. This qualification places some important restrictions on the nature of the set of feasible solutions in the immediate vicinity of $\bar{\mathbf{x}}$. A number of different constraint qualifications of varying complexity have been proposed in the literature. Slater's constraint qualification (1950), although quite restrictive compared to many other such qualifications, has been chosen here because it is particularly simple. For full discussions of various constraint qualifications and their interrelationships, the reader is referred to Bazaraa et al. (2013), Bertsekas (2016), and Cottle and Thapa (2017).

One important result related to convexity, which will be needed below, is stated in the following lemma on separation, which we state here without proof. For theorems on separation, including proofs, the reader is referred to Mangasarian (1969) and Bazaraa et al. (2013).

Lemma 4.8 Let S_1 and S_2 be two disjoint convex sets, and assume that S_2 is open. Then there exists at least one hyperplane $\mathbf{a}\mathbf{x} = b$ where $\mathbf{a} \neq \mathbf{0}, b \in \mathbb{R}$, which separates S_1 and S_2.

Fig. 4.3 Separating
hyperplanes

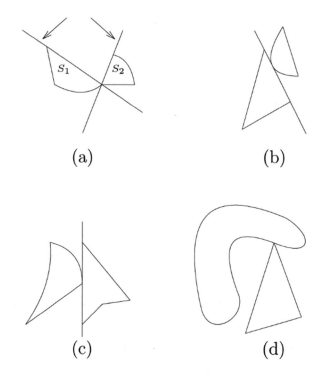

(a)   (b)

(c)   (d)

If two points $\mathbf{x}^1$ and $\mathbf{x}^2$ are given with $\mathbf{x}^1 \in S_1$, $\mathbf{x}^2 \in S_2$, then using the above lemma, one can show that there exist $\mathbf{a}$ and b such that $\mathbf{a}\mathbf{x}^1 \geq b$ and $\mathbf{a}\mathbf{x}^2 < b$, implying that $\mathbf{a}\mathbf{x}^1 > \mathbf{a}\mathbf{x}^2$.

Figure 4.3 shows four different cases. In Fig. 4.3a, infinitely many hyperplanes exist that separate the two convex sets S_1 and S_2, whereas in Fig. 4.3b, the two convex sets can only be separated by a unique hyperplane. If the sets are not both convex, separating hyperplanes may still exist as in Fig. 4.3c but this cannot be assured, as exemplified in Fig. 4.3d.

Let now $\mathbf{y} = [y_0, y_1, \ldots, y_m]^T$ be a vector in $\mathbb{R}^{m+1}$, assume that $\bar{\mathbf{x}}$ is optimal for problem P and define the two sets S_1 and S_2 as follows.

$S_1 = \{\mathbf{y} : \text{there exists some } \mathbf{x} \geq \mathbf{0} \text{ where } y_0 \geq f(\mathbf{x}) \text{ and } y_i \geq g_i(\mathbf{x}) \ \forall \ i = 1, \ldots, m\}$
and
$S_2 = \{\mathbf{y} : y_0 < f(\bar{\mathbf{x}}) \text{ and } y_i < 0 \ \forall \ i = 1, \ldots, m\}.$

It can easily be seen that S_1 as well as S_2 are convex sets, that S_2 is open, and that there exists no point $\mathbf{y}$ which is in both S_1 and S_2. If such a point were to exist, then we would have $\mathbf{y} \in S_1 \cap S_2$, and hence for some $\mathbf{x} \geq \mathbf{0}, f(\bar{\mathbf{x}}) > y_0 \geq f(\mathbf{x})$, i.e., $f(\bar{\mathbf{x}}) > f(\mathbf{x})$. Furthermore, $\mathbf{x}$ would be feasible, which contradicts the optimality of $\bar{\mathbf{x}}$.

From the discussion following Lemma 4.8. on separation as applied to S_1 and S_2, we find that a vector $\mathbf{v} = [v_0, v_1, \ldots, v_m]$ exists such that $\mathbf{v}\mathbf{y}^1 > \mathbf{v}\mathbf{y}^2$ for all $\mathbf{y}^1 \in S_1$ and $\mathbf{y}^2 \in S^2$. Since the components of $\mathbf{y}^2$ may be equal to any large negative number, it follows that $\mathbf{v} \geq \mathbf{0}$. Now for any $\mathbf{x} \geq \mathbf{0}$, let $\mathbf{y}^1 = [f(\mathbf{x}), g_1(\mathbf{x}), g_2(\mathbf{x}), \ldots, g_m(\mathbf{x})]^T$ and $\mathbf{y}^2 = [f(\bar{\mathbf{x}}), 0, 0, \ldots, 0]^T$. Since the strict inequality $\mathbf{v}\mathbf{y}^1 > \mathbf{v}\mathbf{y}^2$ might hold true as a non-strict inequality when both points $\mathbf{y}^1$ and $\mathbf{y}^2$ are located on the border of the sets S_1 and S_2, respectively (note that $\mathbf{y}^2 \notin S_2$), we can write

$$v_0 f(\mathbf{x}) + \sum_{i=1}^{m} v_i g_i(\mathbf{x}) \geq v_0 f(\bar{\mathbf{x}}) \ \forall \ \mathbf{x} \geq \mathbf{0}.$$

This allows us to prove

Lemma 4.9 $v_0 > 0$.

Proof Suppose that the statement is not true. Then since $v_0 \geq 0$, this implies that $v_0 = 0$. In view of the above discussion we would then have $\sum_{i=1}^{m} v_i g_i(\mathbf{x}) \geq 0$. Because Slater's constraint qualification was assumed to be satisfied, there must exist at least one $\mathbf{x}$ such that $g_i(\mathbf{x}) < 0 \ \forall \ i = 1, \ldots, m$. As $\mathbf{v} \geq \mathbf{0}$ but $\mathbf{v} \neq \mathbf{0}$, there exists at least one $k \geq 1$ with $v_k > 0$, which implies that $v_k g_k(\mathbf{x}) < 0$. Therefore, $\sum_{i=1}^{m} v_i g_i(\mathbf{x}) < 0$, which is in contradiction with the inequality just developed. Hence $v_0 > 0$. $\square$

Dividing all terms in the inequality $v_0 f(\mathbf{x}) + \sum_{i=1}^{m} v_i g_i(\mathbf{x}) \geq v_0 f(\mathbf{x})$ by $v_0 > 0$ and defining $\bar{\mathbf{u}} = \dfrac{1}{v_0}[v_1, v_2, \ldots, v_m] \geq \mathbf{0}$ yields the relation

$$L(\mathbf{x}, \bar{\mathbf{u}}) = f(\mathbf{x}) + \bar{\mathbf{u}}\,\mathbf{g}(\mathbf{x}) \geq f(\bar{\mathbf{x}}) \ \forall \ \mathbf{x} \geq \mathbf{0}.$$

Setting $\mathbf{x} := \bar{\mathbf{x}}$, we obtain

$$f(\bar{\mathbf{x}}) + \bar{\mathbf{u}}\mathbf{g}(\bar{\mathbf{x}}) \geq f(\bar{\mathbf{x}}), \ \text{i.e.,} \ \bar{\mathbf{u}}\mathbf{g}(\bar{\mathbf{x}}) \geq 0.$$

However, we must also have $\bar{\mathbf{u}}\mathbf{g}(\bar{\mathbf{x}}) \leq 0$, since $\bar{\mathbf{u}} \geq \mathbf{0}$ and $\mathbf{g}(\bar{\mathbf{x}}) \leq \mathbf{0}$ (by virtue of the feasibility of $\bar{\mathbf{x}}$), consequently $\bar{\mathbf{u}}\mathbf{g}(\bar{\mathbf{x}}) = 0$. Using this fact in the inequality $L(\mathbf{x}, \bar{\mathbf{u}}) \geq f(\bar{\mathbf{x}})$ developed above, we obtain

$$L(\mathbf{x}, \bar{\mathbf{u}}) \geq f(\bar{\mathbf{x}}) = f(\bar{\mathbf{x}}) + \bar{\mathbf{u}}\mathbf{g}(\bar{\mathbf{x}}) = L(\bar{\mathbf{x}}, \bar{\mathbf{u}}) \ \forall \ \mathbf{x} \geq \mathbf{0}, \ \text{or}$$

$$L(\bar{\mathbf{x}}, \bar{\mathbf{u}}) \leq L(\mathbf{x}, \bar{\mathbf{u}}) \ \forall \ \mathbf{x} \geq \mathbf{0},$$

which is the second part of the saddle point criterion. Furthermore, $\mathbf{g}(\bar{\mathbf{x}}) \leq \mathbf{0}$ and $\mathbf{u} \geq \mathbf{0}$ imply that $\mathbf{u}\mathbf{g}(\bar{\mathbf{x}}) \leq 0$. Adding $f(\bar{\mathbf{x}})$ to both sides yields

$$L(\mathbf{x}, \bar{\mathbf{u}}) = f(\bar{\mathbf{x}}) + \mathbf{u}\mathbf{g}(\bar{\mathbf{x}}) \le f(\bar{\mathbf{x}}) = f(\bar{\mathbf{x}}) + \bar{\mathbf{u}}\mathbf{g}(\bar{\mathbf{x}}) \ \forall \ \mathbf{u} \ge \mathbf{0}, \text{ or}$$

$$L(\bar{\mathbf{x}}, \mathbf{u}) \le L(\bar{\mathbf{x}}, \bar{\mathbf{u}}) \ \forall \ \mathbf{u} \ge \mathbf{0},$$

which is the first part of the saddle point criterion. We here therefore proved the following result, which is essentially the converse of Lemma 4.3.

Lemma 4.10 If $\bar{\mathbf{x}}$ is an optimal solution to problem P and the feasible set of P includes at least one point $\mathbf{x}$ with $\mathbf{g}(\mathbf{x}) < \mathbf{0}$, then there exists a vector $\bar{\mathbf{u}}$, such that $(\bar{\mathbf{x}}, \bar{\mathbf{u}})$ is a saddle point for $L(\mathbf{x}, \mathbf{u})$.

Lemmas 4.3 and 4.10 together yield the following

Theorem 4.11 Assume that Slater's constraint qualification holds. Then a point $\bar{\mathbf{x}}$ is an optimal solution for problem P if and only if there exists a vector $\bar{\mathbf{u}} \ge \mathbf{0}$, such that $(\bar{\mathbf{x}}, \bar{\mathbf{u}})$ is a saddle point for $L(\mathbf{x}, \mathbf{u})$.

The development in this section may be illustrated by Fig. 4.4 where for simplicity the assumptions have been omitted.

We end this section by mentioning a generalized version of the Lagrangean function, the *augmented Lagrangean function* $L(\mathbf{x}, \mathbf{u}, r)$ for problems with all equality constraints $\mathbf{g}(\mathbf{x}) = \mathbf{0}$. This function is defined as

$$L(\mathbf{x}, \mathbf{u}, r) = f(\mathbf{x}) + \mathbf{u}\mathbf{g}(\mathbf{x}) + \tfrac{1}{2}r\|\mathbf{g}(\mathbf{x})\|^2,$$

where $r > 0$ is called the *penalty parameter*. One of the reasons for including the penalty term $\tfrac{1}{2}r\|\mathbf{g}(\mathbf{x})\|^2$ is that it tends to "convexify" the Lagrangean near an optimal point, where in the general case it might otherwise be nonconvex. The resulting augmented Langrangean methods can be viewed as combinations of penalty function and local duality methods. We will return to this topic in Chap. 5.

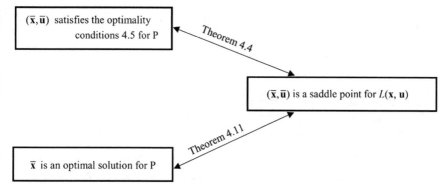

Fig. 4.4 Relations between optimality conditions and the saddle point condition

4.2   Karush-Kuhn-Tucker Optimality Conditions

In this section we describe a set of optimality conditions known in the literature as *Karush-Kuhn-Tucker conditions*. We will derive the relationship between them and the saddle point criterion given in the previous section. For several years, the optimality conditions we will discuss in this section were called Kuhn-Tucker conditions, resulting from the work of H. Kuhn and A. Tucker (1951), which is based on the work by F. John (1948). It was subsequently discovered that such conditions had been studied and developed in a 1939 thesis by W. Karush. Today, these conditions are universally referred to as Karush-Kuhn-Tucker or simply KKT conditions.

For this purpose we assume that the objective function $f(\mathbf{x})$ as well as all the constraints $g_i(\mathbf{x})$ are differentiable functions. Recall the definitions of problem P and the Lagrangean function L, which are restated here for convenience: P: Min $z = f(\mathbf{x})$, s.t. $\mathbf{g}(\mathbf{x}) \leq \mathbf{0}$, $\mathbf{x} \geq \mathbf{0}$, and $L(\mathbf{x}, \mathbf{u}) = f(\mathbf{x}) + \mathbf{u}\mathbf{g}(\mathbf{x})$. The gradient of the Lagrangean function with respect to $\mathbf{x}$ will be denoted by

$$\nabla_{\mathbf{x}} L(\mathbf{x}, \mathbf{u}) = L_{\mathbf{x}}(\mathbf{x}, \mathbf{u}) = \left[\frac{\partial L(\mathbf{x}, \mathbf{u})}{\partial x_1}, \frac{\partial L(\mathbf{x}, \mathbf{u})}{\partial x_2}, \ldots, \frac{\partial L(\mathbf{x}, \mathbf{u})}{\partial x_n} \right]^T,$$

and $L_{\mathbf{u}}$ will be defined accordingly. Note that $L_{\mathbf{u}}(\mathbf{x}, \mathbf{u}) = \mathbf{g}(\mathbf{x})$. We can then write

Definition 4.12 The *Karush-Kuhn-Tucker optimality conditions* for problem P consist of the following six relations.

$$L_{\mathbf{x}}(\bar{\mathbf{x}}, \bar{\mathbf{u}}) \geq \mathbf{0} \tag{4.1}$$

$$\bar{\mathbf{x}}^T L_{\mathbf{x}}(\bar{\mathbf{x}}, \bar{\mathbf{u}}) = 0 \tag{4.2}$$

$$\bar{\mathbf{x}} \geq \mathbf{0} \tag{4.3}$$

$$L_{\mathbf{u}}(\bar{\mathbf{x}}, \bar{\mathbf{u}}) \leq \mathbf{0} \tag{4.4}$$

$$\bar{\mathbf{u}} L_{\mathbf{u}}(\bar{\mathbf{x}}, \bar{\mathbf{u}}) = 0 \tag{4.5}$$

$$\bar{\mathbf{u}} \geq \mathbf{0}. \tag{4.6}$$

Note that with our setup, the Karush-Kuhn-Tucker conditions are "symmetric" in $\bar{\mathbf{x}}$ and $\bar{\mathbf{u}}$. Without the nonnegativity constraint in the problem P, conditions (4.1)–(4.3) would be replaced by the single condition $L_{\mathbf{x}}(\bar{\mathbf{x}}, \bar{\mathbf{u}}) = \mathbf{0}$, see condition (4.12) below.

Lemma 4.13 If $f(\mathbf{x})$ and all $g_i(\mathbf{x})$ are differentiable and there exists a saddle point $(\bar{\mathbf{x}}, \bar{\mathbf{u}})$ of $L(\mathbf{x}, \mathbf{u})$, then $(\bar{\mathbf{x}}, \bar{\mathbf{u}})$ satisfies the Karush-Kuhn-Tucker conditions.

Proof We will proceed by assuming that at least one of the Karush-Kuhn-Tucker conditions is not fulfilled and then show that in this case $(\bar{\mathbf{x}}, \bar{\mathbf{u}})$ would *not* be a saddle point of $L(\mathbf{x}, \mathbf{u})$.

Part 1: Suppose that condition (4.1) and/or (4.2) is violated. Then there exists an $\bar{x}_j = 0$, such that $\left(L_{\mathbf{x}}(\bar{\mathbf{x}}, \bar{\mathbf{u}})\right)_j < 0$, and/or there exists an $\bar{x}_j > 0$, such that $\left(L_{\mathbf{x}}(\bar{\mathbf{x}}, \bar{\mathbf{u}})\right)_j \neq 0$. We consider the following three cases separately.

Case (*i*): $\bar{x}_j = 0$ and $\left(L_{\mathbf{x}}(\bar{\mathbf{x}}, \bar{\mathbf{u}})\right)_j < 0$, which by virtue of the differentiability of $L(\mathbf{x}, \mathbf{u})$ implies that there exists at least one $\tilde{x}_j > 0$, such that $L\left(\bar{x}_1, \bar{x}_2, \ldots, \bar{x}_{j-1}, \tilde{x}_j, \bar{x}_{j+1}, \ldots, \bar{x}_n; \bar{\mathbf{u}}\right) < L(\bar{\mathbf{x}}, \bar{\mathbf{u}})$.

Case (*ii*): $\bar{x}_j > 0$ and $\left(L_{\mathbf{x}}(\bar{\mathbf{x}}, \bar{\mathbf{u}})\right)_j < 0$, which implies that there exists at least one $\tilde{x}_j > \bar{x}_j$, such that $L\left(\bar{x}_1, \bar{x}_2, \ldots, \bar{x}_{j-1}, \tilde{x}_j, \bar{x}_{j+1}, \ldots, \bar{x}_n; \bar{\mathbf{u}}\right) < L(\bar{\mathbf{x}}, \bar{\mathbf{u}})$.

Case (*iii*): $\bar{x}_j > 0$ and $\left(L_{\mathbf{x}}(\bar{\mathbf{x}}, \bar{\mathbf{u}})\right)_j > 0$, which implies that there exists at least one $0 \leq \tilde{x}_j < \bar{x}_j$, such that $L\left(\bar{x}_1, \bar{x}_2, \ldots, \bar{x}_{j-1}, \tilde{x}_j, \bar{x}_{j+1}, \ldots, \bar{x}_n; \bar{\mathbf{u}}\right) < L(\bar{\mathbf{x}}, \bar{\mathbf{u}})$.

To summarize, in all three cases there exists a point $(\tilde{\mathbf{x}}, \bar{\mathbf{u}}) \neq (\bar{\mathbf{x}}, \bar{\mathbf{u}})$ for which the value of the Lagrangean function is smaller. This contradicts the saddle point property of $(\bar{\mathbf{x}}, \bar{\mathbf{u}})$, since the second part of the saddle point criterion stipulates that for a fixed $\bar{\mathbf{u}}$, the Lagrangean function must attain its minimal value precisely at the point $(\bar{\mathbf{x}}, \bar{\mathbf{u}})$.

Part 2: Suppose now that condition (4.4) and/or (4.5) is violated. Then $\bar{u}_i = 0$ exists, such that $\left(L_{\mathbf{u}}(\bar{\mathbf{x}}, \bar{\mathbf{u}})\right)_i > 0$ and/or some $\bar{u}_i > 0$ exists, such that $\left(L_{\mathbf{u}}(\bar{\mathbf{x}}, \bar{\mathbf{u}})\right)_i \neq 0$. As before, three different cases will be examined.

Case (*i*): $\bar{u}_i = 0$ and $\left(L_{\mathbf{u}}(\bar{\mathbf{x}}, \bar{\mathbf{u}})\right)_i = g_i(\bar{\mathbf{x}}) > 0$, which by virtue of the differentiability of $L(\bar{\mathbf{x}}, \bar{\mathbf{u}})$ implies that there exists at least one $\tilde{u}_i > 0$, such that $L\left(\bar{\mathbf{x}}; \bar{u}_1, \bar{u}_2, \ldots, \bar{u}_{i-1}, \tilde{u}_i, \bar{u}_{i+1}, \ldots, \bar{u}_m\right) > L(\bar{\mathbf{x}}, \bar{\mathbf{u}})$.

Case (*ii*): $\bar{u}_i > 0$ and $\left(L_{\mathbf{u}}(\bar{\mathbf{x}}, \bar{\mathbf{u}})\right)_i = g_i(\bar{\mathbf{x}}) > 0$, which implies that there exists at least one $\tilde{u}_i > \bar{u}_i$, such that $L\left(\bar{\mathbf{x}}; \bar{u}_1, \bar{u}_2, \ldots, \bar{u}_{i-1}, \tilde{u}_i, \bar{u}_{i+1}, \ldots, \bar{u}_m\right) > L(\bar{\mathbf{x}}, \bar{\mathbf{u}})$.

Case (*iii*): $\bar{u}_i > 0$ and $\left(L_{\mathbf{u}}(\bar{\mathbf{x}}, \bar{\mathbf{u}})\right)_i = g_i(\bar{\mathbf{x}}) < 0$, which implies that there exists at least one $0 \leq \tilde{u}_i < \bar{u}_i$, such that $L\left(\bar{\mathbf{x}}; \bar{u}_1, \bar{u}_2, \ldots, \bar{u}_{i-1}, \tilde{u}_i, \bar{u}_{i+1}, \ldots, \bar{u}_m\right) > L(\bar{\mathbf{x}}, \bar{\mathbf{u}})$.

Thus, in all the three cases there exists a point different from $(\bar{\mathbf{x}}, \bar{\mathbf{u}})$ for which the value of the Lagrangean function is larger. This fact is in contradiction with the saddle point property of $(\bar{\mathbf{x}}, \bar{\mathbf{u}})$, since the first part of the saddle point criterion requires that for a fixed $\bar{\mathbf{x}}$ the Lagrangean function must attain a maximal value exactly at the point $(\bar{\mathbf{x}}, \bar{\mathbf{u}})$. Therefore, conditions (4.1)–(4.6) must hold and the lemma is proved. $\square$

In order to show that the converse of this lemma also holds true, we make use of the fact that the Lagrangean function $L(\mathbf{x}, \mathbf{u})$ is a convex function of $\mathbf{x}$ and invoke Lemma 1.34. Then one can state and prove the following

Lemma 4.14 If a point $(\bar{\mathbf{x}}, \bar{\mathbf{u}})$ satisfies the Karush-Kuhn-Tucker conditions for problem P with convex functions f and $g_i \ \forall \ i$, then it is also a saddle point of the Langrangean function $L(\mathbf{x}, \mathbf{u})$.

Proof Let the six Karush-Kuhn-Tucker conditions (4.1)–(4.6) be satisfied by the point $(\bar{\mathbf{x}}, \bar{\mathbf{u}})$. With $\bar{\mathbf{u}} \geq \mathbf{0}$, we know that $L(\mathbf{x}, \bar{\mathbf{u}})$ is a convex function of $\mathbf{x}$. Setting $\mathbf{y} := \mathbf{x}$ and $\mathbf{x} := \bar{\mathbf{x}}$ in Lemma 1.34 where $L(\mathbf{x}, \bar{\mathbf{u}})$ corresponds to the convex function $f(\mathbf{x})$, we obtain

$$
\begin{aligned}
L(\mathbf{x}, \bar{\mathbf{u}}) &\geq L(\bar{\mathbf{x}}, \bar{\mathbf{u}}) + (\mathbf{x} - \bar{\mathbf{x}})^T L_{\mathbf{x}}(\bar{\mathbf{x}}, \bar{\mathbf{u}}) \\
&= L(\bar{\mathbf{x}}, \bar{\mathbf{u}}) + \mathbf{x}^T L_{\mathbf{x}}(\bar{\mathbf{x}}, \bar{\mathbf{u}}) - \bar{\mathbf{x}}^T L_{\mathbf{x}}(\bar{\mathbf{x}}, \bar{\mathbf{u}}) \\
&= L(\bar{\mathbf{x}}, \bar{\mathbf{u}}) + \mathbf{x}^T L_{\mathbf{x}}(\bar{\mathbf{x}}, \bar{\mathbf{u}}) \geq L(\bar{\mathbf{x}}, \bar{\mathbf{u}}) \ \forall \ \mathbf{x} \geq \mathbf{0},
\end{aligned}
$$

since from relation (4.2), we have $\bar{\mathbf{x}}^T L_{\mathbf{x}}(\bar{\mathbf{x}}, \bar{\mathbf{u}}) = 0$, and from relation (4.1) we have $L_{\mathbf{x}}(\bar{\mathbf{x}}, \bar{\mathbf{u}}) \geq \mathbf{0}$. This is nothing but the second part of the saddle point criterion. Furthermore, $L(\bar{\mathbf{x}}, \mathbf{u})$ is a linear function of $\mathbf{u}$, so that we obtain

$$
\begin{aligned}
L(\bar{\mathbf{x}}, \mathbf{u}) - L(\bar{\mathbf{x}}, \bar{\mathbf{u}}) &= f(\bar{\mathbf{x}}) + \mathbf{u}\mathbf{g}(\bar{\mathbf{x}}) - [f(\bar{\mathbf{x}}) + \bar{\mathbf{u}}\mathbf{g}(\bar{\mathbf{x}})] = (\mathbf{u} - \bar{\mathbf{u}})\mathbf{g}(\bar{\mathbf{x}}) \\
&= (\mathbf{u} - \bar{\mathbf{u}})L_{\mathbf{u}}(\bar{\mathbf{x}}, \bar{\mathbf{u}}).
\end{aligned}
$$

Hence

$$
\begin{aligned}
L(\bar{\mathbf{x}}, \mathbf{u}) &= L(\bar{\mathbf{x}}, \bar{\mathbf{u}}) + (\mathbf{u} - \bar{\mathbf{u}})L_{\mathbf{u}}(\bar{\mathbf{x}}, \bar{\mathbf{u}}) \\
&= L(\bar{\mathbf{x}}, \bar{\mathbf{u}}) + \mathbf{u}L_{\mathbf{u}}(\bar{\mathbf{x}}, \bar{\mathbf{u}}) - \bar{\mathbf{u}}L_{\mathbf{u}}(\bar{\mathbf{x}}, \bar{\mathbf{u}}) \\
&= L(\bar{\mathbf{x}}, \bar{\mathbf{u}}) + \mathbf{u}L_{\mathbf{u}}(\bar{\mathbf{x}}, \bar{\mathbf{u}}) \leq L(\bar{\mathbf{x}}, \bar{\mathbf{u}}) \ \forall \ \mathbf{u} \geq \mathbf{0},
\end{aligned}
$$

the various steps being justified by relations (4.5) and (4.4) respectively. Since this is simply the first part of the saddle point criterion, the proof is now complete. $\square$

The previous two lemmas can be stated together as

Theorem 4.15 Assume that $f(\mathbf{x})$ and $g_i(\mathbf{x})$, $i = 1, \ldots, m$ are all convex and differentiable. Then $(\bar{\mathbf{x}}, \bar{\mathbf{u}})$ is a point that satisfies the Karush-Kuhn-Tucker conditions of problem P if and only if $(\bar{\mathbf{x}}, \bar{\mathbf{u}})$ is a saddle point of $L(\mathbf{x}, \mathbf{u})$.

Note that no constraint qualification is required in Theorem 4.15. In order to visualize the three main results in this chapter, we provide a diagram in Fig. 4.5, in which, for simplicity, the assumptions have been omitted.

Before proceeding, we emphasize that the development so far has not been using the weakest possible assumptions. For instance, some (but not all) of the implications in Fig. 4.5 hold true even in the absence of convexity; similarly others (but not all) hold true in the absence of a constraint qualification, and/or differentiability.

We are now going to consider the special case in which the objective function is continuously differentiable, convex, and not necessarily linear, whereas the constraints are assumed to be linear, i.e., $\mathbf{g}(\mathbf{x}) = \mathbf{A}\mathbf{x} - \mathbf{b}$, so that problem P becomes

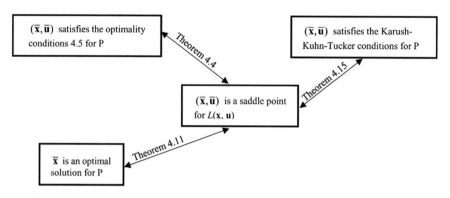

Fig. 4.5 Relations between optimality, KKT, and saddle point conditions

$$P' : \operatorname{Min} z = f(\mathbf{x})$$
$$\text{s.t. } \mathbf{Ax} - \mathbf{b} \le \mathbf{0}$$
$$\mathbf{x} \ge \mathbf{0}.$$

Then $L(\mathbf{x}, \mathbf{u}) = f(\mathbf{x}) + \mathbf{u}g(\mathbf{x}) = f(\mathbf{x}) + \mathbf{u}(\mathbf{Ax} - \mathbf{b})$.

$$\text{Now } \nabla_{\mathbf{x}} g_i(\mathbf{x}) = \begin{bmatrix} \dfrac{\partial g_i(\mathbf{x})}{\partial x_1} \\ \vdots \\ \dfrac{\partial g_i(\mathbf{x})}{\partial x_n} \end{bmatrix} = a_{i\bullet}^T, i = 1, \dots, m.$$

By introducing the n-dimensional column vector $\bar{\mathbf{v}} := L_{\mathbf{x}}(\bar{\mathbf{x}}, \bar{\mathbf{u}})$ and defining the index sets $S := \{i : g_i(\bar{\mathbf{x}}) = 0\}$ and $R := \{j : \bar{x}_j = 0\}$, the Karush-Kuhn-Tucker condition $L_{\mathbf{x}}(\bar{\mathbf{x}}, \bar{\mathbf{u}}) \ge \mathbf{0}$ reads $\bar{\mathbf{v}} \ge \mathbf{0}$, and the condition $\bar{\mathbf{x}}^T L_{\mathbf{x}}(\bar{\mathbf{x}}, \bar{\mathbf{u}}) = 0$ implies $\bar{v}_j = 0 \ \forall \ j \notin R$. The two conditions $\bar{\mathbf{x}} \ge \mathbf{0}$ and $\mathbf{0} \ge L_{\mathbf{u}}(\mathbf{x}, \mathbf{u}) = g(\mathbf{x}) = \mathbf{Ax} - \mathbf{b}$ only serve to establish the feasibility of $\bar{\mathbf{x}}$. In addition, the condition $\bar{\mathbf{u}}L_{\mathbf{u}}(\bar{\mathbf{x}}, \bar{\mathbf{u}}) = 0$ implies that $\bar{u}_i = 0 \ \forall \ i \notin S$ and the last condition $\bar{\mathbf{u}} \ge \mathbf{0}$ gives $\bar{u}_i \ge 0 \ \forall \ i \in S$. Furthermore, a relation between $\bar{\mathbf{v}}$ and $\bar{\mathbf{x}}$ can be established by writing the definition equality $L_{\mathbf{x}}(\bar{\mathbf{x}}, \bar{\mathbf{u}}) = \bar{\mathbf{v}}$ as follows, where $\mathbf{e}_j$ denotes the j-th n-dimensional unit row vector:

$$\nabla_{\mathbf{x}} f(\mathbf{x}) + \sum_{i \in S} \bar{u}_i \nabla_{\mathbf{x}} g_i(\bar{\mathbf{x}}) + \sum_{i \notin S} \bar{u}_i \nabla_{\mathbf{x}} g_i(\bar{\mathbf{x}}) = \sum_{j \in R} \bar{v}_j \mathbf{e}_j^T + \sum_{j \notin R} \bar{v}_j \mathbf{e}_j^T.$$

Since $\bar{u}_i = 0 \ \forall i \notin S$ and $\bar{v}_j = 0 \ \forall j \notin R$, the above relation can be simplified to

$$-\nabla_{\mathbf{x}} f(\bar{\mathbf{x}}) = \sum_{i \in S} \bar{u}_i \nabla_{\mathbf{x}} g_i(\bar{\mathbf{x}}) - \sum_{j \in R} \bar{v}_j \mathbf{e}_j^T.$$

As a consequence, the Karush-Kuhn-Tucker conditions for problem P' can be stated in their equivalent form

$$\bar{v}_j \geq 0 \; \forall j \in R \tag{4.7}$$

$$\bar{v}_j = 0 \; \forall j \notin R \tag{4.8}$$

$$\bar{u}_i = 0 \; \forall i \notin S \tag{4.9}$$

$$\bar{u}_i \geq 0 \; \forall i \in S \tag{4.10}$$

$$-\nabla_{\mathbf{x}} f(\bar{\mathbf{x}}) = \sum_{i \in S} \bar{u}_i \nabla_{\mathbf{x}} g_i(\bar{\mathbf{x}}) - \sum_{j \in R} \bar{v}_j e_j^T \tag{4.11}$$

where $\bar{\mathbf{x}}$ denotes some feasible solution for problem P' (i.e., relations (4.3) and (4.4) must also hold).

It has already been demonstrated in the general convex differentiable case that if $(\bar{\mathbf{x}}, \bar{\mathbf{u}})$ satisfies the Karush-Kuhn-Tucker conditions, then $\bar{\mathbf{x}}$ is an optimal solution for P (Lemmas 4.3 and 4.14). For the converse, however, Slater's constraint qualification had to be satisfied. In the discussion below we will show that such an additional requirement is not necessary for problem P', where all constraints are linear. For this purpose, we introduce an n-dimensional column vector $\mathbf{t}$ of variables $t_1, t_2, \ldots, t_n$ and define the following system of relations I. Recall that $\mathbf{g}(\mathbf{x})$ is now $A\mathbf{x} - \mathbf{b}$, so that $\mathbf{g}(\mathbf{x}) \leq \mathbf{0}$ is $A\mathbf{x} - \mathbf{b} \leq \mathbf{0}$.

$$I: \quad \sum_{j=1}^{n} t_j \left(\nabla_{\mathbf{x}} f(\bar{\mathbf{x}}) \right)_j < 0$$

$$\sum_{j=1}^{n} a_{ij} t_j \leq 0 \; \forall \; i \in S$$

$$t_j \geq 0 \quad \forall \; j \in R.$$

We can then formulate

Lemma 4.16 If $\bar{\mathbf{x}}$ is an optimal solution for P', then system I has no solution $\mathbf{t}$.

Proof We prove this lemma by contradiction. Assume that $\bar{\mathbf{x}}$ is optimal for P' and that a solution $\mathbf{t}$ to I exists. Define a new solution $\tilde{x}_j = \bar{x}_j + t_j \lambda$, $j = 1, \ldots, n$, where $\lambda > 0$ is a constant that is sufficiently small to satisfy $g_i(\tilde{\mathbf{x}}) = \sum_{j=1}^{n} a_{ij} \tilde{x}_j - b_i$ $\leq 0 \; \forall \; i \notin S$ and $\tilde{x}_j \geq 0 \; \forall j \notin R$. Note that since $g_i(\bar{\mathbf{x}}) < 0 \; \forall \; i \notin S$ and $\bar{x}_j > 0 \; \forall j \notin R$, such a λ always exists. Furthermore, we choose λ small enough to ensure that $\mathbf{t}^T \nabla_{\mathbf{x}} f(\mathbf{x}) < 0$ on the entire line segment between $\bar{\mathbf{x}}$ and $\tilde{\mathbf{x}}$. This is possible because of the continuity of $\nabla_{\mathbf{x}} f(\mathbf{x})$ at the point $\bar{\mathbf{x}}$, which, we recall, satisfies system I, specifically $\mathbf{t}^T \nabla_{\mathbf{x}} f(\bar{\mathbf{x}}) < 0$. Since $\sum_{j=1}^{n} a_{ij} \bar{x}_j - b_i \leq 0$ and $\sum_{j=1}^{n} a_{ij} t_j \leq 0 \; \forall \; i \in S$, it follows that

$$g_i(\tilde{\mathbf{x}}) = \sum_{j=1}^{n} a_{ij}\tilde{x}_j - b_i = \sum_{j=1}^{n} a_{ij}\bar{x}_j - b_i + \sum_{j=1}^{n} a_{ij}t_j\lambda \leq 0 \ \forall \ i \in S,$$

which guarantees that $\mathbf{g}(\tilde{\mathbf{x}}) \leq \mathbf{0}$ is satisfied. Now $\bar{x}_j = 0$ and $t_j \geq 0 \ \forall j \in R$, so that $\tilde{x}_j \geq 0 \ \forall j \in R$. But we already had $\tilde{x}_j \geq 0 \ \forall j \notin R$, so that $\tilde{\mathbf{x}} \geq \mathbf{0}$ and conclude that $\tilde{\mathbf{x}}$ is a feasible solution for problem P'. Consider now the objective function $f(\mathbf{x})$, which, by assumption, is continuously differentiable, so that we can apply Theorem 1.15 (linearization). Setting $\lambda := \alpha$, $\mathbf{x} := \tilde{\mathbf{x}}$ and $\tilde{\mathbf{x}} := \bar{\mathbf{x}}$ in the theorem, we can write

$$\begin{aligned} f(\tilde{\mathbf{x}}) &= f(\bar{\mathbf{x}}) + (\tilde{\mathbf{x}} - \bar{\mathbf{x}})^T \nabla_{\mathbf{x}} f(\alpha\tilde{\mathbf{x}} + (1-\alpha)\bar{\mathbf{x}}) \\ &= f(\bar{\mathbf{x}}) + (\bar{\mathbf{x}} + \lambda\mathbf{t} - \bar{\mathbf{x}})^T \nabla_{\mathbf{x}} f(\alpha\tilde{\mathbf{x}} + (1-\alpha)\bar{\mathbf{x}}) \\ &= f(\bar{\mathbf{x}}) + \lambda\mathbf{t}^T \nabla_{\mathbf{x}} f(\alpha\tilde{\mathbf{x}} + (1-\alpha)\bar{\mathbf{x}}). \end{aligned}$$

According to our assumption, $\mathbf{t}^T \nabla_{\mathbf{x}} f(\alpha\tilde{\mathbf{x}} + (1-\alpha)\bar{\mathbf{x}}) < 0$ and hence $\lambda\mathbf{t}^T \nabla_{\mathbf{x}} f(\alpha\tilde{\mathbf{x}} + (1-\alpha)\bar{\mathbf{x}}) < 0$, implying that $f(\tilde{\mathbf{x}}) < f(\bar{\mathbf{x}})$, which violates the assumption that $\bar{\mathbf{x}}$ is an optimal solution for problem P. Thus, if $\bar{\mathbf{x}}$ is optimal, system I cannot have a solution $\mathbf{t}$. $\square$

Recalling that $\mathbf{e}_j$ denotes the j-th n-dimensional unit row vector and defining $\mathbf{d} := -\nabla_{\mathbf{x}} f(\bar{\mathbf{x}})$ and the matrix

$$\tilde{\mathbf{A}} = \begin{bmatrix} \mathbf{a}_{i \bullet} \\ --- \\ -\mathbf{e}_j \end{bmatrix} \quad \begin{matrix} i \in S \\ \\ j \in R \end{matrix}$$

of dimension $[(|S| + |R|) \times n]$, then system I can be rewritten in the equivalent form

$$\begin{aligned} I' : \mathbf{t}^T\mathbf{d} &> 0 \\ \tilde{\mathbf{A}}\mathbf{t} &\leq \mathbf{0}, \end{aligned}$$

which, according to Lemma 4.16 has no solution for an optimal $\bar{\mathbf{x}}$. Now we will invoke a classical result from the duality theory of linear programming called Farkas' Lemma (see Lemma B.14 in Appendix B), which, in the usual notation, states that either the system $\mathbf{A}\mathbf{x} = \mathbf{b} \neq \mathbf{0}$ has a solution $\mathbf{x} \geq \mathbf{0}$, or the system $\mathbf{u}\mathbf{A} \geq \mathbf{0}$, $\mathbf{u}\mathbf{b} < \mathbf{0}$, has a sign-free solution $\mathbf{u}$, but never both.

In our case, Farkas' lemma leads to the result that the system

$$II : \tilde{\mathbf{A}}^T\mathbf{w} = \mathbf{d}$$

has a solution $\bar{\mathbf{w}} = \begin{bmatrix} \bar{\mathbf{u}}^T \\ \overline{\tilde{\mathbf{w}}^T} \end{bmatrix} \geq \mathbf{0}$. Using the above definitions of $\mathbf{d}$ and $\tilde{\mathbf{A}}$, we can write

$$\sum_{i \in S} \bar{u}_i \mathbf{a}_{i\bullet}^T + \sum_{j \in R} -\bar{v}_j \mathbf{e}_j^T = -\nabla_{\mathbf{x}} f(\bar{\mathbf{x}})$$

with $\bar{u}_i \geq 0 \; \forall i \in S$ and $\bar{v}_j \geq 0 \; \forall j \in R$. Since $\nabla_{\mathbf{x}} g_i(\mathbf{x}) = \mathbf{a}_{i\bullet}^T$, the above three relations with $\bar{u}_i = 0 \; \forall i \notin S$ and $\bar{v}_j = 0 \; \forall j \notin R$ are identical to the conditions (4.7)–(4.11). This finally proves

Lemma 4.17 If $\bar{\mathbf{x}}$ is an optimal solution for problem P′, then there exists a vector $\bar{\mathbf{u}}$, such that the point $(\bar{\mathbf{x}}, \bar{\mathbf{u}})$ satisfies the Karush-Kuhn-Tucker conditions.

Since we have already shown that the existence of a saddle point $(\bar{\mathbf{x}}, \bar{\mathbf{u}})$ for $L(\mathbf{x}, \mathbf{u})$ is sufficient for $\bar{\mathbf{x}}$ to be an optimal solution for P′, we can now claim the following equivalences. Note that no constraint qualification is needed.

Theorem 4.18 Assume that in problem P′ the objective function $f(\mathbf{x})$ is convex and differentiable. Then the following three statements are equivalent:

  (i) $(\bar{\mathbf{x}}, \bar{\mathbf{u}})$ is a saddle point of $L(\mathbf{x}, \mathbf{u})$,
 (ii) $(\bar{\mathbf{x}}, \bar{\mathbf{u}})$ satisfies the Karush-Kuhn-Tucker conditions for problem P′, and
(iii) $\bar{\mathbf{x}}$ is an optimal solution of problem P′.

Theorem 4.18 is important when linearly constrained problems with equality constraints are considered, as in those cases Slater's constraint qualification is not satisfied.

Sometimes it may be desirable to obtain some knowledge about the value of the objective function without having to solve the entire problem. In such a case one could search for lower and upper bounds on $f(\mathbf{x})$. Specifically, let $\bar{\mathbf{x}}$ be an optimal solution of problem P and let $\tilde{\mathbf{x}}$ denote any feasible solution to the problem, i.e., $\mathbf{g}(\tilde{\mathbf{x}}) \leq \mathbf{0}$ and $\tilde{\mathbf{x}} \geq \mathbf{0}$. Because of the optimality of the point $\bar{\mathbf{x}}$ we know that $f(\bar{\mathbf{x}}) \leq f(\tilde{\mathbf{x}})$, which provides an upper bound for $f(\tilde{\mathbf{x}})$. Next, any fixed $\mathbf{u}^* \geq \mathbf{0}$ is selected and the problem

$$P_L: \operatorname{Min} z = f(\mathbf{x}) + \mathbf{u}^* \mathbf{g}(\mathbf{x})$$
$$\text{s.t. } \mathbf{x} \geq \mathbf{0}$$

called the *Lagrangean relaxation* of problem P, is solved. If $\hat{\mathbf{x}}$ denotes an optimal solution to the relaxed problem P_L, then

$$f(\hat{\mathbf{x}}) + \mathbf{u}^* \mathbf{g}(\hat{\mathbf{x}}) \leq f(\bar{\mathbf{x}}) + \mathbf{u}^* \mathbf{g}(\bar{\mathbf{x}}).$$

Since $\mathbf{u}^* \geq \mathbf{0}$ and $\mathbf{g}(\bar{\mathbf{x}}) \leq \mathbf{0}$, we can conclude that

$$f(\bar{\mathbf{x}}) + \mathbf{u}^* \mathbf{g}(\bar{\mathbf{x}}) \leq f(\bar{\mathbf{x}})$$

and thus

$$f(\hat{\mathbf{x}}) + \mathbf{u}^*\mathbf{g}(\hat{\mathbf{x}}) \leq f(\bar{\mathbf{x}}),$$

so that we finally obtain $f(\hat{\mathbf{x}}) + \mathbf{u}^*g(\hat{\mathbf{x}}) \leq f(\bar{\mathbf{x}}) \leq f(\tilde{\mathbf{x}})$. This expression provides upper and lower bounds for the optimal value $f(\bar{\mathbf{x}})$. As a numerical illustration, consider the following

Example

$$\text{Min} z = f(x_1, x_2) = x_1^2 - 4x_1 + x_2^2 - 6x_2$$

$$\text{s.t.} \quad x_1 + x_2 \leq 3$$
$$-2x_1 + x_2 \leq 2$$
$$x_1, x_2 \geq 0.$$

Let the feasible solution $\tilde{\mathbf{x}} = [1, 1]^T$ be known; the associated value of the objective function is given by $f(\tilde{\mathbf{x}}) = -8$. Now $\mathbf{u}^* = [2, 2]$ is arbitrarily selected. Then the corresponding Lagrangian relaxation P_1 of problem P reads as

$$P_L: \quad \text{Min} z = x_1^2 - 6x_1 + x_2^2 - 2x_2 - 10$$
$$\text{s.t.} \ x_1, x_2 \geq 0.$$

By setting the first partial derivatives equal to zero, we obtain the linear equations $2x_1 - 6 = 0$ and $2x_2 - 2 = 0$, whose solution is $\hat{\mathbf{x}} = [3, 1]^T$. Hence $f(\hat{\mathbf{x}}) + \mathbf{u}^*\mathbf{g}(\hat{\mathbf{x}})$ $= -20$ and the bounds are established as $-20 \leq f(\bar{\mathbf{x}}) \leq -8$. It is quite obvious that the quality of the upper and lower bounds depends strictly on the choice of $\tilde{\mathbf{x}}$ and $\mathbf{u}^*$. Since the selection of $\tilde{\mathbf{x}}$ and $\mathbf{u}^*$ in this example was not a very fortunate one, the margin between the lower and the upper bound is quite substantial.

Let us digress for a moment and consider how the Lagrangean multiplier vector $\mathbf{u}^*$ might be modified to obtain a tighter lower bound. In some sense, the values of the Lagrangean multipliers u_i can be interpreted as penalties for violating the constraints $g_i(\mathbf{x}) \leq 0$. To see this, assume that $g_i(\hat{\mathbf{x}}) > 0$ for some i, so that the i-th constraint is violated. The term $u_i^* g_i(\hat{\mathbf{x}})$ in the Lagrangean function is then positive, and bigger with a bigger u_i^*, which is disadvantageous, since we are minimizing the Lagrangean function. Intuitively, we may therefore find a better multiplier vector by increasing the value u_i^* for violated constraints and reduce it for constraints that are satisfied with some slack. We will pursue this idea in Chap. 5 when we discuss the so-called *Everett Method*. We should also point out that Lagrangean relaxation is also used in the context of nonlinear duality, which we will cover in Sect. 4.3.

We now extend the scope of our analysis by investigating how certain changes in the structure of problem P will affect the Karush-Kuhn-Tucker conditions. Three particular situations will be examined in detail.

Special Case 1 Suppose that all given variables $\mathbf{x}$ are unrestricted in sign, i.e., $\mathbf{x} \in \mathbb{R}^n$. Then one can replace $\mathbf{x}$ by the difference between two nonnegative vectors

$\mathbf{x}^+$ and $\mathbf{x}^-$, i.e., $\mathbf{x} := \mathbf{x}^+ - \mathbf{x}^-$. For this new equivalent problem the Karush-Kuhn-Tucker condition (4.1) becomes

$$L_{\mathbf{x}^+}\left(\bar{\mathbf{x}}^+ - \bar{\mathbf{x}}^-, \bar{\mathbf{u}}\right) \geq \mathbf{0} \ \text{ and } \ L_{\mathbf{x}^-}\left(\bar{\mathbf{x}}^+ - \bar{\mathbf{x}}^-, \bar{\mathbf{u}}\right) \geq \mathbf{0}.$$

Since

$$
\begin{aligned}
0 \leq L_{\mathbf{x}^+}\left(\bar{\mathbf{x}}^+ - \bar{\mathbf{x}}^-, \bar{\mathbf{u}}\right) &= \frac{\partial L\left(\bar{\mathbf{x}}^+ - \bar{\mathbf{x}}^-, \bar{\mathbf{u}}\right)}{\partial \mathbf{x}^+} = \frac{\partial L\left(\bar{\mathbf{x}}^+ - \bar{\mathbf{x}}^-, \bar{\mathbf{u}}\right)}{\partial \mathbf{x}} \frac{\partial \mathbf{x}}{\partial \mathbf{x}^+} \\
&= \frac{\partial L\left(\bar{\mathbf{x}}^+ - \bar{\mathbf{x}}^-, \bar{\mathbf{u}}\right)}{\partial \mathbf{x}} = L_{\mathbf{x}^+}\left(\bar{\mathbf{x}}^+ - \bar{\mathbf{x}}^-, \bar{\mathbf{u}}\right)
\end{aligned}
$$

and also

$$
\begin{aligned}
0 \leq L_{\mathbf{x}^-}\left(\bar{\mathbf{x}}^+ - \bar{\mathbf{x}}^-, \bar{\mathbf{u}}\right) &= \frac{\partial L\left(\bar{\mathbf{x}}^+ - \bar{\mathbf{x}}^-, \bar{\mathbf{u}}\right)}{\partial \mathbf{x}^-} = \frac{\partial L\left(\bar{\mathbf{x}}^+ - \bar{\mathbf{x}}^-, \bar{\mathbf{u}}\right)}{\partial \mathbf{x}} \frac{\partial \mathbf{x}}{\partial \mathbf{x}^-} \\
&= -\frac{\partial L\left(\bar{\mathbf{x}}^+ - \bar{\mathbf{x}}^-, \bar{\mathbf{u}}\right)}{\partial \mathbf{x}} = -L_{\mathbf{x}^+}\left(\bar{\mathbf{x}}^+ - \bar{\mathbf{x}}^-, \bar{\mathbf{u}}\right)
\end{aligned}
$$

so that

$$L_{\mathbf{x}}\left(\bar{\mathbf{x}}, \bar{\mathbf{u}}\right) = \mathbf{0}. \tag{4.12}$$

Now $\bar{\mathbf{x}}^T L_{\mathbf{x}}\left(\bar{\mathbf{x}}, \bar{\mathbf{u}}\right) = 0$ as long as (4.12) is fulfilled and hence the Karush-Kuhn-Tucker conditions (4.1)–(4.3) are simply replaced by (4.12) when $\mathbf{x} \in \mathbb{R}^n$.

Special Case 2 Assume now that $\mathbf{g}(\mathbf{x}) = \mathbf{0}$ is a set of linear equalities in problem P. Replacing $\mathbf{g}(\mathbf{x}) = \mathbf{0}$ by the two sets of linear inequalities $\mathbf{g}(\mathbf{x}) \leq \mathbf{0}$ and $-\mathbf{g}(\mathbf{x}) \leq \mathbf{0}$ and assigning vectors of nonnegative Langrangean multipliers $\mathbf{u}^+$ and $\mathbf{u}^-$ to the two sets, respectively, the Karush-Kuhn-Tucker condition (4.4) reads as $L_{\mathbf{u}^+}\left(\bar{\mathbf{x}}, \bar{\mathbf{u}}\right) \leq \mathbf{0}$ and $L_{\mathbf{u}^-}\left(\bar{\mathbf{x}}, \bar{\mathbf{u}}\right) \leq \mathbf{0}$. Since $L_{\mathbf{u}^+}\left(\bar{\mathbf{x}}, \bar{\mathbf{u}}\right) = \mathbf{g}(\bar{\mathbf{x}})$ and $L_{\mathbf{u}^-}\left(\bar{\mathbf{x}}, \bar{\mathbf{u}}\right) = -\mathbf{g}(\bar{\mathbf{x}})$, it follows that $\mathbf{g}(\bar{\mathbf{x}}) = \mathbf{0}$. By defining $\mathbf{u} = \mathbf{u}^+ - \mathbf{u}^-$, so that $\mathbf{u} \in \mathbb{R}^m$, we obtain $f(\mathbf{x}) + \mathbf{u}^+ \mathbf{g}(\mathbf{x}) + \mathbf{u}^-(-\mathbf{g}(\mathbf{x})) = f(\mathbf{x}) + (\mathbf{u}^+ - \mathbf{u}^-)\mathbf{g}(\mathbf{x})$ and the Karush-Kuhn-Tucker conditions (4.4)–(4.6) reduce to

$$L_{\mathbf{u}}\left(\bar{\mathbf{x}}, \bar{\mathbf{u}}\right) = \mathbf{0}. \tag{4.13}$$

Special Case 3 All variables are assumed to be unrestricted in sign and all constraints are given as linear equalities. This means that special cases 1 and 2 apply simultaneously. The implication is that the six Karush-Kuhn-Tucker conditions reduce to relations (4.12) and (4.13). This case concerns the traditional Lagrangean multiplier approach from classical differential calculus.

The results derived so far are summarized in Table 4.1.

Table 4.1 Karush-Kuhn-Tucker conditions for various cases

Problem	Karush-Kuhn-Tucker conditions
$f(\mathbf{x})$ convex $g_i(\mathbf{x}) \leq 0$, all convex $\mathbf{x} \geq \mathbf{0}$ (Standard case)	$L_{\mathbf{x}}(\bar{\mathbf{x}}, \bar{\mathbf{u}}) \geq \mathbf{0}, \ \bar{\mathbf{x}}^T L_{\mathbf{x}}(\bar{\mathbf{x}}, \bar{\mathbf{u}}) = 0, \ \bar{\mathbf{x}} \geq \mathbf{0}$ $L_{\mathbf{u}}(\bar{\mathbf{x}}, \bar{\mathbf{u}}) \leq \mathbf{0}, \ \bar{\mathbf{u}} L_{\mathbf{u}}(\bar{\mathbf{x}}, \bar{\mathbf{u}}) = 0, \ \bar{\mathbf{u}} \geq \mathbf{0}$
$f(\mathbf{x})$ convex $g_i(\mathbf{x}) \leq 0$, all convex $\mathbf{x} \in \mathbb{R}^n$ (Special case 1)	$L_{\mathbf{x}}(\bar{\mathbf{x}}, \bar{\mathbf{u}}) = \mathbf{0}$ $L_{\mathbf{u}}(\bar{\mathbf{x}}, \bar{\mathbf{u}}) \leq \mathbf{0}, \ \bar{\mathbf{u}} L_{\mathbf{u}}(\bar{\mathbf{x}}, \bar{\mathbf{u}}) = 0, \ \bar{\mathbf{u}} \geq \mathbf{0}$
$f(\mathbf{x})$ convex $g_i(\mathbf{x}) = 0$, all linear $\mathbf{x} \geq \mathbf{0}$ (Special case 2)	$L_{\mathbf{x}}(\bar{\mathbf{x}}, \bar{\mathbf{u}}) \geq \mathbf{0}, \ \bar{\mathbf{x}}^T L_{\mathbf{x}}(\bar{\mathbf{x}}, \bar{\mathbf{u}}) = 0, \ \bar{\mathbf{x}} \geq \mathbf{0}$ $L_{\mathbf{u}}(\bar{\mathbf{x}}, \bar{\mathbf{u}}) = \mathbf{0}$
$f(\mathbf{x})$ convex $g_i(\mathbf{x}) = 0$, all linear $\mathbf{x} \in \mathbb{R}^n$ (Special case 3)	$L_{\mathbf{x}}(\bar{\mathbf{x}}, \bar{\mathbf{u}}) = \mathbf{0}$ $L_{\mathbf{u}}(\bar{\mathbf{x}}, \bar{\mathbf{u}}) = \mathbf{0}$

A problem of the type described in special case 3 above will now be completely solved in order to illustrate the particular form taken by the Karush-Kuhn-Tucker conditions. A quadratic objective function has been chosen for the simple reason that in this case all the partial derivatives are linear and therefore the problem is easier to solve.

Example Consider the nonlinear programming problem

$$P: \operatorname{Min} z = x_1^2 + 5x_2^2 + 3x_3$$
$$\text{s.t. } x_1 + 4x_2 = 12$$
$$3x_1 + x_2 + x_3 = 9$$
$$x_1, x_2, x_3 \in \mathbb{R}.$$

The corresponding Lagrangean function is

$$L(\mathbf{x}, \mathbf{u}) = x_1^2 + 5x_2^2 + 3x_3 + u_1 x_1 + 4u_1 x_2 - 12u_1 + 3u_2 x_1 + u_2 x_2 + u_2 x_3 - 9u_2,$$

and the Karush-Kuhn-Tucker conditions require that $L_{\mathbf{x}}(\bar{\mathbf{x}}, \bar{\mathbf{u}}) = \mathbf{0}$ and $L_{\mathbf{u}}(\bar{\mathbf{x}}, \bar{\mathbf{u}}) = \mathbf{0}$. Specifically,

$$\frac{\partial L(\mathbf{x}, \mathbf{u})}{\partial x_1} = 2x_1 + u_1 + 3u_2 = 0, \quad \frac{\partial L(\mathbf{x}, \mathbf{u})}{\partial x_2} = 10x_2 + 4u_1 + u_2 = 0,$$

$$\frac{\partial L(\mathbf{x}, \mathbf{u})}{\partial x_3} = 3 + u_2 = 0 \quad \text{and} \quad \frac{\partial L(\mathbf{x}, \mathbf{u})}{\partial u_1} = x_1 + 4x_2 - 12 = 0, \quad \text{as well as}$$

$$\frac{\partial L(\mathbf{x}, \mathbf{u})}{\partial u_2} = 3x_1 + x_2 + x_3 - 9 = 0.$$

The unique solution to this system of simultaneous linear equations is $\bar{x}_1 = 6$, $\bar{x}_2 = 1\frac{1}{2}$, $\bar{x}_3 = -10\frac{1}{2}$, $\bar{u}_1 = -3$, and $\bar{u}_2 = -3$. The optimal value of the objective function is $f(\bar{\mathbf{x}}) = 15\frac{3}{4}$. Selecting another feasible solution, e.g., $\tilde{\mathbf{x}} = [0, 3, 6]^T$, we find a value of the objective function of $f(\tilde{\mathbf{x}}) = 63$, indicating that the above solution $\tilde{\mathbf{x}}$ is a minimal and not a maximal point.

For more complicated problems it may be computationally difficult to derive an optimal solution directly from the Karush-Kuhn-Tucker conditions. Nevertheless, these conditions always give an idea as to the identity of an optimal solution and they also allow for the verification as to whether an available solution actually satisfies optimality.

Due to the fact that in Special Case 3 all the variables are unrestricted in sign and all the constraints are equalities, variable substitution can be used to transform the given constrained problem into an unconstrained problem, which could then be solved by any suitable technique for such models. In particular, the problem above could have been solved by expressing variables x_2 and x_3 in terms of the x_1 by substitution. In this way the objective function can be rewritten as a function of x_1 only. Pursuing this idea, we first find $x_2 = 3 - \frac{1}{4}x_1$ and $x_3 = 6 - \frac{11}{4}x_1$, which, substituted into the objective function, gives $z = \frac{31}{16}x_1^2 - \frac{63}{4}x_1 + 63$. Differentiation and setting the derivative equal to zero yields the unique solution $x_1 = 6$, which is a minimal point for z. Substituting, we obtain $x_2 = 1\frac{1}{2}$ and $x_3 = -10\frac{1}{2}$, in agreement with the solution obtained by using the Karush-Kuhn-Tucker conditions.

The Lagrangean approach, however, also yields values for the multiplier vector $\mathbf{u}$ which can be used to carry out postoptimality analyses. For a general problem of the type in special case 3, let $\tilde{\mathbf{x}}$ be any feasible solution, so that $\mathbf{A}\tilde{\mathbf{x}} = \mathbf{b}$ or $\mathbf{A}\tilde{\mathbf{x}} - \mathbf{b} = \mathbf{0}$. Then the Lagrangean function can be reduced to

$$L(\tilde{\mathbf{x}}, \mathbf{u}) = f(\tilde{\mathbf{x}}) + \mathbf{u}(\mathbf{A}\tilde{\mathbf{x}} - \mathbf{b}) = f(\tilde{\mathbf{x}}),$$

showing that its value and that of the objective function coincide for any value of $\mathbf{u}$ and every feasible solution $\tilde{\mathbf{x}}$. Consider now an isolated change of one of the right-hand side constants, say b_i. Since $\frac{\partial L(\mathbf{x}, \mathbf{u})}{\partial b_i} = -u_i$, we have the following two alternatives.

- $u_i > 0$. This signifies that a small increase (decrease) of b_i by Δb_i results in a decrease (increase) of $L(\mathbf{x}, \mathbf{u})$ and therefore of $f(\mathbf{x})$ by approximately $+u_i\Delta b_i$.
- $u_i < 0$. This indicates that a small increase (decrease) in the value of b_i by Δb_i results in an increase (decrease) of $L(\mathbf{x}, \mathbf{u})$ and hence of $f(\mathbf{x})$ by approximately $-u_i\Delta b_i$.

As a numerical illustration consider the problem above where $\bar{u}_1 = -3$. If for example $b_1 = 12$ is increased by 0.1 to 12.1, then the value of the objective function would increase by approximately 0.3 to 16.05.

In the following we will present two theorems due to Everett (1963) and then, in Chap. 5, an associated algorithm which deals with a Lagrangean multipler method for solving problems where the task is that of allocating scarce resources among an existing variety of possible alternatives. The problem P is assumed to be of the form

$$P: \operatorname{Min} z = f(\mathbf{x})$$
$$\text{s.t. } \mathbf{g}(\mathbf{x}) \le \mathbf{0}$$
$$\mathbf{x} \in \mathbb{R}^n.$$

Consider now two given vectors of nonnegative Lagrangean multipliers denoted by $\tilde{\mathbf{u}}$ and $\hat{\mathbf{u}}$, while $\tilde{\mathbf{x}}$ and $\hat{\mathbf{x}}$ represent the resulting optimal solutions for the respective Lagrangean relaxations

$$\tilde{P}_L: \operatorname*{Min}_{\mathbf{x} \in \mathbb{R}^n} \tilde{z} = f(\mathbf{x}) + \tilde{\mathbf{u}}\mathbf{g}(\mathbf{x}) \quad \text{and} \quad \hat{P}_L: \operatorname*{Min}_{\mathbf{x} \in \mathbb{R}^n} \hat{z} = f(\mathbf{x}) + \hat{\mathbf{u}}\mathbf{g}(\mathbf{x}).$$

Note that $\tilde{\mathbf{x}}$ and $\hat{\mathbf{x}}$ depend on $\tilde{\mathbf{u}}$ and $\hat{\mathbf{u}}$, respectively, i.e., $\tilde{\mathbf{x}} = \tilde{\mathbf{x}}(\tilde{\mathbf{u}})$ and $\hat{\mathbf{x}} = \hat{\mathbf{x}}(\hat{\mathbf{u}})$. Furthermore, neither $\tilde{\mathbf{x}}$ nor $\hat{\mathbf{x}}$ need be feasible for problem P.

Lemma 4.19 If there exists an subscript k, such that

$$g_k(\tilde{\mathbf{x}}) < g_k(\hat{\mathbf{x}}) \quad \text{and} \quad g_i(\tilde{\mathbf{x}}) = g_i(\hat{\mathbf{x}}) \forall i \ne k, \quad \text{then}$$

$$-\tilde{u}_k \le \frac{f(\tilde{\mathbf{x}}) - f(\hat{\mathbf{x}})}{g_k(\tilde{\mathbf{x}}) - g_k(\hat{\mathbf{x}})} \le -\hat{u}_k.$$

Proof Since $\tilde{\mathbf{x}}$ is an optimal solution to problem $\tilde{P}_L$, it follows that

$$f(\tilde{\mathbf{x}}) + \tilde{\mathbf{u}}\mathbf{g}(\tilde{\mathbf{x}}) \le f(\hat{\mathbf{x}}) + \tilde{\mathbf{u}}\mathbf{g}(\hat{\mathbf{x}}),$$

or alternatively,

$$f(\tilde{\mathbf{x}}) - f(\hat{\mathbf{x}}) \le \tilde{u}_k[g_k(\hat{\mathbf{x}}) - g_k(\tilde{\mathbf{x}})] + \sum_{i \ne k} \tilde{u}_i[g_i(\hat{\mathbf{x}}) - g_i(\tilde{\mathbf{x}})].$$

By assumption, $g_i(\tilde{\mathbf{x}}) = g_i(\hat{\mathbf{x}})$ for all $i \ne k$ and therefore the above inequality reduces to

$$f(\tilde{\mathbf{x}}) - f(\hat{\mathbf{x}}) \le \tilde{u}_k[g_k(\hat{\mathbf{x}}) - g_k(\tilde{\mathbf{x}})].$$

Dividing   by $\left[g_k(\hat{\mathbf{x}}) - g_k(\tilde{\mathbf{x}})\right] > 0$ yields $\dfrac{f(\tilde{\mathbf{x}}) - f(\hat{\mathbf{x}})}{g_k(\hat{\mathbf{x}}) - g_k(\tilde{\mathbf{x}})} \le \tilde{u}_k$ or

$\dfrac{f(\tilde{\mathbf{x}}) - f(\hat{\mathbf{x}})}{g_k(\tilde{\mathbf{x}}) - g_k(\hat{\mathbf{x}})} \ge -\tilde{u}_k$, which is first part of the above relation. The second part is shown similarly by starting with the fact that since $\hat{\mathbf{x}}$ is an optimal solution for problem $\hat{P}_L$, then $f(\hat{\mathbf{x}}) + \hat{\mathbf{u}}\mathbf{g}(\hat{\mathbf{x}}) \le f(\tilde{\mathbf{x}}) + \hat{\mathbf{u}}\mathbf{g}(\tilde{\mathbf{x}})$.

This leads to the result that $\dfrac{f(\tilde{\mathbf{x}}) - f(\hat{\mathbf{x}})}{g_k(\tilde{\mathbf{x}}) - g_k(\hat{\mathbf{x}})} \le -\hat{u}_k$ which concludes the proof. □

It is useful to point out that the above result also implies $-\tilde{u}_k \le -\hat{u}_k$ or $\tilde{u}_k \ge \hat{u}_k$. Suppose further that both $\tilde{\mathbf{x}}$ and $\hat{\mathbf{x}}$ are feasible for problem P. Then $\mathbf{g}(\tilde{\mathbf{x}}) \le \mathbf{0}$ and $\mathbf{g}(\hat{\mathbf{x}}) \le \mathbf{0}$ and nonnegative slack variables $S_k(\tilde{\mathbf{x}})$ and $S_k(\hat{\mathbf{x}})$ can then be introduced, such that $g_k(\tilde{\mathbf{x}}) + S_k(\tilde{\mathbf{x}}) = 0$ and $g_k(\hat{\mathbf{x}}) + S_k(\hat{\mathbf{x}}) = 0$. Consequently, $g_k(\tilde{\mathbf{x}}) - g_k(\hat{\mathbf{x}}) = S_k(\hat{\mathbf{x}}) - S_k(\tilde{\mathbf{x}})$ and since by assumption $g_k(\tilde{\mathbf{x}}) < g_k(\hat{\mathbf{x}})$, it follows that $S_k(\tilde{\mathbf{x}}) > S_k(\hat{\mathbf{x}})$. It is now clear that Lemma 4.19 together with the relations $\tilde{u}_k \ge \hat{u}_k$ and $S_k(\tilde{\mathbf{x}}) > S_k(\hat{\mathbf{x}})$ leads to the result that with u_i and $g_i(\mathbf{x})$ fixed for all $i \ne k$, $S_k(\mathbf{x})$ is a monotonically nondecreasing function of u_k. In other words, if for a given solution one keeps the values of u_i and $g_i(\mathbf{x})$ fixed for $i \ne k$, decrease u_k whenever possible and attempt to minimize the Lagrangean function with these new multipliers, then S_k will be no greater than it was in the previous solution.

In addition to this conclusion something can be said about the variation in the objective function. The second part of the relation established in Lemma 4.19, namely

$$\frac{f(\tilde{\mathbf{x}}) - f(\hat{\mathbf{x}})}{g_k(\tilde{\mathbf{x}}) - g_k(\hat{\mathbf{x}})} \le -\hat{u}_k,$$

multiplied by the term $\left[g_k(\tilde{\mathbf{x}}) - g_k(\hat{\mathbf{x}})\right] < 0$ yields

$$f(\tilde{\mathbf{x}}) - f(\hat{\mathbf{x}}) \ge -u_k\left[g_k(\tilde{\mathbf{x}}) - g_k(\hat{\mathbf{x}})\right].$$

By assumption $-u_k$ is nonpositive and the term in brackets is strictly less than zero; hence

$$f(\tilde{\mathbf{x}}) - f(\hat{\mathbf{x}}) \ge 0 \quad \text{or} \quad f(\tilde{\mathbf{x}}) \ge f(\hat{\mathbf{x}}).$$

The interpretation of this statement is important and immediate. A decrease in the value of u_k will not only force $S_k(\mathbf{x})$ not to increase but it will also assure that the value of the objective function in the new solution will not increase, i.e. the new solution cannot possibly be worse than the previous one.

Another interesting result is the following

Theorem 4.20 If $\tilde{\mathbf{x}}$ is an optimal solution to problem $\tilde{P}_L$, then $\tilde{\mathbf{x}}$ is also an optimal solution to the constrained problem

$$\tilde{P} : \text{Min } \tilde{z} = f(\mathbf{x})$$
$$\text{s.t. } \mathbf{g}(\mathbf{x}) \le \mathbf{g}(\tilde{\mathbf{x}})$$
$$\mathbf{x} \in \mathbb{R}^n.$$

Proof By virtue of the optimality of $\tilde{\mathbf{x}}$ for $\tilde{P}_L, f(\tilde{\mathbf{x}}) + \tilde{\mathbf{u}}\mathbf{g}(\tilde{\mathbf{x}}) \le f(\mathbf{x}) + \tilde{\mathbf{u}}\mathbf{g}(\mathbf{x})$ for all $\mathbf{x} \in \mathbb{R}^n$, so that $f(\tilde{\mathbf{x}}) - f(\mathbf{x}) \le \tilde{\mathbf{u}}[\mathbf{g}(\mathbf{x}) - \mathbf{g}(\tilde{\mathbf{x}})]$. Because $\tilde{\mathbf{u}} \ge \mathbf{0}$ by assumption, one can conclude that $f(\tilde{\mathbf{x}}) - f(\mathbf{x}) \le 0$ for all $\mathbf{x} \in \mathbb{R}^n$, such that $\mathbf{g}(\mathbf{x}) \le \mathbf{g}(\tilde{\mathbf{x}})$ and the theorem is proved. $\square$

In practical situations it might be very useful to know that the solution $\tilde{\mathbf{x}}$ to the unconstrained problem $\tilde{P}_L$ is also an optimal solution to the constrained problem $\tilde{P}$, which is nothing but the original problem P with $\mathbf{g}(\tilde{\mathbf{x}})$ instead of $\mathbf{0}$ as right-hand side values of the constraints.

The economic interpretation of the Lagrangean multipliers associated with the solution to a constrained minimization problem is similar to that of so-called *shadow prices* that are associated with the constraints in linear programming models, in that they are related to a particular solution point and correspond to marginal prices, i.e. incremental prices associated with small changes in the requirements expressed by the constraints.

The introduction of slack variables provides an interesting interpretation of the Lagrangean function. Since $g_i(\mathbf{x}) \le 0$ is transformed to $g_i(\mathbf{x}) + S_i(\mathbf{x}) = 0$ or $g_i(\mathbf{x}) = -S_i(\mathbf{x})$, the Lagrangean function can be written as

$$L(\mathbf{x}, \mathbf{u}) = f(\mathbf{x}) - \sum_{i=1}^{m} u_i S_i(\mathbf{x}).$$

Suppose now that the original problem involves minimizing total costs in a production planning model. Then $f(\mathbf{x})$ represents the total production cost and $\mathbf{S}(\mathbf{x})$ stands for the amounts of unused resources such as, for instance, machine hours. Let u_i be the shadow price associated with the i-th resource, i.e., the price for one machine hour. Then $\sum_{i=1}^{m} u_i S_i(\mathbf{x})$ is simply the total amount of money the company would make renting the unused production capacity to some other manufacturer. Under the assumption that all unused resources can be rented at the price of $\mathbf{u}$, then the total cost for our production company is $L(\mathbf{x}, \mathbf{u})$.

We conclude this section by showing how the Karush-Kuhn-Tucker conditions can be utilized to derive the optimality criteria in linear programming problems. Assume that a linear programming problem is given in canonical form:

$$\begin{array}{ll} \text{P: Max} \, z = \mathbf{cx} & \text{P: Min} -z = -\mathbf{cx} \\ \text{s.t. } \mathbf{Ax} \le \mathbf{b} \quad \text{or equivalently} \quad \text{s.t. } \mathbf{Ax} - \mathbf{b} \le \mathbf{0} \\ \quad \mathbf{x} \ge \mathbf{0} & \quad \mathbf{x} \ge \mathbf{0} \end{array}$$

The associated Lagrangean function is $L(\mathbf{x}, \mathbf{u}) = -\mathbf{cx} + \mathbf{uAx} - \mathbf{ub}$ and the corresponding Karush-Kuhn-Tucker conditions can be written as

$$L_\mathbf{x}(\bar{\mathbf{x}}, \bar{\mathbf{u}}) = (-\mathbf{c} + \bar{\mathbf{u}}\mathbf{A})^T \ge \mathbf{0} \quad \text{or} \quad \bar{\mathbf{u}}\mathbf{A} \ge \mathbf{c} \tag{4.14}$$

$$\bar{\mathbf{x}}^T L_\mathbf{x}(\bar{\mathbf{x}}, \bar{\mathbf{u}}) = -\mathbf{c}\bar{\mathbf{x}} + \bar{\mathbf{u}}\mathbf{A}\bar{\mathbf{x}} = 0 \quad \text{or} \quad (\bar{\mathbf{u}}\mathbf{A} - \mathbf{c})\bar{\mathbf{x}} = 0 \tag{4.15}$$

$$L_\mathbf{u}(\bar{\mathbf{x}}, \bar{\mathbf{u}}) = \mathbf{A}\bar{\mathbf{x}} - \mathbf{b} \le \mathbf{0} \quad \text{or} \quad \mathbf{A}\bar{\mathbf{x}} \le \mathbf{b} \tag{4.16}$$

$$\bar{\mathbf{u}} L_\mathbf{u}(\bar{\mathbf{x}}, \bar{\mathbf{u}}) = \bar{\mathbf{u}}\mathbf{A}\bar{\mathbf{x}} - \bar{\mathbf{u}}\mathbf{b} = 0 \quad \text{or} \quad \bar{\mathbf{u}}(\mathbf{A}\bar{\mathbf{x}} - \mathbf{b}) = 0 \tag{4.17}$$

$$\bar{\mathbf{x}} \ge \mathbf{0} \tag{4.18}$$

$$\bar{\mathbf{u}} \ge \mathbf{0} \tag{4.19}$$

Now the inequalities (4.16) and (4.18) represent the conditions for primal feasibility; (4.14) and (4.19) represent the conditions for dual feasibility and the Eqs. (4.15) and (4.17) represent the weak complementary slackness conditions. One can show (see, e.g., Eiselt and Sandblom 2007) that the fulfillment of conditions (4.14)–(4.19) is necessary and sufficient for $\bar{\mathbf{x}}$ and $\bar{\mathbf{u}}$ to represent optimal solutions of the linear programming problem P. Therefore these optimality conditions represent only a special case of a more general situation in which linearity is not a requirement.

4.3 Duality Theory and Postoptimality Analysis

The first papers leading to modern nonlinear duality theory were by Dorn (1960a, b), introducing duality for quadratic programs, i.e., problems with quadratic objective function and linear constraints, and then extending this analysis to general convex objective functions subject to linear constraints. Wolfe (1961) then covered the general convex, differentiable case. The theory on nonlinear duality is now well established and we will refer to this theory as *Lagrangean duality*, since it involves the Lagrangean function. Another approach, often referred to as *conjugate* or *Fenchel duality*, is taken by Rockafellar (1970) who uses the concept of conjugate functions. The two approaches can be shown to be equivalent (Magnanti 1974) and we will restrict our treatment to Lagrangean duality. For an in-depth exposition, see Bertsekas (2016) who covers both, Lagrangean and conjugate, duality. For expository purposes we begin by considering the Wolfe dual problem which is a special case of the more general Lagrangean dual problem.

4.3.1 The Wolfe Dual Problem

As an introduction to the general nonlinear dual problem we begin with Wolfe's approach. Consider the following (primal) nonlinear programming problem:

$$P: \operatorname*{Min}_{\mathbf{x}} z = f(\mathbf{x})$$
$$\text{s.t. } g_i(\mathbf{x}) \leq 0, i = 1, \ldots, m$$
$$\mathbf{x} \in \mathbb{R}^n,$$

where f and g_i, $i = 1, \ldots, m$ are convex and differentiable real-valued functions.

Definition 4.21 Associated with P is the corresponding Wolfe dual problem

$$P_D : \operatorname*{Max}_{\mathbf{x, u}} z_D = L(\mathbf{x, u})$$
$$\text{s.t. } \nabla_{\mathbf{x}} L(\mathbf{x, u}) = \mathbf{0}$$
$$\mathbf{u} \geq \mathbf{0},$$

where $L(\mathbf{x, u})$ as usual denotes the Lagrangean function $L(\mathbf{x, u}) = f(\mathbf{x}) + \sum\limits_{i=1}^{m} g_i(\mathbf{x})$
$= f(\mathbf{x}) + \mathbf{u g}(\mathbf{x})$.

As in the previous section we will use the notation $L_{\mathbf{x}}(\mathbf{x, u})$ for $\nabla_{\mathbf{x}} L(\mathbf{x, u})$ and $L_{\mathbf{u}}(\mathbf{x, u})$ for $\nabla_{\mathbf{u}} L(\mathbf{x, u})$.

The Wolfe dual problem P_D can also be written as

$$P_D: \operatorname*{Max}_{\mathbf{x, u}} z_D = f(\mathbf{x}) + \mathbf{u g}(\mathbf{x})$$
$$\text{s.t. } \nabla f(\mathbf{x}) + \mathbf{u} \nabla \mathbf{g}(\mathbf{x}) = \mathbf{0}$$
$$\mathbf{u} \geq \mathbf{0},$$

where $\nabla \mathbf{g}(\mathbf{x})$ is the $[m \times n]$-dimensional matrix

$$\nabla \mathbf{g}(\mathbf{x}) = [(\nabla \mathbf{g}(\mathbf{x})_{ij}] = \frac{\partial g_i}{\partial x_j}.$$

Note that the primal problem consists of minimizing a function of n variables, i.e., $\mathbf{x} \in \mathbb{R}^n$, whereas the Wolfe dual problem consists of maximizing a function of $n + m$ variables, i.e., $(\mathbf{x, u}) \in \mathbb{R}^{m+n}$. We will refer to $\hat{\mathbf{x}}$ as a primal feasible solution, if $\hat{\mathbf{x}}$ satisfies the constraints of P, and to $(\tilde{\mathbf{x}}, \tilde{\mathbf{u}})$ as a dual feasible solution, if $(\tilde{\mathbf{x}}, \tilde{\mathbf{u}})$ satisfies the constraints of P_D. Just as for linear programming one can obtain the following

Theorem 4.22 (The Weak Wolfe Duality Theorem) If $\hat{\mathbf{x}}$ is primal feasible and $(\tilde{\mathbf{x}}, \tilde{\mathbf{u}})$ is dual feasible, then

$$L(\tilde{\mathbf{x}}, \tilde{\mathbf{u}}) \leq f(\hat{\mathbf{x}}).$$

Proof As f is convex, Theorem 1.34 implies that

$$
\begin{aligned}
f(\hat{\mathbf{x}}) - f(\tilde{\mathbf{x}}) &\geq (\hat{\mathbf{x}} - \tilde{\mathbf{x}})^T \nabla f(\tilde{\mathbf{x}}) \\
&= -(\hat{\mathbf{x}} - \tilde{\mathbf{x}})^T \sum_{i=1}^{m} u_i \nabla g_i(\tilde{\mathbf{x}}) \qquad \text{(according to the dual feasibility of } (\tilde{\mathbf{x}}, \tilde{\mathbf{u}})) \\
&\geq -\sum_{i=1}^{m} \tilde{u}_i (g_i(\hat{\mathbf{x}}) - g_i(\tilde{\mathbf{x}})) \qquad \text{(according to the convexity of } g_i \text{ and again due} \\
&\hspace{6cm} \text{to Theorem 1.34)} \\
&\geq \sum_{i=1}^{m} \tilde{u}_i g_i(\tilde{\mathbf{x}}) \qquad \text{(according to the primal feasibility of } \hat{\mathbf{x}} \text{ and as } \tilde{\mathbf{u}} \geq \mathbf{0}).
\end{aligned}
$$

Rearranging, we obtain

$$
f(\hat{\mathbf{x}}) \geq f(\tilde{\mathbf{x}}) + \sum_{i=1}^{m} \tilde{u}_i g_i(\tilde{\mathbf{x}}), \text{ or}
$$
$$
f(\hat{\mathbf{x}}) \geq L(\tilde{\mathbf{x}}, \tilde{\mathbf{u}}),
$$

which completes the proof. $\square$

The above result can be strengthened by the following

Theorem 4.23 (The Strong Wolfe Duality Theorem) If the primal constraint set satisfies Slater's constraint qualification and if $\bar{\mathbf{x}}$ is an optimal solution to the primal problem P, then there exists some $\bar{\mathbf{u}}$ such that $(\bar{\mathbf{x}}, \bar{\mathbf{u}})$ is an optimal solution to the Wolfe dual problem P_D, and the extrema are equal, i.e., $f(\bar{\mathbf{x}}) = L(\bar{\mathbf{x}}, \bar{\mathbf{u}})$ or simply $\bar{z} = \bar{z}_D$.

Proof The convexity (with respect to $\mathbf{x}$) of f, g_1, $\ldots$, g_m implies the convexity of L (with respect to $\mathbf{x}$ for any given $\mathbf{u}$, and with respect to $\mathbf{u}$ for any given $\mathbf{x}$) for all $\mathbf{u} \geq \mathbf{0}$. Furthermore, the Wolfe dual problem P_D has feasible solutions due to the constraint qualification and the existence of an optimal solution to the primal problem P; this follows from Lemmas 4.10 and 4.13. Assume now that $(\tilde{\mathbf{x}}, \tilde{\mathbf{u}})$ and $(\hat{\mathbf{x}}, \tilde{\mathbf{u}})$ are dual feasible solutions. The convexity of L then implies the following (due to Theorem 1.34):

$$
\begin{aligned}
L(\tilde{\mathbf{x}}, \tilde{\mathbf{u}}) - L(\hat{\mathbf{x}}, \tilde{\mathbf{u}}) &\geq (\tilde{\mathbf{x}} - \hat{\mathbf{x}})^T L_{\mathbf{x}}(\hat{\mathbf{x}}, \tilde{\mathbf{u}}) = \mathbf{0} \\
L(\hat{\mathbf{x}}, \tilde{\mathbf{u}}) - L(\tilde{\mathbf{x}}, \tilde{\mathbf{u}}) &\geq (\hat{\mathbf{x}} - \tilde{\mathbf{x}})^T L_{\mathbf{x}}(\tilde{\mathbf{x}}, \tilde{\mathbf{u}}) = \mathbf{0}.
\end{aligned}
$$

Therefore $L(\tilde{\mathbf{x}}, \tilde{\mathbf{u}}) = L(\hat{\mathbf{x}}, \tilde{\mathbf{u}})$, so that for any $\tilde{\mathbf{u}}$ the function $L(\mathbf{x}, \tilde{\mathbf{u}})$ does not depend on $\mathbf{x}$, as long as $(\mathbf{x}, \tilde{\mathbf{u}})$ is dual feasible. Again using Lemmas 4.10 and 4.13, we find that there exists a $\bar{\mathbf{u}}$, such that $(\bar{\mathbf{x}}, \bar{\mathbf{u}})$ is dual feasible, and furthermore

$$L(\bar{\mathbf{x}},\bar{\mathbf{u}}) = \max_{\mathbf{u}\geq 0} L(\mathbf{x},\mathbf{u}) \qquad\qquad (\text{according to the saddle point property of } (\bar{\mathbf{x}},\bar{\mathbf{u}}))$$

$$\geq \max_{\mathbf{u}} \{L(\bar{\mathbf{x}},\mathbf{u}) : (\bar{\mathbf{x}},\mathbf{u}) \text{ is dual feasible}\} \quad (\text{dual feasibility requires } \mathbf{u}\geq 0)$$

$$= \max_{\mathbf{x},\mathbf{u}} \{L(\mathbf{x},\mathbf{u}) : (\mathbf{x},\mathbf{u}) \text{ is dual feasible}\} \quad (L(\mathbf{x},\mathbf{u}) \text{ is independent of } \mathbf{x}\,\text{for dual}$$

$$\qquad\qquad\qquad\qquad\qquad\qquad\qquad\qquad \text{feasible } (\mathbf{x},\mathbf{u}))$$

$$\geq L(\bar{\mathbf{x}},\bar{\mathbf{u}}) \qquad\qquad\qquad\qquad\qquad ((\bar{\mathbf{x}},\bar{\mathbf{u}}) \text{ is dual feasible}).$$

Consequently, $(\bar{\mathbf{x}}, \bar{\mathbf{u}})$solves the dual problem P_D. Finally, since $\sum_{i=1}^{m} \bar{u}_i g_i(\bar{\mathbf{x}}) = 0$, by virtue of the complementarity condition (b) of Theorem 4.4,

$$L(\bar{\mathbf{x}}, \bar{\mathbf{u}}) = f(\bar{\mathbf{x}}) + \sum_{i=1}^{m} \bar{u}_i g_i(\bar{\mathbf{x}}) = f(\bar{\mathbf{x}}),$$

which completes the proof. $\square$

From the weak Wolfe duality theorem we obtain $f(\bar{\mathbf{x}}) - L(\bar{\mathbf{x}}, \bar{\mathbf{u}}) \geq 0$, whereas the above strong Wolfe duality theorem asserts that the difference $f(\bar{\mathbf{x}}) - L(\bar{\mathbf{x}}, \bar{\mathbf{u}})$, called the *duality gap*, is actually zero.

Example   Consider the primal problem

$$\text{P: } \min_{\mathbf{x}} z = x_1^2 + x_2^2$$
$$\text{s.t. } -x_1 - x_2 + 3 \leq 0$$
$$x_1, x_2 \in \mathbb{R}.$$

Geometrically, P can be represented as in Fig. 4.6, where the concentric circles are level curves for the objective function. By inspection we find that the optimal solution to P is $(\bar{x}_1, \bar{x}_2) = \left(\dfrac{3}{2}, \dfrac{3}{2}\right)$ with $\bar{z} = \dfrac{9}{2}$.

Fig. 4.6 Level curves and optimal solution of a quadratic optimization problem

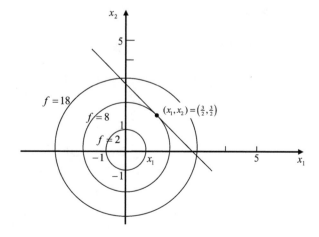

Furthermore, the Wolfe dual problem is

$$P_D: \underset{\mathbf{x},u}{\text{Max}}\ L(\mathbf{x},u) = x_1^2 + x_2^3 + u(-x_1 - x_2 + 3)$$
$$\text{s.t.}\ \ 2x_1 - u = 0$$
$$2x_2 - u = 0$$
$$u \geq 0.$$

Simplifying the dual problem, we obtain $u = 2x_1 = 2x_2 \geq 0$, so that

$$L(\mathbf{x},u) = x_1^2 + x_1^2 + 2x_1(-2x_1 + 3) = -2(x_1^2 - 3x_1) = -2\left(x_1 - \frac{3}{2}\right)^2 + \frac{9}{2}.$$

Thus $\bar{z}_D = \dfrac{9}{2}$ for $(\bar{x}_1, \bar{x}_2, \bar{u}) = \left(\dfrac{3}{2}, \dfrac{3}{2}, 3\right)$. As expected, $\bar{z}_D = \bar{z}$ and one can also see

that the other results of the weak and strong Wolfe duality theorems hold true.

To give a geometric interpretation of the Wolfe dual problem in general, we will use the following approach (see e.g., Bertsekas, 2016 or Bazaraa et al., 2013). Set $\mathbf{y}^1 = \mathbf{g}(\mathbf{x})$ and $y^2 = f(\mathbf{x})$ and consider the set G in the $(m + 1)$-dimensional $(\mathbf{y}^1, y^2)$-space where

$$G = \left\{ (\mathbf{y}^1, y^2) \in \mathbb{R}^m \times \mathbb{R} : \mathbf{y}^1 = \mathbf{g}(\mathbf{x}) \text{ and } y^2 = f(\mathbf{x}) \text{ for some } \mathbf{x} \in \mathbb{R}^n \right\}.$$

In other words, G is the image of the $\mathbf{x}$-space under the map $(\mathbf{g}(\mathbf{x}), f(\mathbf{x}))$. One can show that due to the convexity of $f(\mathbf{x})$ and $\mathbf{g}(\mathbf{x})$, the set G will also be convex. Geometrically, the primal problem P can now be interpreted as finding a point in $\mathbb{R}^{m+1}$ with $\mathbf{y}^1 \leq \mathbf{0}$, belonging to G and with minimal y^2-intercept. This minimal y^2-point occurs at the intersection of this boundary with the y^2-axis, see Fig. 4.7.

For the Wolfe dual problem P_D, we will now first show that dual feasible points correspond to points $(\mathbf{y}^1, y^2)$ in G where y^2 is minimal for any given $\mathbf{y}^1$, i.e., the lower boundary of the set G. To see this, assume that $\hat{\mathbf{u}} \geq \mathbf{0}$ and $\hat{\mathbf{y}}^1 = \mathbf{g}(\hat{\mathbf{x}})$ are given. Dual feasibility requires that $L_\mathbf{x}(\hat{\mathbf{x}}, \hat{\mathbf{u}}) = \mathbf{0}$, which implies that $\hat{\mathbf{x}}$ minimizes $L(\mathbf{x}, \hat{\mathbf{u}})$ due to

Fig. 4.7 Geometric interpretation of the Wolfe dual

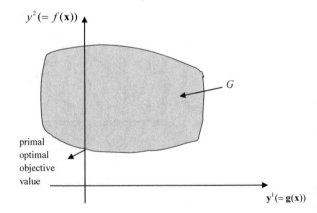

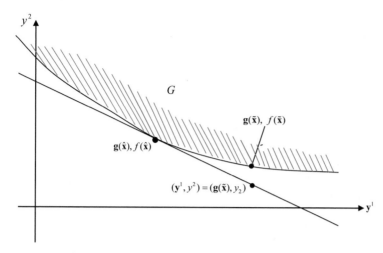

Fig. 4.8 Supporting hyperplane to the set G

the convexity of $L(\mathbf{x}, \hat{\mathbf{u}})$ for any fixed $\hat{\mathbf{u}} \geq \mathbf{0}$. Then $\hat{y}^2 = f(\hat{\mathbf{x}})$ must be minimal in G, since otherwise $\hat{y}^2$ could be reduced by keeping $\hat{\mathbf{y}}^1$ constant (thus reducing $f(\mathbf{x})$ while keeping $\mathbf{g}(\mathbf{x})$ constant) and hence $\hat{\mathbf{u}}\mathbf{g}(\mathbf{x}) = \hat{\mathbf{u}}\hat{\mathbf{y}}^1$ would also be constant. This would decrease $f(\mathbf{x}) + \mathbf{u}\mathbf{g}(\mathbf{x}) = L(\mathbf{x}, \hat{\mathbf{u}})$, violating the restriction $L_{\mathbf{x}}(\hat{\mathbf{x}}, \hat{\mathbf{u}}) = \mathbf{0}$. Furthermore, given a particular dual feasible point $(\hat{\mathbf{x}}, \hat{\mathbf{u}})$, the hyperplane $y^2 - f(\hat{\mathbf{x}})$ $+ \hat{\mathbf{u}}(\mathbf{y}^1 - \mathbf{g}(\hat{\mathbf{x}})) = y^2 + \hat{\mathbf{u}}\mathbf{y}^1 - L(\hat{\mathbf{x}}, \hat{\mathbf{u}}) = 0$ is a supporting hyperplane to G. To show this, take $(\mathbf{y}^1, y^2)$ on this hyperplane with $\mathbf{y}^1 \neq \mathbf{g}(\hat{\mathbf{x}})$. If $\tilde{\mathbf{x}}$ exists, such that $\mathbf{y}^1 = \mathbf{g}(\tilde{\mathbf{x}})$, i.e., $(\mathbf{g}(\mathbf{x}), f(\mathbf{x})) \in G$, we must show that $f(\tilde{\mathbf{x}}) \geq y^2$, see Fig. 4.8.

This holds true, because

$$
\begin{aligned}
f(\tilde{\mathbf{x}}) - y^2 &= f(\tilde{\mathbf{x}}) + \hat{\mathbf{u}}\mathbf{y}^1 - L(\hat{\mathbf{x}}, \hat{\mathbf{u}}) && \\
&= f(\tilde{\mathbf{x}}) + \hat{\mathbf{u}}\mathbf{g}(\tilde{\mathbf{x}}) - L(\hat{\mathbf{x}}, \hat{\mathbf{u}}) && \left(\text{as } \tilde{\mathbf{u}}\mathbf{y}^1 = \hat{\mathbf{u}}\mathbf{g}(\tilde{\mathbf{x}}) \text{ by assumption}\right) \\
&= L(\tilde{\mathbf{x}}, \hat{\mathbf{u}}) - L(\hat{\mathbf{x}}, \hat{\mathbf{u}}) && (\text{by definition}) \\
&\geq (\tilde{\mathbf{x}} - \hat{\mathbf{x}})^T L_{\mathbf{x}}(\hat{\mathbf{x}}, \hat{\mathbf{u}}) && (\text{by convexity of } L) \\
&= \mathbf{0} && \left(\text{due to dual feasibility of } (\hat{\mathbf{x}}, \hat{\mathbf{u}})\right).
\end{aligned}
$$

Therefore the Wolfe dual problem of maximizing $L(\mathbf{x}, \mathbf{u}) = y^2 + \mathbf{u}\mathbf{y}^1$ is that of finding the maximum y^2-intercept of any supporting hyperplane $L(\mathbf{x}, \mathbf{u}) = y^2 + \mathbf{u}\mathbf{y}^1$ of G with $\mathbf{u} \geq \mathbf{0}$.

As an example, consider the previous numerical problem

$$
\begin{aligned}
\text{P}: \ &\underset{x_1, x_2}{\text{Min}} \ z = x_1^2 + x_2^2 \\
&\text{s.t.} -x_1 - x_2 + 3 \leq 0 \\
&x_1, x_2 \in \mathbb{R}.
\end{aligned}
$$

Setting $y^2 = f(\mathbf{x}) = x_1^2 + x_2^2$ and $y^1 = g(\mathbf{x}) = -x_1 - x_2 + 3$ and starting with the trivial inequality $(x_1 - x_2)^2 \geq 0 \ \forall \ x_1, x_2 \in \mathbb{R}$, we obtain $0 \leq (x_1 - x_2)^2 = 2(x_1^2 + x_2^2) - (x_1 + x_2)^2 = 2y^2 - (3 - y^1)^2$. It follows that P can be written as

$$P' : \underset{y^1, y^2}{\text{Min}} \ z = y^2$$
$$\text{s.t. } y^1 \leq 0$$
$$(3 - y^1)^2 - 2y^2 \leq 0$$
$$y^1, y^2 \in \mathbb{R},$$

where the nonnegativity of y^2 follows from the second constraint. The Wolfe dual problem P_D can then be expressed as

$$P'_D : \underset{y^1, y^2, u}{\text{Max}} \ z = uy^1 + y^2 \quad (= f(\mathbf{x}) + ug(\mathbf{x}) = L(\mathbf{x}, u))$$
$$\text{s.t. } y^1 = 3 - u \qquad\qquad (\text{from } y^1 = g(\mathbf{x}) = -x_1 - x_2 + 3 \text{ and } 2x_1 = 2x_2 = u)$$
$$y^1 \leq 3 \qquad\qquad\qquad (\text{via } u \geq 0)$$
$$y^2 = \tfrac{1}{2}u^2 \qquad\qquad\quad \left(\text{as } y^2 = x_1^2 + x_2^2 \text{ and } 2x_1 = 2x_2 = u\right)$$
$$y^1, y^2 \in \mathbb{R}$$
$$u \in \mathbb{R}_+.$$

The set G in the (y^1, y^2)-plane becomes

$$G = \left\{ (y^1, y^2) \in \mathbb{R}^2 : y^1 = -x_1 - x_2 + 3, \ y^2 = x_1^2 + x_2^2, \text{for some } (x_1, x_2) \in \mathbb{R}^2 \right\},$$

so that G is the image of the (x_1, x_2)-plane under the map $(-x_1 - x_2 + 3, x_1^2 + x_2^2)$. From the above discussion it follows that $G = \{(y^1, y^2) \in \mathbb{R}^2 : (3 - y^1)^2 \leq 2y^2\}$, see Fig. 4.9.

The primal problem P' of finding a point in G with minimum ordinate y^2 in the left half-plane $y^1 \leq 0$ has then the unique optimal solution $\bar{y}^1 = 0, \bar{y}^2 = \tfrac{9}{2}$. The dual problem P'_D is that of finding the maximum y_2-axis intercept $L(\mathbf{x}, u)$ of any supporting hyperplane $L(\mathbf{x}, u) = y^2 + uy^1$ of G, with $u \geq 0$. In this case, the supporting hyperplanes are lines tangent to G and $L(\mathbf{x}, u) = y^2 + uy^1$ is the equation of a straight line in the (y^1, y^2)-plane with a slope of $-u$ and y^2-intercept $L(\mathbf{x}, u)$. The slope of a tangent to the set G is obtained by elementary differential calculus; it is $y^1 - 3$, which then equals $-u$. Then $u \geq 0$ requires that $y^1 \leq 3$ and the dual problem P'_D consists of finding the highest y^2-axis intercept for any tangent to the set G to the left of $y^1 = 3$. From Fig. 4.9 we see that this maximum intercept is $\bar{y}^2 = \tfrac{9}{2}$, which confirms the result of the strong Wolfe duality Theorem 4.23.

We will now show that if the primal problem P happens to be a linear programming problem, its Wolfe dual will be identical to the linear programming dual. Thus Wolfe duality is a true extension of linear programming duality to the convex differentiable nonlinear case. Specifically, let the primal problem P be a linear programming problem in canonical form, i.e.

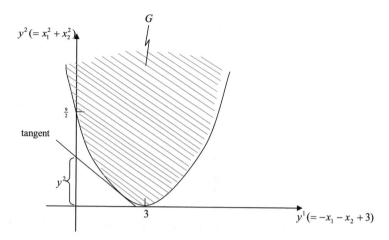

Fig. 4.9 Mapping the set G into the (y_1, y_2)-plane

$$P: \text{Max } z = \mathbf{cx}$$
$$\text{s.t. } \mathbf{Ax} \leq \mathbf{b}$$
$$\mathbf{x} \geq \mathbf{0},$$

which transforms into:

$$P': \underset{\mathbf{x}}{\text{Min}} -z = -\mathbf{cx}$$
$$\text{s.t. } \mathbf{Ax} - \mathbf{b} \leq \mathbf{0}$$
$$-\mathbf{x} \leq \mathbf{0}$$
$$\mathbf{x} \in \mathbb{R}^n.$$

Then associate dual variables (row vectors) $\mathbf{u}$ and $\mathbf{v}$ with the structural and nonnegativity constraints, respectively. Note that these two types of constraints together form the constraint $\mathbf{g(x)} \leq \mathbf{0}$. Using the resulting Lagrangean function $L(\mathbf{x}, \mathbf{u}, \mathbf{v}) = -\mathbf{cx} + \mathbf{u(Ax - b)} + \mathbf{v(-x)}$, the Wolfe dual becomes

$$P_D: \underset{\mathbf{x, u, v}}{\text{Max}} \ z_D = -\mathbf{cx} + \mathbf{uAx} - \mathbf{ub} - \mathbf{vx}$$
$$\text{s.t. } -\mathbf{c} + \mathbf{uA} - \mathbf{v} = \mathbf{0}$$
$$\mathbf{u, v} \geq \mathbf{0}$$
$$\mathbf{x} \in \mathbb{R}^n.$$

Because of the requirement that $-\mathbf{c} + \mathbf{uA} - \mathbf{v} = \mathbf{0}$, the objective function in P_D can be simplified to $z_D = (-\mathbf{c} + \mathbf{uA} - \mathbf{v})\mathbf{x} - \mathbf{ub} = -\mathbf{ub}$, so that the Wolfe dual can be written as

$$P_D' : \underset{\mathbf{x,u,v}}{\text{Max}} \ z_D = -\mathbf{ub}$$
$$\text{s.t.} - \mathbf{c} + \mathbf{uA} - \mathbf{v} = \mathbf{0}$$
$$\mathbf{u,v} \geq \mathbf{0}$$
$$\mathbf{x} \in \mathbb{R}^n.$$

Furthermore, $-\mathbf{c} + \mathbf{uA} = \mathbf{v} \geq \mathbf{0}$, so that the constraint set can be written as $-\mathbf{c} + \mathbf{uA} \geq \mathbf{0}$ or $\mathbf{uA} \geq \mathbf{c}$. Transforming P_D' into a minimization problem yields

$$P_D'' : \underset{\mathbf{u}}{\text{Min}} -z_D = \mathbf{ub}$$
$$\text{s.t.} \ \mathbf{uA} \geq \mathbf{c}$$
$$\mathbf{u} \geq \mathbf{0},$$

which is the well known dual linear programming problem in canonical form.

4.3.2   The Lagrangean Dual Problem

In this section the Lagrangean dual problem will be discussed. The Lagrangean dual is more general than the Wolfe dual since it does not require differentiability. Let as usual $L(\mathbf{x}, \mathbf{u})$ denote the Lagrangean function

$$L(\mathbf{x},\mathbf{u}) = f(\mathbf{x}) + \sum_{i=1}^{m} u_i g_i(\mathbf{x}) = f(\mathbf{x}) + \mathbf{ug}(\mathbf{x})$$

and consider a (primal) nonlinear programming problem

$$P: \underset{x}{\text{Min}} \ z = f(\mathbf{x})$$
$$\text{s.t.} \ g_i(\mathbf{x}) \leq 0, \quad i = 1, \ldots, m$$
$$\mathbf{x} \in \mathbb{R}^n.$$

Definition 4.24 Associated with P is the Lagrangean dual problem

$$P_D : \underset{\mathbf{u} \geq \mathbf{0}}{\text{Max}} \ z_D = \underset{\mathbf{x}}{\inf} \ \{L(\mathbf{x},\mathbf{u})\}$$
$$\mathbf{x} \in \mathbb{R}^n, \mathbf{u} \in \mathbb{R}^m.$$

For the definition of infimum and its relation to minimum, see e.g., Bertsekas (2016) or Bazaraa et al. (2013). Sometimes, $\underset{\mathbf{x}}{\inf} \ \{L(\mathbf{x},\mathbf{u})\}$ will be denoted by $h(\mathbf{u})$, so that the Lagrangean dual problem reads as

$$P_D : \underset{u \geq 0}{\text{Max}} \ z_D = h(\mathbf{u})$$
$$\mathbf{u} \in \mathbb{R}^m.$$

One can also form a partial Lagrangean dual with respect to only some of the m constraints $g_i(\mathbf{x}) \leq 0$, $i = 1, \ldots, m$. For instance, the partial Lagrangean dual of P with respect to the k constraints $g_i(\mathbf{x}) \leq 0$, $i = 1, \ldots, k$, where $1 \leq k < m$, is given by

$$P_D^P : \underset{u \geq 0}{\text{Max}} \ z_D = \underset{\substack{\mathbf{x}:g_i(\mathbf{x}) \leq 0 \\ i=k+1,\ldots,m}}{\inf} \left\{ f(\mathbf{x}) + \sum_{i=1}^k u_i g_i(\mathbf{x}) \right\}.$$

Note that convexity assumptions have not yet been introduced. If the infimum is taken over an empty set, its value is assumed to be $+\infty$. If the set of feasible values $L(\mathbf{x}, \mathbf{u})$ is unbounded from below, the infimum is taken to be $-\infty$. Unlike the linear case, this nonlinear duality formulation is not symmetric. One way to express the primal and dual problems so that a certain symmetry results would be to consider the problems

$$P' : \underset{u \geq 0}{\sup} \left\{ \underset{\mathbf{x}}{\inf} \ L(\mathbf{x}, \mathbf{u}) \right\} \quad \text{and} \quad P'' : \underset{\mathbf{x}}{\inf} \left\{ \underset{u \geq 0}{\sup} \ L(\mathbf{x}, \mathbf{u}) \right\}.$$

Problem P′ corresponds to the Lagrangean dual problem P_D with "Max" replaced by "sup." As $\underset{u \geq 0}{\sup} \ L(\mathbf{x}, \mathbf{u}) = +\infty$ for $\mathbf{x}$, such that $g_i(\mathbf{x}) > 0$ for at least one subscript i (i.e., violating at least one constraint), we need only consider those $\mathbf{x}$ in P″ for which $\mathbf{g}(\mathbf{x}) \leq \mathbf{0}$. Therefore, we have

$$\underset{\mathbf{x}}{\inf} \left\{ \underset{u \geq 0}{\sup} \ L(\mathbf{x}, \mathbf{u}) \right\} = \underset{\mathbf{x}}{\inf} \left\{ \underset{u \geq 0}{\sup} \ L(\mathbf{x}, \mathbf{u}) : \mathbf{g}(\mathbf{x}) \leq \mathbf{0} \right\},$$

and, since $\underset{u \geq 0}{\sup} \ \{f(\mathbf{x}) + \mathbf{u}\mathbf{g}(\mathbf{x})\} = f(\mathbf{x})$ for $\mathbf{u} \geq \mathbf{0}$ and $\mathbf{g}(\mathbf{x}) \leq \mathbf{0}$, we can conclude that

$$\underset{\mathbf{x}}{\inf} \left\{ \underset{u \geq 0}{\sup} \ L(\mathbf{x}, \mathbf{u}) \right\} = \underset{\mathbf{x}}{\inf} \ \{f(\mathbf{x}) : \mathbf{g}(\mathbf{x}) \leq \mathbf{0}\}.$$ This corresponds to the usual primal

problem P only with "Min" replaced by "inf."

The symmetry would be enhanced by retaining the nonnegativity condition for $\mathbf{x}$ as in the previous two sections. Symmetry may also be obtained using conjugate duality, see Rockafellar (1970). Cottle (1963) considered symmetric Lagrangean duality for quadratic programs. Now the general Lagrangean dual of two general problem types will be discussed.

(a) Let the primal problem P be the usual linear programming problem

$$P: \text{Max } z = \mathbf{cx}$$
$$\text{s.t. } \mathbf{Ax} \leq \mathbf{b}$$
$$\mathbf{x} \geq \mathbf{0},$$

which can be rewritten as

$$P^*: \text{Min} - z = -\mathbf{cx}$$
$$\text{s.t. } \mathbf{Ax} - \mathbf{c} \leq \mathbf{0}$$
$$\mathbf{x} \geq \mathbf{0}$$

and let us formulate the Lagrangean dual with respect to the structural constraints.

The Lagrangean function then becomes $L(\mathbf{x}, \mathbf{u}) = -\mathbf{cx} + \mathbf{u}(\mathbf{Ax} - \mathbf{b})$ and the resulting partial Lagrangean dual is

$$\mathbf{P_D^{*P}} : \text{Max} -z_D = \inf_{\mathbf{x} \geq \mathbf{0}} \left\{ \left(-\mathbf{cx} + \mathbf{u}(\mathbf{Ax} - \mathbf{b}) \right) \right\}.$$

Since $\inf_{\mathbf{x} \geq \mathbf{0}} \{-\mathbf{c} + \mathbf{u}(\mathbf{Ax} - \mathbf{b})\} = \inf_{\mathbf{x} \geq \mathbf{0}} \{(\mathbf{uA} - \mathbf{c})\mathbf{x} - \mathbf{ub}\}$ takes the value of $-\mathbf{ub}$, if $\mathbf{uA} - \mathbf{c} \geq \mathbf{0}$ and $-\infty$ otherwise, the Lagrangean dual is then

$$\mathbf{P_D^{*P}} : \text{Max} -z_D = -\mathbf{ub}$$
$$\mathbf{uA} \geq \mathbf{c}$$
$$\mathbf{u} \geq \mathbf{0}$$

or simply

$$P_D: \text{Min } z_D = \mathbf{ub}$$
$$\text{s.t. } \mathbf{uA} \geq \mathbf{c}$$
$$\mathbf{u} \geq \mathbf{0},$$

which is the regular linear programming dual.

(b) Let now the primal problem be

$$P: \quad \underset{\mathbf{x}}{\text{Min }} z = f(\mathbf{x})$$
$$\text{s.t. } g_i(\mathbf{x}) \leq 0, i = 1, \ldots, m$$
$$\mathbf{x} \in \mathbb{R}^n,$$

where f and g_i, $i = 1, \ldots, m$ are convex differentiable real-valued functions. The Lagrangean dual problem is then

$$P_D^* : \underset{u \geq 0}{\text{Max}} \ z_D = \underset{x}{\inf} \ \left\{ f(\mathbf{x}) + \sum_{i=1}^{m} u_i g_i(\mathbf{x}) \right\} = \underset{x}{\inf} \ \{ L(\mathbf{x}, \mathbf{u}) \}.$$

For any given $\mathbf{u} \geq \mathbf{0}$, the Lagrangean function L is convex and differentiable in $\mathbf{x}$ and hence its infimum is attained when $L_x(\mathbf{x}, \mathbf{u}) = \mathbf{0}$. The Lagrangean dual therefore becomes

$$P_D^* : \underset{u \geq 0}{\text{Max}} \ z_D = \{ L(\mathbf{x}, \mathbf{u}) : L_x(\mathbf{x}, \mathbf{u}) = \mathbf{0} \}$$

or simply

$$P_D: \underset{\mathbf{x}, \mathbf{u}}{\text{Max}} \ z_D = L(\mathbf{x}, \mathbf{u})$$
$$\text{s.t.} \ L_x(\mathbf{x}, \mathbf{u}) = \mathbf{0}$$
$$\mathbf{u} \geq \mathbf{0},$$

which is precisely the Wolfe dual.

Although both the primal and dual problems may have optimal solutions, their respective objective function values need not be equal in the general case. The difference is referred to as the *duality gap*. One can show that although for continuous convex problems duality gaps do not exist, problems with nonconvex objective functions and/or nonconvex constraints may exhibit duality gaps, as shown in the following example.

Example Consider the primal problem

$$P : \text{Min} \ z = -x_1^3 + 3x_1 + x_2$$
$$\text{s.t.} \ x_1 \leq 0$$
$$x_1 \leq 3$$
$$x_2 \geq 0.$$

Although the constraint $x_1 \leq 3$ is redundant in the formulation, it is retained for reasons to become clear below. The unique optimal solution to the primal problem is $\bar{x}_1 = -1, \bar{x}_2 = 0$, yielding an objective function value $\bar{z} = f(\bar{x}_1, \bar{x}_2) = f(-1, 0) = -2$. The graph of the function $f(x_1, 0)$ is displayed in Fig. 4.10.

Now we formulate the partial Lagrangean dual problem P_D^P with respect to the first constraint.

$$P_D^P : \underset{u \geq 0}{\text{Max}} \ z_D = \underset{\substack{x_1 \leq 3 \\ x_2 \geq 0}}{\inf} \ \{ -x_1^3 + 3x_1 + x_2 + ux_1 \} = \underset{\substack{x_1 \leq 3 \\ x_2 \geq 0}}{\inf} \ \{ L^P(x_1, x_2, u) \}.$$

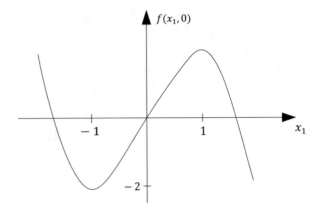

Fig. 4.10 Objective for the numerical example with $x_2 = 0$

Here, L^P denotes the partial Lagrangean function with respect to only the first constraint. Since $x_2 \geq 0$ appears just once with positive sign and separate from all other variables, the infimum of $L^P(x_1, x_2, u)$ will be obtained for $x_2 = 0$. Hence we need to study in some detail the function

$$L^P(x_1, 0, u) = -x_1^3 + (3 + u)x_1 \text{ for } x_1 \leq 3.$$

Using differential calculus this function can be shown to have two local minimal points, one for $x_1 = -\sqrt{1 + u/3}$ with a function value of $-2\sqrt{(1 + u/3)^3}$ and one for $x_1 = 3$ with a function value of $3(u - 6)$; see Fig. 4.11, where the value of u is arbitrarily set at $u = 3$.
Therefore, $h(u) = \inf_{\substack{x_1 \leq 3 \\ x_2 \geq 0}} L(x_1, x_2, u) = \inf_{\substack{x_1 \leq 3 \\ x_2 \geq 0}} \{-x_1^3 + 3x_1 + ux_1 + x_2\}$

$$= \inf_{x_1 \leq 3} L^P(x_1, 0, u)$$

is equal to the smaller of the two local minima, so that the partial Langrangean dual problem P_D^P is reduced to

$$P_D^{P*} : \underset{u \geq 0}{\text{Max}} \, z_D = \min\left(-2\sqrt{\left(1 + \frac{1}{3}u\right)^3}, 3(u - 6)\right).$$

This corresponds to finding the intersection of the two functions in Fig. 4.12, where the function $h(u)$ is marked in boldface. The intersection is found by solving the equation

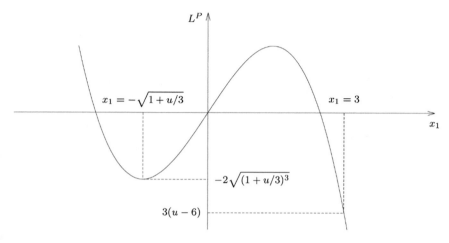

Fig. 4.11 Partial Lagrangean function of the numerical example

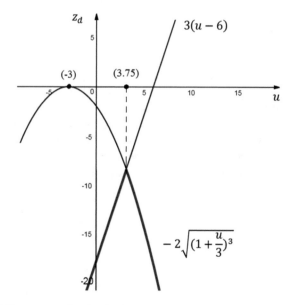

Fig. 4.12 Partial Lagrangean function in the space of the dual variable

$$-2\left(1+\frac{1}{3}\bar{u}\right)^{\frac{3}{2}} = 3(\bar{u} - 6),$$

which has the unique solution $\bar{u} = 3\frac{3}{4}$, resulting in $\bar{z}_D = -6\frac{3}{4}$. With $\bar{z} = -2 > -6\frac{3}{4} = \bar{z}_D$, a duality gap exists. Using set G in the (y_1, y_2)-plane to illustrate the duality gap in this problem, Fig. 4.13 is obtained.

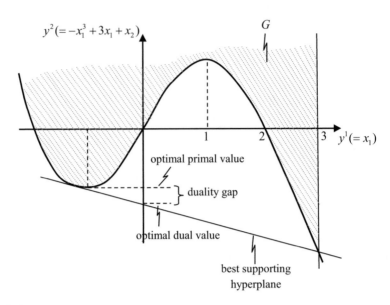

Fig. 4.13 Duality gap

In this figure, the best supporting hyperplane of G is that with the highest y^2-intercept, which corresponds to the optimal objective function value of the dual problem. Note that in this example the optimal objective function value of the primal problem does not correspond to a point on the y^2-axis as in Fig. 4.7.

Returning to our general discussion, some of the main results of nonlinear duality theory will now be stated in general form. To avoid becoming too technical, several proofs have been omitted; for details the reader is referred to Bertsekas (2016) or Bazaraa et al. (2013). Recall the definition of the primal problem and the Lagrangean dual problem

$$\text{P: } \underset{\mathbf{x}}{\text{Min}}\ z = f(\mathbf{x}) \qquad \text{P}_\text{D}: \underset{\mathbf{u} \geq 0}{\text{Max}}\ z_D = \underset{\mathbf{x}}{\inf} L(\mathbf{x}, \mathbf{u})$$
$$\text{s.t.}\, \mathbf{g}(\mathbf{x}) \leq \mathbf{0} \qquad\qquad \text{s.t.}\ \ \mathbf{x} \in \mathbb{R}^n, \mathbf{u} \in \mathbb{R}^m.$$
$$\mathbf{x} \in \mathbb{R}^n.$$

As before, letting $\underset{\mathbf{x}}{\inf} L(\mathbf{x}, \mathbf{u})$ be denoted by $h(\mathbf{u})$, then the Lagrangean dual problem can also be written as

$$\text{P}_\text{D}: \underset{\mathbf{u} \geq 0}{\text{Max}}\ z_D = h(\mathbf{u}).$$
$$\text{s.t.}\, \mathbf{u} \in \mathbb{R}^m.$$

We can then state

Lemma 4.25 The function $h(\mathbf{u})$ is a concave function of $\mathbf{u} \geq \mathbf{0}$.

Proof Select some arbitrary $\lambda \in]0, 1[$. For any $\mathbf{u}^1, \mathbf{u}^2 \geq \mathbf{0}$, we then obtain

$$
\begin{aligned}
L(\mathbf{x}, \lambda\mathbf{u}^1 + (1 - \lambda)\mathbf{u}^2) &= f(\mathbf{x}) + (\lambda\mathbf{u}^1 + (1 - \lambda)\mathbf{u}^2)\mathbf{g}(\mathbf{x}) = \\
&= \lambda(f(\mathbf{x}) + \mathbf{u}^1\mathbf{g}(\mathbf{x})) + (1 - \lambda)(f(\mathbf{x}) + \mathbf{u}^2\mathbf{g}(\mathbf{x})) = \\
&= \lambda L(\mathbf{x}, \mathbf{u}^1) + (1 - \lambda)L(\mathbf{x}, \mathbf{u}^2),
\end{aligned}
$$

which expresses the fact that $L(\mathbf{x}, \mathbf{u})$ is a linear function of $\mathbf{u}$ for given $\mathbf{x}$. Using this result we obtain

$$
\begin{aligned}
h(\lambda\mathbf{u}^1 + (1 - \lambda)\mathbf{u}^2) &= \inf_{\mathbf{x}} \left\{ L\big(\mathbf{x}, \lambda\mathbf{u}^1 + (1 - \lambda)u^2 \big\} \right. \\
&= \inf_{\mathbf{x}} \left\{ \lambda L(\mathbf{x}, \mathbf{u}^1) + (1 - \lambda)L(\mathbf{x}, \mathbf{u}^2) \right\} \\
&\geq \inf_{\mathbf{x}} \left\{ \lambda L(\mathbf{x}, \mathbf{u}^1) \right\} + \inf_{\mathbf{x}} \left\{ (1 - \lambda)L(\mathbf{x}, \mathbf{u}^2) \right\} \\
&= \lambda \inf_{\mathbf{x}} \left\{ L(\mathbf{x}, \mathbf{u}^1) \right\} + (1 - \lambda) \inf_{\mathbf{x}} \left\{ L(\mathbf{x}, \mathbf{u}^2) \right\} \\
&= \lambda h(\mathbf{u}^1) + (1 - \lambda)h(\mathbf{u}^2).
\end{aligned}
$$

The inequality sign above stems from the fact that for any functions $\Phi(\mathbf{x})$ and $\Psi(\mathbf{x})$, we have $\Phi(\mathbf{x}) \geq \inf \{\Phi(\mathbf{x})\}$ and $\Psi(\mathbf{x}) \geq \inf \{\Psi(\mathbf{x})\}$, so that $\Phi(\mathbf{x}) + \Psi(\mathbf{x}) \geq \inf \{\Phi(\mathbf{x})\} + \inf \{\Psi(\mathbf{x})\}$, implying that $\inf\{\Phi(\mathbf{x}) + \Psi(\mathbf{x})\} \geq \inf \{\Phi(\mathbf{x})\} + \inf \{\Psi(\mathbf{x})\}$. Having shown that $h(\lambda\mathbf{u}^1 + (1 - \lambda)\mathbf{u}^2) \geq \lambda h(\mathbf{u}^1) + (1 - \lambda)h(\mathbf{u}^2)$ for any $\lambda \in]0, 1[$ and for any $\mathbf{u}^1, \mathbf{u}^2 \geq \mathbf{0}$, we can conclude that the function h is concave (see Definition 1.29). $\square$

Definition 4.26 A point $(\bar{\mathbf{x}}, \bar{\mathbf{u}})$ is said to satisfy the *optimality conditions* for a problem P if all of the following four conditions are satisfied.

(a) $\bar{\mathbf{x}}$ minimizes $L(\mathbf{x}, \bar{\mathbf{u}})$,
(b) $\bar{\mathbf{u}}\mathbf{g}(\bar{\mathbf{x}}) = \mathbf{0}$
(c) $\mathbf{g}(\bar{\mathbf{x}}) \leq \mathbf{0}$, and
(d) $\bar{\mathbf{u}} \geq \mathbf{0}$.

Note that these optimality conditions closely resemble those given in Definition 4.5; the only difference is that here $\mathbf{x} \geq \mathbf{0}$ is not required. The definition of an optimal Lagrangean multiplier vector is modified accordingly. Theorem 4.4, on the equivalence between points satisfying the saddle point criteria and the optimality conditions, also holds with obvious modifications. The same is true for Lemma 4.10, Theorem 4.11, Lemmas 4.13 and 4.14, as well as Theorem 4.15.

Theorem 4.27 (Weak Lagrangean Duality Theorem) If $\tilde{\mathbf{x}}$ is a primal feasible solution and $\tilde{\mathbf{u}}$ is a dual feasible solution, then $h(\tilde{\mathbf{u}}) \leq f(\tilde{\mathbf{x}})$.

Proof By definition

$$h(\tilde{\mathbf{u}}) = \inf_{\mathbf{x}} \{f(\mathbf{x}) + \tilde{\mathbf{u}}\mathbf{g}(\mathbf{x})\} \leq f(\tilde{\mathbf{x}}) + \tilde{\mathbf{u}}\mathbf{g}(\tilde{\mathbf{x}}) \leq f(\tilde{\mathbf{x}}),$$

where the last inequality follows because $\tilde{\mathbf{u}} \geq \mathbf{0}$ and $\mathbf{g}(\tilde{\mathbf{x}}) \leq \mathbf{0}$ imply $\tilde{\mathbf{u}}\mathbf{g}(\tilde{\mathbf{x}}) \leq 0$. $\square$

The weak Lagrangean duality theorem states that any duality gap $f(\tilde{\mathbf{x}}) - h(\tilde{\mathbf{u}})$ must be nonnegative. It is useful, because a dual feasible solution will therefore provide a lower bound on the optimal objective function value of the primal; similarly a primal feasible solution will provide an upper bound on the optimal objective function value of the Lagrangean dual.

For the following lemma, recall that $\bar{\mathbf{u}}$ is an optimal Lagrangean multiplier vector for problem P if there exists an $\bar{\mathbf{x}}$ such that $(\bar{\mathbf{x}}, \bar{\mathbf{u}})$ satisfies the optimality conditions.

Lemma 4.28 If $\hat{\mathbf{u}}$ is an optimal Lagrangean multiplier vector for problem P, then $(\bar{\mathbf{x}}, \hat{\mathbf{u}})$ will satisfy the optimality conditions (a) – (d) for *every* primal optimal $\bar{\mathbf{x}}$.

Proof Assume that $\hat{\mathbf{x}}$ is associated with $\hat{\mathbf{u}}$, so that $(\hat{\mathbf{x}}, \hat{\mathbf{u}})$ satisfy the optimality conditions. Then $\hat{\mathbf{u}}\mathbf{g}(\hat{\mathbf{x}}) = 0$ and $\hat{\mathbf{u}}\mathbf{g}(\bar{\mathbf{x}}) \leq 0$, and furthermore $L(\hat{\mathbf{x}}, \hat{\mathbf{u}}) \leq L(\bar{\mathbf{x}}, \hat{\mathbf{u}})$, because $\hat{\mathbf{x}}$ minimizes $L(\mathbf{x}, \hat{\mathbf{u}})$. Consequently, we have

$$f(\bar{\mathbf{x}}) \leq f(\hat{\mathbf{x}}) = f(\hat{\mathbf{x}}) + \hat{\mathbf{u}}\mathbf{g}(\hat{\mathbf{x}}) = L(\hat{\mathbf{x}}, \hat{\mathbf{u}}) \leq L(\bar{\mathbf{x}}, \hat{\mathbf{u}}) = f(\bar{\mathbf{x}}) + \hat{\mathbf{u}}\mathbf{g}(\bar{\mathbf{x}}) \leq f(\bar{\mathbf{x}}).$$

Thus all relations must be satisfied as equalities which proves that the optimality conditions (a) and (b) hold for $(\bar{\mathbf{x}}, \hat{\mathbf{u}})$. Condition (c) holds because $\bar{\mathbf{x}}$ is feasible for problem P and (d) holds because $\hat{\mathbf{u}} \geq \mathbf{0}$. $\square$

Lemma 4.28 asserts that an optimal Lagrangean multiplier vector for problem P is associated with P itself rather than with any particular primal optimal solution to P. We can also relate the optimality conditions for P to the optimality for P and its Lagrangean dual P_D as well as to the absence of a duality gap. Specifically, we can formulate and prove the following

Theorem 4.29 A point $(\bar{\mathbf{x}}, \bar{\mathbf{u}})$ satisfies the optimality conditions (a)–(d) for problem P if and only if all of the three following conditions are satisfied.

 (i) $\bar{\mathbf{x}}$ is an optimal solution for P,
 (ii) $\bar{\mathbf{u}}$ is an optimal solution for P_D,
(iii) $\bar{z} = \bar{z}_D$.

Proof Allowing $+\infty$ and $-\infty$ as objective function values, optimal objective function values of P and P_D will always exist, regardless of the existence of feasible or unbounded solutions. Assume now that the optimality conditions are satisfied. Then (i) follows directly from Theorem 4.4 and Lemma 4.3, in a form modified to allow any sign-free $\mathbf{x} \in \mathbb{R}^n$. To prove (ii), we observe that the following relations hold for all $\mathbf{u} \geq \mathbf{0}$, since $\mathbf{ug}(\bar{\mathbf{x}}) \leq 0$ due to optimality condition (c), and $\bar{\mathbf{u}}\mathbf{g}(\bar{\mathbf{x}}) = 0$ via optimality condition (b):

$$h(\mathbf{u}) = \inf_{\mathbf{x}} \{f(\mathbf{x}) + \mathbf{ug}(\mathbf{x})\} \leq f(\bar{\mathbf{x}}) + \mathbf{ug}(\bar{\mathbf{x}}) \leq f(\bar{\mathbf{x}}) + \bar{\mathbf{u}}\mathbf{g}(\bar{\mathbf{x}})$$
$$= \inf_{\mathbf{x}} \{f(\mathbf{x}) + \bar{\mathbf{u}}\mathbf{g}(\mathbf{x})\} = h(\bar{\mathbf{u}}).$$

This means that $\bar{\mathbf{u}}$ is an optimal solution for P_D and hence (ii) is proved. We also found that $f(\bar{\mathbf{x}}) = h(\bar{\mathbf{u}})$, and thus condition (iii) holds.

Conversely, assume that conditions (i), (ii) and (iii) are all satisfied. Then $\bar{\mathbf{u}}\mathbf{g}(\bar{\mathbf{x}}) \leq 0$, so that

$$h(\bar{\mathbf{u}}) = \inf_{\mathbf{x}} \{f(\mathbf{x}) + \bar{\mathbf{u}}\mathbf{g}(\mathbf{x})\} \leq f(\bar{\mathbf{x}}) + \bar{\mathbf{u}}\mathbf{g}(\bar{\mathbf{x}}) \leq f(\bar{\mathbf{x}}).$$

Because $h(\bar{\mathbf{u}}) = f(\bar{\mathbf{x}})$ according to (iii), all relations in the above expression must be satisfied as equalities. This proves (a) and (b) of the optimality conditions. Finally (c) and (d) hold because of the feasibility of $\bar{\mathbf{x}}$ and $\bar{\mathbf{u}}$ for P and P_D, respectively. $\square$

Note that (iii) implies that the duality gap is zero.

This will for now end our discussion of duality. For further developments in this area, the reader is referred to the detailed coverage in Bertsekas (2016) or Bazaraa et al. (2013), see also Luenberger and Ye (2008). However, we will return below to the treatment of duality with some additional results.

4.3.3 Perturbations, Stability and Computational Aspects

The reader will have noticed that the concept of convexity has not been used so far in our discussion of Lagrangean duality. This requirement is now needed in order to obtain the convexity of what is known as the *perturbation function*.

Definition 4.30 The *perturbation function* v of a given convex problem P is defined as

$$v(\mathbf{b}) = \inf_{\mathbf{x}} \{f(\mathbf{x}) : \mathbf{g}(\mathbf{x}) \leq \mathbf{b}\}.$$

One can show that v is a convex function on $\mathbb{R}^m$. Clearly, $v(\mathbf{0})$ equals the optimal objective function value $\bar{z}$ of problem P. The behavior of $v(\mathbf{b})$ for other values of $\mathbf{b}$ are of interest for a sensitivity analysis of P with respect to right-hand side perturbations of the constraints. Although the function v is convex, it need not be everywhere differentiable, as the following example will demonstrate.

Fig. 4.14 Perturbation
function

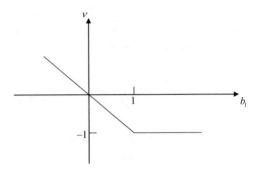

Example Consider the single-variable problem

$$P: \operatorname{Min} x$$
$$\text{s.t.} - x \leq 0$$
$$-x - 1 \leq 0$$
$$x \in \mathbb{R}.$$

Then the perturbation function $v(\mathbf{b}) = v(b_1, b_2)$ is

$$v(b_1, b_2) = \inf_x \{x : -x \leq b_1, -x - 1 \leq b_2\} \quad \text{or}$$

$$v(b_1, b_2) = \max\{-b_1, -1 - b_2\}.$$

For $b_2 = 0$, we obtain

$$v(b_1, 0) = \max\{-b_1, -1\} = \begin{cases} -b_1, & \text{if } b_1 \leq 1 \\ -1, & \text{if } b_1 > 1 \end{cases}.$$

The function $v(b_1, 0)$ is displayed in Fig. 4.14.

Although $v(b_1, 0)$ is convex, it is not differentiable at $b_1 = 1$. As v is not necessarily differentiable everywhere, a gradient may not always be defined. In the definition below the concept of a gradient will therefore be generalized.

Definition 4.31 Given any convex function $\theta(\mathbf{y})$ of $\mathbf{y} \in \mathbb{R}^n$. An n-vector $\mathbf{t}$ is said to be a *subgradient* of θ at $\hat{\mathbf{y}}$, if

$$\theta(\mathbf{y}) \geq \theta(\hat{\mathbf{y}}) + (\mathbf{y} - \hat{\mathbf{y}})^T \mathbf{t} \quad \forall \mathbf{y} \in \mathbb{R}^n.$$

The inequality sign is reversed if concave functions are considered. A subgradient is a generalization of the gradient (see Definition 1.13) in the sense that for a differentiable convex function, a subgradient at any point will coincide with the gradient of the function at this point. We also note that the subgradient definition inequality is identical to the necessary and sufficient condition for the convexity of a

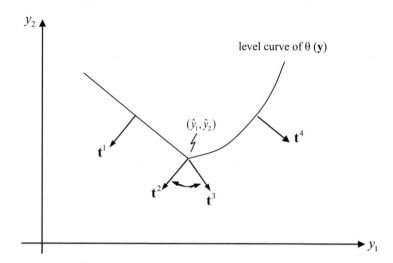

Fig. 4.15 Subgradients of a function

differentiable function (Theorem 1.34 with $\mathbf{t} = \nabla\theta(\mathbf{y})$). For other treatments of subgradients the reader is referred to Bazaraa et al. (2013), Bertsekas (2016) or Sun and Yuan (2006). As far as convex functions are concerned, the subgradient concept is a strict generalization of the gradient only at "corners" or other nondifferentiable points of nonsmooth functions.

In Fig. 4.15, the subgradients $\mathbf{t}^1$ and $\mathbf{t}^4$ are unique and hence gradients, whereas at $(\hat{y}_1, \hat{y}_2)$ both $\mathbf{t}^2$ and $\mathbf{t}^3$ (and in fact any linear convex combination of $\mathbf{t}^2$ and $\mathbf{t}^3$) are subgradients. Denote the *subdifferential*, which is defined as the *set* of all subgradients of θ at the point $(\hat{y}_1, \hat{y}_2)$ by $\underline{\nabla}\theta(\hat{y}_1, \hat{y}_2)$; by abuse of language we will sometimes also use $\underline{\nabla}\theta$ to denote an *individual* subgradient.

Example For the above perturbation function (with $\mathbf{y}$ instead of $\mathbf{b}$), $\nu(y_1, y_2) = \max\{-y_1, -1 - y_2\}$, we display the level curve $\nu = -2$ in Fig. 4.16.

We will now show that for $(\hat{y}_1, \hat{y}_2) = (-1, -1)$ a subgradient of the perturbation function of ν is $\mathbf{t} = [-1, 0]^T$. On one hand $\nu(\mathbf{y}) - \nu(\hat{\mathbf{y}}) = \nu(y_1, y_2) - \nu(-1, -1) = \max\{-y_1, -1 - y_2\} - \max\{1, 0\} = \max\{-y_1, -1 - y_2\} - 1 = \max\{-1 - y_1, -2 - y_2\}$, and, on the other hand, $(\mathbf{y} - \hat{\mathbf{y}})^T \mathbf{t} = [y_1 + 1,$ $y_2 + 1] \begin{bmatrix} -1 \\ 0 \end{bmatrix} = -1 - y_1$, so that $\nu(\mathbf{y}) - \nu(\hat{\mathbf{y}}) \geq -1 - y_1 = (\mathbf{y} - \hat{\mathbf{y}})^T \mathbf{t}$. In fact, because ν is differentiable in the vicinity of $[\hat{y}_1, \hat{y}_2] = [-1, -1]$, where it takes the value $\nu = -y_1$, the subgradient $\mathbf{t} = [-1, 0]^T$ is unique and coincides with the gradient $\nabla\nu$ of ν at the point $(-1, -1)$. Similarly, $[0, -1]^T$ is the unique subgradient of ν at $(\hat{y}_1, \hat{y}_2) = (3, -1)$, in the vicinity of which $\nu = -1 - y_2$, so that the subgradient is identical to the gradient of ν at this point.

Fig. 4.16 Level curve for
the perturbation function

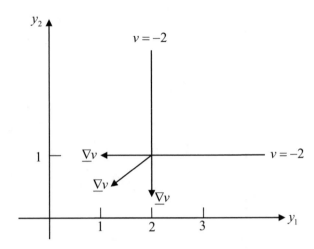

At the point $(\hat{y}_1, \hat{y}_2) = (2, 1)$, where $v(\hat{y}_1, \hat{y}_2) = \max\{-\hat{y}_1, -1 - \hat{y}_2\} = \max\{-2, -2\} = -2$, both $[-1, 0]^T$ and $[0, -1]^T$ are subgradients of the perturbation function v, as well as for example $[-\frac{1}{2}, -\frac{1}{2}]^T$. To see this, calculate

$$(\mathbf{y} - \hat{\mathbf{y}})^T \mathbf{t} = [y_1 - 2, y_2 - 1]\begin{bmatrix} -\frac{1}{2} \\ -\frac{1}{2} \end{bmatrix} = \frac{1}{2}(-y_1 + 2 - y_2 + 1) = \frac{1}{2}(-y_1 - y_2 + 3)$$

and also

$$v(\mathbf{y}) - v(\hat{\mathbf{y}}) = v(y_1, y_2) - (-2) = \max\{-y_1, -1 - y_2\} + 2$$
$$= \max\{2 - y_1, 1 - y_2\} \geq \frac{1}{2}(3 - y_1 - y_2)$$

(using the fact that $\max\{\alpha, \beta\} = \frac{1}{2}(\alpha + \beta) + \frac{1}{2}|\alpha - \beta| \geq \frac{1}{2}(\alpha + \beta) \; \forall \; \alpha, \beta \in \mathbb{R}$). Therefore, $v(\mathbf{y}) \geq v(\hat{\mathbf{y}}) + (\mathbf{y} - \hat{\mathbf{y}})^T \mathbf{t}$, so that $\mathbf{t} = [-\frac{1}{2}, -\frac{1}{2}]^T$ is a subgradient of v at the point $(2, 1)$. In fact, every $\underline{\nabla} v$ of the form $\underline{\nabla} v = \lambda_1[-1, 0]^T + \lambda_2[0, -1]^T$, where $\lambda_1 \geq 0, \lambda_2 \geq 0$ and $\lambda_1 + \lambda_2 = 1$, i.e. every linear convex combination of $[-1, 0]^T$ and $[0, -1]^T$ can be shown to be a subgradient of v at the point $(2, 1)$.

An interesting relationship between the perturbed and unperturbed problem is provided by the following theorem due to Everett (1963), whose method will be considered in the next chapter. Recall the optimality conditions of Definition 4.26.

Theorem 4.32 If the point $(\bar{\mathbf{x}}, \bar{\mathbf{u}})$ satisfies the optimality conditions for problem P, then $\bar{\mathbf{x}}$ is also an optimal solution to the perturbed problem P^b given by:

$$P^b: \text{Min } z^b = f(\mathbf{x})$$
$$\text{s.t. } \mathbf{g}(\mathbf{x}) \leq \mathbf{b}$$
$$\mathbf{x} \in \mathbb{R}^n,$$

where $\mathbf{b} \in \mathbb{R}^m$ is any right-hand side vector, such that $g_i(\bar{\mathbf{x}}) = b_i$ if $\bar{u}_i > 0$ and g_i $(\bar{\mathbf{x}}) \le b_i$ if $\bar{u}_i = 0$.

Proof Define $\mathbf{g}^b(\mathbf{x}) := \mathbf{g}(\mathbf{x}) - \mathbf{b}$ and $L^b(\mathbf{x}, \mathbf{u}) := f(\mathbf{x}) + \mathbf{u}\mathbf{g}^b(\mathbf{x})$. By assumption $(\bar{\mathbf{x}}, \bar{\mathbf{u}})$ satisfies the optimality conditions for P. Thus $\bar{\mathbf{x}}$ minimizes $L(\mathbf{x}, \bar{\mathbf{u}})$, i.e.,

$$L(\bar{\mathbf{x}}, \bar{\mathbf{u}}) \le L(\mathbf{x}, \bar{\mathbf{u}}) \quad \forall \mathbf{x} \in \mathbb{R}^n$$

$$\text{or } f(\bar{\mathbf{x}}) + \bar{\mathbf{u}}\mathbf{g}(\bar{\mathbf{x}}) \le f(\mathbf{x}) + \bar{\mathbf{u}}\mathbf{g}(\mathbf{x}) \quad \forall \mathbf{x} \in \mathbb{R}^n.$$

Subtracting $\bar{\mathbf{u}}\mathbf{b}$ from both sides yields

$$f(\bar{\mathbf{x}}) + \bar{\mathbf{u}}\mathbf{g}(\bar{\mathbf{x}}) - \bar{\mathbf{u}}\mathbf{b} \le f(\mathbf{x}) + \bar{\mathbf{u}}\mathbf{g}(\mathbf{x}) - \bar{\mathbf{u}}\mathbf{b} \quad \forall \mathbf{x} \in \mathbb{R}^n$$

or simply

$$L^b(\bar{\mathbf{x}}, \bar{\mathbf{u}}) \le L^b(\mathbf{x}, \bar{\mathbf{u}}) \quad \forall \mathbf{x} \in \mathbb{R}^n,$$

which is optimality condition (a) for the problem P^b (see Definition 4.26). Furthermore, $g_i(\bar{\mathbf{x}}) = b_i$ for $\bar{u}_i > 0$ and $g_i(\bar{\mathbf{x}}) \le b_i$ for $\bar{u}_i = 0$ imply $\bar{u}_i(g_i(\bar{\mathbf{x}}) - b_i) = 0$ $\forall$ $i = 1, \ldots, m$, so that $\bar{\mathbf{u}}\mathbf{g}^b(\bar{\mathbf{x}}) = \bar{\mathbf{u}}(\mathbf{g}(\bar{\mathbf{x}}) - \mathbf{b}) = \mathbf{0}$. Finally, $\mathbf{g}^b(\bar{\mathbf{x}}) = \mathbf{g}(\bar{\mathbf{x}}) - \mathbf{b} \le \mathbf{0}$ and $\bar{\mathbf{u}} \ge \mathbf{0}$, so that optimality conditions (b), (c) and (d) are also satisfied for P^b. Using Theorem 4.29, we then conclude that $\bar{\mathbf{x}}$ is optimal for P^b. $\square$

Theorem 4.32 is important for postoptimality analysis, in that it specifies how much the right-hand side may be perturbed without affection the optimal solution. For a further discussion of perturbations and postoptimality, see e.g., Bazaraa et al. (2013).

Definition 4.33 The convex problem P is said to be *stable* if a real number $s > 0$ exists such that its perturbation function $v(\mathbf{y})$ satisfies the inequality

$$\frac{v(\mathbf{0}) - v(\mathbf{y})}{\|\mathbf{y}\|} \le s \; \forall \mathbf{y} \in \mathbb{R}^m \backslash \{\mathbf{0}\}.$$

In other words, stability requires that the perturbation function v does not decrease infinitely steeply in any direction. Using the notion of stability the following important theorem can now be formulated; for a proof see Bazaraa et al. (2013).

Theorem 4.34 Suppose that an optimal solution exists for problem P. Then

(i) An optimal multiplier vector $\bar{\mathbf{u}}$ exists is and only if and only if P is stable.
(ii) $\bar{\mathbf{u}}$ is an optimal multiplier vector if and only if $-\bar{\mathbf{u}}$ is a subgradient of the perturbation function

$$v(\mathbf{y}) \text{ at } \mathbf{y} = \mathbf{0}.$$

Stability is a regularity condition similar to a constraint qualification. However, the stability condition involves not only the constraints but the objective function as well. It is intuitively clear that both the objective function f and the constraints g should be involved. For instance, one can construct examples where f is so "well-behaved" that the problem is stable no matter what g is selected. Conversely, with a sufficiently "ill-behaved" g, the problem will not be stable for any f (see Geoffrion 1971). Now the strong Lagrangean duality theorem can be stated.

Theorem 4.35 (Strong Lagrangean Duality) Assume that the problem P is stable. Then the following four statements are true.

(i) The Lagrangean dual problem P_D has an optimal solution.
(ii) The problems P and P_D have equal optimal objective function values.
(iii) $\bar{\mathbf{u}}$ is optimal for P_D if and only if $-\bar{\mathbf{u}}$ is a subgradient of the perturbation function $v(\mathbf{y})$ at $\mathbf{y} = \mathbf{0}$.
(iv) Suppose that $\bar{\mathbf{u}}$ solves P_D. Then $\bar{\mathbf{x}}$ solves P if and only if $(\bar{\mathbf{x}}, \bar{\mathbf{u}})$ satisfy the optimality conditions for problem P.

It should be pointed out that the existence of at least one optimal solution to problem P has not been assumed in the above theorem. Finally we state, without proof, the following converse Lagrangean duality theorem which gives sufficient conditions for the stability of problem P.

Theorem 4.36 (Converse Lagrangean Duality) Assume that f and g are continuous functions such that $\bar{\mathbf{u}}$ is optimal for P_D, $\bar{\mathbf{x}}$ is the unique minimal point of $L(\mathbf{x}, \bar{\mathbf{u}})$, and $g(\bar{\mathbf{x}}) \leq \mathbf{0}$. Then the following three statements are true.

(i) $\bar{\mathbf{x}}$ is the unique solution to problem P;
(ii) $\bar{\mathbf{u}}$ is an optimal multiplier vector for P;
(iii) P is stable.

This section will be concluded by considering some computational aspects for solving the Lagrangean dual problem P_D. We also show some cases where it is advantageous to solve the dual problem rather than the primal problem directly.

Lemma 4.37 Consider the problem P and assume that $\bar{\mathbf{u}} \geq \mathbf{0}$. If $\bar{\mathbf{x}}$ minimizes $L(\mathbf{x}, \bar{\mathbf{u}})$, i.e., if $h(\bar{\mathbf{u}}) = L(\bar{\mathbf{x}}, \bar{\mathbf{u}}) = \inf_{\mathbf{x}} L(\mathbf{x}, \bar{\mathbf{u}})$, then $g(\bar{\mathbf{x}})$ is a subgradient of h at $\bar{\mathbf{u}}$.

Proof By definition, $h(\mathbf{u}) = \inf_{\mathbf{x}} L(\mathbf{x}, \mathbf{u}) \leq L(\mathbf{x}, \mathbf{u}) = f(\mathbf{x}) + \mathbf{u}g(\mathbf{x}) \quad \forall \mathbf{x} \in \mathbb{R}^n$. Inserting $\mathbf{x} = \bar{\mathbf{x}}$, we obtain $h(\mathbf{u}) \leq f(\bar{\mathbf{x}}) + \mathbf{u}g(\bar{\mathbf{x}})$. However, by assumption $h(\bar{\mathbf{u}}) = L(\bar{\mathbf{x}}, \bar{\mathbf{u}}) = f(\bar{\mathbf{x}}) + \bar{\mathbf{u}}g(\bar{\mathbf{x}})$ Subtracting this from the previous inequality we find $h(\mathbf{u}) - h(\bar{\mathbf{u}}) \leq f(\bar{\mathbf{x}}) + \mathbf{u}g(\bar{\mathbf{x}}) - [f(\bar{\mathbf{x}}) + \bar{\mathbf{u}}g(\bar{\mathbf{x}})]$, i.e., $h(\mathbf{u}) \leq h(\bar{\mathbf{u}}) + (\mathbf{u} - \bar{\mathbf{u}})g(\bar{\mathbf{x}})$ $\forall \mathbf{u} \geq \mathbf{0}$. As h is concave (by virtue of Lemma 4.25), this shows that $g(\bar{\mathbf{x}})$ is a subgradient of h at $\bar{\mathbf{u}}$, see Definition 4.31.□

Therefore each time the dual function $h(\mathbf{u})$ is evaluated, a subgradient of h at this point can also easily be obtained. Although h need not be differentiable, we might nevertheless apply one of the gradient techniques of Sect. 2.2 to maximize h by using the subgradient $\mathbf{g}(\mathbf{x})$ instead of the gradient. In practice, this approach tends to be quite efficient, although convergence cannot be guaranteed. Another approach is to use algorithms specifically designed for nondifferentiable optimization; some of these will be covered in the following chapters.

Considering computational advantages by solving the Lagrangean dual rather than the primal problem, let us first point out that the result of the weak Lagrangean duality theorem (Theorem 4.27) can be used to assess how near-optimal a proposed solution is; furthermore it can provide bounds for the optimal value of the objective function. Another important point is that by carefully selecting the constraints with respect to which the Lagrangean dual is formed, it may be possible to obtain a dual problem which is more convenient to solve than the primal. This may be particularly advantageous when the primal problem is structured. We will conclude this chapter by an example [based on an idea due to Erlander (1981); see also Svanberg (1982)] for which the dual problem is computationally much easier to solve than the primal.

Example Consider the problem

$$\text{P}: \quad \underset{\mathbf{x}}{\text{Min}} \ z = f(\mathbf{x}) = \sum_{j=1}^{n} \frac{c_j}{x_j}$$
$$\text{s.t.} \ \mathbf{Ax} = \mathbf{b}$$
$$\mathbf{x} \geq \mathbf{0},$$

where $c_j \geq 0, j = 1, \ldots, n$. If only one of the coefficients c_j is nonzero, problem P is of the fractional programming type and can be transformed into a linear programming problem (see, e.g., Eiselt and Sandblom 2007), but if two or more of the coefficients c_j are positive, this is no longer possible. Assume for now that all c_j are positive. For $\mathbf{x} > \mathbf{0}$, the objective function $f(\mathbf{x})$ is then convex and nonnegative, and assuming that the feasible region is bounded, problem P will possess a finite optimal solution point $\bar{\mathbf{x}} > \mathbf{0}$. The Lagrangean dual problem P_D with respect to all structural constraints is then

$$\text{P}_\text{D}: \quad \underset{\mathbf{u}}{\text{Max}} \ z_D = \underset{\mathbf{x} \geq \mathbf{0}}{\inf} L(\mathbf{x}, \mathbf{u}) = \underset{\mathbf{x} \geq \mathbf{0}}{\inf} \{f(\mathbf{x}) + \mathbf{u}(\mathbf{Ax} - \mathbf{b})\},$$

where the vector $\mathbf{u}$ is not restricted in sign since it is associated with the primal equality constraints. The fact that by assumption an optimal solution $\bar{\mathbf{x}}$ exists for problem P, together with Theorems 4.4, 4.18, and 4.29 leads to the conclusion that there exists a vector $\bar{\mathbf{u}}$, such that

$$f(\bar{\mathbf{x}}) = \underset{\mathbf{x} \geq \mathbf{0}}{\inf} \{f(\mathbf{x}) + \bar{\mathbf{u}}(\mathbf{Ax} - \mathbf{b})\} = h(\bar{\mathbf{u}}).$$

Now $\bar{\mathbf{u}}\mathbf{A} \geq \mathbf{0}$ because otherwise the Lagrangean expression in brackets would be unbounded from below. We will even show below that $\bar{\mathbf{u}}\mathbf{A} > \mathbf{0}$, since otherwise the optimality of $\bar{\mathbf{x}}$ for problem P would be contradicted. As the Lagrangean function $L(\mathbf{x}, \bar{\mathbf{u}}) = f(\mathbf{x}) + \bar{\mathbf{u}}(A\mathbf{x} - \mathbf{b})$ attains its minimum at $\bar{\mathbf{x}} > \mathbf{0}$, it follows that

$$L_{\mathbf{x}}(\bar{\mathbf{x}}, \bar{\mathbf{u}}) = \nabla f(\bar{\mathbf{x}}) + A^T \bar{\mathbf{u}}^T = \mathbf{0} \quad \text{or} \quad (\nabla f(\bar{\mathbf{x}}))^T = -\bar{\mathbf{u}}A.$$

However, $\nabla f(\mathbf{x}) = \nabla \sum_{j=0}^{n} c_j / x_j = \begin{bmatrix} -c_1/x_1^2 \\ -c_2/x_2^2 \\ \vdots \\ -c_n/x_n^2 \end{bmatrix}$ so that $-\dfrac{c_j}{\bar{x}_j^2} = -(\bar{\mathbf{u}}A)_j$ or simply

$\bar{x}_j = \sqrt{\dfrac{c_j}{(\bar{\mathbf{u}}A)_j}}$ for $j = 1, \ldots, n$, where we must have $\bar{\mathbf{u}}A > \mathbf{0}$. Furthermore,

$-f(\bar{\mathbf{x}}) = (\nabla f(\bar{\mathbf{x}}))^T \bar{\mathbf{x}} = -\bar{\mathbf{u}}A\bar{\mathbf{x}}$, so that $L(\bar{\mathbf{x}}, \bar{\mathbf{u}}) = f(\bar{\mathbf{x}}) + \bar{\mathbf{u}}(A\bar{\mathbf{x}} - \mathbf{b}) = 2\bar{\mathbf{u}}A\bar{\mathbf{x}} - \bar{\mathbf{u}}$ $\mathbf{b}$. By considering $\mathbf{u}$ in general, such that $\mathbf{u}A > \mathbf{0}$, the Lagrangean dual problem P_D becomes

$$P_D' : \underset{\mathbf{x}, \mathbf{u}}{\text{Max}} \ z_D = 2\mathbf{u}A\mathbf{x} - \mathbf{u}\mathbf{b}$$

$$\text{s.t. } x_j = \sqrt{\dfrac{c_j}{(\mathbf{u}A)_j}}, j = 1, \ldots, n$$

$$\mathbf{u}A > \mathbf{0},$$

where, implicitly, due to the equality constraints, $\mathbf{x} > \mathbf{0}$ must hold. By substituting the expression for $\mathbf{x}$ in the objective function and by realizing that $\mathbf{u}A > \mathbf{0}$ can be considered implicitly satisfied, an unconstrained maximization problem with variables $\mathbf{u}$ results; its optimal solution $\bar{\mathbf{u}}$ yields the optimal solution $\bar{\mathbf{x}}$ to the original problem P, given by the expressions

$$\bar{x}_j = \sqrt{\dfrac{c_j}{(\mathbf{u}A)_j}}, j = 1, \ldots, n.$$

If any of the coefficients c_j are zero, so that the corresponding variable x_j is missing from the objective function, one might attempt to eliminate such a variable from the constraint equations. This approach will work as long as the nonnegativity of the eliminated variable will be assured.

Example As a numerical illustration consider the problem

$$P: \underset{x_1, x_2}{\text{Min}} \ z = \frac{2}{x_1} + \frac{3}{x_2}$$
$$\text{s.t. } 4x_1 + x_2 = 9$$
$$x_1, x_2 \geq 0.$$

With $A = [4, 1]$, the Lagrangean dual problem P_D' becomes

$$P'_D : \quad \underset{\mathbf{x}, u}{\text{Max}} \, z_D = 2u[4, 1] \begin{bmatrix} x_1 \\ x_2 \end{bmatrix} - 6u$$

$$\text{s.t. } x_1 = \sqrt{\frac{2}{4u}}$$

$$x_2 = \sqrt{\frac{3}{u}}$$

$$u[4, 1] > \mathbf{0}.$$

Substituting, we obtain

$$z_D = 8u\sqrt{\frac{2}{4u}} + 2u\sqrt{\frac{3}{u}} - 6u = \left(4\sqrt{2} + 2\sqrt{3}\right)\sqrt{u} - 6u.$$

The Lagrangean dual objective function value z_D is maximized for $\sqrt{\bar{u}} = \dfrac{2\sqrt{2} + \sqrt{3}}{6},$ so that we can conclude that

$$\bar{x}_1 = \frac{6}{4 + \sqrt{6}} \approx 0.9303, \text{ and}$$

$$\bar{x}_2 = \frac{18}{3 + 2\sqrt{6}} \approx 2.2788.$$

Similar techniques can be used for linearly constrained problems with entropy type objective functions, such as

$$\sum_{j=1}^{n} c_j / x_j^\alpha \text{ for } \alpha > 1, \quad \sum_{j=1}^{n} c_j \log x_j, \text{ or } \sum_{j=1}^{n} c_j x_j^\alpha \log x_j \quad \text{for } \alpha > 0.$$

References

Bazaraa MS, Sherali HD, Shetty CM (2013) *Nonlinear programming: theory and algorithms.* (3rd ed.) Wiley, New York

Bertsekas DP (2016) *Nonlinear programming.* (3rd ed.) Athena Scientific, Belmont, MA

Cottle RW (1963) Symmetric dual quadratic programs. *Quarterly of Applied Mathematics* **21**: 237-243

Cottle RW, Thapa MN (2017) *Linear and nonlinear optimization.* Springer-Verlag, Berlin-Heidelberg-New York

Dorn WS (1960a) Duality in quadratic programming. *Quarterly of Applied Mathematics* **18**: 155-162

Dorn WS (1960b) A duality theorem for convex programs. *IBM Journal of Research and Development* **4**: 407-413

Eiselt HA, Sandblom C-L (2007) *Linear programming and its applications.* Springer-Verlag, Berlin-Heidelberg

Erlander S (1981) Entropy in linear programs. *Mathematical Programming* **21**: 137-151

Everett H (1963) Generalized Lagrange multiplier method for solving problems of optimum allocation of resources. *Operations Research* **11**: 399-417

Geoffrion AM (1971) Duality in nonlinear programming: a simplified applications-oriented development. *SIAM Review* **13** 1-37

John F (1948) Extremum problems with inequalities as side constraints. pp. 187-204 in: Friedrichs KO, Neugebauer OE, Stoker JJ (eds.) *Studies and Essays, Courant Anniversary Volume.* Wiley Interscience, New York

Karush W (1939): *Minima of functions of several variables with inequalities as side conditions.* MS Thesis, Department of Mathematics, University of Chicago, IL

Kuhn HW, Tucker AW (1951) Nonlinear programming. Pages 481-492 in Neyman J (ed.) *Proceedings of the Second Berkeley Symposium on Mathematical Statistics and Probability.* University of California Press, Berkeley, CA

Luenberger DL, Ye Y (2008) *Linear and nonlinear programming.* (3rd ed.) Springer-Verlag, Berlin-Heidelberg-New York

Magnanti TL (1974) Fenchel und Lagrange duality are equivalent. *Mathematical Programming* **7**: 253-258

Mangasarian OL (1969) *Nonlinear programming.* McGraw-Hill, New York

Rockafellar RT (1970) *Convex analysis.* University Press, Princeton, NJ

Slater M (1950) *Lagrange multipliers revisited: a contribution to nonlinear programming.* Cowles Commission Discussion Paper, Mathematics 403, Yale University, New Haven, CT

Sun W, Yuan Y (2006) *Optimization theory and methods. Nonlinear programming.* Springer-Verlag, Berlin-Heidelberg-New York

Svanberg K (1982) An algorithm for optimum structural design using duality. *Mathematical Programming Study* **20**: 161-177

Wolfe P (1961) A duality theorem for non-linear programming. *Quarterly of Applied Mathematics* **19**: 239-244

Chapter 5
General Solution Methods for Constrained Optimization

In Chap. 1 we dealt with optimality conditions for unconstrained problems, and then we described a number of algorithms for such problems in Chap. 2. The previous chapter discussed optimality conditions for constrained problems, and in parallel fashion, we will now turn to algorithms for such optimization problems. This chapter will cover some general approaches for constrained optimization, and the following three chapters describe some of the many algorithms for specific types of problems.

Whereas unconstrained optimization algorithms usually start from some initial point, the situation in the constrained case is quite different. Most of the methods that we will encounter require an initial solution that is feasible, i.e., one that satisfies all given constraints. Such an initial feasible point may be difficult to find; and then there is, of course, the case, in which there exists no feasible solution at all. Therefore, a number of methods require a preliminary search method that finds an initial feasible solution point, from which the optimization procedure commences. Borrowing from linear programming terminology, we will call such search a *Phase 1* procedure, whereas, starting with an initial feasible solution in Phase 1, the process of finding an optimal point is referred to as *Phase 2*. Although depending on the specific algorithm under consideration, a way of finding an initial feasible solution may naturally present itself, we will here present a general Phase 1 procedure, which can be used in conjunction with any constrained optimization algorithm that needs a feasible starting point. Assume that the problem under consideration is

$$P: \ \text{Min} \ z = f(\mathbf{x})$$
$$\text{s.t.} \ g_i(\mathbf{x}) \leq 0, \ i = 1, \ \ldots, \ m$$
$$\mathbf{x} \in \mathbb{R}^n.$$

© Springer Nature Switzerland AG 2019
H. A. Eiselt, C.-L. Sandblom, *Nonlinear Optimization*, International Series in
Operations Research & Management Science 282,
https://doi.org/10.1007/978-3-030-19462-8_5

A General Phase 1 Procedure

Step 1 Solve the unconstrained problem

$$P_1: \text{Min } z_1 = g_1(\mathbf{x})$$
$$\mathbf{x} \in \mathbb{R}^n$$

by any appropriate technique. Stop as soon as an $\mathbf{x}^1$, such that $g_1(\mathbf{x}^1) \leq 0$ has been found. If no such $\mathbf{x}^1$ can be found, we stop; the problem P has no feasible solution. Otherwise, set $k := 1$ and go to Step 2.

Step 2 Is $k = m$?

If yes: Stop, an initial feasible solution for P has been found.
If no: Go to Step 3.

Step 3 Using any suitable constrained optimization technique, solve the problem

$$P_{k+1}: \text{Min } z_{k+1} = g_{k+1}(\mathbf{x})$$
$$\text{s.t. } g_i(\mathbf{x}) \leq 0, \ i = 1, \ \ldots, \ k,$$

using $\mathbf{x}^k$ as a starting feasible point. Stop as soon as an $\mathbf{x}^{k+1}$ with $g_{k+1}(\mathbf{x}^{k+1}) \leq 0$ has been found. If no such $\mathbf{x}^{k+1}$ can be found, the problem P has no feasible solution and we stop. Otherwise set $k := k + 1$ and go to Step 2.

If an interior feasible point is required rather than just a feasible point, then the above procedure is simply modified by replacing "≤ 0" by "< 0" in Steps 1 and 3.

5.1 Barrier and Penalty Methods

A major category of algorithms is formed by the *barrier* and *penalty function methods* which are described below. We first discuss the principle underlying the barrier function methods. For that purpose, consider the convex programming problem

$$P: \text{Min } z = f(\mathbf{x})$$
$$g_i(\mathbf{x}) \leq 0, \ i = 1, \ \ldots, \ m$$
$$\mathbf{x} \in \mathbb{R}^n,$$

where the nonnegativity constrains, if any, are expressed explicitly as structural constraints. The idea is to convert P into a corresponding unconstrained problem with an optimal solution near to that of the original problem. For $\mathbf{x}$ such that $g^i(\mathbf{x}) < 0, \ i = 1, \ldots, m$, and for $r > 0$, the *barrier function* $B(\mathbf{x}, r)$ is defined as

Fig. 5.1 Objective function
of the numerical example on
its domain

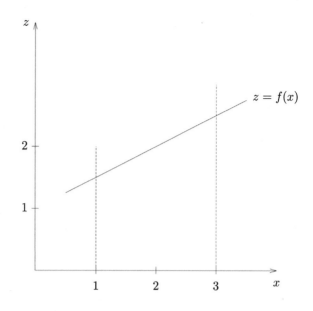

$$B(\mathbf{x},\ r) = f(\mathbf{x}) - r \sum_{i=1}^{m} \frac{1}{g_i(\mathbf{x})}.$$

The terms $-\dfrac{r}{g_i(\mathbf{x})}$ are called the *boundary repulsion* terms and force $\mathbf{x}$ to stay
within the feasible region so that unconstrained techniques become applicable. As a
simple numerical illustration we consider the problem

$$\begin{aligned} &\text{P:}\quad \text{Min }\ z = f(x) = \tfrac{1}{2}x + 1 \\ &\text{s.t. }\ g_1(x) = x - 3 \leq 0 \\ &\qquad\quad g_2(x) = -x + 1 \leq 0 \\ &\qquad\quad x \in \mathbb{R}, \end{aligned}$$

the graph of which is shown in Fig. 5.1.

If $r = 0.1$ is selected, the barrier function becomes

$$B(x,\ 0.1) = f(x) - \frac{0.1}{g_1(x)} - \frac{0.1}{g_2(x)} = \tfrac{1}{2}x + 1 - x - \frac{0.1}{x-3} - \frac{0.1}{-x+1},$$

which is graphically illustrated in Fig. 5.2.

Returning to the general case, we can see that starting from any interior point of
the feasible region and trying to perform unconstrained minimization of the auxiliary
function $B(\mathbf{x}, r)$, the boundary repulsion terms have the effect that the point $\mathbf{x}$ will
always remain inside the feasible region. The approach then consists in solving a
sequence of unconstrained minimization problems with successively smaller $r > 0$,
such that the generated sequence of optimal points to the unconstrained problems

Fig. 5.2 Barrier function

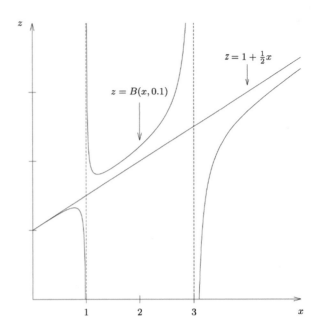

will converge to the solution of the original constrained problem P. Here we have used boundary repulsion terms of the type $-\dfrac{r}{g_i(\mathbf{x})}$, other types are also possible as shown later in this section.

In penalty function methods, *penalty* terms instead of boundary repulsion terms are added in order to represent large penalties for violating any constraint. For example, a penalty function could be

$$P(\mathbf{x},\ r) = f(\mathbf{x}) + r \sum_{i=1}^{m} \left[\max\{0,\ g_i(\mathbf{x})\}\right]^2$$

for a given $r > 0$. Using $r = 10$, the penalty function for our numerical example above would be

$$P(x,\ 10) = \tfrac{1}{2}x + 1 + 10[\max\{0,\ x - 3\}]^2 + 10[\max\{0,\ -x + 1\}]^2,$$

where the penalty terms force x to stay within a close vicinity of the feasible region, so that unconstrained techniques can be used. For our previous numerical example, the function $P(x, 10)$ is shown graphically in Fig. 5.3.

For the equation-constrained problem with $\mathbf{g}(\mathbf{x}) = \mathbf{0}$, a natural penalty function would be $P_{eq}(\mathbf{x}, r) = f(\mathbf{x}) + r\|\mathbf{g}(\mathbf{x})\|^2$.

The approach for penalty function methods is the same as for barrier function methods except that $r > 0$ is successively chosen larger and larger. It is perhaps

Fig. 5.3 Penalty function

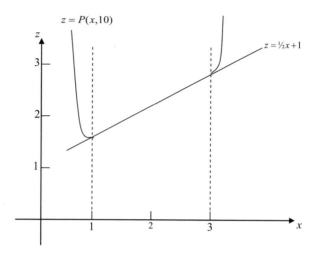

$z = P(x,10)$

$z = \frac{1}{2}x + 1$

surprising to find out that penalty functions exist, for which even finite values of the penalty parameter r will lead to solutions of the unconstrained problem that exactly coincide with an optimal solution to the original problem. This is the case with the penalty function $P(\mathbf{x}, \ r) = f(\mathbf{x}) + r \sum_{i=1}^{m} \max\{0, \ g_i(\mathbf{x})\}$, for which the following result holds true.

Theorem 5.1 Suppose that the point $(\bar{\mathbf{x}}, \ \bar{\mathbf{u}})$ satisfies the optimality conditions 4.26 for the original problem P. Then for values of the parameter r in the above penalty function $P(\mathbf{x}, r)$ that satisfy $r > \max\{\bar{u}_1, \ \bar{u}_2, \ \dots, \ \bar{u}_m\}$, $\bar{\mathbf{x}}$ will also be a minimal point of $P(\mathbf{x}, r)$.

Penalty functions, which have solutions that are exactly the same as those of the original problems are called *exact penalty functions*. The function $P(\mathbf{x}, r)$ is such a function in view of Theorem 5.1. The advantage of an exact penalty function is that if the value of the parameter r is properly chosen, only one minimization needs to be carried out. The drawback is that $P(\mathbf{x}, r)$ is not differentiable. For a proof of the theorem and further discussion, see Bazaraa et al. (2013), Bertsekas (2016), and Luenberger and Ye (2008).

In general one can regard barrier methods as *inner approximation methods* in the sense that the auxiliary problems have feasible regions that approximate from inside the feasible region of the original problem. Similarly, penalty methods can be interpreted as *outer approximation methods*, since they approximate the feasible region from the outside. By convention, the term "penalty function methods" has become accepted to mean "penalty and barrier function methods" although this is a somewhat misleading terminology.

The first algorithm of the barrier and penalty function type to be presented here is the *Sequential Unconstrained Minimization Technique*, or *SUMT* for short. This method is due to Fiacco and McCormick (1964a, b) and is a solution technique using barrier functions. Following an idea of Carroll (1961), Fiacco and McCormick (1964a, b) have devised an algorithm using the barrier function

$$B(\mathbf{x}, \ r) = f(\mathbf{x}) - r \sum_{i=1}^{m} \frac{1}{g_i(\mathbf{x})}$$

to solve the original problem P. This approach is both simple and intuitive: for a given positive strictly decreasing sequence of numbers r_k, $k = 1, 2, \ldots$ such that $\lim_{k \to \infty} r_k = 0$, one solves the auxiliary unconstrained barrier function problems

$$P^B : \underset{\mathbf{x} \in \mathbb{R}^n}{\text{Min}} z^B = B(\mathbf{x}, \ r^k), \ k = 1, 2, \ \ldots,$$

denoting the respective optimal solutions by $\mathbf{x}^k$. For each unconstrained problem, the minimization process starts using the optimal point of the previous problem; thus we are assured that the generated points stay in the interior of the feasible region. In this way a sequence $\mathbf{x}^k$, $k = 1, 2, \ldots$ of points, feasible for the original problem, is obtained. Given certain assumptions we will then show that $\lim_{k \to \infty} f(\mathbf{x}^k) = f(\bar{\mathbf{x}}) = \bar{z}$, where $\bar{\mathbf{x}}$ denotes an optimal solution and $\bar{z}$ symbolizes the optimal function value of the original problem P. Consequently, the given constrained minimization problem has been transformed into a sequence of unconstrained problems. The boundary repulsion term $r_k \sum_{i=1}^{m} \frac{1}{g_i(\mathbf{x})}$ serves to ensure the feasibility of the points $\mathbf{x}^k$.

The *SUMT* method has been further developed and found to be numerically efficient. An extensive treatment is available in Fiacco and McCormick (1968). In order to prove convergence of the *SUMT* method the following lemma is needed.

Lemma 5.2 If a real-valued function $h(\mathbf{x}) < 0$ is convex, then so is $-\dfrac{1}{h(\mathbf{x})}$.

Proof The function $f(t) = t^{-1}$ is convex for all $t > 0$, since its second derivative $f''(t) = 2t^{-3}$ is positive for all $t > 0$. Due to the convexity of $h(\mathbf{x})$,

$$h(\lambda_1 \mathbf{x}^1 + \lambda_2 \mathbf{x}^2) \leq \lambda_1 h(\mathbf{x}^1) + \lambda_2 h(\mathbf{x}^2) < 0$$

for all $\mathbf{x}^1, \mathbf{x}^2 \in \mathbb{R}^n$ and for all $\lambda_1, \lambda_2 \geq 0$, such that $\lambda_1 + \lambda_2 = 1$. Therefore,

$$-\frac{1}{h(\lambda_1 \mathbf{x}^1 + \lambda_2 \mathbf{x}^2)} \leq -\frac{1}{\lambda_1 h(\mathbf{x}^1) + \lambda_2 h(\mathbf{x}^2)} \leq -\frac{\lambda_1}{h(\mathbf{x}^1)} - \frac{\lambda_2}{h(\mathbf{x}^2)},$$

where the last inequality follows because of the convexity of $\dfrac{1}{t}$ when $t > 0$; we conclude that $-\dfrac{1}{h(\mathbf{x})}$ is convex. $\qquad \square$

This result implies that if $f(\mathbf{x})$ and all $g_i(\mathbf{x})$ are convex, then for $r > 0$, the barrier function $B(\mathbf{x}, r)$ is convex with respect to the variable $\mathbf{x} \in \mathbb{R}^n$. A minimal point can therefore be found by applying any of the multidimensional unconstrained optimization techniques discussed in Chap. 2.

Let now $S := \{\mathbf{x} : g_i(\mathbf{x}) \leq 0, i = 1, \ldots, m\}$ be the feasible set and let $S^0 := \{\mathbf{x} : g_i(\mathbf{x}) < 0, i = 1, \ldots, m\}$. In other words, $S^0 \subseteq S$ is the subset of the feasible region, for which the barrier function $B(\mathbf{x}, r^k)$ is defined. By assuming that

(i) $f, g_1, g_2, \ldots, g_m$ are all convex and twice continuously differentiable functions,
(ii) S^0 is not empty, and
(iii) the set $\{\mathbf{x} : f(\mathbf{x}) \leq a, \mathbf{x} \in \mathbb{R}^n\}$ is bounded for all $a \in \mathbb{R}$,

one can show that there exists an $\bar{\mathbf{x}} \in \mathbb{R}^n$, such that $\inf_{\mathbf{x} \in S} f(\mathbf{x}) = f(\bar{\mathbf{x}}) = \bar{z}$. Furthermore, from the above discussion we know that for any fixed $r > 0$, the assumption (i) implies the convexity of $B(\mathbf{x}, r)$ for $\mathbf{x} \in S^0$. Then, if at least one of the f and g_i functions is strictly convex, so is $B(\mathbf{x}, r)$.

Theorem 5.3 Let the assumptions (i), (ii), and (iii) be satisfied, assume that $r_k > 0$ and that and B is strictly convex. Then for each $k = 1, 2, \ldots$, there exists a unique $\mathbf{x}^k \in S^0$, such that $\inf_{\mathbf{x} \in S^0} B(\mathbf{x}, r_k) = B(\mathbf{x}^k, r_k)$ and $\nabla_{\mathbf{x}} B(\mathbf{x}^k, r_k) = \mathbf{0}$.

Proof Due to the assumptions and the theory of convex functions (see e.g., Rockafellar 1970), the function $B(\mathbf{x}, r)$ is known to be bounded from below on S^0. Since, however, S^0 is an open set we must prove that the infimum of $B(\mathbf{x}, r_k)$ is actually attained on S^0. Recall that a continuous function will attain its infimum on a compact set (see e.g., Bertsekas 2016; Cottle and Thapa 2017) but that this need not be true for open sets, as exemplified by the trivial one-dimensional case where $f(x) = x$, which has infimum zero over the open set $\{x : x > 0\}$. Now take any $\tilde{\mathbf{x}} \in S^0$, set $z_k := B(\tilde{\mathbf{x}}, r_k)$, and define the sets $\tilde{S}_0 := \{\mathbf{x} : f(\mathbf{x}) \leq z_k, \mathbf{x} \in S\}$ and $\tilde{S}_i := \left\{ \mathbf{x} : -\dfrac{r_k}{g_i(\mathbf{x})} \leq z_k - \bar{z}, \mathbf{x} \in S^0 \right\}$ for $i = 1, \ldots, m$. Then the sets $\tilde{S}_i$ are closed for $i = 0, 1, \ldots, m$, the set $\tilde{S}_0$ is bounded, and

$$-\frac{r_k}{g_i(\tilde{\mathbf{x}})} \leq -r_k \sum_{i=1}^{m} \frac{1}{g_i(\tilde{\mathbf{x}})} = z_k - f(\tilde{\mathbf{x}}) \leq z_k - \bar{z},$$

so that $\tilde{\mathbf{x}} \in \tilde{S}_i, i = 1, \ldots, m$. Furthermore, $\tilde{S} = \bigcap_{i=0}^{m} \tilde{S}_i \neq \varnothing$ must be compact and $\tilde{S} \subseteq S^0$. Since the function $B(\mathbf{x}, r_k)$ is continuous on $\tilde{S}$, it follows that $\inf_{\mathbf{x} \in \tilde{S}} B(\mathbf{x}, r_k)$ must be attained on $\tilde{S}$. However, the inclusion $\tilde{S} \subseteq S^0$ implies that $\inf_{\mathbf{x} \in \tilde{S}} B(\mathbf{x}, r_k) \geq \inf_{\mathbf{x} \in S^0} B(\mathbf{x}, r_k)$, and if $S^0 \ni \mathbf{x} \notin \tilde{S}$, then there exists at least one $i, 0 \leq i \leq m$, such that $\mathbf{x} \notin \tilde{S}_i$. If $i \geq 1$, then, after some manipulation, we obtain $B(\mathbf{x}, r_k) > f(\mathbf{x}) + z_k - \bar{z} > z_k$, while if $i = 0$, then

$$B(\mathbf{x}, \ r_k) = f(\mathbf{x}) - r_k \sum_{i=1}^{m} \frac{1}{g_i(\mathbf{x})} > z_k - r_k \sum_{i=1}^{m} \frac{1}{g_i(\mathbf{x})} > z_k.$$

Hence, $S^0 \ni \mathbf{x} \notin \tilde{S}$ implies that $B(\mathbf{x}, \ r_k) > z_k$. Therefore, $\inf_{\mathbf{x} \in \tilde{S}} B(\mathbf{x}, \ r_k) = \inf_{\mathbf{x} \in S^0} B(\mathbf{x}, \ r_k)$ and even the latter infimum is attained at the same point of S^0. Because of the strict convexity of $B(\mathbf{x}, \ r_k)$, this point is uniquely determined. This infimal point is denoted by $\mathbf{x}^k$ and as $B(\mathbf{x}, \ r_k)$ is also continuously differentiable over S^0, the conclusion is that $\nabla_{\mathbf{x}} B(\mathbf{x}^k, \ r_k) = \mathbf{0}$. □

The results of the above theorem can be sharpened as shown in the following

Theorem 5.4 If the assumptions of Theorem 5.3 are satisfied, and if r_k is a strictly decreasing sequence with limit zero, then

(i) $\lim_{k \to \infty} B(\mathbf{x}^k, \ r_k) = f(\bar{\mathbf{x}}) = \bar{z}$,

(ii) $\lim_{k \to \infty} f(\mathbf{x}^k) = \bar{z}$, and

(iii) $\lim_{k \to \infty} r_k \sum_{i=1}^{m} \frac{1}{g_i(\mathbf{x})} = 0$,

with the convergence being monotonically decreasing in (i) and (ii).

Proof For any given $\varepsilon > 0$ there exists at least one $\tilde{\mathbf{x}} \in S^0$, such that $f(\tilde{\mathbf{x}}) < \bar{z} + \frac{1}{2}\varepsilon$ and k' such that $r_{k'} < \frac{\varepsilon}{2m} \left\{ \min_i |g_i(\tilde{\mathbf{x}})| \right\}$. It follows that for $k > k'$ we must have

$$\begin{aligned}
\bar{z} &\leq \inf_{\mathbf{x} \in S^0} B(\mathbf{x}, \ r_k) = B(\mathbf{x}^k, \ r_k) && \text{(via Theorem 5.3)} \\
&\leq B(\mathbf{x}^{k'}, r_k) && \text{(via the definition of } \mathbf{x}^k) \\
&< B(\mathbf{x}^{k'}, r_{k'}) && \text{(as } B(\mathbf{x}, \ r_k) \text{ is strictly decreasing with } k) \\
&\leq B(\tilde{\mathbf{x}}, \ r_{k'}) && \text{(by definition of } \mathbf{x}^{k'}) \\
&= f(\tilde{\mathbf{x}}) - r_{k'} \sum_{i=1}^{m} \frac{1}{g_i(\tilde{\mathbf{x}})} < \bar{z} + \frac{\varepsilon}{2} + \frac{\varepsilon}{2} && \text{(by assumption)} \\
&= \bar{z} + \varepsilon
\end{aligned}$$

and (i) is proved as well as the fact that $B(\mathbf{x}^{k_1}, \ r_{k_1}) > B(\mathbf{x}^{k_2}, \ r_{k_2})$ for any $k_1 < k_2$. Furthermore, $\bar{z} \leq f(\mathbf{x}^k) \leq B(\mathbf{x}^k, \ r_k)$, which proves (ii) if we let $k \to \infty$. Finally parts (i) and (ii) imply (iii) and only the monotonic convergence in (ii) is left to be demonstrated. We showed above that if $k_1 < k_2$, then $B(\mathbf{x}^{k_2}, \ r_{k_2}) \leq B(\mathbf{x}^{k_1}, \ r_{k_2})$, and by definition of $\mathbf{x}^{k_1}$, we obtain $B(\mathbf{x}^{k_1}, \ r_{k_1}) \leq B(\mathbf{x}^{k_2}, \ r_{k_1})$. Writing these inequalities explicitly yields $f(\mathbf{x}^{k_2}) - r_{k_2} \sum_{i=1}^{m} \frac{1}{g_i(\mathbf{x}^{k_2})} \leq f(\mathbf{x}^{k_1}) - r_{k_2} \sum_{i=1}^{m} \frac{1}{g_i(\mathbf{x}^{k_1})}$ and $f(\mathbf{x}^{k_1})$ $-r_{k_1} \sum_{i=1}^{m} \frac{1}{g_i(\mathbf{x}^{k_1})} \leq f(\mathbf{x}^{k_2}) - r_{k_1} \sum_{i=1}^{m} \frac{1}{g_i(\mathbf{x}^{k_2})}$. Multiplying the first inequality by r_{k_1} and the second by r_{k_2} and adding results in

$$r_{k_1} f\left(\mathbf{x}^{k_2}\right) + r_{k_2} f\left(\mathbf{x}^{k_1}\right) \le r_{k_1} f\left(\mathbf{x}^{k_1}\right) + r_{k_2} f\left(\mathbf{x}^{k_2}\right), \quad \text{or}$$

$$\left(r_{k_1} - r_{k_2}\right)\left(f\left(\mathbf{x}^{k_2}\right) - f\left(\mathbf{x}^{k_1}\right)\right) \le 0,$$

which, because $k_1 < k_2$, so that $r_{k_1} - r_{k_2} > 0$, implies that $f\left(\mathbf{x}^{k_2}\right) \le f\left(\mathbf{x}^{k_1}\right)$ and the theorem is finally proved.   □

Note that nothing can be said about the monotonicity of the convergence in (iii) for the sum of the boundary repulsion terms. Based on the above discussion, a solution procedure can now be formulated in the following algorithmic form. Recall that the problem under consideration is

$$\begin{aligned} \text{P:} \quad &\text{Min } z = f(\mathbf{x}) \\ &\text{s.t. } g_i(\mathbf{x}) \le 0, \ i = 1, \ \ldots, \ m \\ &\mathbf{x} \in \mathbb{R}^n. \end{aligned}$$

The Sequential Unconstrained Minimization Technique (SUMT)

Step 0 (Phase 1) Disregarding the constraints of P, find a minimal point for the unconstrained problem. Is this point feasible?

If yes: Stop, an optimal solution to problem P has been found.
If no: Go to Step 1.

Step 1 Start with an initial interior feasible point $\mathbf{x}^0$ and set $k: = 1$.

Step 2 Use any unconstrained minimization technique to find the minimum of the barrier function

$$B(\mathbf{x}, \ r_k) = f(\mathbf{x}) - r_k \sum_{i=1}^{m} \frac{1}{g_i(\mathbf{x})},$$

where, for instance, $r_k = 10^{1-k}$. Start the minimization process from the point x^{k-1} and denote the resulting optimal point by $\mathbf{x}^k$.

Step 3 Does the current solution satisfy some stop criterion?

If yes: Stop, an optimal solution to P has been found.
If no: Set $k: = k + 1$ and to go Step 2.

In the algorithm above we have employed the sequence $\{r_k\}_{k=1}^{\infty} = \{10^{1-k}\}_{k=1}^{\infty}$, which empirically has been shown to yield satisfactory results; see Fiacco and McCormick (1968). A slower converging r_k sequence may unnecessarily delay the convergence of the SUMT algorithm whereas a faster converging r_k sequence may create numerical instability. A suitable stop criterion resulting from duality considerations will be developed after the numerical example below. The interior feasible point required to initiate the algorithm can be obtained using the Phase 1 procedure which was described in the beginning of this chapter.

Example The *SUMT* method is now illustrated by carrying out four iterations for the problem

$$P: \operatorname{Min} z = 3x_1^2 - 2x_1x_2 + 2x_2^2 - 26x_1 - 8x_2$$
$$\text{s.t. } x_1 + 2x_2 \le 6$$
$$x_1 - x_2 \ge 1$$
$$x_1,\ x_2 \ge 0.$$

All the requirements for using the *SUMT* method are met. The barrier function $B(\mathbf{x},\ r)$ is $B(x_1,\ x_2,\ r) = 3x_1^2 - 2x_1x_2 + 2x_2^2 - 26x_1 - 8x_2 - r\left[\dfrac{1}{x_1 + 2x_2 - 6} + \dfrac{1}{-x_1 + x_2 + 1} + \dfrac{1}{-x_1} + \dfrac{1}{-x_2}\right]$. In Step 0 we find that the unconstrained minimal point turns out to be $\hat{\mathbf{x}} = [\hat{x}_1,\ \hat{x}_2]^T = [6,\ 5]^T$, which is infeasible when we take the constraints into consideration. In Step 1, we arbitrarily pick $\mathbf{x}^0 = [x_1^0,\ x_2^0]^T = [3,\ 1]^T$ as an initial interior feasible point. Then, for $k = 1$, we set $r^1 := 10^{k-1} = 10^{1-1} = 1$ and minimize the barrier function $B(x_1, x_2, 1)$, starting out from $\mathbf{x}^0$, using the Newton minimization method described in Sect. 2.2.7. The result is $\mathbf{x}^1 = [x_1^2,\ x_2^1]^T \approx [3.691,\ 0.953]^T$ with $B(x_1^2,\ x_2^1,\ 1) \approx -63.56$. Continuing in the same fashion, the successive three iterations yield

$$\min B(x_1,\ x_2,\ r_1) = B(x_1^2,\ x_2^2,\ 10^{-1}) \approx B(3.750,\ 1.059,\ 10^{-1}) \approx -68.55,$$

$$\min B(x_1,\ x_2,\ r_2) = B(x_1^3,\ x_2^3,\ 10^{-2}) \approx B(3.769,\ 1.095,\ 10^{-2})$$
$$\approx -69.73,\ \text{ and}$$

$$\min B(x_1,\ x_2,\ r_3) = B(x_1^4,\ x_2^4,\ 10^{-3}) \approx B(3.775,\ 1.106,\ 10^{-3}) \approx -70.07.$$

Having carried out four iterations with the *SUMT* method, we stop here. The exact optimal solution is $\bar{\mathbf{x}} = [\bar{x}_1,\ \bar{x}_2]^T = [3\frac{7}{9},\ 1\frac{1}{9}]^T \approx [3.778,\ 1.111]^T$, which is quite close to $\mathbf{x}^4$. The solution points generated above are displayed in Fig. 5.4.

It turns out that for establishing upper and lower bounds on the optimal function value $\bar{z}$, some duality properties of the *SUMT* method can be obtained. The Lagrangean function of problem P is $L(\mathbf{x},\ \mathbf{u}^k) = f(\mathbf{x}) + \sum_{i=1}^{m} u_i^k g_i(\mathbf{x})$, which has the gradient $\nabla_{\mathbf{x}} L(\mathbf{x}, \mathbf{u}^k) = \nabla f(\mathbf{x}) + \sum_{i=1}^{m} u_i^k g_i(\mathbf{x})$. Inserting $u_i^k = \dfrac{r_k}{[g_i(\mathbf{x}^k)]^2} > 0, i = 1, \ldots, m$, we obtain $\nabla_{\mathbf{x}} L(\mathbf{x}^k, \mathbf{u}^k) = \nabla f(\mathbf{x}^k) + \sum_{i=1}^{m} \dfrac{r_k}{(g_i(\mathbf{x}^k))^2} \nabla g_i(\mathbf{x}^k) = \nabla f(\mathbf{x}^k) - r_k \sum_{i=1}^{m} \nabla \left[\dfrac{1}{g_i(\mathbf{x}^k)}\right] = \nabla_{\mathbf{x}} B(\mathbf{x}^k, r_k)$. But as $\nabla_{\mathbf{x}} B(\mathbf{x}^k,\ r_k) = \mathbf{0}$ due to Theorem 5.3, it follows that $(\mathbf{x}^k,\ \mathbf{u}^k)$ is

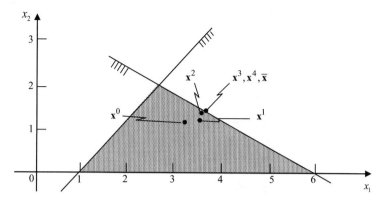

Fig. 5.4 Iterative solutions with the *SUMT* method

Lagrangean dual feasible, and at each iteration the *SUMT* algorithm will generate points feasible for both the primal and the Lagrangean dual problems. According to the weak Lagrangean duality result of Theorem 4.27 we can then conclude that $f(\mathbf{x}^k)+r_k\sum_{i=1}^{m}\frac{1}{g_i(\mathbf{x}^k)}\leq\bar{z}\leq f(\mathbf{x}^k)$, which provides upper and lower bounds for the optimal function value $\bar{z}$. In other words, $r_k\sum_{i=1}^{m}\frac{1}{g_i(\mathbf{x}^k)}=f(\mathbf{x}^k)-B(\mathbf{x}^k,r_k)$ is an estimate of how far $f(\mathbf{x}^k)$ is from the optimal value $\bar{z}=f(\bar{\mathbf{x}})$. Hence the above relation provides a useful stop criterion for the algorithm, since from part (iii) of Theorem 5.4 it is known that this expression approaches 0 when $k \to \infty$.

So far we have proved that the successive values of the objective function converge toward an optimal function *value*, but nothing has been said about the sequence of solution *points* $\mathbf{x}^0, \mathbf{x}^1, \mathbf{x}^2, \ldots$ The question is if it will converge, and, if so, to an optimal solution $\bar{\mathbf{x}}$ of the original problem. This aspect has been investigated by Sandblom (1974) with the outcome that such a sequence need not be convergent. However, if the optimal solution $\bar{\mathbf{x}}$ is unique, then it can in fact be shown that $\mathbf{x}^k \to \bar{\mathbf{x}}$ as $k \to \infty$. We state and prove this in the following

Theorem 5.5 If the assumptions (i), (ii), and (iii) of Theorem 5.3 are satisfied and if $\bar{\mathbf{x}}$ is the unique optimal solution, then the sequence $\mathbf{x}^k$, $k = 0, 1, 2, \ldots$ is convergent with limit $\bar{\mathbf{x}}$.

Proof From Theorem 5.4 we know that $f(\mathbf{x}^k) \to f(\bar{\mathbf{x}})$ as $k \to \infty$. Suppose that $\mathbf{x}^k \nrightarrow \bar{\mathbf{x}}$ as $k \to \infty$, i.e., for some $\varepsilon > 0$ we would have $\|\bar{\mathbf{x}} - \mathbf{x}^k\| \geq \varepsilon$ for infinitely many k. As $\mathbf{x}^k$, $k = 0, 1, 2, \ldots$ belongs to the set $\{\mathbf{x} : f(\mathbf{x}) \leq f(\mathbf{x}^0), \mathbf{x} \in S^0\}$ which is closed and bounded, a convergent subsequence $\mathbf{x}^{k_v}$, $v = 1, 2, \ldots$ exists with $\|\bar{\mathbf{x}} - \mathbf{x}^{k_v}\| \geq \varepsilon$ for all v. Denote the limit of this subsequence by $\mathbf{x}^*$. As $\|\bar{\mathbf{x}} - \mathbf{x}^{k_v}\| \geq \varepsilon$ we conclude that $\|\bar{\mathbf{x}} - \mathbf{x}^*\| \geq \varepsilon$. But $f(\mathbf{x}^k) \to f(\bar{\mathbf{x}})$ as $k \to \infty$

implies that $f(\mathbf{x}^{k_\nu}) \to f(\bar{\mathbf{x}})$ as $\nu \to \infty$. From the continuity of f it follows that $f(\mathbf{x}^{k_\nu}) \to f(\mathbf{x}^*)$ as $\nu \to \infty$. Hence $f(\bar{\mathbf{x}}) = f(\mathbf{x}^*)$ which by the uniqueness assumption is a contradiction and the result $\mathbf{x}^k \to \bar{\mathbf{x}}$ as $k \to \infty$ is proved.   □

Fiacco and McCormick (1966) have extended the *SUMT* algorithm to incorporate equality constraints, i.e., problems of the type

$$P: \quad \text{Min } z = f(\mathbf{x})$$
$$\text{s.t. } g_i(\mathbf{x}) \le 0, \ i = 1, \ \ldots, \ m$$
$$h_\ell(\mathbf{x}) = 0, \ell = 1, \ \ldots, \ p.$$

Then the barrier function is defined as

$$B(\mathbf{x}, \ r) = f(\mathbf{x}) - r \sum_{i=1}^{m} \frac{1}{g_i(\mathbf{x})} + r^{-\frac{1}{2}} \sum_{\ell=1}^{p} (h_\ell(\mathbf{x}))^2$$

With assumptions similar to those previously considered, corresponding results are obtained.

Another extension is the *slacked SUMT*, known as *SLUMT*, which is also due to Fiacco and McCormick (1967a). Contrary to the methods examined until now it is not necessary to start it with an interior feasible point. Instead of forming the barrier function $B(\mathbf{x}, r)$ as before and then proceeding from an element $\mathbf{x}$ of the set S^0, in *SLUMT* we define

$$B(\mathbf{x}, \ \mathbf{t}, \ r) := f(\mathbf{x}) + r^{-1} \sum_{i=1}^{m} [g_i(\mathbf{x}) + t_i]^2$$

where t_i serves the same purpose as slack or artificial variables do in linear programming. For a given sequence r_k, $k = 0, 1, 2, \ldots$ with $r_k \to 0$ as $k \to \infty$, the problem then consists in minimizing $B(\mathbf{x}, \mathbf{t}, r_k)$ for all possible values of $\mathbf{t} \ge \mathbf{0}$ and $\mathbf{x}$. The advantage with this algorithm is that no Phase 1 procedure is necessary. A drawback is that the new barrier function has to be minimized with respect to $(n + m)$ variables instead of only n as before.

A third modification is *SUMT* without parameters. Fiacco and McCormick (1967b) suggest starting from some $\mathbf{x}^0 \in S^0$, then setting $k := 0$ and minimizing, with respect to $\mathbf{x}$, the barrier function

$$Q(\mathbf{x}, \ \mathbf{x}^k) = \frac{1}{f(\mathbf{x}^k) - f(\mathbf{x})} - \sum_{i=1}^{m} \frac{1}{g_i(\mathbf{x})},$$

which will automatically be done over the set $\{\mathbf{x} : f(\mathbf{x}) < f(\mathbf{x}^k), \ g_i(\mathbf{x}) < 0, \ i = 1, \ldots, m\}$. The solution to this auxiliary problem is called $\mathbf{x}^{k+1}$ and the minimization process is then repeated with $k := k + 1$. It can be shown that the sequence $\mathbf{x}^k$, $k = 0$, 1, 2, ... thus obtained will be such that $f(\mathbf{x}^k) \to f(\bar{\mathbf{x}})$ as $k \to \infty$.

Finally, the *logarithmic barrier method*, originally due to the Norwegian economist K. R. Frisch (1956), uses logarithmic boundary repulsion terms $-r \ln(-g_i(\mathbf{x}))$ to force $\mathbf{x}$ to stay within the feasible region. Since $\ln(t)$ approaches $-\infty$ as $t > 0$ approaches 0, the effect of the boundary repulsion terms is to ensure that $g_i(\mathbf{x})$ stays negative. The logarithmic barrier function $LB(\mathbf{x}, r)$ is then defined as

$$LB(\mathbf{x}, \ r) = f(\mathbf{x}) - r \sum_{i=1}^{m} \ln\left(-g_i(\mathbf{x})\right),$$

and the method proceeds, just like the *SUMT* method, by sequentially minimizing the barrier function $LB(\mathbf{x}, r)$ with respect to $\mathbf{x}$, for decreasing preselected values of the positive parameter r. For details, see Bertsekas (2016) or Luenberger and Ye (2008).

The logarithmic barrier method underlies a class of interior point methods for linear programming. The problem to be solved is then

$$\begin{aligned} \mathrm{P:} \quad &\mathrm{Max} \ \ z = \mathbf{c}\mathbf{x} \\ &\mathrm{s.t.} \ \ \mathbf{A}\mathbf{x} = \mathbf{b} \\ &\quad \ \ \mathbf{x} \geq \mathbf{0}, \end{aligned}$$

where $\mathbf{c}$ is a $[1 \times n]$-dimensional row vector, $\mathbf{x} \in \mathbb{R}^n$ as usual, $\mathbf{A}$ is an $[m \times n]$-dimensional matrix, and $\mathbf{b}$ is an $[m \times 1]$-dimensional column vector. In order to keep all variables x_j strictly positive in the solution procedure, logarithmic barrier terms are added to the objective function (remember that we are now maximizing, rather than minimizing), resulting in the barrier problem

$$\begin{aligned} \mathrm{P^B:} \quad &\mathrm{Max} \ \ z_B = \mathbf{c}\mathbf{x} + r \sum_{j=1}^{n} \ln x_j \\ &\mathrm{s.t.} \ \ \mathbf{A}\mathbf{x} = \mathbf{b}, \end{aligned}$$

where the nonnegativity constraints, being redundant, have been dropped. This is now a maximization problem with a strictly concave objective function and linear equation constraints; from the Karush-Kuhn-Tucker optimality conditions developed in the previous chapter, one can show that the necessary and sufficient conditions for an optimal solution $\bar{\mathbf{x}}$ to the problem $\mathrm{P^B}$ is the existence of a row vector $\bar{\mathbf{u}} \in \mathbb{R}^m$ such that

$$\begin{aligned} &\mathbf{A}\bar{\mathbf{x}} = \mathbf{b} \\ &\left(\bar{\mathbf{u}}A\right)_j - r/\bar{x}_j = c_j \\ &\bar{\mathbf{x}} > \mathbf{0}. \end{aligned}$$

This nonlinear system can now be solved by the Newton-Raphson method, see Sect. 2.2.7. Repeating the procedure with preselected values of the positive parameter approaching zero, an optimal solution to the initial problem P can be obtained. The whole procedure has been given the lengthy name of the *Primal-Dual Newton Step Interior Point Method of Linear Programming*. Although it employs nonlinear techniques, it is still considered part of the field of linear programming and it will not be described here; we refer readers to Dantzig and Thapa (2003), and Eiselt and Sandblom (2007).

5.2   Everett's Lagrangean Multiplier Method

Having covered various conditions for optimality in Chap. 4, a method based on generalized Lagrangean multipliers will now be described. The technique is due to Everett (1963) and it is introduced together with the corresponding algorithm and an example. Let us again consider our inequality-constrained problem

$$\text{P: Min } z = f(\mathbf{x})$$
$$\text{s.t. } g_i(\mathbf{x}) \leq 0, \ i = 1, \ \ldots, \ m$$
$$\mathbf{x} \in \mathbb{R}^n,$$

where we also use the vector notation $\mathbf{g}(\mathbf{x}) \leq \mathbf{0}$ for the structural constraints. Refer now to Lemma 4.19 and Theorem 4.20 of Sect. 4.2, as well as our discussion how to interpret the associated dual variables or Lagrangean multipliers $\mathbf{u}$ as shadow costs or penalties for violating the associated constraints. Assume that for a given Lagrange multiplier vector $\tilde{\mathbf{u}} \geq \mathbf{0}$ with $\tilde{\mathbf{u}} \neq \mathbf{0}$ we minimize the associated Lagrangean function $\tilde{z}_L = f(\mathbf{x}) + \tilde{\mathbf{u}}\mathbf{g}(\mathbf{x})$, obtaining the solution $\tilde{\mathbf{x}}$. From the discussion in Chap. 4, it follows that it is advantageous to increase the values of $\tilde{u}_i$ for violated constraints, i.e., those constraints with $g_i(\tilde{\mathbf{x}}) > 0$, and to reduce any $\tilde{u}_i > 0$ for satisfied constraints. In this way, we can consider the Lagrangean multipliers $\tilde{u}_i$ as penalties for violating the constraints. The Lagrangean multiplier method of Everett performs these modifications in a systematic form to obtain a heuristic method, which we will now describe in algorithmic form. We recall that the problem to be solved is formulated as

$$\text{P: Min } z = f(\mathbf{x})$$
$$\text{s.t. } \mathbf{g}(\mathbf{x}) \leq \mathbf{0}$$
$$\mathbf{x} \in \mathbb{R}^n.$$

Let some stop criterion (e.g., the Karush-Kuhn-Tucker optimality conditions or some preselected maximal iteration count) be given, as well as an initial preselected nonzero Lagrangean multiplier vector $\tilde{\mathbf{u}} \geq \mathbf{0}$.

Everett's Lagrangean Multiplier Method

Step 0 Disregarding the constraints of P, find a minimal point for the resulting unconstrained problem. Is this point feasible?

If yes: Stop, an optimal solution to P has been found.
If no: Go to Step 1.

Step 1 Solve the problem $\tilde{P}_L$: $\underset{x \in \mathbb{R}^n}{\text{Min}}\ \tilde{z}_L = f(x) + \tilde{u}g(x)$ and let $\tilde{x}$ be an optimal solution of $\tilde{P}_L$. Determine $S_i(\tilde{x}) = -g_i(\tilde{x})$, $i = 1, 2, \ldots, m$.

Step 2 Is $S_i(\tilde{x}) \geq 0\ \forall i$?

If yes: Go to Step 3.
If no: Go to Step 5.

Step 3 Is the stop criterion satisfied?

If yes: Stop, an optimal or near-optimal solution of P has been found.
If no: Go to Step 4.

Step 4 Select some index k such that $\tilde{u}_k > 0$ and some $\varepsilon > 0$, such that $\tilde{u}_k > \varepsilon > 0$. Set $\tilde{u}_k := \tilde{u}_k - \varepsilon$ and go to Step 1.

Step 5 Select some index ℓ such that $S_\ell(\tilde{x}) < 0$ and some $\varepsilon > 0$. Set $\tilde{u}_\ell := \tilde{u}_\ell + \varepsilon$ and go to Step 1.

It may be computationally advantageous to select k in Step 4, such that $S_k(\tilde{x}) = \max\{S_i(\tilde{x}) : \tilde{u}_i > 0\}$. Using the Everett method, we will now solve the following

Example Consider the problem

$$P:\ \text{Min } z = x_1^2 - 4x_1 + x_2^2 - 6x_2$$
$$\text{s.t. } x_1 + x_2 \leq 3$$
$$-2x_1 + x_2 \leq 2$$
$$x_1,\ x_2 \in \mathbb{R}.$$

The Lagrangean function is

$$L(x,\ u) = x_1^2 - 4x_1 + x_2^2 - 6x_2 + u_1 x_1 + u_1 x_2 - 3u_1 - 2u_2 x_1 + u_2 x_2 - 2u_2.$$

Starting with the arbitrary Lagrangean multipliers $\tilde{u}_1 = \tilde{u}_2 = 5$, the optimal solution of the Lagrangean relaxation which consists in minimizing $L(x, \tilde{u})$ with respect to x is $\tilde{x}_1 = 4\frac{1}{2}$ and $\tilde{x}_2 = -2$, so that $f(\tilde{x}) = 18\frac{1}{4}$. In this solution the slack variables are $S_1(\tilde{\mathbf{x}}) = \frac{1}{2}$ and $S_2(\tilde{\mathbf{x}}) = 13$, and therefore we will decrease $\tilde{u}_2$. The results of the iterations are as follows.

Iteration 1: $\tilde{\mathbf{u}} = [5, 5]$; $\tilde{\mathbf{x}} = [4\frac{1}{2}, -2]^T$; $f(\tilde{\mathbf{x}}) = 18\frac{1}{4}$; $S_1(\tilde{\mathbf{x}}) = \frac{1}{2}$; $S_2(\tilde{\mathbf{x}}) = 13$.

The value of $\tilde{u}_2$ will be decreased, arbitrarily choose $\varepsilon = 2$.

Iteration 2: $\tilde{\mathbf{u}} = [5, 3]$; $\tilde{\mathbf{x}} = [2\frac{1}{2}, -1]^T$; $f(\tilde{\mathbf{x}}) = 3\frac{1}{4}$; $S_1(\tilde{\mathbf{x}}) = 1\frac{1}{2}$, $S_2(\tilde{\mathbf{x}}) = 8$.

The value of $\tilde{u}_2$ will be decreased further, use $\varepsilon = 3$.

Iteration 3: $\tilde{\mathbf{u}} = [5, 0]$; $\tilde{\mathbf{x}} = [-\frac{1}{2}, \frac{1}{2}]^T$; $f(\tilde{\mathbf{x}}) = -\frac{1}{2}$; $S_1(\tilde{\mathbf{x}}) = 3$, $S_2(\tilde{\mathbf{x}}) = \frac{1}{2}$. The value of $\tilde{u}_1$ will be decreased, use $\varepsilon = 2$.

Iteration 4: $\tilde{\mathbf{u}} = [3, 0]$; $\tilde{\mathbf{x}} = [\frac{1}{2}, 1\frac{1}{2}]^T$; $f(\tilde{\mathbf{x}}) = -8\frac{1}{2}$; $S_1(\tilde{\mathbf{x}}) = 1$, $S_2(\tilde{\mathbf{x}}) = 1\frac{1}{2}$.

The value of $\tilde{u}_1$ will be further decreased, use $\varepsilon = 2$ again.

Iteration 5: $\tilde{\mathbf{u}} = [1, 0]$; $\tilde{\mathbf{x}} = [1\frac{1}{2}, 2\frac{1}{2}]^T$; $f(\tilde{\mathbf{x}}) = -12\frac{1}{2}$; $S_1(\tilde{\mathbf{x}}) = -1$,

$$S_2(\tilde{\mathbf{x}}) = 2\frac{1}{2}.$$

In the transition from iteration 4 to iteration 5 the value of $\tilde{u}_1$ was decreased too much, causing S_1 to be negative (i.e., -1), so that the first constraint is violated. Therefore, in the next step $\tilde{u}_1$ will have to be increased again; in particular, it will have to be chosen from the interval $[1, 3]$.

Iteration 6: Selecting $\varepsilon = 1$, we obtain $\tilde{\mathbf{u}} = [2, 0]$; $\tilde{\mathbf{x}} = [1, 2]^T$; $f(\tilde{\mathbf{x}}) = -11$;

$$S_1(\tilde{\mathbf{x}}) = 0, \ S_2(\tilde{\mathbf{x}}) = 2.$$

At this stage $\tilde{u}_1$ cannot be decreased any further, as this would cause both $\tilde{x}_1$ and $\tilde{x}_2$ to increase and then result in a negative value for the slack variable $S_1(\tilde{\mathbf{x}})$ and, on the other hand, neither can $\tilde{u}_2$ be decreased since the nonnegativity condition $\tilde{\mathbf{u}} \geq \mathbf{0}$ must be satisfied. Thus, iteration 6 yields the optimal solution

$$\bar{x}_1 = 1, \ \bar{x}_2 = 2, \ \bar{z} = -11, \ \bar{u}_1 = 2, \ \bar{u}_2 = 0,$$

and our problem is solved.

5.3   Augmented Lagrangean Methods

In this section we will consider problems with equation constraints, i.e., of the type

$$\text{P: Min } z = f(\mathbf{x})$$
$$\text{s.t. } \mathbf{g}(\mathbf{x}) = \mathbf{0}$$
$$\mathbf{x} \in \mathbb{R}^n.$$

Our standard formulation with inequality constraints $\mathbf{g}(\mathbf{x}) \leq \mathbf{0}$ can be brought into equation form by adding sign-free slack variables A_i to obtain $\mathbf{g}(\mathbf{x}) + A_i^2 = \mathbf{0}$, $i = 1, \ldots, m$, where A_i^2 will provide the nonnegative slack required.

Definition 5.6 Associated with problem P is the *augmented Lagrangean function L* $(\mathbf{x}, \mathbf{u}, r)$ given by

$$L(\mathbf{x}, \mathbf{u}, r) := f(\mathbf{x}) + \mathbf{u}\mathbf{g}(\mathbf{x}) + \tfrac{1}{2}r\|\mathbf{g}(\mathbf{x})\|^2,$$

where the row vector $\mathbf{u} \in \mathbb{R}^m$ is referred to as the *Lagrangean multiplier* or *dual variable*, and the positive scalar r is called the *penalty parameter*.

Now consider a modification of P, i.e., the problem $P_\mathbf{u}$ given by

$$P_\mathbf{u}: \text{Min } z_\mathbf{u} = f(\mathbf{x}) + \mathbf{u}\mathbf{g}(\mathbf{x})$$
$$\text{s.t. } \mathbf{g}(\mathbf{x}) = \mathbf{0}$$
$$\mathbf{x} \in \mathbb{R}^n.$$

It is clear that problems P and $P_\mathbf{u}$ have the same optimal solution points and the same objective function values for all feasible solutions, since adding combinations of the equation constraints to the objective function will not change its values. Assume now that $\bar{\mathbf{x}}$ is an optimal solution to problem P with an associated optimal multiplier vector $\bar{\mathbf{u}}$, so that $\nabla f(\bar{\mathbf{x}}) + \sum_{i=1}^{m} \bar{u}_i \nabla g_i(\bar{\mathbf{x}}) = \mathbf{0}$. Taking the gradient of the augmented Lagrangean function $L(\mathbf{x}, \mathbf{u}, r)$ of problem P, we obtain

$$\nabla_\mathbf{x} L(\mathbf{x}, \mathbf{u}, r) = \nabla f(\mathbf{x}) + \sum_{i=1}^{m} u_i \nabla g_i(\mathbf{x}) + r \sum_{i=1}^{m} g_i(\mathbf{x}) \nabla g_i(\mathbf{x}),$$

and from the above discussion we can conclude that at optimum,

$$\nabla_\mathbf{x} L(\bar{\mathbf{x}}, \bar{\mathbf{u}}, r) = \mathbf{0}.$$

This means that $L(\mathbf{x}, \mathbf{u}, r)$ is an exact penalty function for problem P.

Example Consider the problem

$$P: \quad \text{Min } z = 3x_1^2 - 2x_1x_2 + 2x_2^2 - 26x_1 - 8x_2$$
$$\text{s.t. } x_1 + 2x_2 = 6$$
$$x_1, \ x_2 \in \mathbb{R}.$$

This problem has the unique optimal solution $(\bar{x}_1, \bar{x}_2, u) = \left(3\frac{7}{9}, 1\frac{1}{9}, 5\frac{5}{9}\right)$, which can be verified, e.g., by the Karush-Kuhn-Tucker conditions. Its augmented Lagrangean function is

$$L(x_1, \ x_2, \ u, \ r) = 3x_1^2 - 2x_1x_2 + 2x_2^2 - 26x_1 - 8x_2 + u(x_1 + 2x_2 - 6)$$
$$+ \frac{1}{2}r(x_1 + 2x_2 - 6)^2,$$

so that determining $\nabla_{\mathbf{x}}L(\mathbf{x}, u, r)$, we obtain

$$\begin{cases} \dfrac{\partial L}{\partial x_1} = 6x_1 - 2x_2 - 26 + u + r(x_1 + 2x_2 - 6) \\ \dfrac{\partial L}{\partial x_2} = -2x_1 + 4x_2 - 8 + 2u + 2r(x_1 + 2x_2 - 6). \end{cases}$$

Inserting, we find $\nabla_{\mathbf{x}}L\left(3\dfrac{7}{9}, \ 1\dfrac{1}{9}, \ 5\dfrac{5}{9}, \ r\right) = \mathbf{0}$.

One can envisage a variety of strategies for finding an optimal solution $\bar{\mathbf{x}}$ to our original problem P using the augmented Lagrangean function. We will demonstrate an approach, where we switch between updating estimates $\mathbf{x}^k$ of an optimal solution and estimates $\mathbf{u}^k$ of an optimal Lagrangean multiplier. Specifically, given the multiplier $\mathbf{u}^k$ and the parameter r, we find $\mathbf{x}^k$ as the unconstrained minimal point of the augmented Lagrangean $L\left(\mathbf{x}, \mathbf{u}^k, r\right) = f(\mathbf{x}) + \mathbf{u}^k\mathbf{g}(\mathbf{x}) + \dfrac{r}{2}\|\mathbf{g}(\mathbf{x})\|^2$. Then we update the multiplier $\mathbf{u}^k$ via the expression

$$\mathbf{u}^{k+1} := \mathbf{u}^k + r\mathbf{g}\left(\mathbf{x}^k\right),$$

which makes intuitive sense considering u_i as prices relating to the constraint functions g_i. This argument can be given a more rigorous explanation; see, e.g., Luenberger and Ye (2008). We are now ready to describe the procedure formally. Recall that the problem to be solved is

$$\begin{aligned} \text{P:} \quad & \text{Min } z = f(\mathbf{x}) \\ & \text{s.t. } \mathbf{g}(\mathbf{x}) = \mathbf{0} \\ & \mathbf{x} \in \mathbb{R}^n. \end{aligned}$$

The process is initialized with a preselected value of the penalty parameter r, a tentative solution point $\mathbf{x}^1$, which is not necessarily feasible, and a tentative Lagrangean multiplier (dual variable) $\mathbf{u}^1$. Let some stop criterion be given.

An Augmented Lagrangean Algorithm

Step 0 Disregarding the constraints of P, find a minimal point for the unconstrained problem. Is this point feasible?

If yes: Stop, an optimal solution to problem P has been found.
If no: Set $k: = 1$ and go to Step 1.

Step 1 Use any unconstrained optimization technique to find the minimal point of the augmented Lagrangean function $L(\mathbf{x}, \mathbf{u}^k, r) = f(\mathbf{x}) + \mathbf{u}^k \mathbf{g}(\mathbf{x}) + \frac{1}{2}r\|\mathbf{g}(\mathbf{x})\|^2$, denoting the solution by $\mathbf{x}^{k+1}$.

Step 2 Update the Lagrangean multiplier $\mathbf{u}^k$ according to the formula

$$\mathbf{u}^{k+1} := \mathbf{u}^k + r\mathbf{g}(\mathbf{x}^k).$$

Step 3 Is the stop criterion satisfied?

If yes: Stop with the current solution $(\mathbf{x}^{k+1}, \mathbf{u}^{k+1})$.
If no: Set $k := k + 1$ and go to Step 1.

Example Consider the problem

$$\text{P:} \quad \text{Min } z = 3x_1^2 - 2x_1x_2 + 2x_2^2 - 26x_1 - 8x_2$$
$$\text{s.t. } x_1 + 2x_2 = 6$$
$$x_1, \ x_2 \in \mathbb{R}.$$

Using the augmented Lagrangean method, and arbitrarily starting with $(\mathbf{x}^1, \mathbf{u}^1, r) = (x_1^1, x_2^1, u^1, r) = (3, 2, 5, 2)$, we will now carry out three iterations of the algorithm. Note that the initial solution is not feasible. In Step 0, we find that the solution point $\hat{\mathbf{x}}$ to the unconstrained problem is $\hat{\mathbf{x}} = [\hat{x}_1, \hat{x}_2]^T = [6, 5]^T$, which is not feasible, when the constraint is taken into consideration, so we move to Step 1. The augmented Lagrangean function is now $L(\mathbf{x}, u^1, r) = L(x_1, x_2, u^1, r) = L(x_1, x_2, 5, 2) = 3x_1^2 - 2x_1x_2 + 2x_2^2 - 26x_1 - 8x_2 + 5(x_1 + 2x_2 - 6) + (x_1 + 2x_2 - 6)^2$, from which we conclude that

$$\begin{cases} \dfrac{\partial L}{\partial x_1} = 8x_1 + 2x_2 - 33 \\ \dfrac{\partial L}{\partial x_2} = 2x_1 + 12x_2 - 22. \end{cases}$$

Solving $\dfrac{\partial L}{\partial x_1} = 0 = \dfrac{\partial L}{\partial x_2}$, we find that $x_1 = 88/23 \approx 3.82609$ and $x_2 = 55/46 \approx 1.19565$. We set $(x_1^2, x_2^2) := (3.83, 1.20)$ and move to Step 2, updating the Lagrangean multiplier as

$$u^2 := u^1 + r(x_1^2 + 2x_2^2 - 6) = 5 + 2(3.83 + 2.40 - 6) = 5.46.$$

We take this result to the next iteration. In Step 1, minimizing the augmented Lagrangean function $L(x_1, x_2, u^2, r) = L(x_1, x_2, 5.46, 2)$. We find

$$\begin{cases} \dfrac{\partial L}{\partial x_1} = 8x_1 + 2x_2 - 32.54 \\ \dfrac{\partial L}{\partial x_2} = 2x_1 + 12x_2 - 21.08 \end{cases}$$

and solving $\dfrac{\partial L}{\partial x_1} = 0 = \dfrac{\partial L}{\partial x_2}$, we find that $x_1 \approx 3.786087$ and $x_2 \approx 1.125652$.

Setting $\left(x_1^3,\ x_2^3\right) := (3.786,\ 1.126)$, we obtain in Step 2 $u^3 := u^2 + r\left(x_1^3 + 2x_2^3 - 6\right)$ $= 5.46 + 2(3.786 + (2)(1.126) - 6) = 5.536$.

Finally, in the third iteration, minimizing $L(x_1, x_2, u^3, r) = L(x_1, x_2, 5.536, 2)$, we proceed as in the previous iterations, by solving $\dfrac{\partial L}{\partial x_1} = 0 = \dfrac{\partial L}{\partial x_2}$ with the result $\left(x_1^4,\ x_2^4\right) \approx (3.7789, 1.1167)$. Updating the Lagrangean multiplier u, we find $u^4 := u^3 + r\left(x_1^4 + 2x_2^4 - 6\right) = 5.536 + 2(3.7789 + (2)(1.1167) - 6) = 5.5606$.

Having carried out three complete iterations, we stop here with the solution $\left(x_1^4,\ x_2^4,\ u^4\right) = (3.7789, 1.1167, 5.5606)$. Comparing this solution with the true optimum $(\bar{x}_1, \bar{x}_2, u) = \left(3\dfrac{7}{9}, 1\dfrac{1}{9}, 5\dfrac{5}{9}\right) \approx (3.7778, 1.1111, 5.5556)$, we see that we are quite close. One should also note that our solution $\mathbf{x}^4$ is near-feasible.

Readers will have noticed that we have not discussed the value of the penalty parameter r, nor have we suggested any procedure for updating it. Besides pushing the solution towards feasibility, higher values of r will also tend to "convexify" the augmented Lagrangean function via the penalty term $\dfrac{r}{2} \|\mathbf{g}(\mathbf{x})\|^2$. In our numerical example above, this was not an issue since the Lagrangean function was already convex before the penalty term was added, but in the general case one might have to increase the value of r iteratively above some value to achieve convexity, even if this were only required locally around the optimal solution. Although it is intuitively clear that a sufficiently large value of the penalty parameter r will ensure convexity, this can be proved rigorously; see, e.g., Luenberger and Ye (2008) and Bertsekas (2016). The reason for requiring the augmented Lagrangean to be convex is that most unconstrained minimization techniques require convexity. This is also an issue in consideration of duality aspects; again, we refer to the aforementioned references.

Finally, we should briefly mention the *sequential* (or *successive*) *linear programming* approach (*SLP*), in which linear programming problems are successively generated and solved, as linearizations of the objective function and the constraints of a nonlinear programming problem. Bounds or trust region constraints are added as needed. Convergence of those methods has proved to be fast for optimal solutions that are at or near a vertex of the feasible region, but slow otherwise. For details, readers are referred to Bazaraa et al. (2013). In *sequential* (or *successive*) *quadratic programming* (*SQP*) approaches, quadratic programming problems are successively generated by forming quadratic approximations of the Lagrangean function, to be maximized over linear approximations of the constraints. *SQP* methods are extensively treated by Sun and Yuan (2006); see also Bertsekas (2016) and Cottle and Thapa (2017).

References

Bazaraa MS, Sherali HD, Shetty CM (2013) *Nonlinear programming: theory and algorithms.* (3rd ed.) Wiley, New York

Bertsekas DP (2016) *Nonlinear programming.* (3rd ed.) Athena Scientific, Belmont, MA

Carroll CW (1961) The created response surface technique for optimizing nonlinear restrained systems. *Operations Research* **9**: 169-184.

Cottle RW, Thapa MN (2017) *Linear and nonlinear optimization.* Springer-Verlag, Berlin-Heidelberg-New York

Dantzig GB, Thapa MN (2003) *Linear programming 2: theory and extensions.* Springer, New York, Berlin, Heidelberg

Eiselt HA, Sandblom C-L (2007) *Linear programming and its applications.* Springer-Verlag, Berlin-Heidelberg

Everett H (1963) Generalized Lagrange multiplier method for solving problems of optimum allocation of resources. *Operations Research* **11**: 399-417

Fiacco AV, McCormick GP (1964a) Computational algorithm for the sequential unconstrained minimization technique for nonlinear programming. *Management Science* **10**: 601-617

Fiacco AV, McCormick GP (1964b) The sequential unconstrained minimization technique for nonlinear programming-a primal-dual method. *Management Science* **10**: 360-366

Fiacco AV, McCormick GP (1966) Extensions of SUMT for nonlinear programming equality constraints and extrapolation. *Management Science* **12**: 816-828

Fiacco AV, McCormick GP (1967a) The sequential unconstrained minimization technique (SUMT) without parameters. *Operations Research* **15**: 820-827

Fiacco AV, McCormick GP (1967b) The slacked unconstrained minimization technique for convex programming. *The SIAM Journal of Applied Mathematics* **15**: 505-515

Fiacco AV, McCormick GP (1968) *Nonlinear programming.* J. Wiley & Sons, New York

Frisch KR (1956) La résolution des problèmes de programme lineaire pour la méthode du potential logarithmique. *Cahiers du seminaire d'Économetrie* **4**: 7-20

Luenberger DL, Ye Y (2008) *Linear and nonlinear programming.* (3rd ed.) Springer-Verlag, Berlin-Heidelberg-New York

Rockafellar RT (1970) *Convex analysis.* University Press, Princeton, NJ

Sandblom C-L (1974) On the convergence of SUMT. *Mathematical Programming* **6**: 360-364

Sun W, Yuan Y (2006) *Optimization theory and methods. Nonlinear programming.* Springer-Verlag, Berlin-Heidelberg-New York

Chapter 6
Methods for Linearly Constrained Problems

In this and the following two chapters, several algorithms for solving nonlinear constrained optimization problems will be described. First the most special of all constrained nonlinear programming problems is considered, namely the quadratic programming problem for which the objective function is convex and quadratic and the constraints are linear. In the second section of the chapter methods for the more general problem of optimizing a differentiable convex function subject to linear constraints are discussed. Although every convex quadratic programming problem could be solved also by these more general methods, it is generally preferable to employ quadratic programming methods when possible. As a general principle it is advisable to use more specialized techniques for more specialized problems. Consequently, for a given problem one should select a method (covered in the previous, this, or the next two chapters) from a box as high up in Table 6.1 as possible. The third section considers problems in which the objective function is quadratic, but concave, a difficult case.

An important special case of linearly constrained problems are those with *fractional* objective function $f(\mathbf{x})$, i.e., in which $f(\mathbf{x}) = \dfrac{c_0 + \mathbf{cx}}{d_0 + \mathbf{dx}}$, where c_0 and d_0 are scalars and $\mathbf{c}$ and $\mathbf{d}$ are n-vectors. Such problems can be converted to linear programming problems and will therefore not be treated in this book. For details, see Eiselt and Sandblom (2007).

6.1 Problems with Quadratic Objective

Optimization problems that arise in a managerial context will often contain quadratic terms in their model formulations. If the quadratic terms appear in the objective function only, and the optimization is done subject to linear constraints, a *quadratic programming* problem is given. In Chap. 3 we showed that quadratic programming

© Springer Nature Switzerland AG 2019
H. A. Eiselt, C.-L. Sandblom, *Nonlinear Optimization*, International Series in
Operations Research & Management Science 282,
https://doi.org/10.1007/978-3-030-19462-8_6

Table 6.1 Nonlinear optimization methods

Objective function	Constraints	Algorithm	Sections
Convex quadratic	Linear	Wolfe's Method Lemke's Complementary Pivot Method	6.1.1 6.1.2
Convex differentiable	Linear	Frank-Wolfe Method Gradient Projection Method	6.2.1 6.2.2
Concave quadratic	Linear	Concave quadratic progamming	6.3
Linear	Nonlinear (convex and differentiable)	Cutting plane	7.1
Convex differentiable		Generalized Reduced Gradient Method (*GRG*) Sequential Unconstrained Minimization Technique (*SUMT*)	7.2.1 5.1
Concave	Convex	Simplicial Branch-and-Bound Method	7.3
Convex nondifferentiable	Nonlinear (convex and non-differentiable)	Shor-Polyak Subgradient Algorithm	7.2.2
Posynomial	–	Unconstrained posynomial geometric programming	8.3.1
Posynomial	Posynomial	Constrained posynomial geometric programming	8.3.2
Signomial	Signomial	Constrained signomial geometric programming	8.4

models are common in such diverse areas as curve fitting, portfolio analysis, macroeconomic and production planning, and so forth. Several algorithms have been developed for solving quadratic programming problems. After an introductory discussion, we present a technique due to Wolfe (1959), which is an adaptation of the simplex method for linear programming. As for several other quadratic programming algorithms, Wolfe's method is based on the Karush-Kuhn-Tucker optimality conditions. Then in the next section, the so-called *linear complementarity problem* is defined, followed by a description of how it can be solved by the *complementary pivot algorithm* due to Lemke (1968). It is then shown how the quadratic programming problem can be formulated as a linear complementarity problem, and consequently solved by the complementary pivot algorithm.

Consider the quadratic programming problem in canonical form

$$P: \text{ Min } z = f(\mathbf{x}) = \mathbf{x}^T \mathbf{Q} \mathbf{x} + \mathbf{c}\mathbf{x} + d_0$$
$$\text{s.t. } \mathbf{A}\mathbf{x} \leq \mathbf{b}$$
$$\mathbf{x} \geq \mathbf{0},$$

where $\mathbf{x}$ and $\mathbf{b}$ are n- and m-dimensional column vectors, respectively, $\mathbf{c}$ is an n-dimensional row-vector, $\mathbf{Q}$ is an $[n \times n]$-dimensional symmetric matrix, $\mathbf{A}$ is an $[m \times n]$-dimensional matrix and d_0 is a scalar which will be omitted in the following discussion since it does not affect the process of optimization. Any quadratic programming problem can be written in this form (suitably transformed, if necessary). For instance, the quadratic programming problem

$$P: \text{Min } z = f(x_1, \ x_2) = 3x_1^2 - 2x_1x_2 + 2x_2^2 - 26x_1 - 8x_2$$
$$\text{s.t. } x_1 + 2x_2 \le 6$$
$$x_1 - x_2 \ge 1$$
$$x_1, \ x_2 \ge 0$$

can be written in the above standard form with

$$\mathbf{Q} = \begin{bmatrix} 3 & -1 \\ -1 & 2 \end{bmatrix}, \ \mathbf{c} = [-26, \ -8], \ d_0 = 0, \ \mathbf{A} = \begin{bmatrix} 1 & 2 \\ -1 & 1 \end{bmatrix}, \ \mathbf{b} = \begin{bmatrix} 6 \\ -1 \end{bmatrix}.$$

If no further requirements on the matrix $\mathbf{Q}$ are specified, the level sets of the objective function always have the shape of multidimensional conic sections such as ellipsoids, hyperboloids, paraboloids or mixed forms such as hyperbolical paraboloids. Most of the techniques available in the literature of quadratic programming are restricted to the solution of problems in which for given values of the objective function z, the level sets are concentric ellipsoids or, as a special case, hyperspheres. Some techniques also allow the degenerate case corresponding to a paraboloid. In these cases, the problem P is said to belong to the class of convex quadratic programming models. From the discussion in Sect. 1.3 it will be clear that a convex quadratic programming model has a matrix $\mathbf{Q}$, which is positive definite, or positive semidefinite in the degenerate case. Furthermore, if a quadratic programming problem is formulated in matrix form with a nonsymmetric matrix $\mathbf{Q}$, then $\mathbf{Q}$ can be replaced with $\frac{1}{2}(\mathbf{Q} + \mathbf{Q}^T)$, which is symmetric and generates the same quadratic function. We should mention that problems with objective functions of the type $\left(\sum_{j=1}^{n} q_j x_j \right)^2$ can be handled by linear programming methods. The minimal point for $\left(\sum_{j=1}^{n} q_j x_j \right)^2$ is the same as the minimal point for the smallest of the minimal points for $\sum_{j=1}^{n} q_j x_j$ and $- \sum_{j=1}^{n} q_j x_j$, so all we need to do is solve two linear programming problems and pick the solution with the smaller objective function value.

Consider now a quadratic programming problem with a strictly convex objective function $f(\mathbf{x})$, i.e., one with a positive definite matrix $\mathbf{Q}$. Any level set of this objective function (i.e. points with identical objective function value) will then be an ellipsoid. The smaller the objective function, the smaller will be the corresponding ellipsoid. The unconstrained minimum is then located in the center, common to all these ellipsoids. This allows us to formulate the following graphical solution technique for quadratic optimization problems.

The Graphical Solution Technique

Step 1 Graph all constraints and identify the set of feasible solutions, i.e., the polytope $\{\mathbf{x} : \mathbf{Ax} \leq \mathbf{b}, \ \mathbf{x} \geq \mathbf{0}\}$.

Step 2 Determine the constrained minimal solution, i.e., the point $\hat{\mathbf{x}}$ at which $\nabla f(\hat{\mathbf{x}}) = \mathbf{0}$.

Step 3 Does the unconstrained minimal point belong to the feasible region?

If yes: Stop, the unconstrained minimal point is an optimal solution to P.
If no: Go to Step 4.

Step 4 Select the smallest ellipsoid that has at least one point in common with the given polytope.

The resulting point is then a constrained optimal solution. Before illustrating such a geometric procedure with a numerical example, we will first consider what can be said about the constrained optimal solution in general. Without loss of generality we assume that the quadratic objective function defines a class of concentric hyperspheres. Denoting by $\hat{\mathbf{x}}$ the unconstrained and by $\bar{\mathbf{x}}$ the constrained minimal points, we can then distinguish three different situations as depicted by the graphs (a), (b), and (c) of Fig. 6.1.

As can be seen from Fig. 6.1, the constrained optimal solution may occur in the interior of the given polytope as in case (a); on a binding hyperplane as in case (b); or at an extreme point as in case (c). In other words, any point of the given polytope is a candidate for the constrained optimum, in contrast to the situation in linear programming where at least one of the optimal solutions is situated at an extreme point of the given polytope and interior points can never be optimal. This implies that, excluding cases involving degeneracy, at most n constraints can be binding at an optimal point while in linear programming they were exactly n.

It is also important to note that if an interior point is the constrained optimum, then this point must also be the unconstrained optimum. This fact leads to the

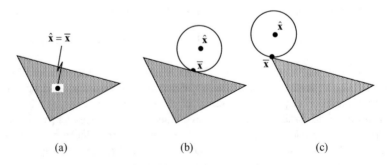

(a)   (b)   (c)

Fig. 6.1 Optimal points for a quadratic objective with linear constraints

suggestion that the unconstrained optimum is to be determined first and subsequently one checks if it satisfies the given set of constraints. When this happens to be so, then the optimal solution has already been determined. In quadratic optimization this is particularly simple because the relation $\nabla f(\mathbf{x}) = \mathbf{0}$ amounts to a system of n simultaneous linear equations with n variables, the solution of which can be found by using for instance the Gauss-Jordan pivoting technique (see Appendix D). This procedure avoids lengthy calculations if it should turn out that an interior point is optimal.

Example To illustrate the graphical solution technique described above, consider the quadratic programming problem

$$\text{P: Min } z = f(x_1, \ x_2) = x_1^2 + x_2^2 - 16x_1 - 10x_2$$
$$\text{s.t. } -2x_1 + 4x_2 \le 8$$
$$x_1 + x_2 \le 6$$
$$x_2 \le 3$$
$$x_1, \ x_2 \ge 0.$$

It can easily be seen that the unconstrained optimum is located at the point $(\hat{x}_1, \ \hat{x}_2) = (8, \ 5)$, which is not feasible for the constrained problem.

In Fig. 6.2 various circles representing different levels for the value of the objective function are displayed, together with the straight lines corresponding to the three linear structural constraints of the problem. Note that the circle with radius $\sqrt{89 - 64\frac{1}{2}} = \sqrt{24\frac{1}{2}}$ touches on the given polyhedron at just one point, namely $\bar{\mathbf{x}} = [4\frac{1}{2}, \ 1\frac{1}{2}]$ and no smaller circle has a point in common with the

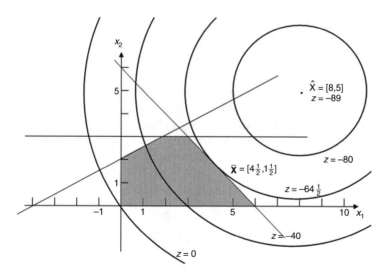

Fig. 6.2 Level curves for the numerical example

polyhedron. Thus, the point $\bar{\mathbf{x}}$ constitutes the unique constrained optimal solution for problem P. This corresponds to the situation (b) in Fig. 6.1, in which the optimal point is located not at an extreme point, but on the boundary of the feasible region.

6.1.1   Wolfe's Method

Based on the Karush-Kuhn-Tucker optimality conditions, Wolfe (1959) devised an algorithm for the quadratic programming problem, which uses the simplex method for linear programming, albeit in a slightly modified form. For a description of the simplex method for linear programming, readers are referred to Appendix D. Recall that the general quadratic programming problem P is formulated as

$$P : \text{Min } z = f(\mathbf{x}) = \mathbf{x}^T \mathbf{Q} \mathbf{x} + \mathbf{c}\mathbf{x}$$
$$\text{s.t. } \mathbf{A}\mathbf{x} \le \mathbf{b}$$
$$\mathbf{x} \ge \mathbf{0}.$$

The associated Lagrangean function is $L(\mathbf{x}, \mathbf{u}) = \mathbf{x}^T \mathbf{Q}\mathbf{x} + \mathbf{c}\mathbf{x} + \mathbf{u}\mathbf{A}\mathbf{x} - \mathbf{u}\mathbf{b}$, so that the Karush-Kuhn-Tucker optimality conditions for this linearly constrained problem (see Definition 4.12) are:

$$L_{\mathbf{x}}\big(\bar{\mathbf{x}},\bar{\mathbf{u}}\big) = 2\mathbf{Q}\bar{\mathbf{x}} + \mathbf{c}^T + \mathbf{A}^T\bar{\mathbf{u}}^T \ge \mathbf{0},$$
$$\bar{\mathbf{x}}^T L_{\mathbf{x}}\big(\bar{\mathbf{x}},\bar{\mathbf{u}}\big) = 2\bar{\mathbf{x}}^T\mathbf{Q}\bar{\mathbf{x}} + \bar{\mathbf{x}}^T\mathbf{c}^T + \bar{\mathbf{x}}^T\mathbf{A}^T\bar{\mathbf{u}}^T = 2\bar{\mathbf{x}}^T\mathbf{Q}\bar{\mathbf{x}} + \mathbf{c}\bar{\mathbf{x}} + \bar{\mathbf{u}}\mathbf{A}\bar{\mathbf{x}} = 0,$$
$$\bar{\mathbf{x}} \ge \mathbf{0},$$
$$L_{\mathbf{u}}\big(\bar{\mathbf{x}},\bar{\mathbf{u}}\big) = \mathbf{A}\bar{\mathbf{x}} - \mathbf{b} \le \mathbf{0},$$
$$\bar{\mathbf{u}}L_{\mathbf{u}}\big(\bar{\mathbf{x}},\bar{\mathbf{u}}\big) = \bar{\mathbf{u}}\mathbf{A}\bar{\mathbf{x}} - \bar{\mathbf{u}}\mathbf{b} = 0, \quad \text{and}$$
$$\bar{\mathbf{u}} \ge \mathbf{0}.$$

From Theorem 4.18 we know that if the objective function f is convex and differentiable (which in the case of quadratic optimization means that $\mathbf{Q}$ is positive definite or positive semidefinite), then these Karush-Kuhn-Tucker conditions are necessary and sufficient for optimality. Defining $\mathbf{v}^T := L_{\mathbf{x}}(\mathbf{x}, \mathbf{u}) = 2\mathbf{Q}\mathbf{x} + \mathbf{c}^T + \mathbf{A}^T\mathbf{u}^T \ge \mathbf{0}$ and $\mathbf{y} := -L_{\mathbf{u}}(\mathbf{x}, \mathbf{u}) = \mathbf{b} - \mathbf{A}\mathbf{x} \ge \mathbf{0}$ leads to the system of relations

$$I: -2\mathbf{Q}\mathbf{x} - \mathbf{A}^T\mathbf{u}^T + \mathbf{v}^T = \mathbf{c}^T \tag{6.1}$$

$$\mathbf{A}\mathbf{x} + \mathbf{y} = \mathbf{b} \tag{6.2}$$

$$\mathbf{v}\mathbf{x} = 0, \ \mathbf{u}\mathbf{y} = 0 \tag{6.3a}$$

$$\mathbf{x}, \ \mathbf{v}, \ \mathbf{u}, \ \mathbf{y} \ge \mathbf{0}.$$

Note that because of the nonnegativity requirements, the two relations (6.3a) can be written as the single relation (6.3b) $\mathbf{v}\mathbf{x} + \mathbf{u}\mathbf{y} = 0$.

From the above discussion it follows that $\bar{\mathbf{x}}$ is an optimal solution to the quadratic programming problem if and only if $\bar{\mathbf{u}}$, $\bar{\mathbf{v}}$, and $\bar{\mathbf{y}}$ exist, such that $(\bar{\mathbf{x}},\bar{\mathbf{u}},\bar{\mathbf{v}},\bar{\mathbf{y}})$ satisfies the system I. Without the nonlinear equations $\mathbf{vx} = 0$ and $\mathbf{uy} = 0$ (usually called the *complementarity* or *orthogonality* conditions, which express the fact that the surplus variable $\mathbf{v}$ is orthogonal to $\mathbf{x}$, and $\mathbf{u}$ is orthogonal to the slack variables $\mathbf{y}$), system I is simply a system of simultaneous linear equations. As long as the matrix $\mathbf{Q}$ is positive definite, a modified simplex procedure for the solution of problem P can now be devised. The case in which $\mathbf{Q}$ is not positive definite will be considered at the end of this section.

In order to initialize the technique, artificial variables w_k^+ and w_k^- are introduced in the constraints (6.1). Furthermore, artificial surplus variables w_k^0 are subtracted from the left-hand side of constraints (6.2), but only where required (i.e., where $b_k < 0$). After these preparations have been carried out, the system of constraints with the exclusion of the complementarity conditions becomes

$$II: -2\mathbf{Q}\mathbf{x} - \mathbf{A}^T\mathbf{u}^T + \mathbf{v}^T + \mathbf{w}^+ - \mathbf{w}^- = \mathbf{c}^T$$
$$\mathbf{A}\mathbf{x} + \mathbf{y} - \mathbf{w}^0 = \mathbf{b}$$
$$\mathbf{x}, \mathbf{u}, \mathbf{v}, \mathbf{y}, \mathbf{w}^+, \mathbf{w}^-, \mathbf{w}^0 \geq \mathbf{0}.$$

Following Wolfe's approach, system II can be solved in two phases.

Phase 1 The sum $\mathbf{ew}^0$ of the artificial variables w_k^0 is minimized (where $\mathbf{e}$ is a summation vector of appropriate dimension) with the additional restriction that $\mathbf{u}$ and $\mathbf{v}$ stay out of the basis (i.e., have a value of zero) in this phase. At the end of this phase, there are two possible results: either $\mathbf{ew}^0 > 0$ or $\mathbf{ew}^0 = 0$. In case $\mathbf{ew}^0 > 0$, some of the given constraints $\mathbf{Ax} \leq \mathbf{b}$, $\mathbf{x} \geq \mathbf{0}$ are incompatible, so that problem P has no feasible solution. If $\mathbf{ew}^0 = 0$, we proceed to Phase 2, dropping the artificial variables $\mathbf{w}^0$ from consideration. The variables $\mathbf{w}^+$ and $\mathbf{w}^-$ are combined into one vector, i.e., $\mathbf{w}^+ - \mathbf{w}^- = \mathbf{Dw}$, where $\mathbf{D}$ is a diagonal matrix with elements 1 or -1 along the main diagonal, depending on whether w_k^+ or w_k^- is in the basis at this time. The system of constraints, with the exclusion of the complementarity conditions, is therefore at the end of Phase 1

$$III: -2\mathbf{Q}\mathbf{x} - \mathbf{A}^T\mathbf{u}^T + \mathbf{v}^T + \mathbf{Dw} = \mathbf{c}^T$$
$$\mathbf{A}\mathbf{x} + \mathbf{y} = \mathbf{b}$$
$$\mathbf{x}, \mathbf{u}, \mathbf{v}, \mathbf{y}, \mathbf{w} \geq \mathbf{0},$$

where $\mathbf{u} = \mathbf{0}$ and $\mathbf{v} = \mathbf{0}$. Hence the complementarity conditions are automatically satisfied.

Phase 2 The sum $\mathbf{ew}$ of the artificial variables w_k is minimized. At the end of this phase, there are again two possible results: we either have $\mathbf{ew} > 0$ or $\mathbf{ew} = 0$. Theorem 6.1 below will demonstrate that under certain assumptions this phase will always end with $\mathbf{ew} = 0$.

Until now, the complementarity conditions $\mathbf{vx} = 0$ and $\mathbf{uy} = 0$ have been disregarded. A very simple rule can be applied in addition to the regular simplex method in such a way that the complementarity conditions are always satisfied in the two phases described above. This rule can be stated as follows.

Complementarity Rule A variable x_j (or v_j) can enter the basis only if either v_j (or x_j) is currently a nonbasic variable or if v_j (x_j) leaves the basis at the same time that x_j (v_j) enters the basis. Similarly, we ensure that the variables u_i and y_i cannot be in the basis simultaneously. The modification of this rule in the case of degeneracy, i.e., with zero-valued basic variables, is obvious. In the following we assume that this rule is applied in carrying out the iterations in Phase 2. With the successful completion of Phase 2, an optimal solution to problem P will have been found.

Theorem 6.1 Assume that the matrix $\mathbf{Q}$ in problem P is positive definite and symmetric and that feasible solutions exist. Then Phase 1 will end with $\mathbf{ew}^0 = 0$ and Phase 2 will end with $\mathbf{ew} = 0$, yielding an optimal solution to P.

Proof Due to the existence of a feasible solution for P, $\mathbf{ew}^0$ will be zero at the end of Phase 1, following the above discussion, i.e., $\mathbf{w}^0 = \mathbf{0}$. Assume now that Phase 2 was completed with $(\hat{\mathbf{x}},\hat{\mathbf{u}},\hat{\mathbf{v}},\hat{\mathbf{y}},\hat{\mathbf{w}})$ and that the solution is nondegenerate in the sense that either $\hat{x}_j$ or $\hat{v}_j$ is zero, but not both, and either $\hat{y}_i$ or $\hat{u}_i$ is zero, but not both. Consider now the problem

$$\hat{\mathrm{P}}: \mathrm{Min} \ \mu = \mathbf{ew}$$

$$\text{s.t.} - 2\mathbf{Qx} - \mathbf{A}^T\mathbf{u}^T + \mathbf{v}^T + \mathbf{Dw} = \mathbf{c}^T \qquad (6.4)$$

$$\mathbf{Ax} + \mathbf{y} = \mathbf{b} \qquad (6.5)$$

$$\hat{\mathbf{v}}\mathbf{x} + \hat{\mathbf{u}}\mathbf{y} + \mathbf{v}\hat{\mathbf{x}} + \hat{\mathbf{u}}\mathbf{y} \leq 0 \qquad (6.6)$$

$$\mathbf{x}, \mathbf{u}, \mathbf{v}, \mathbf{y}, \mathbf{w} \geq \mathbf{0}. \qquad (6.7)$$

Note that problem $\hat{\mathrm{P}}$ is a linear programming problem with variables $\mathbf{x}$, $\mathbf{u}$, $\mathbf{v}$, $\mathbf{y}$, $\mathbf{w}$ and that $\hat{\mathbf{v}}, \hat{\mathbf{y}}, \hat{\mathbf{x}}$, and $\hat{\mathbf{u}}$ serve as coefficients in the linear constraint (6.6). Due to the nondegeneracy assumption, the constraint (6.6) is satisfied if and only if $\hat{\mathbf{v}}\hat{\mathbf{x}} = 0$ and $\hat{\mathbf{u}}\hat{\mathbf{y}} = 0$, so that $(\hat{\mathbf{x}},\hat{\mathbf{u}},\hat{\mathbf{v}},\hat{\mathbf{y}},\hat{\mathbf{w}})$ is an optimal solution to $\hat{\mathrm{P}}$ with objective function value $\hat{\mu} = \mathbf{e}\hat{\mathbf{w}}$. Using duality theory, we will now show that $\hat{\mu} = 0$. Defining row vectors $\boldsymbol{\alpha} \in \mathbb{R}^n$, $\boldsymbol{\beta} \in \mathbb{R}^m$ and $\gamma \in \mathbb{R}$ as dual variables associated with the respective primal constraints, the dual problem of $\hat{\mathrm{P}}$ is

$$\hat{\mathrm{P}}_\mathrm{D} : \mathrm{Min} \ \nu = \boldsymbol{\alpha}\mathbf{c}^T + \boldsymbol{\beta}\mathbf{b}$$

$$\text{s.t.} - 2\boldsymbol{\alpha}\mathbf{Q} + \boldsymbol{\beta}\mathbf{A} + \gamma\hat{\mathbf{v}} \geq \mathbf{0} \qquad (6.8)$$

$$-\boldsymbol{\alpha}\mathbf{A}^T + \gamma\hat{\mathbf{y}}^T \geq \mathbf{0} \qquad (6.9)$$

$$\boldsymbol{\alpha} + \gamma \hat{\mathbf{x}}^T \geq \mathbf{0} \tag{6.10}$$

$$\boldsymbol{\beta} + \gamma \hat{\mathbf{u}} \geq \mathbf{0} \tag{6.11}$$

$$\boldsymbol{\alpha} \mathbf{D} \geq -\mathbf{e} \tag{6.12}$$

$$\boldsymbol{\alpha} \in \mathbb{R}^n, \ \boldsymbol{\beta} \in \mathbb{R}^m, \ \gamma \geq 0.$$

Since the problem possesses a finite optimal solution, strong duality of linear programming implies that the dual problem $\hat{P}_D$ has a finite optimal solution, which will be denoted by $(\hat{\boldsymbol{\alpha}}, \hat{\boldsymbol{\beta}}, \hat{\gamma})$. Postmultiplying (6.8) by the transpose of (6.10) yields

$$-2\hat{\boldsymbol{\alpha}}\mathbf{Q}\hat{\boldsymbol{\alpha}}^T + \hat{\boldsymbol{\beta}}\mathbf{A}\hat{\boldsymbol{\alpha}}^T + \hat{\gamma}\hat{\mathbf{v}}\hat{\boldsymbol{\alpha}}^T - 2\hat{\gamma}\hat{\boldsymbol{\alpha}}\mathbf{Q}\hat{\mathbf{x}} + \hat{\gamma}\hat{\boldsymbol{\beta}}\mathbf{A}\hat{\mathbf{x}} + \hat{\gamma}^2\hat{\mathbf{v}}\hat{\mathbf{x}} \geq 0 \tag{6.13}$$

We will now demonstrate that the last four terms on the left-hand side of relation (6.13) cancel out to zero. First, the two primal constraints (6.6) and (6.7) imply that $\hat{\mathbf{v}}\hat{\mathbf{x}} = 0$. Secondly, due to weak complementary slackness (cf. relations (4.15) and (4.17) at the very end of Sect. 4.2) for $\mathbf{v}$ and its associated dual constraint (6.10), we obtain $\hat{\mathbf{v}}\hat{\boldsymbol{\alpha}}^T + \gamma\hat{\mathbf{v}}\hat{\mathbf{x}} = 0$, or $\hat{\mathbf{v}}\hat{\boldsymbol{\alpha}}^T = 0$, since $\hat{\mathbf{v}}\hat{\mathbf{x}} = 0$. Thirdly, weak complementary slackness for $\mathbf{x}$ and its associated dual constraint (6.8) yields $-2\hat{\boldsymbol{\alpha}}\mathbf{Q}\hat{\mathbf{x}} + \hat{\boldsymbol{\beta}}\mathbf{A}\hat{\mathbf{x}} + \gamma\hat{\mathbf{v}}$ $\hat{\mathbf{x}} = 0$ or $-2\hat{\boldsymbol{\alpha}}\mathbf{Q}\hat{\mathbf{x}} + \hat{\boldsymbol{\beta}}\mathbf{A}\hat{\mathbf{x}} = 0$. Using these results, relation (6.13) reduces to

$$-2\hat{\boldsymbol{\alpha}}\mathbf{Q}\hat{\boldsymbol{\alpha}}^T + \hat{\boldsymbol{\beta}}\mathbf{A}\hat{\boldsymbol{\alpha}}^T \geq 0. \tag{6.14}$$

In order to show that $\hat{\boldsymbol{\beta}}\mathbf{A}\hat{\boldsymbol{\alpha}}^T$ (or rather its transpose $\hat{\boldsymbol{\alpha}}\mathbf{A}^T\hat{\boldsymbol{\beta}}^T$) is nonpositive, postmultiply (6.9) by the transpose of (6.11), which results in

$$-\hat{\boldsymbol{\alpha}}\mathbf{A}^T\hat{\boldsymbol{\beta}}^T + \hat{\gamma}\hat{\mathbf{y}}^T\hat{\boldsymbol{\beta}}^T + \hat{\gamma}\left(-\hat{\boldsymbol{\alpha}}\mathbf{A}^T\hat{\mathbf{u}}^T + \hat{\gamma}\hat{\mathbf{y}}^T\hat{\mathbf{u}}^T\right) \geq 0. \tag{6.15}$$

First, due to weak complementary slackness for $\mathbf{y}$ and its associated dual constraint (6.11), we obtain $(\hat{\boldsymbol{\beta}} + \hat{\gamma}\hat{\mathbf{u}})\hat{\mathbf{y}} = \hat{\boldsymbol{\beta}}\hat{\mathbf{y}} + \hat{\gamma}\hat{\mathbf{u}}\hat{\mathbf{y}} = \mathbf{0}$, from which we conclude that $\hat{\boldsymbol{\beta}}\hat{\mathbf{y}} = 0$, since $\hat{\mathbf{u}}\hat{\mathbf{y}} = 0$ (via the same argument that shows that $\hat{\mathbf{v}}\hat{\mathbf{x}} = 0$). Secondly, weak complementary slackness for $\mathbf{u}$ and its associated dual constraint (6.9) yields $(-\hat{\boldsymbol{\alpha}}\mathbf{A}^T + \hat{\gamma}\hat{\mathbf{y}}^T)\hat{\mathbf{u}}^T = -\hat{\boldsymbol{\alpha}}\mathbf{A}^T\hat{\mathbf{u}}^T + \hat{\gamma}\hat{\mathbf{y}}^T\hat{\mathbf{u}}^T = 0$, so that the expression in brackets in relation (6.15) vanishes. Hence (6.15) reduces to

$$-\hat{\boldsymbol{\alpha}}\mathbf{A}^T\hat{\boldsymbol{\beta}}^T \geq 0. \tag{6.16}$$

Now, relations (6.14) and (6.16) together imply that $2\hat{\boldsymbol{\alpha}}\mathbf{Q}\hat{\boldsymbol{\alpha}}^T \leq 0$ and as $\mathbf{Q}$ was assumed to be positive definite, it follows that $\hat{\boldsymbol{\alpha}} = \mathbf{0}$. Hence the dual constraint (6.12) is satisfied as a strict inequality, and it follows from weak complementary slackness that the primal variables $\mathbf{w}$ associated with these constraints must be zero at optimality, i.e., $\hat{\mathbf{w}} = \mathbf{0}$. Therefore, Phase 2 must end with $\mathbf{e}\hat{\mathbf{w}} = 0$, and the proof is complete. $\square$

The above discussion has demonstrated that if the system $\mathbf{Ax} \le \mathbf{b}$, $\mathbf{x} \ge \mathbf{0}$, has at least one feasible solution and the matrix $\mathbf{Q}$ of the given quadratic objective function $f(\mathbf{x}) = \mathbf{x}^T \mathbf{Q}\mathbf{x} + \mathbf{c}\mathbf{x}$ is positive definite, then the proposed method, known as the *short form* of Wolfe's method, will find an optimal solution to the problem P. The *long form* of this method is slightly more general, since it requires $\mathbf{Q}$ only to be positive semidefinite. For a treatment of the long form, readers are referred to Künzi et al. (1966).

The short form of Wolfe's method will now be described in algorithmic form. Recall that the problem under consideration is

$$P: \text{Min } z = \mathbf{x}^T \mathbf{Q}\mathbf{x} + \mathbf{c}\mathbf{x} + d_0$$
$$\text{s.t. } \mathbf{Ax} \le \mathbf{b}$$
$$\mathbf{x} \ge \mathbf{0},$$

where the square matrix $\mathbf{Q}$ is required to be symmetric and positive definite, and refer to the systems *II* and *III* in the discussion above.

Wolfe's Method (Short Form)

Step 0 Disregarding the constraints of problem P, find a minimal point for the unconstrained problem. Is this point feasible?

If yes: Stop, an optimal solution to problem P has been found.
If no: Go to Step 1.

Step 1 Set up the initial tableau as follows:

x	u	v	y	$\mathbf{w}^+$	$\mathbf{w}^-$	$\mathbf{w}^0$	1
A	0	0	I	0	0	$-\mathbf{I}$	b
$-2\mathbf{Q}$	$-\mathbf{A}^T$	I	0	I	$-\mathbf{I}$	0	$\mathbf{c}^T$
Phase 2 artificial objective function: Min ew							μ
Phase 1 artificial objective function: Min $\mathbf{ew}^0$							

Artificial variables $\mathbf{w}^0$ are only introduced in rows where necessary, i.e., for negative right-hand side values. Multiplying by (-1) where necessary ensures that all right-hand side values are nonnegative.

Step 2 (Phase 1) Using the primal simplex method, minimize the sum of artificial variables $\mathbf{w}^0$ (if any), while satisfying the additional restriction that $\mathbf{u}$ and $\mathbf{v}$ stay out of the basis in this phase.

Step 3 Is the minimal sum $\mathbf{ew}^0$ positive?

If yes: Stop, the problem P has no feasible solution.
If no: Go to Step 4.

Step 4 (Phase 2) Continuing with the tableau resulting from Phase 1, drop the $\mathbf{w}^0$ columns (if any), combine the $\mathbf{w}^+$ and $\mathbf{w}^-$ variables into one vector $\mathbf{w}$, i.e., set $\mathbf{Dw} := \mathbf{w}^+ - \mathbf{w}^-$, where $\mathbf{D}$ is a diagonal matrix with elements $+1$ or -1, and minimize the sum $\mathbf{ew}$ of the components of the artificial variable $\mathbf{w}$. Use the primal simplex method with the additional rule that a column is pivot-eligible, only if it has a negative entry in the $\mathbf{ew}$ objective function row and its complementary variable is either nonbasic or leaves the basis in this step. At minimum, $\mathbf{e\bar{w}}$ will be zero, and the corresponding variables $\bar{\mathbf{x}}$ constitute an optimal solution to the problem P.

Note that the tableau in Step 1 corresponds to System *II* with the two matrix equations in reverse order.

Example Consider the quadratic programming problem

$$P : \text{Min } z = 3x_1^2 - 2x_1x_2 + 2x_2^2 - 26x_1 - 8x_2$$
$$\text{s.t. } x_1 + 2x_2 \leq 6$$
$$x_1 - -x_2 \geq 1$$
$$x_1, x_2 \geq 0,$$

which is graphically represented in Fig. 6.3.

Note that for the various values of z, the objective function defines various ellipsoids, all of which are centered at the point $\hat{\mathbf{x}} = [6, 5]^T$. Since this point is not in the feasible polytope, Wolfe's method proceeds beyond Step 0 in its search for an optimal solution. For the given problem, we multiply the second constraint by (-1) and define the following matrices: $\mathbf{Q} = \begin{bmatrix} 3 & -1 \\ -1 & 2 \end{bmatrix}$, $\mathbf{c} = [-26 \ -8]$,

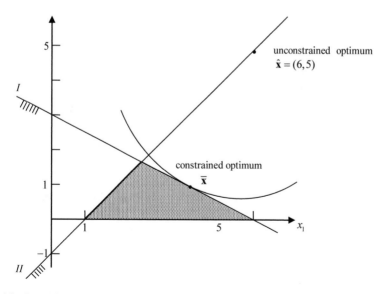

Fig. 6.3 Constrained and unconstrained optimum for the numerical example

$\mathbf{A} = \begin{bmatrix} 1 & 2 \\ -1 & 1 \end{bmatrix}$, and $\mathbf{b} = \begin{bmatrix} 6 \\ -1 \end{bmatrix}$. Furthermore, we will use the acronym "Ph1" to denote the artificial objective function in phase 1, while "Ph2" denotes the artificial objective function in phase 2. The initial tableau is then
T^1:

Basis	x_1	x_2	u_1	u_2	v_1	v_2	y_1	y_2	w_1^+	w_2^+	w_1^-	w_2^-	w_1^0	1
y_1	1	2	0	0	0	0	1	0	0	0	0	0	0	6
w_1^0	①	-1	0	0	0	0	0	-1	0	0	0	0	1	1
w_1^-	6	-2	1	-1	-1	0	0	0	-1	0	1	0	0	26
w_2^-	-2	4	2	1	0	-1	0	0	0	-1	0	1	0	8
Ph2	-4	-2	-3	0	1	1	0	0	0	0	0	0	0	-34
Ph1	-1	1	0	0	0	0	0	1	0	0	0	0	0	-1

The objective function in tableau T^1 is $\mathbf{e}\mathbf{w}^0 = w_1^0$, which is to be minimized. The appropriate pivot element is circled. Then x_1 is chosen as entering variable, which is allowed, since its complementary variable v_1 is currently nonbasic. The artificial variable w_1^0 will therefore leave the basis and thus the variable w_1^0 and its column can be discarded. Performing a pivot operation, we obtain the next tableau T^2, which represents a basic feasible solution, indicating the end of Phase 1 and the beginning of Phase 2. For brevity, we omit the leftmost basis column, and, according to the procedure in Step 4, the variables, w_1^+, w_2^+, w_1^-, and w_2^- are combined into w_1 and w_2.
T^2:

	x_1	x_2	u_1	u_2	v_1	v_2	y_1	y_2	w_1	w_2	1
	0	③	0	0	0	0	1	1	0	0	5
	1	-1	0	0	0	0	0	-1	0	0	1
	0	4	1	-1	-1	0	0	6	1	0	20
	0	2	2	1	0	-1	0	-2	0	1	10
Ph2	0	-6	-3	0	1	1	0	-4	0	0	-30

The objective function in tableau T^2 is $\mathbf{e}\mathbf{w} = w_1 + w_2$, which is to be minimized. Now x_2 is chosen as the entering variable, which is permitted, since v_2 is currently nonbasic, and y_1 leaves the basis in this step. The resulting transformed tableau T^3 is then

T^3:

x_1	x_2	u_1	u_2	v_1	v_2	y_1	y_2	w_1	w_2	1
0	1	0	0	0	0	$\frac{1}{3}$	$\frac{1}{3}$	0	0	$\frac{5}{3}$
1	0	0	0	0	0	$\frac{1}{3}$	$-\frac{2}{3}$	0	0	$\frac{8}{3}$
0	0	1	-1	-1	0	$-\frac{4}{3}$	$\frac{14}{3}$	1	0	$\frac{40}{3}$
0	0	②	1	0	-1	$-\frac{2}{3}$	$-\frac{8}{3}$	0	1	$\frac{20}{3}$
Ph2	0	-3	0	1	1	2	-2	0	0	-20

Now u_1 is chosen as the entering variable (y_1 is not in the basis), and the artificial variable w_2 leaves the basis, its column can therefore be deleted. The resulting tableau is then T^4.

T^4:

x_1	x_2	u_1	u_2	v_1	v_2	y_1	y_2	w_1	1
0	1	0	0	0	0	$\frac{1}{3}$	$\frac{1}{3}$	0	$\frac{5}{3}$
1	0	0	0	0	0	$\frac{1}{3}$	$-\frac{2}{3}$	0	$\frac{8}{3}$
0	0	0	$-\frac{3}{2}$	-1	$\frac{1}{2}$	-1	⑥	1	10
0	0	1	$\frac{1}{2}$	0	$-\frac{1}{2}$	$-\frac{1}{3}$	$-\frac{4}{3}$	0	$\frac{10}{3}$
Ph2	0	0	$\frac{3}{2}$	1	$-\frac{1}{2}$	1	-6	0	-10

Here, v_2 cannot be selected as the new entering variable since x_2 is currently in the basis. Thus the variable y_2 is chosen as the entering variable (its complementary variable u_2 is not in the basis), and the artificial variable w_1 leaves the basis, its column can therefore be deleted. As the new tableau T^5 does not include any further artificial variables, it represents an optimal solution.

T^5:

x_1	x_2	u_1	u_2	v_1	v_2	y_1	y_2	1
0	1	0	$\frac{1}{12}$	$\frac{1}{18}$	$-\frac{1}{36}$	$\frac{7}{18}$	0	$\frac{10}{9}$
1	0	0	$-\frac{1}{6}$	$-\frac{1}{9}$	$\frac{1}{18}$	$\frac{2}{9}$	0	$\frac{34}{9}$
0	0	0	$-\frac{1}{4}$	$-\frac{1}{6}$	$\frac{1}{12}$	$-\frac{1}{6}$	1	$\frac{5}{3}$
0	0	1	$\frac{1}{6}$	$-\frac{2}{9}$	$-\frac{7}{18}$	$-\frac{5}{9}$	0	$\frac{50}{9}$
Ph2	0	0	0	0	0	0	0	0

The unique optimal solution $\bar{\mathbf{x}} = \begin{bmatrix} \bar{x}_1, & \bar{x}_2 \end{bmatrix}^T$ to problem P can now be read off Tableau T^5; it is $\bar{x}_1 = 3\frac{7}{9} \approx 3.778$, $\bar{x}_2 = 1\frac{1}{9} \approx 1.111$. That $\mathbf{x} = \begin{bmatrix} 3\frac{7}{9}, & 1\frac{1}{9} \end{bmatrix}^T$ is indeed an optimal solution to P can easily be checked by verifying that this $\bar{\mathbf{x}}$ together with $\bar{\mathbf{u}} = \begin{bmatrix} 5\frac{5}{9}, & 0 \end{bmatrix}$, $\bar{\mathbf{v}} = [0, 0]$, and $\mathbf{y} = \begin{bmatrix} 0, & 1\frac{2}{3} \end{bmatrix}^T$ (all from Tableau T^5) satisfy the Karush-Kuhn-Tucker conditions expressed in system I.

As mentioned above, the short form of Wolfe's method requires the matrix $\mathbf{Q}$ of the objective function to be positive definite. If $\mathbf{Q}$ happens to be merely positive semidefinite, other methods must be used, e.g., the long form of Wolfe's method, or the method of Beale (1955, 1959). Another very simple approach, suggested by Charnes and Cooper (1961), is to slightly perturb $\mathbf{Q}$ to make it positive definite. The idea can be described as follows. Assume that $\mathbf{Q}$ is positive semidefinite and consider the perturbed matrix $\mathbf{Q} + \varepsilon\mathbf{I}$, where $\varepsilon > 0$ is some small preselected number and $\mathbf{I}$ is the identity matrix of appropriate dimension. The perturbed matrix $\mathbf{Q} + \varepsilon\mathbf{I}$ is identical to $\mathbf{Q}$, except that all elements along the main diagonal have been increased by an amount ε. As $\mathbf{x}^T\mathbf{I}\mathbf{x} > 0 \;\; \forall \; \mathbf{x} \neq \mathbf{0}$, it follows that $\mathbf{x}^T(\mathbf{Q} + \varepsilon\mathbf{I})\mathbf{x} = \mathbf{x}^T\mathbf{Q}\mathbf{x} + \varepsilon\mathbf{x}^T\mathbf{I}\mathbf{x} \geq \varepsilon\mathbf{x}^T\mathbf{I}\mathbf{x} = \varepsilon\|\mathbf{x}\|^2 > 0 \;\; \forall \;\; \mathbf{x} \neq \mathbf{0}$, so that $\mathbf{Q} + \varepsilon\mathbf{I}$ is positive definite. For numerical purposes, it is, of course, desirable to select the perturbation small enough so as not to alter in a significant way the resulting optimal solution. On the other hand, the perturbation should not be so small that numerical instability is caused.

Because Wolfe's method uses simplex tableaus, postoptimality analyses for quadratic programming problems can be performed as for linear programming. Among other methods for quadratic programming, the algorithm of Frank and Wolfe (1956) has to be mentioned. Although this method is labeled as a quadratic programming technique, it actually admits any convex differentiable objective function and is therefore more general. The method will be treated in a later section of this chapter. Several other quadratic programming algorithms are described in Künzi et al. (1966).

6.1.2   The Linear Complementarity Problem

In this section the linear complementarity problem and an algorithm for its solution will be described. We then show how the quadratic programming problem can be formulated and therefore also solved as a linear complementarity problem. The approach is illustrated numerically by using the same example as in the previous section.

Definition 6.2 Assume that $\mathbf{M}$ and $\mathbf{q}$ are given $[n \times n]$- and $[n \times 1]$-dimensional matrices, respectively. The *linear complementarity problem* (or *LCP* for short) is then to find vectors $\mathbf{w}$ and $\mathbf{z}$ such that

$$P^{LCP}: \quad Mz + q = w$$
$$z^T w = 0$$
$$z, w \geq 0.$$

The notation in Definition 6.2 adheres to that generally accepted in the linear complementarity literature; it deviates slightly from our usual notation elsewhere in this book. The relationship $z^T w = 0$ is called the *complementarity condition*, and if it is satisfied, z and w are said to be *complementary* and each a *complement* of the other. From a geometrical point of view, complementary vectors are orthogonal to each other. Note that the linear complementarity problem is linear except for the complementarity condition which is nonlinear.

We will now show that linear and quadratic programming problems can be converted into linear complementarity problems. First consider a linear programming problem P^L in canonical form and its linear programming dual P_D^L:

$$P^L: \text{Max}\, z = cx \qquad\qquad P_D^L: \text{Min}\, z_D = ub$$
$$\text{s.t. } Ax \leq b \qquad\qquad\qquad \text{s.t. } uA \geq c$$
$$x \geq 0 \qquad\qquad\qquad\qquad u \geq 0.$$

Based on P^L and P_D^L the following linear system can be established:

$$P^*: \quad Ax \leq b$$
$$x \geq 0$$
$$uA \geq c$$
$$u \geq 0$$
$$cx - ub = 0.$$

From Theorem B.15 in the Appendix, or, alternatively, combining Theorems 4.18 and 4.23 from Chap. 4, we know that $(\bar{x}, \bar{u})$ is a solution to system P^* if and only if $\bar{x}$ is an optimal solution to P^L and $\bar{u}$ is an optimal solution to P_D^L. By introducing a vector of nonnegative slack variables y such that $Ax + y = b$ and a vector of nonnegative excess variables v such that $uA - v = c$, the constraint $cx - ub = 0$ becomes $(uA - c)x - u(Ax - b) = 0$ or $vx + uy = 0$, so that the system P^* is transformed into

$$IV: \quad Ax + y = b \qquad\qquad\qquad (6.17)$$

$$uA - v = c \qquad\qquad\qquad (6.18)$$

$$vx + uy = 0 \qquad\qquad\qquad (6.19)$$

$$x, u, y, v \geq 0.$$

Note the resemblance of this system with system I of the previous section. Equations (6.17) and (6.19) are identical to (6.2) and (6.3b), respectively. By defining

$$\mathbf{M} = \begin{bmatrix} \mathbf{0} & -\mathbf{A} \\ \mathbf{A}^T & \mathbf{0} \end{bmatrix}, \ \mathbf{q} = \begin{bmatrix} \mathbf{b} \\ -\mathbf{c}^T \end{bmatrix}, \ \mathbf{z} = \begin{bmatrix} \mathbf{u}^T \\ \mathbf{x} \end{bmatrix}, \ \text{and} \ \mathbf{w} = \begin{bmatrix} \mathbf{y} \\ \mathbf{v}^T \end{bmatrix},$$

system IV can be rewritten as

$$\mathrm{P}^{\mathrm{LCP}}: \quad \mathbf{Mz} + \mathbf{q} = \mathbf{w}$$
$$\mathbf{z}^T \mathbf{w} = 0$$
$$\mathbf{z}, \mathbf{w} \geq \mathbf{0},$$

which is precisely the linear complementarity problem LCP in Definition 6.2 above. Note that $\mathbf{M}$ here is an $[m + n]$-dimensional square skew-symmetric matrix, i.e., $\mathbf{M}^T = -\mathbf{M}$, see Definition A.5 in the Appendix. For any not necessarily nonnegative $\mathbf{z} = \begin{bmatrix} \mathbf{u}^T \\ \mathbf{x} \end{bmatrix}$, we find that

$$\mathbf{z}^T \mathbf{Mz} = [\mathbf{u}, \ \mathbf{x}^T] \begin{bmatrix} \mathbf{0} & -\mathbf{A} \\ \mathbf{A}^T & \mathbf{0} \end{bmatrix} \begin{bmatrix} \mathbf{u}^T \\ \mathbf{x} \end{bmatrix} = [\mathbf{x}^T \mathbf{A}^T, \ -\mathbf{uA}] \begin{bmatrix} \mathbf{u}^T \\ \mathbf{x} \end{bmatrix} = \mathbf{x}^T \mathbf{A}^T \mathbf{u}^T - \mathbf{uAx}$$
$$= (\mathbf{uAx})^T - \mathbf{uAx} = 0,$$

because $(\mathbf{uAx})^T = \mathbf{uAx}$ is a real number. Consequently, $\mathbf{M}$ is merely a positive semidefinite matrix but not positive definite.

Having shown that any linear programming problem in canonical form together with its dual and the optimality condition $\mathbf{c}\bar{\mathbf{x}} = \bar{\mathbf{u}}\mathbf{b}$ can be transformed into a linear complementarity problem, we now consider the quadratic programming problem

$$\mathrm{P}^{\mathrm{Q}}: \ \mathrm{Min} \ \mathbf{x}^T \mathbf{Qx} + \mathbf{cx}$$
$$\mathbf{Ax} \leq \mathbf{b}$$
$$\mathbf{x} \geq \mathbf{0}.$$

In Sect. 6.1.1 we showed that solving the problem P^{Q} is equivalent to finding a solution to system I, which is repeated here for convenience.

$$I: -2\mathbf{Qx} - \mathbf{A}^T \mathbf{u}^T + \mathbf{v}^T = \mathbf{c}^T \tag{6.1}$$

$$\mathbf{Ax} + \mathbf{y} = \mathbf{b} \tag{6.2}$$

$$\mathbf{vx} + \mathbf{uy} = 0 \tag{6.3b}$$

$$\mathbf{x}, \mathbf{u}, \mathbf{v}, \mathbf{y} \geq \mathbf{0}.$$

By defining $\mathbf{M} = \begin{bmatrix} \mathbf{0} & -\mathbf{A} \\ \mathbf{A}^T & 2\mathbf{Q} \end{bmatrix}$, $\mathbf{q} = \begin{bmatrix} \mathbf{b} \\ \mathbf{c}^T \end{bmatrix}$, $\mathbf{z} = \begin{bmatrix} \mathbf{u}^T \\ \mathbf{x} \end{bmatrix}$, and $\mathbf{w} = \begin{bmatrix} \mathbf{y} \\ \mathbf{v}^T \end{bmatrix}$, we see
that system I can be written in the form of the linear complementarity problem LCP.
Note the close resemblance between the linear and quadratic programming formulations in LCP form. An alternative way to express system I in LCP form is by
defining

$$\mathbf{M} = \begin{bmatrix} 2\mathbf{Q} & \mathbf{A}^T \\ -\mathbf{A} & \mathbf{0} \end{bmatrix}, \quad \mathbf{q} = \begin{bmatrix} \mathbf{c}^T \\ \mathbf{b} \end{bmatrix}, \quad \mathbf{z} = \begin{bmatrix} \mathbf{x} \\ \mathbf{u}^T \end{bmatrix}, \quad \text{and} \quad \mathbf{w} = \begin{bmatrix} \mathbf{v}^T \\ \mathbf{y} \end{bmatrix}.$$

The first way is useful for stressing the similarity of the linear and quadratic
programming formulations in LCP form; the second way will be used below in the
numerical illustration after the description of Lemke's complementarity pivot algo-
rithm. Using the first way, and for any not necessarily nonnegative $\mathbf{z} = \begin{bmatrix} \mathbf{u}^T \\ \mathbf{x} \end{bmatrix}$ we
find that

$$\mathbf{z}^T \mathbf{M} \mathbf{z} = [\mathbf{u}, \mathbf{x}^T] \begin{bmatrix} \mathbf{0} & -\mathbf{A} \\ \mathbf{A}^T & 2\mathbf{Q} \end{bmatrix} \begin{bmatrix} \mathbf{u}^T \\ \mathbf{x} \end{bmatrix} = [\mathbf{x}^T \mathbf{A}^T, \; -\mathbf{u}\mathbf{A} + 2\mathbf{x}^T \mathbf{Q}] \begin{bmatrix} \mathbf{u}^T \\ \mathbf{x} \end{bmatrix}$$
$$= \mathbf{x}^T \mathbf{A}^T \mathbf{u}^T - \mathbf{u}\mathbf{A}\mathbf{x} + 2\mathbf{x}^T \mathbf{Q}\mathbf{x} = 2\mathbf{x}^T \mathbf{Q}\mathbf{x} \geq 0,$$
$$\text{again}$$
$$\text{because } \mathbf{x}^T \mathbf{A}^T \mathbf{u}^T = \mathbf{u}\mathbf{A}\mathbf{x},$$

so that $\mathbf{M}$ is again merely a positive semidefinite matrix, as long as $\mathbf{Q}$ is positive
definite or only positive semidefinite.

In the following we describe a complementary pivot method due to Lemke (1968)
for solving the linear complementary problem. The convergence of this method is
guaranteed if the matrix $\mathbf{M}$ satisfies certain properties, for instance being positive
definite or having positive diagonal elements and nonnegative elements otherwise.
As these properties are not satisfied in the linear complementarity problems we
obtained above from linear or quadratic programming, weaker conditions on $\mathbf{M}$ are
desirable. Such conditions are outlined in the following

Definition 6.3 If a square matrix $\mathbf{M}$ satisfies the conditions

(i) $\mathbf{z}^T \mathbf{M}\mathbf{z} \geq 0 \; \forall \, \mathbf{z} \geq \mathbf{0}$, and
(ii) $(\mathbf{M} + \mathbf{M}^T)\mathbf{z} = 0$, if $\mathbf{z}^T \mathbf{M}\mathbf{z} = 0$ and $\mathbf{z} \geq \mathbf{0}$,

then $\mathbf{M}$ is said to be *copositive-plus*. If $\mathbf{M}$ satisfies only (i), it is said to be *copositive*.

For the linear complementarity problem resulting from a linear programming
problem, we find that $\mathbf{M} = \begin{bmatrix} \mathbf{0} & -\mathbf{A} \\ \mathbf{A}^T & \mathbf{0} \end{bmatrix}$, which implies that $\mathbf{z}^T \mathbf{M}\mathbf{z} = 0 \; \forall \; \mathbf{z} \in \mathbb{R}^{m+n}$,
and hence (i) is satisfied. As furthermore the matrix $\mathbf{M}$ is skew-symmetric, i.e.,
$\mathbf{M}^T = -\mathbf{M}$, (ii) is also trivially satisfied, and hence $\mathbf{M}$ is copositive-plus. For
quadratic programming problems, $\mathbf{M} = \begin{bmatrix} \mathbf{0} & -\mathbf{A} \\ \mathbf{A}^T & 2\mathbf{Q} \end{bmatrix}$ and it is then easy to verify

that $\mathbf{z}^T\mathbf{Mz} \geq 0$ $\forall$ $\mathbf{z} \in \mathbb{R}^{m+n}$, so that (i) is satisfied. To show that (ii) holds, assume that $\mathbf{z} \geq \mathbf{0}$, i.e., $\mathbf{x} \geq \mathbf{0}$ and $\mathbf{u} \geq \mathbf{0}$. Straightforward matrix calculations then show that as $\mathbf{z}^T\mathbf{Mz} = 2\mathbf{x}^T\mathbf{Qx}$ and $\mathbf{Q}$ is assumed to be positive definite, $\mathbf{z}^T\mathbf{Mz} = 0$ requires that $\mathbf{x} = \mathbf{0}$. Consequently,

$$\left(\mathbf{M} + \mathbf{M}^T\right)\mathbf{z} = \left[\begin{bmatrix} \mathbf{0} & -\mathbf{A} \\ \mathbf{A}^T & 2\mathbf{Q} \end{bmatrix} + \begin{bmatrix} \mathbf{0} & -\mathbf{A} \\ \mathbf{A}^T & 2\mathbf{Q} \end{bmatrix}^T\right]\begin{bmatrix} \mathbf{u}^T \\ \mathbf{x} \end{bmatrix} = \begin{bmatrix} \mathbf{0} & \mathbf{0} \\ \mathbf{0} & 4\mathbf{Q} \end{bmatrix}\begin{bmatrix} \mathbf{u}^T \\ \mathbf{0} \end{bmatrix} = \begin{bmatrix} \mathbf{0} \\ \mathbf{0} \end{bmatrix},$$

so that condition (ii) is satisfied; $\mathbf{M}$ is therefore copositive-plus. In general, one can show that any positive semidefinite matrix is copositive-plus.

Consider now the linear complementarity problem

$$\mathrm{P}^{\mathrm{LCP}} : \mathbf{Mz} + \mathbf{q} = \mathbf{w} \tag{6.20a}$$

$$\mathbf{z}^T\mathbf{w} = 0 \tag{6.21}$$

$$\mathbf{z}, \ \mathbf{w} \geq \mathbf{0}. \tag{6.22a}$$

If $\mathbf{q} \geq \mathbf{0}$, then the solution $\mathbf{z} = \mathbf{0}$, $\mathbf{w} = \mathbf{q}$ is immediately available. If, however, at least one component of $\mathbf{q}$ is negative, a single nonnegative artificial variable A is introduced in the system, so that

$$\mathbf{Mz} + \mathbf{q} + A\mathbf{e} = \mathbf{w} \tag{6.20b}$$

where $\mathbf{e}$ is the column summation vector of appropriate dimension. Relation (6.22a) then becomes

$$\mathbf{z}, \mathbf{w} \geq \mathbf{0}, \ A \geq 0, \tag{6.22b}$$

and a solution $A = \min_i \{q_i\}$, $\mathbf{z} = \mathbf{0}$, $\mathbf{w} = \mathbf{q} + A\mathbf{e}$ is readily available.

Definition 6.4 A solution $(\mathbf{z}, \mathbf{w})$ to the linear complementarity problem (6.20a), (6.21), (6.22a) is said to be a *complementary basic feasible solution* if for each j, either z_j or w_j is basic.

Definition 6.5 A solution $(\mathbf{z}, \mathbf{w})$ to the system (6.20b), (6.21), (6.22b) is said to be an *almost complementary basic feasible solution*, if

   (i) there exists an index k, such that neither z_k nor w_k is basic
  (ii) for each $j \neq k$, either z_j or w_j is basic
 (iii) the artificial variable A is in the basis.

The idea underlying the complementary pivot algorithm is to move from one almost complementary basic feasible solution to another until either a complementary basic feasible solution is obtained or a solution is found, which indicates unboundedness for the *LCP* with artificial variable A (or, equivalently, infeasibility for the *LCP* without A). The latter case is called *ray termination*, as a ray in the direction of unboundedness is then found, analogous to finding an unbounded solution in the simplex tableau for linear programming. The difference here is that in this case no feasible solution to the linear complementarity problem exists. One can show (see Lemke 1968; Cottle and Dantzig 1968; or Bazaraa et al. 2013) that if **M** is copositive-plus and barring degeneracy, the complementary pivot algorithm terminates in a finite number of steps. Eaves (1971) has extended the validity of this result to a more general class of matrices; see also Karamardian (1972) who describes several different conditions on **M**.

We are now ready to state the Lemke complementary pivot method in algorithmic form. Recall that the linear complementarity problem (*LCP*) is to find a solution **z**, **w** to the system

$$P^{LCP}: \mathbf{Mz} + \mathbf{q} = \mathbf{w}$$
$$\mathbf{z}^T \mathbf{w} = 0$$
$$\mathbf{z}, \ \mathbf{w} \geq \mathbf{0},$$

where **M** is a given square copositive-plus matrix and **q** is a given column matrix.

Lemke's Complementary Pivot Algorithm

Step 0 Is $\mathbf{q} \geq \mathbf{0}$?

If yes: Stop; $(\mathbf{z}, \mathbf{w}) = (\mathbf{0}, \mathbf{q})$ is a solution to the *LCP*.
If no: Go to Step 1.

Step 1 Set up the initial tableau as follows

z	**w**	A	1
$-\mathbf{z}$	**I**	$-\mathbf{e}$	**q**

Find a pivot in the column for the artificial variable A and the r-th row, where r is determined by $q_r = \min_i \{q_i\}$. Note that w_r will leave the basis. Perform a regular tableau transformation with this pivot.

Step 2 Is $A = 0$?

If yes: Stop, the current solution is a complementary basic solution.
If no: Go to Step 3.

Step 3 Suppose that z_s (or w_s) has left the basis in the preceding iteration. Then w_s (or z_s) is selected as next pivot column. Is the pivot column nonpositive?

If yes: Stop with ray termination. No feasible solution exists for the *LCP*.
If no: Go to Step 4.

Step 4 Select a pivot in the pivot column according to the minimum ratio criterion of the primal simplex method. With this pivot, perform one tableau transformation and go to Step 2.

Example The complementary pivot algorithm is now used to solve the linear complementarity problem with

$$\mathbf{M} = \begin{bmatrix} 6 & -2 & 1 & -1 \\ -2 & 4 & 2 & 1 \\ -1 & -2 & 0 & 0 \\ 1 & -1 & 0 & 0 \end{bmatrix} \quad \text{and} \quad \mathbf{q} = \begin{bmatrix} -26 \\ -8 \\ 6 \\ -1 \end{bmatrix}.$$

This corresponds to solving the quadratic programming problem of Sect. 6.1.1.

$$P^Q : \quad \text{Min } z = 3x_1^2 - 2x_1x_2 + 2x_2^2 - 26x_1 - 8x_2$$
$$\text{s.t. } x_1 + 2x_2 \le 6$$
$$x_1 - x_2 \ge 1$$
$$x_1, x_2 \ge 0,$$

where $\mathbf{M} = \begin{bmatrix} 2\mathbf{Q} & \mathbf{A}^T \\ -\mathbf{A} & \mathbf{0} \end{bmatrix}$, $\mathbf{q} = \begin{bmatrix} \mathbf{c}^T \\ \mathbf{b} \end{bmatrix}$, with $\mathbf{Q} = \begin{bmatrix} 3 & -1 \\ -1 & 2 \end{bmatrix}$, $\mathbf{A} = \begin{bmatrix} 1 & 2 \\ -1 & 1 \end{bmatrix}$, $\mathbf{c} = [-26, \quad -8]$, and $\mathbf{b} = \begin{bmatrix} 6 \\ -1 \end{bmatrix}$ as before; as well as $\mathbf{z} = \begin{bmatrix} \mathbf{x} \\ \mathbf{u}^T \end{bmatrix}$, and $\mathbf{w} = \begin{bmatrix} \mathbf{v}^T \\ \mathbf{y} \end{bmatrix}$, and then solving system I, which expresses the Karush-Kuhn-Tucker optimality conditions for quadratic programming (see the previous section).

As $\mathbf{q}$ contains negative elements, we must introduce the artificial variable A, so that the initial tableau is

T^1:

Basis	z_1	z_2	z_3	z_4	w_1	w_2	w_3	w_4	A	1
w_1	-6	2	-1	1	1	0	0	0	(-1)	-26
w_2	2	-4	-2	-1	0	1	0	0	-1	-8
w_3	1	2	0	0	0	0	1	0	-1	6
w_4	-1	1	0	0	0	0	0	1	-1	-1

Note the close resemblance between this tableau and the first tableau for solving the same quadratic programming problem by Wolfe's method from the previous section. Some columns are missing here, the two top rows have been exchanged with the two bottom rows and the signs have also been changed. This resemblance will stay throughout the iterations. As $\min_{i}\{q_i\}$ is attained for the w_1 row, w_1 will leave the basis as A enters it. Performing a pivot operation, the next tableau is T^2:

z_1	z_2	z_3	z_4	w_1	w_2	w_3	w_4	A	1
6	−2	1	−1	−1	0	0	0	1	26
⑧	−6	−1	−2	−1	1	0	0	0	18
7	0	1	−1	−1	0	1	0	0	32
5	−1	1	−1	−1	0	0	1	0	25

Here, z_1 is the entering variable. The minimum ratio of elements on the right-hand side to those in the z_1 column is obtained for the w_2 row; hence w_2 leaves the basis and z_2 will be the next entering variable. Performing a tableau transformation, we obtain the next tableau: T^3:

z_1	z_2	z_3	z_4	w_1	w_2	w_3	w_4	A	1
0	5/2	7/4	1/2	−1/4	−3/4	0	0	1	25/2
1	−3/4	−1/8	−1/4	−1/8	1/8	0	0	0	9/4
0	㉑/4	15/8	3/4	−1/8	−7/8	1	0	0	65/4
0	11/4	13/8	1/4	−3/8	−5/8	0	1	0	55/4

The leaving variable in T^3 is w_3 and consequently, z_3 will be the next entering variable. The next tableau is therefore T^4:

z_1	z_2	z_3	z_4	w_1	w_2	w_3	w_4	A	1
0	0	6/7	1/7	−4/21	−1/3	−10/21	0	1	100/21
1	0	1/7	1/7	−1/7	0	1/7	0	0	32/7
0	1	5/14	1/7	−1/42	−1/6	4/21	0	0	65/21
0	0	9/14	−1/7	−13/42	−1/6	−11/21	1	0	110/21

The leaving variable in T^4 is the artificial variable A so that in the next tableau T^5 a complementary basic feasible solution to the *LCP* will be found. The tableau is: T^5:

z_1	z_2	z_3	z_4	w_1	w_2	w_3	w_4	A	1
0	0	1	1/6	–2/9	–7/18	–5/9	0	7/6	50/9
1	0	0	5/42	1/9	1/18	2/9	0	–1/6	34/9
0	1	0	1/12	1/63	–1/36	7/18	0	–5/12	10/9
0	0	0	–1/4	–1/6	1/12	–1/6	1	–3/4	5/3

Reading from Tableau T_5, we have $\bar{\mathbf{z}} = \left[\bar{z}_1, \bar{z}_2, \bar{z}_3, \bar{z}_4\right]^T = \left[\bar{x}_1, \bar{x}_2, \bar{u}_1, \bar{u}_2\right]^T = \left[3\frac{7}{9}, 1\frac{1}{9}, 5\frac{5}{9}, 0\right]^T$ and $\bar{\mathbf{w}} = \left[\bar{w}_1, \bar{w}_2, \bar{w}_3, \bar{w}_4\right]^T = \left[\bar{v}_1, \bar{v}_2, \bar{y}_1, \bar{y}_2\right]^T = \left[0, 0, 0, 1\frac{2}{3}\right]^T$. The optimal solution to the quadratic programming problem is therefore $\bar{x}_1 = 3\frac{7}{9} \approx 3.778$, $\bar{x}_2 = 1\frac{1}{9} \approx 1.111$, in agreement with the result obtained by the Wolfe method. As before, we also find $\bar{\mathbf{u}} = \left[5\frac{5}{9}, 0\right]$, $\bar{\mathbf{v}} = [0,0]$, and $\bar{\mathbf{y}} = \left[0, 1\frac{2}{3}\right]^T$.

The similarity of Lemke's and Wolfe's methods in the quadratic programming context is not coincidental since both are based on primal simplex tableau transformations with an added rule assuring complementarity. Generalizations of Lemke's method due to Eaves (1978) as well as Talman and van der Heyden (1981) have been established, allowing starting points other than the origin. A special class of linear complementary problems can be solved as linear programs; see Mangasarian (1976), Cottle and Pang (1978), and Aganagić (1984). Finally, nonlinear complementarity problems of the type

$$\mathbf{g}(\mathbf{z}) = \mathbf{w}$$
$$\mathbf{z}^T\mathbf{w} = 0$$
$$\mathbf{z}, \mathbf{w} \geq \mathbf{0}$$

with a nonlinear vector-valued function $\mathbf{g}$ have been studied by Karamardian (1969a, b) as extensions to the linear complementarity problem. For general treatments of various aspects of the complementarity problem, see Balinski and Cottle (1978), Cottle et al. (2009), Bazaraa et al. (2013), and Cottle and Thapa (2017).

6.2 Problems with Differentiable Convex Objective

Having considered quadratic programming problems (i.e. problems with quadratic objective functions and linear constraints) in Sect. 6.1, the discussion is now extended to problems with a general differentiable convex objective function. Two methods will be described in this section. First we consider the famous method of Frank and Wolfe (1956) for solving problems of the type

$$P: \text{Min } z = f(\mathbf{x})$$
$$\text{s.t. } \mathbf{Ax} \leq \mathbf{b}$$
$$\mathbf{x} \in \mathbb{R}^n,$$

where the objective function $f(\mathbf{x})$ is assumed to be convex and differentiable. Then in Sect. 6.2.2 we discuss the *gradient projection method* due to Rosen (1960). That method applies to problems of the type

$$P: \text{Min } z = f(\mathbf{x})$$
$$\text{s.t. } \mathbf{Ax} \leq \mathbf{b}$$
$$\tilde{\mathbf{A}}\mathbf{x} = \tilde{\mathbf{b}}$$
$$\mathbf{x} \in \mathbb{R}^n,$$

where again the objective function $f(\mathbf{x})$ is assumed to be convex and differentiable. Note that the above two problem types are equivalent since the equation constraints $\tilde{\mathbf{A}}\mathbf{x} = \tilde{\mathbf{b}}$ can be written as $\tilde{\mathbf{A}}\mathbf{x} \leq \tilde{\mathbf{b}}$ and $-\tilde{\mathbf{A}}\mathbf{x} \leq -\tilde{\mathbf{b}}$.

6.2.1 The Method of Frank and Wolfe

Although the method of Frank and Wolfe (1956) is referred to as an algorithm for quadratic programming, it actually admits any convex differentiable objective function. In the general case this method may numerically be not as efficient as more recent methods, but because of its conceptual simplicity and appeal we will describe it here.

Consider the problem

$$P: \text{Min } z = f(\mathbf{x})$$
$$\text{s.t. } \mathbf{Ax} \leq \mathbf{b}$$
$$\mathbf{x} \in \mathbb{R}^n.$$

Assume that the objective function $f(\mathbf{x})$ is convex and differentiable and suppose for the time being that the feasible region $\{\mathbf{x} \in \mathbb{R}^n : \mathbf{Ax} \leq \mathbf{b}\}$ is bounded. Assume furthermore that a feasible point $\mathbf{x}^k$ has been generated, for instance by a Phase 1 procedure. The k-th iteration of the algorithm consists of two steps. First, we solve a linear programming problem P^ℓ obtained from P by linearizing $f(\mathbf{x})$ around the point $\mathbf{x}^k$, i.e.,

$$P^\ell: \operatorname*{Min}_{\mathbf{y}} \ z^\ell = \left[\nabla f(\mathbf{x}^k)\right]^T \mathbf{y}$$
$$\text{s.t. } \mathbf{Ay} \le \mathbf{b}$$
$$\mathbf{y} \in \mathbb{R}^n.$$

If the objective function $f(\mathbf{x})$ happens to be linear, then the problems P and P^ℓ are identical (apart from a possible additive constant in the objective function). An optimal solution point of P^ℓ, which by assumption must be finite, is denoted by $\mathbf{y}^k$. In the second step the best interpolation point on the line segment between $\mathbf{x}^k$ and $\mathbf{y}^k$ is selected. Since the feasible region of problem P is convex, and as $\mathbf{x}^k$ and $\mathbf{y}^k$ both belong to this feasible region by construction, it follows that any point on the line segment is also feasible. We now perform the interpolation search

$$P^\lambda: \operatorname*{Min}_{0 \le \lambda \le 1} z^\lambda = f\left(\lambda \mathbf{x}^k + (1 - \lambda)\mathbf{y}^k\right),$$

and denote by λ^k an optimal solution to this one-dimensional search problem. Recall that one-dimensional search methods were covered in Sect. 2.1. The resulting point $\mathbf{x}^{k+1}$ is then defined by $\mathbf{x}^{k+1} := \lambda^k \mathbf{x}^k + (1 - \lambda^k)\mathbf{y}^k$, which we know will be feasible for problem P. Then we test for optimality of $\mathbf{x}^{k+1}$, and if it is not optimal we proceed to the next iteration. The Frank-Wolfe method can now be described in algorithmic form as follows. Recall that the objective function $f(\mathbf{x})$ is assumed to be convex and differentiable, and the feasible region $\{\mathbf{x} \in \mathbb{R}^n : \mathbf{Ax} \le \mathbf{b}\}$ to be bounded. Note that boundedness can always be assured by adding upper and lower bound constraints for all variables, e.g., $-\mathbf{e}M \le \mathbf{x} \le \mathbf{e}M$, where $\mathbf{e}$ is a column vector of ones and M is a suitably large number. If $\mathbf{x} \in \mathbb{R}^n_+$ is required, then the addition of the single constraint $\mathbf{x} \le \mathbf{e}M$ will be sufficient to ensure boundedness (note that Step 2 will have to be modified accordingly). The boundedness requirement can actually be replaced by a weaker condition, which will be discussed below.

The Frank-Wolfe Method

Step 0 Disregarding the constraints, find a minimal point for the unconstrained problem. Is this point feasible?
If yes: Stop, an optimal solution to problem P is found.
If no: Go to Step 1.

Step 1 Use a Phase 1 procedure to locate an initial feasible point. Does such a point exist?

If yes: Denote such a point by $\mathbf{x}^1$, set $k := 1$ and go to Step 2.
If no: Stop, the problem has no feasible solution.

Step 2 Evaluate the objective function gradient $\nabla f(\mathbf{x}^k)$ and solve the linear programming problem

$$\mathbf{P}^{\ell}: \underset{\mathbf{y}}{\text{Min}}\, z^{\ell} = \left[\nabla f(\mathbf{x}^k)\right]^T \mathbf{y}$$
$$\text{s.t. } \mathbf{Ay} \leq \mathbf{b}$$
$$\mathbf{y} \in \mathbb{R}^n.$$

Denote an optimal solution to $\mathbf{P}^{\ell}$ by $\mathbf{y}^k$.

Step 3 Perform a one-dimensional interpolation search, i.e. solve the problem

$$\mathbf{P}^{\lambda}: \underset{0 \leq \lambda \leq 1}{\text{Min}}\, z^{\lambda} = f\left(\lambda \mathbf{x}^k + (1 - \lambda)\mathbf{y}^k\right),$$

denote the resulting optimal λ by λ^k, and set $\mathbf{x}^{k+1} := \lambda^k \mathbf{x}^k + (1 - \lambda^k)\mathbf{y}^k$.

Step 4 Does $\mathbf{x}^{k+1}$ satisfy some stop criterion?

If yes: Stop, $\mathbf{x}^{k+1}$ is the desired point.
If no: Let $k := k + 1$ and go to Step 2.

The stop criterion in Step 4 could be

$$\left|\frac{f\left(\mathbf{x}^{k+1}\right) - f\left(\mathbf{x}^k\right)}{\max\{\varepsilon_1, |f(\mathbf{x}^{k+1})|, |f(\mathbf{x}^k)|\}}\right| \leq \varepsilon_2 \quad \text{or} \quad \left\|\frac{\mathbf{x}^{k+1} - \mathbf{x}^k}{\max\{\varepsilon_3, \|\mathbf{x}^{k+1}\|, \|\mathbf{x}^k\|\}}\right\| \leq \varepsilon_4,$$

where ε_1, ε_2, ε_3, and ε_4 are predetermined numbers; these two criteria measure the relative changes in $f(\mathbf{x})$ and $\mathbf{x}$, respectively, from one iteration to the next. An additional stop criterion could be $\|\nabla f(\mathbf{x}^{k+1})\| \leq \varepsilon_5$ with a predetermined small ε_5. Moreover, since the feasible region of the linear programming problem in Step 2 is always the same, the optimal solution $\mathbf{y}^k$ from one iteration can be used as an initial solution for the linear programming problem in the next iteration.

We will now illustrate the method on the previously used

Example Perform three iterations with the Frank-Wolfe method for the problem

$$\mathbf{P}: \text{Min } z = 3x_1^2 - 2x_1x_2 + 2x_2^2 - 26x_1 - 8x_2$$
$$\text{s.t. } x_1 + 2x_2 \leq 6$$
$$x_1 - x_2 \geq 1$$
$$x_1, x_2 \geq 0.$$

From the previous section on quadratic programming we know that the unconstrained minimum for this problem is attained at the point $[\hat{x}_1, \hat{x}_2]^T = [6, 5]^T$ which is infeasible. Therefore we go to Step 1 and start with the feasible arbitrarily chosen point $\mathbf{x}^1 = [3, 1]^T$.

Computing $\nabla f(\mathbf{x}) = \begin{bmatrix} 6x_1 - 2x_2 - 26 \\ -2x_1 + 4x_2 - 8 \end{bmatrix}$ we find that $\nabla f(\mathbf{x}^1) = \begin{bmatrix} -10 \\ -10 \end{bmatrix}$. The linear programming problem P^ℓ in Step 2 therefore becomes

$$P^\ell: \text{ Min } z^\ell = -10y_1 - 10y_2$$
$$\text{s.t. } y_1 + 2y_2 \le 6$$
$$y_1 - y_2 \ge 1$$
$$y_1, \, y_2 \ge 0,$$

which has the unique optimal solution $\mathbf{y}^1 = [6, 0]^T$. Now $z^\lambda = f(\lambda \mathbf{x}^1 + (1 - \lambda)\mathbf{y}^1) = f(3\lambda + 6(1 - \lambda), \, \lambda) = f(6 - 3\lambda, \, \lambda) = 35\lambda^2 - 50\lambda - 48$, which is minimized, e.g., by differential calculus, with the result that $\lambda^1 = 5/7$. Hence $\mathbf{x}^2 = \frac{5}{7}\mathbf{x}^1 + \frac{2}{7}\mathbf{y}^1 = \frac{5}{7}[3, \, 1]^T + \frac{2}{7}[6, \, 0]^T = \left[\frac{27}{7}, \frac{5}{7}\right]^T \approx [3.86, \, 0.71]^T$ and we go to the second iteration.

The objective function gradient at $\mathbf{x}^2 = \left[\frac{27}{7}, \frac{5}{7}\right]^T$ is now $\nabla f(\mathbf{x}^2) = \begin{bmatrix} 30/7 \\ -90/7 \end{bmatrix}$, so that the linear programming problem in Step 2 is

$$P^\ell: \text{ Min } z^\ell = \frac{30}{7}y_1 - \frac{90}{7}y_2$$
$$\text{s.t. } y_1 + 2y_2 \le 6$$
$$y_1 - y_2 \ge 1$$
$$y_1, \, y_2 \ge 0,$$

which has the unique optimal solution $\mathbf{y}^2 = \left[\frac{8}{3}, \frac{5}{3}\right]^T \approx [2.67, 1.67]^T$. Now $f(\lambda \mathbf{x}^2 + (1 - \lambda)\mathbf{y}^2) = f\left(\frac{25\lambda + 56}{21}, \frac{-20\lambda + 35}{21}\right) = \frac{1}{21}(175\lambda^2 - 200\lambda - 1358)$, which is minimized for $\lambda^2 = 4/7$. Hence $\mathbf{x}^3 = \frac{4}{7}\mathbf{x}^2 + \frac{3}{7}\mathbf{y}^2 = \frac{4}{7}\left[\frac{27}{7}, \frac{5}{7}\right]^T + \frac{3}{7}\left[\frac{8}{3}, \frac{5}{7}\right]^T = \left[\frac{164}{49}, \frac{55}{49}\right]^T \approx [3.35, 1.12]^T$, and we go to the third iteration. The objective function gradient at $\mathbf{x}^3 = \left[\frac{164}{49}, \frac{55}{49}\right]^T$ is $\nabla f(\mathbf{x}^3) = \begin{bmatrix} -400/49 \\ -500/49 \end{bmatrix}$, so that the linear programming problem is now

$$P^\ell: \text{ Min } z^\ell = -\frac{400}{49}y_1 - \frac{500}{49}y_2$$
$$\text{s.t. } y_1 + 2y_2 \le 6$$
$$y_1 - y_2 \ge 1$$
$$y_1, \, y_2 \ge 0,$$

which has the unique optimal solution $\mathbf{y}^3 = [6, 0]^T$. Now $f(\lambda \mathbf{x}^3 + (1 - \lambda)\mathbf{y}^3) \approx f(\lambda[3.35, \, 1.12]^T + (1 - \lambda)[6, \, 0]^T) = f(6 - 2.65\lambda, \, 1.12\lambda) = 29.5123\lambda^2 - 48.9\lambda - 48,$

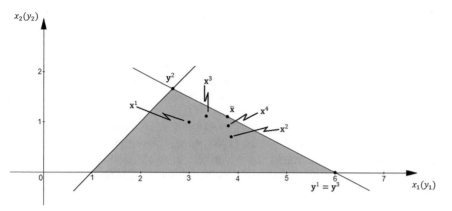

Fig. 6.4 Iterative solutions with the Frank-Wolfe method

which is minimized for $\lambda^3 \approx 0.828$, and we conclude that $\mathbf{x}^4 = 0.828\mathbf{x}^3 + 0.172\mathbf{y}^3 \approx [3.80, 0.93]^T$.

We stop here, having carried out three complete iterations of the method. The true optimal solution is $\bar{\mathbf{x}} = \left[3\frac{7}{9}, 1\frac{1}{9} \right]^T \approx [3.78, 1.11]^T$, so that $\mathbf{x}^4$ is quite close to the optimal point. Figure 6.4 displays the progression of the solution points $\mathbf{x}^k$ and $\mathbf{y}^k$.

Returning to the general discussion we note that if the objective function $f(\mathbf{x})$ happens to be linear, Step 0 gives an unbounded solution and, if a feasible solution exists, an optimal solution will then always be reached by solving the linear programming problem in Step 2 of the first iteration; this is, in fact, the original problem P itself.

As indicated above, the requirement that the feasible region be bounded can be replaced by a weaker assumption, namely that for any given feasible point $\tilde{\mathbf{x}}$, the set $\left\{ \left[\nabla f(\tilde{\mathbf{x}}) \right]^T \mathbf{y} : \mathbf{A}\mathbf{y} \leq \mathbf{b} \right\}$ should be bounded from below. Frank and Wolfe (1956) also indicate how the boundedness requirement can be dropped altogether with appropriate modifications in the algorithm; they also give a proof of the convergence of the method. In his seminal work on convergence theory for nonlinear programming algorithms, Wolfe (1970) proposes a modified version of the Frank-Wolfe method and considers the convergence rates for the original and the modified methods. The Frank-Wolfe method has been generalized to the case where the constraints are nonlinear as well, see Holloway (1974).

6.2.2 The Gradient Projection Method for Linear Constraints

A major class of algorithms for constrained nonlinear optimization is the category of *feasible direction methods*. For such a method, an iteration can be described as follows. From a feasible point a step is taken in a so-called *improving feasible direction* which is such that a sufficiently small move in this direction will improve the value of the objective function and at the same time attempt to preserve—or almost preserve—feasibility. In some methods, the step taken may be such that one leaves the feasible region. A recovery step is then taken so that a feasible point is again reached and the next iteration can begin. Thus in a feasible direction method, we alternate between determining the step direction and the step length for each subsequent move.

The first method of feasible directions is credited to Zoutendijk (1960); it was later modified by Topkis and Veinott (1967) so that convergence is guaranteed. In 1972 Wolfe gave an example for which Zoutendijk's method converges to a nonoptimal point. For a description of Zoutendijk's method, Wolfe's counterexample and the Topkis and Veinott modification, we refer to the original works or to the detailed discussion in Bazaraa et al. (2013); see also Bertsekas (2016) and Luenberger and Ye (2008). One can interpret the Frank-Wolfe method of the previous subsection as some kind of feasible direction procedure. Another algorithm in this class is the *gradient projection method* by Rosen (1960, 1961) which we will now describe.

The problem to be considered for the gradient projection method reads as follows.

$$
\begin{aligned}
\text{P: } \text{Min } & z = f(\mathbf{x}) \\
\text{s.t. } & \mathbf{Ax} \leq \mathbf{b} \\
& \tilde{\mathbf{A}}\mathbf{x} = \tilde{\mathbf{b}} \\
& \mathbf{x} \in \mathbb{R}^n,
\end{aligned}
$$

where $f \colon \mathbb{R}^n \to \mathbb{R}$ denotes a convex differentiable function, $\mathbf{x} \in \mathbb{R}^n$, and $\mathbf{A}, \mathbf{b}, \tilde{\mathbf{A}}, \tilde{\mathbf{b}}$ are $[m \times n]$-, $[m \times 1]$-, $[r \times n]$-, and $[r \times 1]$-dimensional matrices, respectively. If $m = 0$, $\mathbf{A}$ and $\mathbf{b}$ will be vacuous, so that there are no inequality constraints, and if $r = 0$, $\tilde{\mathbf{A}}$ and $\tilde{\mathbf{b}}$ are vacuous, so that there are no equality constraints.

Assume that now we have arrived at $\mathbf{x}^k$, which is a feasible point. If the objective function gradient $\nabla f(\mathbf{x}^k) = \mathbf{0}$, then $\mathbf{x}^k$ is optimal; if $\nabla f(\mathbf{x}^k) \neq \mathbf{0}$, the direction of steepest descent is $-\nabla f(\mathbf{x}^k)$, and f is then to be minimized along this direction, subject to the restriction that feasibility is retained.

Should $\mathbf{x}^k$ happen to be a point in the interior of the feasible region, then $-\nabla f(\mathbf{x}^k)$ is a feasible direction, i.e., it would be possible to decrease the value of $f(\mathbf{x})$ by taking a step of strictly positive length in the direction $-\nabla f(\mathbf{x}^k)$. If, however, $\mathbf{x}^k$ is a point on the boundary of the feasible regions, then $-\nabla f(\mathbf{x}^k)$ need not necessarily be a feasible direction. If it is not, then any step in this direction, no matter how small, would lead outside the feasible region. These possibilities are illustrated in Fig. 6.5, where the polytope indicates the feasible region.

Here $\mathbf{x}^4$ is an interior point so that $-\nabla f(\mathbf{x}^4)$ is a feasible direction; all the other points indicated are boundary points. We can see that $-\nabla f(\mathbf{x}^1)$ and $-\nabla f(\mathbf{x}^2)$ are also feasible directions, but that $-\nabla f(\mathbf{x}^3)$, $-\nabla f(\mathbf{x}^5)$ and $-\nabla f(\mathbf{x}^6)$ are not.

The basic principle of the gradient projection method consists in moving along the negative gradient if this is a feasible direction. If it is not, the negative gradient is projected onto the boundary of the given polytope; if the result is a feasible direction

Fig. 6.5 Negative gradients at various feasible points

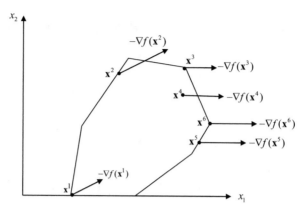

Fig. 6.6 Gradient projections

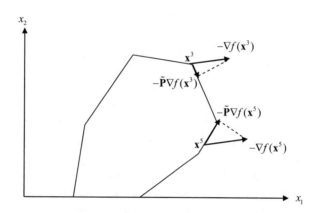

one moves in this direction. If, however, the result of the projection is not a feasible direction, then one can show that the current solution must necessarily be optimal. For the time being only the direction of the move and not its length will be considered.

In order to obtain feasible directions for the points in Fig. 6.5 we need to project the gradients at $\mathbf{x}^3$ and $\mathbf{x}^5$ onto the respective binding hyperplane; no projection of the gradient at $\mathbf{x}^6$ will make it a feasible direction and therefore $\mathbf{x}^6$ is an optimal point. Below we will show that the projections are achieved by premultiplying $-\nabla f(\mathbf{x})$ by a suitable *projection matrix* $\tilde{\mathbf{P}}$. The projection results for $\mathbf{x}^3$ and $\mathbf{x}^5$ are indicated in Fig. 6.6.

Definition 6.6 A matrix $\tilde{\mathbf{P}}$ is called a *projection matrix*, if $\tilde{\mathbf{P}} = \tilde{\mathbf{P}}^T$ and $\tilde{\mathbf{P}}\tilde{\mathbf{P}} = \tilde{\mathbf{P}}$.

If $\tilde{\mathbf{P}}$ is any projection matrix, then $\mathbf{x}^T\tilde{\mathbf{P}}\mathbf{x} = \mathbf{x}^T\tilde{\mathbf{P}}\tilde{\mathbf{P}}\mathbf{x} = \mathbf{x}^T\tilde{\mathbf{P}}^T\tilde{\mathbf{P}}\mathbf{x} = \left\|\tilde{\mathbf{P}}\mathbf{x}\right\|^2 \geq 0$ for all $\mathbf{x} \in \mathbb{R}^n$, so that $\tilde{\mathbf{P}}$ is positive semidefinite. Furthermore, one can easily check that $\tilde{\mathbf{P}}$ is a projection matrix if and only if $\mathbf{I} - \tilde{\mathbf{P}}$ is also a projection matrix. One can also show that a projection matrix maps the whole space $\mathbb{R}^n$ onto a hyperplane in $\mathbb{R}^n$.

Example Consider the matrix $\tilde{\mathbf{P}}_1 = \begin{bmatrix} 1 & 0 \\ 0 & 0 \end{bmatrix}$. We find that $\tilde{\mathbf{P}}_1^T = \begin{bmatrix} 1 & 0 \\ 0 & 0 \end{bmatrix} = \tilde{\mathbf{P}}_1$ and

$\tilde{\mathbf{P}}_1 \tilde{\mathbf{P}}_1 = \begin{bmatrix} 1 & 0 \\ 0 & 0 \end{bmatrix} \begin{bmatrix} 1 & 0 \\ 0 & 0 \end{bmatrix} = \begin{bmatrix} 1 & 0 \\ 0 & 0 \end{bmatrix} = \tilde{\mathbf{P}}_1$, so that $\tilde{\mathbf{P}}_1$ is a projection matrix.

Similarly one can show that $\tilde{\mathbf{P}}_2 = \begin{bmatrix} 0 & 0 \\ 0 & 1 \end{bmatrix}$ and $\tilde{\mathbf{P}}_3 = \begin{bmatrix} \dfrac{1}{5} & \dfrac{2}{5} \\ \dfrac{2}{5} & \dfrac{4}{5} \end{bmatrix}$ are also projection

matrices. Furthermore we find that

$$\tilde{\mathbf{P}}_1 \mathbf{x} = \begin{bmatrix} 1 & 0 \\ 0 & 0 \end{bmatrix} \begin{bmatrix} x_1 \\ x_2 \end{bmatrix} = \begin{bmatrix} x_1 \\ 0 \end{bmatrix} \forall \ \mathbf{x} \in \mathbb{R}^2$$

$$\tilde{\mathbf{P}}_2 \mathbf{x} = \begin{bmatrix} 0 & 0 \\ 0 & 1 \end{bmatrix} \begin{bmatrix} x_1 \\ x_2 \end{bmatrix} = \begin{bmatrix} 0 \\ x_2 \end{bmatrix} \forall \ \mathbf{x} \in \mathbb{R}^2$$

$$\tilde{\mathbf{P}}_3 \mathbf{x} = \begin{bmatrix} \dfrac{1}{5} & \dfrac{2}{5} \\ \dfrac{2}{5} & \dfrac{4}{5} \end{bmatrix} \begin{bmatrix} x_1 \\ x_2 \end{bmatrix} = \begin{bmatrix} \dfrac{1}{5}x_1 + \dfrac{2}{5}x_2 \\ \dfrac{2}{5}x_1 + \dfrac{4}{5}x_2 \end{bmatrix} \forall \ \mathbf{x} \in \mathbb{R}^2 .$$

In other words, $\tilde{\mathbf{P}}_1$ projects the $\mathbb{R}^2$-space onto the x_1-axis, and $\tilde{\mathbf{P}}_2$ projects the $\mathbb{R}^2$-space onto the x_2-axis. By defining $\mathbf{y}:=\tilde{\mathbf{P}}_3\mathbf{x}$ we find $y_1 = \dfrac{1}{5}x_1 + \dfrac{2}{5}x_2$ and $y_2 = \dfrac{2}{5}x_1 + \dfrac{4}{5}x_2$, so that $2y_1 - y_2 = 0$, from which follows that $\tilde{\mathbf{P}}_3$ projects the $\mathbb{R}^2$-space onto the hyperplane (straight line) $2x_1 - x_2 = 0$.

To show how the gradient projection approach works mathematically, recall that the nonlinear programming problem P considered here was expressed as

$$\text{P: Min } z = f(\mathbf{x})$$
$$\text{s.t. } \mathbf{A}\mathbf{x} \leq \mathbf{b}$$
$$\tilde{\mathbf{A}}\mathbf{x} = \tilde{\mathbf{b}}$$
$$\mathbf{x} \in \mathbb{R}^n .$$

Any nonnegativity constraints are supposed to be included in the first set of inequality constraints. According to the previous discussion, the direction of steepest descent is $-\nabla f(\mathbf{x}^k)$ for any feasible $\mathbf{x}^k$, but this need not be a feasible direction if $\mathbf{x}^k$ is on the border of the feasible region. This occurs for instance if the matrix $\mathbf{A}$ is vacuous and it could also occur if one or more of the inequality constraints happened to be satisfied with equality at $\mathbf{x}^k$. In that case, assume that

$$\mathbf{A}^= \mathbf{x}^k = \mathbf{b}^=$$
$$\mathbf{A}^< \mathbf{x}^k < \mathbf{b}^<$$
$$\tilde{\mathbf{A}} \mathbf{x}^k = \tilde{\mathbf{b}},$$

where

$$\begin{bmatrix} \mathbf{A}^= \\ \mathbf{A}^< \end{bmatrix} = \mathbf{A} \text{ and } \begin{bmatrix} \mathbf{b}^= \\ \mathbf{b}^< \end{bmatrix} = \mathbf{b},$$

after rearranging the rows, if necessary. We can then form the new matrix

$$\mathbf{M} := \begin{bmatrix} \mathbf{A}^= \\ \tilde{\mathbf{A}} \end{bmatrix} \text{ and define}$$

$$\tilde{\mathbf{P}} := \mathbf{I} - \mathbf{M}^T \left(\mathbf{M}\mathbf{M}^T \right)^{-1} \mathbf{M},$$

with $\mathbf{M}$ assumed to have no more rows than columns and also to have full rank so as to make $\mathbf{M}\mathbf{M}^T$ nonsingular. If $\mathbf{M}$ is not of full rank, degeneracy occurs and a sufficient number of rows will have to be deleted from $\mathbf{M}$ to establish the full rank property. When $\mathbf{M}$ is vacuous, i.e., when there are no equality constraints and all constraints are satisfied with strict inequality, then $\tilde{\mathbf{P}}$ is taken to be the identity matrix $\mathbf{I}$.

To understand the logic behind the expression for $\tilde{\mathbf{P}}$, consider Fig. 6.7, where $\mathbf{x}^k$ is the current feasible point, located on the boundary of the feasible region with $\mathbf{M}\mathbf{x}^k = \begin{bmatrix} \mathbf{A}^= \\ \tilde{\mathbf{A}} \end{bmatrix} \mathbf{x}^k = \begin{bmatrix} \mathbf{b}^= \\ \tilde{\mathbf{b}} \end{bmatrix}$.

The negative gradient of f at $\mathbf{x}^k$, i.e., $-\nabla f(\mathbf{x}^k)$ can be decomposed into two orthogonal components $-\nabla f(\mathbf{x}^k) = \mathbf{s}^k + \mathbf{M}^T\mathbf{u}^k$, where $\mathbf{s}^k$ belongs to the subspace spanned by the active constraints, and $\mathbf{M}^T\mathbf{u}^k$ is some linear combination of the normals to the active constraints. We must then also have $\mathbf{M}\mathbf{s}^k = \mathbf{0}$. It follows that $-\mathbf{M}\nabla f(\mathbf{x}^k) = \mathbf{M}\mathbf{s}^k + \mathbf{M}\mathbf{M}^T\mathbf{u}^k = \mathbf{M}\mathbf{M}^T\mathbf{u}^k$, and premultiplying by $(\mathbf{M}\mathbf{M}^T)^{-1}$, we obtain $-(\mathbf{M}\mathbf{M}^T)^{-1}\mathbf{M}\nabla f(\mathbf{x}^k) = \mathbf{u}^k$. Inserting this expression for $\mathbf{u}^k$ into the

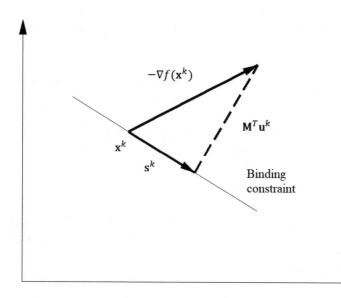

Fig. 6.7 Projection of a gradient onto a binding constraint

decomposition formula for the negative gradient yields $-\nabla f(\mathbf{x}^k) = \mathbf{s}^k + \mathbf{M}^T\mathbf{u}^k = \mathbf{s}^k - \mathbf{M}^T(\mathbf{M}\mathbf{M}^T)^{-1}\mathbf{M}\nabla f(\mathbf{x}^k)$, from which $\mathbf{s}^k = -\nabla f(\mathbf{x}^k) + \mathbf{M}^T(\mathbf{M}\mathbf{M}^T)^{-1}\mathbf{M}\nabla f(\mathbf{x}^k)$
$= -\left(\mathbf{I} - \mathbf{M}^T(\mathbf{M}\mathbf{M}^T)^{-1}\mathbf{M}\right)\nabla f(\mathbf{x}^k) = -\tilde{\mathbf{P}}\nabla f(\mathbf{x}^k)$. Finally, since $\mathbf{s}^k$ is orthogonal to
$\nabla f(\mathbf{x}^k) + \mathbf{s}^k$ (see Fig. 6.7), we have $(\mathbf{s}^k)^T(\nabla f(\mathbf{x}^k) + \mathbf{s}^k) = 0$, so that $(\mathbf{s}^k)^T\nabla f$
$(\mathbf{x}^k) = -(\mathbf{s}^k)^T\mathbf{s}^k = -\|\mathbf{s}^k\| \leq 0$, implying that $\mathbf{s}^k$ is an improving feasible direction
if $\mathbf{s}^k \neq \mathbf{0}$. For more details, see Luenberger and Ye (2008), Bazaraa et al. (2013), or
the early work of Rosen (1960). The matrix $\mathbf{M}^T(\mathbf{M}\mathbf{M}^T)^{-1}$ is often called the *Moore-Penrose pseudoinverse* or the *generalized inverse* of $\mathbf{M}$ (see Moore 1935; Penrose
1955; or Ben-Israel and Greville 2003).

Now, if $\mathbf{s}^k = \mathbf{0}$ and $\mathbf{M}$ is vacuous, $\mathbf{0} = \mathbf{s}^k = -\tilde{\mathbf{P}}\nabla f(\mathbf{x}^k) = -\nabla f(\mathbf{x}^k)$, so that $\mathbf{x}^k$
must be optimal. If $\mathbf{s}^k = \mathbf{0}$ and $\mathbf{M}$ is not vacuous, we consider the vector $\mathbf{u}^k$, which is
actually the Lagrangean multiplier vector associated with the constraints, and partition it according to which of the constraints its components are associated with, i.e.,
we define $\mathbf{u}^k(\mathbf{M}) := \begin{bmatrix} \mathbf{u}^k(\mathbf{A}^=) \\ \mathbf{u}^k(\tilde{\mathbf{A}}) \end{bmatrix}$. Note that $\mathbf{u}^k$ is here a column vector, whereas we
usually employ Lagrangean multipliers in row vector form. Also note that $\mathbf{u}^k$ here
only contains components associated with constraints that are tight (active) at the
current point $\mathbf{x}^k$. Remembering that $-\nabla f(\mathbf{x}^k) = \mathbf{s}^k + \mathbf{M}^T\mathbf{u}^k$, we find, with $\mathbf{s}^k = \mathbf{0}$, that
$\nabla f(\mathbf{x}^k) + \mathbf{M}^T\mathbf{u}^k = \mathbf{0}$, so that with $\mathbf{u}^k(\mathbf{A}^=) \geq \mathbf{0}$, i.e., the Lagrangean multipliers
associated with the tight inequality constraints being nonnegative, the Karush-Kuhn-Tucker conditions for optimality are satisfied at $\mathbf{x}^k$. If $\mathbf{u}^k(\mathbf{A}^=) \geq \mathbf{0}$ does not
hold, some negative component $\mathbf{u}^k(\mathbf{A}^=)$ is selected (typically the "most negative"
one) and the corresponding row of $\mathbf{A}^=$ is deleted. Then $\mathbf{M}$ is updated accordingly.

Finally, we consider the case with $\mathbf{s}^k = -\tilde{\mathbf{P}}\nabla f(\mathbf{x}^k) \neq \mathbf{0}$ and proceed to move into
this direction as far as possible. This move will be limited by the constraints, which
are currently not tight, i.e., $\mathbf{A}^<\mathbf{x}^k < \mathbf{b}^<$. To identify which constraint will be violated
first, we define the slack $\mathbf{b}^* := \mathbf{b}^< - \mathbf{A}^<\mathbf{x}^k > \mathbf{0}$ and $\mathbf{s}^* := \mathbf{A}^<\mathbf{s}^k$. Then we compute the
maximal step length α^* determined by

$$\alpha^* := \begin{cases} \min_j \left\{ b_j^*/s_j^* : s_j^* > 0 \right\}, & \text{if some component of } \mathbf{s}^* \text{ is positive} \\ +\infty & \text{otherwise} \end{cases}$$

A one-dimensional bounded search is now performed in the improving feasible
direction $\mathbf{s}^k$, but limited by the maximal step length α^*, i.e., we solve

$$P^\alpha : \min_{0 \leq \alpha \leq \alpha^*} z^\alpha = f(\mathbf{x}^k + \alpha\mathbf{s}^k).$$

Denoting an optimal solution to P^α by α^k, we then set $\mathbf{x}^{k+1} := \mathbf{x}^k + \alpha^k\mathbf{s}^k$ and go to
the next iteration. The gradient projection method for the minimization problem P as
formulated above can now be formally stated.

Recall that the problem under consideration is

$$\text{P: Min } z = f(\mathbf{x})$$
$$\text{s.t. } \mathbf{A}\mathbf{x} \le \mathbf{b}$$
$$\tilde{\mathbf{A}}\mathbf{x} \le \tilde{\mathbf{b}}$$
$$\mathbf{x} \in \mathbb{R}^n,$$

The Gradient Projection Method for Linear Constraints

Step 0 Disregarding the constraints, find a minimal point for the unconstrained problem. Is this point feasible?

If yes: Stop, an optimal solution to problem P has been found.
If no: Go to Step 1.

Step 1 Start with an initial feasible point $\mathbf{x}^1$ (which can be found for instance by Phase 1 of the two phase method of linear programming) and set $k := 1$.

Step 2 Provide the partitions $\mathbf{A} = \begin{bmatrix} \mathbf{A}^= \\ \mathbf{A}^< \end{bmatrix}$ and $\mathbf{b} = \begin{bmatrix} \mathbf{b}^= \\ \mathbf{b}^< \end{bmatrix}$ such that $\mathbf{A}^=\mathbf{x}^k = \mathbf{b}^=$ and $\mathbf{A}^<\mathbf{x}^k < \mathbf{b}^<$.

Step 3 Set $\mathbf{M} := \begin{bmatrix} \mathbf{A}^= \\ \tilde{\mathbf{A}} \end{bmatrix}$ and set $\tilde{\mathbf{P}} := \begin{cases} \mathbf{I}, & \text{if } \mathbf{M} \text{ is vacuous,} \\ \mathbf{I} - \mathbf{M}^T (\mathbf{M}\mathbf{M}^T)^{-1}\mathbf{M} & \text{otherwise} \end{cases}$, and set $\mathbf{s}^k = -\tilde{\mathbf{P}}\nabla f(\mathbf{x}^k)$. Is $\mathbf{s}^k = \mathbf{0}$?

If yes: Go to Step 4.
If no: Go to Step 7.

Step 4 Is $\mathbf{M}$ vacuous?

If yes: Stop, $\mathbf{x}^k$ is an optimal solution.
If no: Go to Step 5.

Step 5 Let $\mathbf{u}^k(\mathbf{M}) = \begin{bmatrix} \mathbf{u}^k(\mathbf{A}^=) \\ \mathbf{u}^k(\tilde{\mathbf{A}}) \end{bmatrix} = -(\mathbf{M}\mathbf{M}^T)^{-1}\mathbf{M}\nabla f(\mathbf{x}^k)$. Is $\mathbf{u}^k(\mathbf{A}^=) \ge 0$?

If yes: Stop, $\mathbf{x}^k$ is an optimal solution.
If no: Go to Step 6.

Step 6 Select some negative component from the vector $\mathbf{u}^k(\mathbf{A}^=)$ and delete the corresponding row from $\mathbf{A}^=$. Go to Step 2.

Step 7 Set $\mathbf{b}^* = \mathbf{b}^< - \mathbf{A}^<\mathbf{x}^k$, $\mathbf{s}^* := \mathbf{A}^<\mathbf{s}^k$ and

$$\alpha^* := \begin{cases} \min_j \left\{ \dfrac{b_j^*}{s_j^*} : s_j^* > 0 \right\}, & \text{if } \mathbf{s}^* \not\le \mathbf{0} \\ +\infty & \text{otherwise} \end{cases}$$

and solve the following one-dimensional minimization search problem

$$P^\alpha: \operatorname*{Min}_{\alpha} \; z^\alpha = f\left(\mathbf{x}^k + \lambda \mathbf{s}^k\right)$$
$$\text{s.t. } 0 \leq \alpha \leq \alpha^*.$$

Denote an optimal solution of P^α by α^k, set $\mathbf{x}^{k+1} := \mathbf{x}^k + \alpha^k \mathbf{s}^k$, $k: = k + 1$ and go to Step 2.

Rosen's original paper gives a recursive method for updating the projection matrix $\tilde{\mathbf{P}}$ in Step 3 which considerably reduces the computational effort. It also indicates how to deal with the degenerate case of a singular $\mathbf{MM}^T$ matrix. In Step 7, the limits on the step length for the search imposed by α^* will ensure the feasibility of $\mathbf{x}^{k+1}$ for the next iteration. Note that if problem P happens to be a linear programming problem, the gradient projection method generates a sequence of solution points which is identical to that of the primal simplex method with the steepest unit ascent/descent rule for the pivot column selection. We will now show how the gradient projection method works numerically. For comparative purposes the illustration will be the same as that used in Sect. 6.1 on quadratic programming and in the previous section.

Example Consider the problem

$$P: \operatorname{Min} \; z = 3x_1^2 - 2x_1x_2 + 2x_2^2 - 26x_1 - 8x_2$$
$$\text{s.t.} x_1 + 2x_2 \leq 6$$
$$x_1 - x_2 \geq 1$$
$$x_1, x_2 \geq 0.$$

The gradient of the objective function $f(\mathbf{x})$ is

$$\nabla f(\mathbf{x}) = \begin{bmatrix} 6x_1 - 2x_2 - 26 \\ -2x_1 + 4x_2 - 8 \end{bmatrix}.$$

The unconstrained minimal point $[\hat{x}_1, \hat{x}_2]^T = [6, \; 5]^T$ is infeasible so that we proceed to Step 1. Start with the feasible arbitrarily chosen point $\mathbf{x}^1 = [x_1^1, x_2^1]^T = [3,1]^T$. There are no equality constraints and all constraints are satisfied as strict inequalities; hence $\mathbf{A}^=$ is vacuous. We find:

$$\mathbf{A}^< = \begin{bmatrix} 1 & 2 \\ -1 & 1 \\ -1 & 0 \\ 0 & -1 \end{bmatrix}, \quad \mathbf{b}^< = \begin{bmatrix} 6 \\ -1 \\ 0 \\ 0 \end{bmatrix}, \quad \text{and } \mathbf{x}^1 = [x_1^1, \; x_2^1]^T, \text{ so that}$$

$$x_1^1 + 2x_2^1 = 3 + 2(1) = 5 < 6$$
$$-x_1^1 + x_2^1 = -3 + 1 = -2 < -1$$
$$-x_1^1 = -3 < 0$$
$$-x_2^1 = -1 < 0.$$

It follows that in Step 3, the matrix $\mathbf{M}$ is vacuous, so that $\mathbf{P} := \mathbf{I} = \begin{bmatrix} 1 & 0 \\ 0 & 1 \end{bmatrix}$.

Therefore the search direction is $\mathbf{s}^1 = -\tilde{\mathbf{P}}\nabla f(\mathbf{x}^1) = -\nabla f(\mathbf{x}^1)$

$$= -\begin{bmatrix} 6x_1^1 - 2x_2^1 - 26 \\ -2x_1^1 + 4x_2^1 - 8 \end{bmatrix} = \begin{bmatrix} 10 \\ 10 \end{bmatrix} \neq \begin{bmatrix} 0 \\ 0 \end{bmatrix}, \text{ so that we go to Step 7.}$$

Now, $\mathbf{b}^* = \mathbf{b}^< - \mathbf{A}^< \mathbf{x}^1 = \begin{bmatrix} 6 \\ -1 \\ 0 \\ 0 \end{bmatrix} - \begin{bmatrix} 1 & 2 \\ -1 & 1 \\ -1 & 0 \\ 0 & -1 \end{bmatrix} \begin{bmatrix} 3 \\ 1 \end{bmatrix} = \begin{bmatrix} 1 \\ 1 \\ 3 \\ 1 \end{bmatrix}$ and

$$\mathbf{s}^* = \mathbf{A}^< \mathbf{s}^1 = \begin{bmatrix} 1 & 2 \\ -1 & 1 \\ -1 & 0 \\ 0 & -1 \end{bmatrix} \begin{bmatrix} 10 \\ 10 \end{bmatrix} = \begin{bmatrix} 30 \\ 0 \\ -10 \\ -10 \end{bmatrix} \nleq 0, \text{ so that } \alpha^* = \frac{b_1^*}{s_1^*} = \frac{1}{30}.$$

Consider now problem P^α and minimize the function $f(\mathbf{x}^1 + \alpha \mathbf{s}^1) = f(x_1^1 + 10\alpha, \, x_2^1 + 10\alpha) = f(3 + 10\alpha, \, 1 + 10\alpha)$ subject to $0 \le \alpha \le \alpha^*$.

Furthermore, $\dfrac{df}{d\alpha} = 10\dfrac{\partial f}{\partial x_1} + 10\dfrac{\partial f}{\partial x_2} = 10(6x_1 - 2x_2 - 26) + 10(-2x_1 + 4x_2 - 8) = 40x_1 + 20x_2 - 340 = 600\alpha - 200$. Hence $f(3 + 10\alpha, \, 1 + 10\alpha)$ is strictly decreasing for $0 \le \alpha \le \alpha^* = \dfrac{1}{30}$ and the minimum value is attained for $\alpha = \dfrac{1}{30}$, i.e., $\alpha^1 = \dfrac{1}{30}$. Now $\mathbf{x}^2 = \mathbf{x}^1 + \alpha^1 \mathbf{s}^1$ and specifically

$$x_1^2 = x_1^1 + \alpha^1 s_1^1 = 3 + \frac{1}{30}(10) = \frac{10}{3} \approx 3.33 \text{ and}$$

$$x_2^2 = x_2^1 + \alpha^1 s_2^1 = 1 + \frac{1}{30}(10) = \frac{4}{3} \approx 1.33.$$

Since $\mathbf{x}^2 = \begin{bmatrix} \dfrac{10}{3}, & \dfrac{4}{3} \end{bmatrix}^T$, we set $k := 2$ and go to Step 2 of the next iteration. First we determine $\mathbf{A}^=$ and $\mathbf{A}^<$. Due to the fact that

$$x_1^2 + 2x_2^2 = \frac{10}{3} + \frac{8}{3} = 6$$
$$-x_1^2 + x_2^2 = -\frac{10}{3} + \frac{4}{3} = -\frac{6}{3} < 1$$
$$-x_1^2 = -\frac{10}{3} < 0$$
$$-x_2^2 = -\frac{4}{3} < 0,$$

only the first constraint is binding so that $\mathbf{A}^= = [1, 2]$ and $b^= = 6$. Furthermore, $\mathbf{M} = [1, 2]$ and thus $\tilde{\mathbf{P}} = \mathbf{I} - \mathbf{M}^T \left(\mathbf{M}\mathbf{M}^T\right)^{-1} \mathbf{M} = \begin{bmatrix} 1 & 0 \\ 0 & 1 \end{bmatrix} - \begin{bmatrix} 1 \\ 2 \end{bmatrix} \left[[1,2] \begin{bmatrix} 1 \\ 2 \end{bmatrix} \right]^{-1}$

$$[1,2] = \begin{bmatrix} 1 & 0 \\ 0 & 1 \end{bmatrix} - \frac{1}{5} \begin{bmatrix} 1 & 2 \\ 2 & 4 \end{bmatrix} = \begin{bmatrix} \dfrac{4}{5} & -\dfrac{2}{5} \\ -\dfrac{2}{5} & \dfrac{1}{5} \end{bmatrix}.$$

Next, $\mathbf{s}^2 = -\tilde{\mathbf{P}}\nabla f(\mathbf{x}^2) = - \begin{bmatrix} \dfrac{4}{5} & -\dfrac{2}{5} \\ -\dfrac{2}{5} & \dfrac{1}{5} \end{bmatrix} \begin{bmatrix} 6x_1^2 - 2x_2^2 - 26 \\ -2x_1^2 + 4x_2^2 - 8 \end{bmatrix} =$

$$= - \begin{bmatrix} \dfrac{4}{5} & -\dfrac{2}{5} \\ -\dfrac{2}{5} & \dfrac{1}{5} \end{bmatrix} \begin{bmatrix} 6\left(\dfrac{10}{3}\right) - 2\left(\dfrac{4}{3}\right) - 26 \\ -2\left(\dfrac{10}{3}\right) + 4\left(\dfrac{4}{3}\right) - 8 \end{bmatrix} = \begin{bmatrix} \dfrac{16}{5} \\ -\dfrac{8}{5} \end{bmatrix} \neq \begin{bmatrix} 0 \\ 0 \end{bmatrix}.$$

We now proceed to Step 7, where the problem P^α is solved, i.e., we minimize the function $f(\mathbf{x}^2 + \alpha\mathbf{s}^2) = f\left(x_1^2 + \dfrac{16}{5}\alpha, \; x_2^2 - \dfrac{8}{5}\alpha\right) = f\left(\dfrac{10}{3} + \dfrac{16}{5}\alpha, \; \dfrac{4}{3} - \dfrac{8}{5}\alpha\right)$, subject to $0 \leq \alpha \leq \alpha^*$, where α^* is determined by first calculating

$$\mathbf{b}^* = \mathbf{b}^< - \mathbf{A}^<\mathbf{x}^2 = \begin{bmatrix} -1 \\ 0 \\ 0 \end{bmatrix} - \begin{bmatrix} -1 & 1 \\ -1 & 0 \\ 0 & -1 \end{bmatrix} \begin{bmatrix} \dfrac{10}{3} \\ \dfrac{4}{3} \end{bmatrix} = \begin{bmatrix} \dfrac{1}{10} \\ \dfrac{10}{3} \\ \dfrac{4}{3} \end{bmatrix}, \text{ and}$$

$$\mathbf{s}^* = \mathbf{A}^<\mathbf{s}^2 = \begin{bmatrix} -1 & 1 \\ -1 & 0 \\ 0 & -1 \end{bmatrix} \begin{bmatrix} \dfrac{16}{5} \\ -\dfrac{8}{5} \end{bmatrix} = \dfrac{8}{5}\begin{bmatrix} -3 \\ -2 \\ 1 \end{bmatrix} \nleq 0, \quad \text{so that } \alpha^* = b_3^*/s_3^* = \dfrac{4}{3}\Big/\dfrac{8}{5} = \dfrac{5}{6}.$$

Moreover, $\dfrac{df}{d\alpha} = \dfrac{8}{5}\left[2\dfrac{\partial f}{\partial x_1} - \dfrac{\partial f}{\partial x_2}\right] = \dfrac{8}{5}[2(6x_1 - 2x_2 - 26) - (-2x_1 + 4x_2 - 8)]$

$= \dfrac{8}{5}(14x_1 - 8x_2 - 44) = \dfrac{8}{5}\left[14\left(\dfrac{10}{3} + \dfrac{16}{5}\alpha\right) - 8\left(\dfrac{4}{3} - \dfrac{8}{5}\alpha\right) - 44\right] = \dfrac{8}{5}\left(\dfrac{288}{5}\alpha - 8\right).$

It follows that $f\left(\dfrac{10}{3} + \dfrac{16}{5}\alpha, \; \dfrac{4}{3} - \dfrac{8}{5}\alpha\right)$ has a minimum for $\alpha = \dfrac{40}{288} = \dfrac{5}{36}$, which lies in the interval $\left[0, \dfrac{5}{6}\right]$, i.e., $\alpha^2 = \dfrac{5}{36}$.

With $\mathbf{x}^3 = \mathbf{x}^2 + \alpha^2\mathbf{s}^2$, we obtain

$$x_1^3 = x_1^2 + \alpha^2 s_1^2 = \dfrac{10}{3} + \dfrac{5}{36}\left(\dfrac{16}{5}\right) = \dfrac{34}{9}$$

$$x_2^3 = x_2^2 + \alpha^2 s_2^2 = \dfrac{4}{3} + \dfrac{5}{36}\left(-\dfrac{8}{5}\right) = \dfrac{10}{9},$$

i.e., $\mathbf{x}^3 = \begin{bmatrix} \dfrac{34}{9}, & \dfrac{10}{9} \end{bmatrix}^T$. Set $k: = 3$ and go to Step 2 of the next iteration.

Again determine $\mathbf{A}^=$ and $\mathbf{A}^<$. Now

$$
\begin{aligned}
x_1^3 + 2x_2^3 &= \frac{34}{9} + \frac{20}{9} = 6, \\
-x_1^3 + x_2^3 &= \frac{34}{9} + \frac{10}{9} = -\frac{8}{3} < -1 \\
-x_1^3 \quad &= -\frac{34}{9} < 0 \\
-x_2^3 &= -\frac{10}{9} < 0.
\end{aligned}
$$

We note that only the first constraint is binding, so that $\mathbf{A}^= = [1, 2]$ and $\mathbf{b}^= = 6$.

Furthermore, $\mathbf{M} = [1, 2]$ and $\tilde{\mathbf{P}} = \begin{bmatrix} \frac{4}{5} & -\frac{2}{5} \\ -\frac{2}{5} & \frac{1}{5} \end{bmatrix}$ as in the previous iteration.

Therefore, $\mathbf{s}^3 = -\tilde{\mathbf{P}}\nabla f(\mathbf{x}^3) = - \begin{bmatrix} \frac{4}{5} & -\frac{2}{5} \\ -\frac{2}{5} & \frac{1}{5} \end{bmatrix} \begin{bmatrix} 6x_1^3 - 2x_2^3 - 26 \\ -2x_1^3 + 4x_1^3 - 8 \end{bmatrix} = - \begin{bmatrix} \frac{4}{5} & -\frac{2}{5} \\ -\frac{2}{5} & \frac{1}{5} \end{bmatrix}$

$\begin{bmatrix} 6\left(\frac{34}{9}\right) - 2\left(\frac{10}{9}\right) - 26 \\ -2\left(\frac{34}{9}\right) + 4\left(\frac{10}{9}\right) - 8 \end{bmatrix} = \begin{bmatrix} 0 \\ 0 \end{bmatrix}$. Since $\mathbf{M}$ is not vacuous, we go to Step 5, and

since $\tilde{\mathbf{A}}$ is vacuous, $\mathbf{u}^3(\mathbf{M}) = \mathbf{u}^3(\mathbf{A}^=) = -\left(\mathbf{M}\mathbf{M}^T\right)^{-1}\mathbf{M}\nabla f(\mathbf{x}^3)$

$= -\left([1, 2]\begin{bmatrix} 1 \\ 2 \end{bmatrix}\right)^{-1}[1, 2]\begin{bmatrix} -\frac{50}{9} \\ \frac{100}{9} \end{bmatrix} = -\frac{1}{5}\left(-\frac{250}{9}\right) = \frac{50}{9} > 0$. As $\mathbf{u}^3(\mathbf{A}^=) \geq \mathbf{0}$,

the algorithm terminates with an optimal solution. It is $\bar{\mathbf{x}} = \mathbf{x}^3 = \begin{bmatrix} \frac{34}{9}, \frac{10}{9} \end{bmatrix}^T \approx$

$[3.78, 1.11]^T$, which is the same solution that was found using the Frank-Wolfe method in the previous section. The progression of the points $\mathbf{x}^k$ is shown in Fig. 6.8.

Fig. 6.8 Iterative solutions with Rosen's gradient projection method

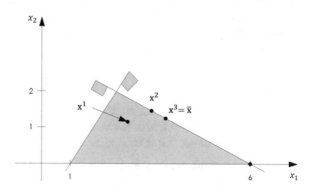

An example for which the gradient projection method cycles is given by Powell (1981). Rosen (1961) generalized the gradient projection method to allow nonlinear constraints.

We can regard the gradient projection method as belonging to a set of methods that treat constraints, depending on whether or not they are binding at the current iteration. If such is the case, we talk about *active set methods*. This aspect of the gradient projection method, as well as its convergence properties, are further explored in Luenberger and Ye (2008) and Bertsekas (2016), see also Cottle and Thapa (2017).

6.3   Problems with Nonconvex Quadratic Objective

In Sect. 6.1 we considered the convex quadratic programming problem with linear constraints, i.e.,

$$\text{P: } \underset{\mathbf{x}}{\text{Min}}\, z = f(\mathbf{x}) = \tfrac{1}{2}\mathbf{x}^T \mathbf{Q} \mathbf{x} + \mathbf{c}\mathbf{x} + d_0 \tag{6.23}$$

$$\text{s.t. } \mathbf{A}\mathbf{x} \le \mathbf{b}$$
$$\mathbf{x} \ge \mathbf{0},$$

and where the symmetric $[n \times n]$-dimensional matrix $\mathbf{Q}$ is positive semidefinite, causing $f(\mathbf{x})$ to be a convex function. We will now discuss the cases where $\mathbf{Q}$ is negative semidefinite or indefinite. Assume that $\mathbf{Q} \ne \mathbf{0}$.

Definition 6.7 If the square matrix $\mathbf{Q}$ in (6.23) is negative semidefinite (negative definite), problem P is called a *concave (strictly concave) quadratic minimization problem*. If $\mathbf{Q}$ is indefinite, problem P is called an *indefinite quadratic minimization problem*.

If the feasible region $\{\mathbf{x} \ge \mathbf{0}\colon \mathbf{A}\mathbf{x} \le \mathbf{b}\}$ is bounded and nonempty, we know that an optimal solution point for P in the convex case could be anywhere in the feasible region, i.e., at an extreme point, at a nonextreme boundary point, or at an interior point. In contrast, for strictly concave problems P, optimality will only occur at an extreme point and for indefinite problems at a boundary point, extreme or nonextreme. This can be summarized in

Theorem 6.8 If the quadratic programming problem P has an optimal solution, then at least one optimal solution point will be

- anywhere in the feasible region, if P is convex,
- at an extreme point, if P is concave, and
- at a boundary point, if P is indefinite.

For a proof of this theorem and other results in this section, see Floudas (2000) and Horst et al. (2000). As an illustration of the strictly concave case, take the "upside down" parabola $f(\mathbf{x}) = -\sum_{j=1}^{n} x_j^2 = -||\mathbf{x}||^2$, where $\mathbf{x} \in \mathbb{R}^n$, and consider the problem

$$P: \text{Min } z = f(\mathbf{x}) = -\sum_{j=1}^{n} x_j^2$$
$$\text{s.t. } -1 \leq x_j \leq 1, \; j = 1, \ldots, n.$$

The feasible region for this problem is a hypercube with side length 2, centered at the origin and axis parallel. One can easily demonstrate that f is minimized, with optimal objective function value $\bar{z} = -n$, at each of the 2^n corners of the feasible hypercube, which are *all* strict global minima!

If a problem is concave, but not strictly concave, there may be optimal points that are located on the boundary of the feasible region, but which are not extreme points. An example is the following problem:

$$P: \text{Min } z = f(\mathbf{x}) = -(x_1 - 1)^2$$
$$\text{s.t. } x_1 + x_2 \leq 1$$
$$x_1, \; x_2 \geq 0,$$

which has all its global minimal points on the line segment $(\bar{x}_1, \bar{x}_2) = (0, \; t)$ for all $t \in [0, 1]$. For $t = 0$ and $t = 1$, we obtain extreme points, while for $0 < t < 1$, we obtain non-extreme boundary points of the feasible region.

As an application of quadratic programming, we may refer to the portfolio selection problem, described in Sect. 3.5. Depending on the values of the elements in the covariance matrix $\mathbf{V} = (\sigma_{ij})$, the problem may then be either convex, concave, or indefinite. The optimization model concerning production planning developed in Sect. 3.9 is also quadratic. As yet another application, consider the following simple

Example A monopolist produces and sells two products in a region. The marketed quantities of the products are x_1 and x_2, respectively. The unit prices of the two products are p_1 and p_2. In particular, let the price functions of the firm be given by

$$p_1 = a_0 - a_1 x_1 + a_2 x_2 \text{ and}$$
$$p_2 = b_0 + b_1 x_1 - b_2 x_2,$$

where all coefficients are nonnegative. These functions include substitutional effects expressed by the coefficients a_1 and b_2, respectively, as well as complementarity effects between the two products, expressed by the coefficients a_2 and b_1, respectively. The marginal production costs of the two products are given by the coefficients c_1 and c_2, respectively, and there are production capacities that may be expressed by the constraints $x_1 + 2x_2 \leq 14$ and $3x_1 + 2x_2 \leq 26$, respectively. The firm wants to maximize its profits that can be expressed as $\Pi = -z = p_1 x_1 + p_2 x_2 - c_1 x_1 - c_2 x_2 = (a_0 - a_1 x_1 + a_2 x_2)x_1 + (b_0 + b_1 x_1 - b_2 x_2)x_2 - c_1 x_1 - c_2 x_2 = -a_1 x_1^2$

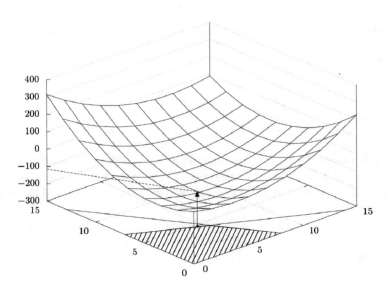

Fig. 6.9 Level curve of a quadratic optimization problem

$+(a_2 + b_1)x_1x_2 - b_2x_2^2 + (a_0 - c_1)x_1 + (b_0 - c_2)x_2$. If $(a_0,\ a_1,\ a_2) = (20,\ 2,\ 1)$, $(b_0,\ b_1,\ b_2) = (30,\ 1,\ 3)$, and $(c_1,\ c_2) = (5,\ 6)$, we then obtain the model

$$\text{P: Min } z = 2x_1^2 - 2x_1x_2 + 3x_2^2 - 15x_1 - 24x_2$$
$$\text{s.t. } x_1 + 2x_2 \leq 14$$
$$3x_1 + 2x_2 \leq 26$$
$$x_1,\ x_2 \geq 0.$$

This is a quadratic programming problem, and in terms of the standard form expressed in relation (6.23), we have $\mathbf{Q} = \begin{bmatrix} 4 & -2 \\ -2 & 6 \end{bmatrix}$, $\mathbf{c} = [-15,\ -24]$, $d_0 = 0$, $\mathbf{A} = \begin{bmatrix} 1 & 2 \\ 3 & 2 \end{bmatrix}$, and $\mathbf{b} = \begin{bmatrix} 14 \\ 26 \end{bmatrix}$. Since the matrix $\mathbf{Q}$ is positive definite, the problem P is a convex quadratic programming problem. Its unique optimal solution is located at the point $(\bar{x}_1, \bar{x}_2) = \left(\tfrac{76}{15},\ \tfrac{67}{15}\right) \approx (5.07, 4.47)$, which is a nonextreme boundary point. This is shown in Fig. 6.9.

Assume now that the production capacities double from 14 and 26 to 28 and 52, respectively. The unique optimal solution is then $(\bar{x}_1,\ \bar{x}_2) = (6.9, 6.3)$, which is an interior point, see Fig. 6.10.

If, on the other hand, the production capacities were 15 and 25, respectively, the unique optimal solution would be at the extreme point $(\bar{x}_1, \bar{x}_2) = (5, 5)$, see Fig. 6.11.

This demonstrates the statement made above, viz., that a convex quadratic programming problem may have its optimal solution at an extreme point, at a nonextreme boundary point, or at an interior point.

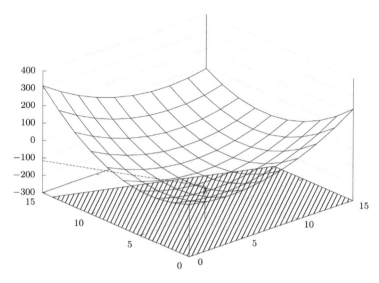

Fig. 6.10 3-D plot of a quadratic optimization problem with double capacities

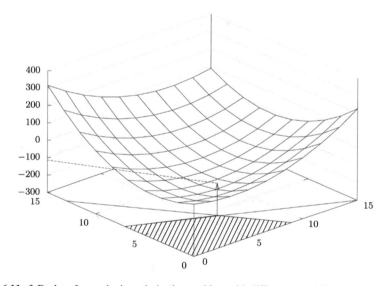

Fig. 6.11 3-D plot of a quadratic optimization problem with different capacities

Let us now return to the original model, but with changed coefficients, so that $a_1 = 1$ instead of 2, $a_2 = 2$ instead of 1, and $b_1 = 2$ instead of 1, reflecting less substitutional and more complementarity effects. We then obtain $\mathbf{Q} = \begin{bmatrix} 2 & -4 \\ -4 & 6 \end{bmatrix}$, so that the quadratic programming problem P is now indefinite. The unique optimal solution is now at the extreme point $(\bar{x}_1, \bar{x}_2) = (6, 4)$, see Fig. 6.12. If the capacity

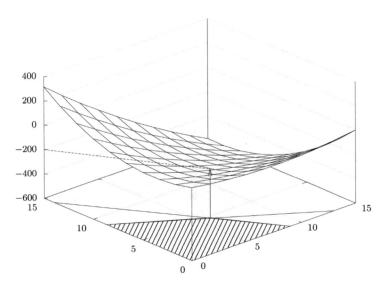

Fig. 6.12 3-D plot of an indefinite quadratic optimization problem

resource of the first constraint is increased from 14 to 16 units, the unique optimal solution becomes $(\bar{x}_1, \bar{x}_2) = (296/55, 271/55) \approx (5.38, 4.93)$, which is a nonextreme boundary point, at which the first constraint is loose and the second is tight, see Fig. 6.13.

This shows that for an indefinite quadratic programming problem, both extreme points and nonextreme boundary points can occur as optimal solutions.

Finally, let us assume that in the original model $a_1 = a_2 = b_1 = -1$ and $b_2 = -3$, so that both products exhibit the luxury commodity characteristic of having positively sloping demand functions, i.e., the higher the price, the higher the demand. We then have $\mathbf{Q} = \begin{bmatrix} -2 & 2 \\ 2 & -6 \end{bmatrix}$, which is negative definite, so that we have a strictly concave quadratic programming problem. According to the theory, the corresponding optimal solution must therefore be located at an extreme point of the set of feasible solutions. Solving P, we find the unique optimal solution $(\bar{x}_1, \bar{x}_2) = (0, 7)$, which is indeed an extreme point, as predicted and shown in Fig. 6.14.

Since the general quadratic programming problem is **NP**-hard (see, e.g., Horst et al. 2000), the exact solution method we will present here could possibly take an exponential number of steps and time to converge. This exact method is based on the Karush-Kuhn-Tucker conditions, which are used to identify an optimal solution. Specifically, we consider the problem

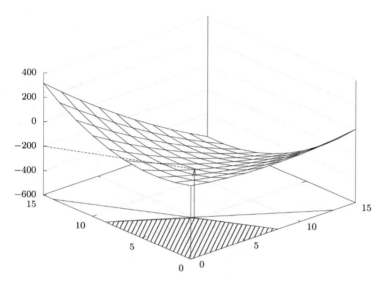

Fig. 6.13 3-D plot of an indefinite quadratic optimization problem with different capacities

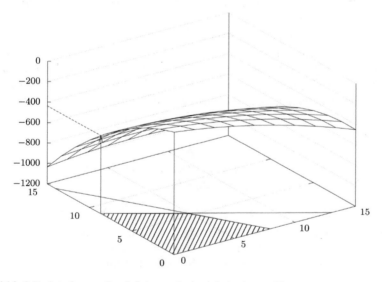

Fig. 6.14 3-D plot of a negative definite quadratic optimization problem

$$\text{P: } \underset{\mathbf{x}}{\text{Min }} z = f(\mathbf{x}) = \tfrac{1}{2}\mathbf{x}^T\mathbf{Q}\mathbf{x} + \mathbf{c}\mathbf{x}$$
$$\text{s.t. } \mathbf{A}\mathbf{x} \leq \mathbf{b}$$
$$\mathbf{x} \in \mathbb{R}^n,$$

where $\mathbf{Q}$ is symmetric and negative definite, negative semidefinite, or indefinite, and where any nonnegativity constraints are considered to be included in the m constraints $\mathbf{A}\mathbf{x} \leq \mathbf{b}$. The associated Lagrangean function is

$$L(\mathbf{x}, \mathbf{u}) = \frac{1}{2}\mathbf{x}^T\mathbf{Q}\mathbf{x} + \mathbf{c}\mathbf{x} + \mathbf{u}(\mathbf{A}\mathbf{x} - \mathbf{b}),$$

so that the Karush-Kuhn-Tucker conditions for optimality of P are

$$\left. \begin{array}{l} \mathbf{Q}\bar{\mathbf{x}} + \mathbf{c}^T + \mathbf{A}^T\bar{\mathbf{u}}^T = \mathbf{0} \\ \mathbf{A}\bar{\mathbf{x}} \leq \mathbf{b} \\ \bar{\mathbf{u}}(\mathbf{A}\bar{\mathbf{x}} - \mathbf{b}) = 0 \\ \bar{\mathbf{u}} \geq \mathbf{0} \end{array} \right\}. \tag{6.24}$$

From Theorem 6.8 we know that optimality must occur at a boundary point, so that there is an index set $I \subseteq \{1, 2, \ldots, m\}$ for the active (tight) constraints $\mathbf{A}_I\bar{\mathbf{x}} = \mathbf{b}_I$ and $\bar{I} = \{1, 2, \ldots, m\}\backslash I$ for the inactive (loose) constraints $\mathbf{A}_{\bar{I}}\bar{\mathbf{x}} < \mathbf{b}_{\bar{I}}$, where the submatrices $\mathbf{A}_I$ and $\mathbf{b}_I$ correspond to active rows, and $\mathbf{A}_{\bar{I}}$ and $\mathbf{b}_{\bar{I}}$ correspond to rows of inactive constraints. The Lagrangean multiplier row vector $\bar{\mathbf{u}}$ is correspondingly decomposed into $\bar{\mathbf{u}}_I$ for active/tight constraints and $\bar{\mathbf{u}}_{\bar{I}}$ for inactive/loose constraints, respectively. The complementarity conditions $\bar{\mathbf{u}}(\mathbf{A}\bar{\mathbf{x}} - \mathbf{b}) = 0$ then imply that $\bar{\mathbf{u}}_{\bar{I}} = \mathbf{0}$, so that system (6.24) reduces to the linear system

$$\left. \begin{array}{l} \mathbf{Q}\bar{\mathbf{x}} + \mathbf{c}^T + \mathbf{A}^T\bar{\mathbf{u}}^T = \mathbf{0} \\ \mathbf{A}_I\bar{\mathbf{x}} = \mathbf{b}_I, \quad \mathbf{A}_{\bar{I}}\bar{\mathbf{x}} < \mathbf{b}_{\bar{I}} \\ \bar{\mathbf{u}}_I \geq \mathbf{0}, \quad \bar{\mathbf{u}}_{\bar{I}} = \mathbf{0}, \end{array} \right\}. \tag{6.25}$$

For each possible choice of I, we will now have to find a solution to the system (6.25) and compute the corresponding objective function value z. The smallest of these z-values will therefore identify the optimal solution(s) to P. Note that if the feasible region $\{\mathbf{x}: \mathbf{A}\mathbf{x} \leq \mathbf{b}\}$ is unbounded, it could happen that no finite optimal solution exists.

Example Consider the problem

$$\text{P: } \text{Min } z = x_1^2 - 4x_1x_2 + 3x_2^2 - 15x_1 - 24x_2$$
$$\text{s.t. } x_1 + 2x_2 \leq 16$$
$$3x_1 + 2x_2 \leq 26$$
$$x_1, x_2 \geq 0,$$

which is the indefinite case of the monopolist production example above, but with $b_1 = 16$ instead of 14, see Fig. 6.12. With $\mathbf{Q} = \begin{bmatrix} 2 & -4 \\ -4 & 6 \end{bmatrix}$, $\mathbf{c} = [-15, -24]$,

$$\mathbf{A} = \begin{bmatrix} 1 & 2 \\ 3 & 2 \\ -1 & 0 \\ 0 & -1 \end{bmatrix}, \text{ and } \mathbf{b} = \begin{bmatrix} 16 \\ 26 \\ 0 \\ 0 \end{bmatrix} \text{ we obtain the system (6.24) as}$$

$$
\begin{aligned}
&2x_1 - 4x_2 - 15 + u_1 + 3u_2 - u_3 = 0 \\
&-4x_1 + 6x_2 - 24 + 2u_1 + 2u_2 - u_4 = 0 \\
&x_1 + 2x_2 \le 16 \\
&3x_1 + 2x_2 \le 26 \\
&-x_1 \le 0 \\
&-x_2 \le 0 \\
&u_1(x_1 + 2x_2 - 16) + u_2(3x_1 + 2x_2 - 26) - u_3x_1 - u_4x_2 = 0 \\
&u_1, u_2, u_3, u_4 \ge 0.
\end{aligned}
$$

Trying with $I = \{1\}$ so that $\bar{I} = \{2, 3, 4\}$, we obtain the system (6.25) as

$$
\begin{aligned}
&2x_1 - 4x_2 - 15 + u_1 = 0 \\
&-4x_1 + 6x_2 - 24 + 2u_1 = 0 \\
&x_1 + 2x_2 = 16 \\
&3x_1 + 2x_2 < 26 \\
&-x_1 < 0 \\
&-x_2 < 0 \\
&u_1 \ge 0.
\end{aligned}
$$

The three equations yield the solution $(x_1, x_2, u_1) = \left(\dfrac{118}{15}, \dfrac{61}{15}, \dfrac{233}{15} \right)$, which violates the constraint $3x_1 + 2x_2 < 26$. Therefore, the Karush-Kuhn-Tucker conditions do not have a solution with the first constraint being active and the other three constraints being loose.

Trying instead with $I = \{2\}$ so that $\bar{I} = \{1, 3, 4\}$, we obtain the system (6.25) as

$$
\begin{aligned}
&2x_1 - 4x_2 - 15 + 3u_2 = 0 \\
&-4x_1 + 6x_2 - 24 + 2u_2 = 0 \\
&x_1 + 2x_2 \qquad < 16 \\
&3x_1 + 2x_2 \qquad = 26 \\
&-x_1 \qquad < 0 \\
&-x_2 \qquad < 0 \\
&u_2 \ge 0.
\end{aligned}
$$

The three equations yield $(x_1, x_2, u_2) = \left(\dfrac{296}{55}, \dfrac{271}{55}, \dfrac{878}{55} \right) \approx (5.38, 4.93, 15.96)$, which satisfies the three inequalities. With the Karush-Kuhn-Tucker conditions satisfied, this might therefore be an optimal solution; the corresponding objective function value is $\bar{z} = -\dfrac{11{,}179}{55} \approx -203.25$.

For $I = \{3\}$, we obtain $\bar{I} = \{1,\ 2,\ 4\}$ and the system

$$
\begin{aligned}
2x_1 - 4x_2 - 15 - u_3 &= 0 \\
-4x_1 + 6x_2 - 24 &= 0 \\
x_1 + 2x_2 &< 16 \\
3x_1 + 2x_2 &< 26 \\
-x_1 &= 0 \\
-x_2 &< 0 \\
u_3 &\geq 0,
\end{aligned}
$$

which yields $(x_1, x_2, u_3) = (0, 4, -31)$, violating $u_3 \geq 0$, so that the Karush-Kuhn-Tucker conditions are not satisfied. Similarly, for $I = \{4\}$ we obtain $\bar{I} = \{1,\ 2,\ 3\}$ and the system

$$
\begin{aligned}
2x_1 - 4x_2 - 15 &= 0 \\
-4x_1 + 6x_2 - 24 - u_4 &= 0 \\
x_1 + 2x_2 &< 16 \\
3x_1 + 2x_2 &< 26 \\
-x_1 &< 0 \\
-x_2 &= 0 \\
u_3 &\geq 0,
\end{aligned}
$$

from which we obtain $(x_1, x_2, u_4) = \left(\dfrac{15}{2}, 0,\ -54\right)$, which violates $u_4 \geq 0$.

Continuing in this fashion, we try $I = \{1, 2\}$, so that $\bar{I} = \{3,\ 4\}$ and

$$
\begin{aligned}
2x_1 - 4x_2 - 15 + u_1 + 3u_2 &= 0 \\
-4x_1 + 6x_2 - 24 + 2u_1 + 2u_2 &= 0 \\
x_1 + 2x_2 &= 16 \\
3x_1 + 2x_2 &= 26 \\
-x_1 &< 0 \\
-x_2 &< 0 \\
u_1,\ u_2 &\geq 0,
\end{aligned}
$$

where the four equations have the unique solution $(x_1,\ x_2,\ u_1,\ u_2) = \left(5,\ \dfrac{11}{2},\ -\dfrac{21}{4},\ \dfrac{43}{4}\right)$, which violates the nonnegativity of u_1.

Next, $I = \{1, 3\}$ gives
$(x_1, x_2, u_1, u_3) = (0, 8, -12, -59)$, (violating the nonnegativity of u_1 and u_3),
$I = \{1, 4\}$ gives
$(x_1, x_2, u_1, u_4) = (16, 0, -17, -122)$, (violating the nonnegativity of u_1 and u_4),
$I = \{2, 3\}$ gives
$(x_1, x_2, u_2, u_3) = (0, 13, -27, -148)$, (violating the nonnegativity of u_2 and u_3),

$I = \{2, 4\}$ gives

$(x_1, x_2, u_2, u_4) = \left(\dfrac{26}{3}, 0, -\dfrac{7}{9}, -\dfrac{542}{9}\right)$, (violating the nonnegativity of u_2 and u_4),

$I = \{3, 4\}$ gives

$(x_1, x_2, u_3, u_4) = (0, 0, -15, -24)$, (violating the nonnegativity of u_3 and u_4).

The choices $I = \{1, 2, 3\}, \{1, 2, 4\}, \{1, 3, 4\}, \{2, 3, 4\}$, and $\{1, 2, 3, 4\}$ will all lead right away to incompatible constraints. For instance, $I = \{1, 2, 3\}$, so that constraints 1, 2, and 3 are active/tight, leads to the following system of equations:

$$\begin{aligned}
x_1 + 2x_2 &= 16 \\
3x_1 + 2x_2 &= 26 \\
x_1 &= 0,
\end{aligned}$$

which is clearly incompatible. All possibilities are now exhausted. Only the case with $I = \{2\}$ as the single active constraint and $\bar{I} = \{1, 3, 4\}$ as the inactive/loose constraints lead to a solution of the Karush-Kuhn-Tucker conditions. The global optimal solution to our problem is therefore $(\bar{x}_1, \bar{x}_2, \bar{z}) \approx (5.38, 4.93, -203.25)$ with $(\bar{u}_1, \bar{u}_2, \bar{u}_3, \bar{u}_4) \approx (0, 15.96, 0, 0)$.

The attractive feature of the above exact procedure for solving quadratic programming problems is that the Karush-Kuhn-Tucker conditions reduce to systems of linear equations, due to the fact that the objective function $f(\mathbf{x})$ is quadratic and the constraints are linear. The drawback of the procedure derives from the fact that there is an exponentially large number of sets I of active constraints.

Concerning approximate methods for solving nonconvex quadratic programming problems, the concave case with negative definite or negative semidefinite matrix $\mathbf{Q}$ will be dealt with in the next chapter in Sect. 7.3, covering general concave minimization. The case with an indefinite matrix $\mathbf{Q}$ is a special case of D.C. programming, which will be covered in Sect. 7.4. For a full treatment of the subject area, useful references are Floudas and Visweswaran (1995) and Horst et al. (2000).

References

Aganagić M (1984) Newton's method for linear complementarity problems. *Mathematical Programming* **28**: 349-362

Balinski ML, Cottle RW (eds.) (1978) Complementarity and fixed point problems. *Mathematical Programming Study* **7**, North Holland, Amsterdam

Bazaraa MS, Sherali HD, Shetty CM (2013) *Nonlinear programming: theory and algorithms.* (3rd ed.) Wiley, New York

Beale EML (1955) On minimizing a convex function subject to linear inequalities. *Journal of the Royal Statistical Society* **17B**: 173-184

Beale EML (1959) On quadratic programming. *Naval Research Logistics Quarterly* **6**: 227-243

Ben-Israel A, Greville T (2003) *Generalized inverses.* Springer-Verlag, Berlin-Heidelberg-New York

Bertsekas DP (2016) *Nonlinear programming.* (3rd ed.) Athena Scientific, Belmont, MA

Charnes A, Cooper WW (1961) *Management models and industrial applications of linear programming.* Wiley, New York

Cottle RW, Dantzig GB (1968) Complementary pivot theory of mathematical programming. *Linear Algebra and its Applications* **1**: 103-125

Cottle RW, Pang JS (1978) On solving linear complementarity problems as linear programs. *Mathematical Programming Study* **7**: 88-107

Cottle RW, Pang JS, Stone RE (2009) *The linear complementarity problem.* Classics in Applied Mathematics, SIAM, Philadelphia

Cottle RW, Thapa MN (2017) *Linear and nonlinear optimization.* Springer-Verlag, Berlin-Heidelberg-New York

Eaves BC (1971) The linear complementarity problem. *Management Science* **17**: 612-634

Eaves BC (1978) Computing stationary points. *Mathematical Programming Study* **7**: 1-14

Eiselt HA, Sandblom C-L (2007) *Linear programming and its applications.* Springer-Verlag, Berlin-Heidelberg

Floudas CA (2000) *Deterministic global optimization. Theory, methods, and applications.* Springer Science + Business Media, Dordrecht

Floudas CA, Visweswaran V (1995) Quadratic optimization. In: Horst R, Pardalos PM (eds.) *Handbook of global optimization.* Kluwer, Boston, MA

Frank M, Wolfe P (1956) An algorithm for quadratic programming. *Naval Research Logistics Quarterly* **3**: 95-110

Holloway CA (1974) An extension of the Frank and Wolfe method of feasible directions. *Mathematical Programming* **6**: 14-27

Horst R, Pardalos PM, Thoai NV (2000) *Introduction to global optimization*, vol. 2. Kluwer, Dordrecht, The Netherlands

Karamardian S (1969a) The nonlinear complementarity problem with applications, part I. *Journal of Optimization Theory and Applications* **4**: 87-98

Karamardian S (1969b) The nonlinear complementarity Problem with Applications, part II. *Journal of Optimization Theory and Applications* **4**: 167-181

Karamardian S (1972) The complementarity problem. *Mathematical Programming* **2**: 107-129

Künzi HP, Krelle W, Oettli W (1966) *Nonlinear programming.* Blaisdell Publ. Co., MA

Lemke CE (1968) On complementary pivot theory. pp. 95-114 in Dantzig GB, Veinott AF (eds.) *Mathematics of the Decision Sciences, Part 1.* American Mathematical Society, Providence, RI

Luenberger DL, Ye Y (2008) *Linear and nonlinear programming.* (3rd ed.) Springer-Verlag, Berlin-Heidelberg-New York

Mangasarian OL (1976) Linear complementarity problems solvable by a single linear program. *Mathematical Programming* **10**: 263-270

Moore EH (1935) *General analysis, part I.* Memoranda of the American Philosophical Society, Vol. I

Penrose R (1955) A generalized inverse for matrices. *Proceedings of the Cambridge Philosophical Society* **51**: 406-413

Powell MJD (1981) An example of cycling in a feasible point algorithm. *Mathematical Programming* **20**: 353-357

Rosen JB (1960) The gradient projection method for non-linear programming: part I, linear constraints. *Journal of SIAM* **8**: 181-217

Rosen JB (1961) The gradient projection method for non-linear programming: part II. *Journal of SIAM* **9**: 514-532

Talman D, Van der Heyden L (1981) *Algorithms for the linear complementarity problem which allow an arbitrary starting point.* Cowles Foundation Discussion Paper No. 600. Yale University, New Haven, CT

Topkis DM, Veinott AF (1967) On the convergence of some feasible direction algorithms for nonlinear programming. *SIAM Journal on Control* **5/2**: 268-279

Wolfe P (1959) The simplex method for quadratic programming. *Econometrica* **27**: 382-398

Wolfe P (1970) Convergence theory in nonlinear programming. pp. 1-36 in Abadie J (ed.) *Integer and nonlinear programming.* North Holland, Amsterdam

Zoutendijk G (1960) *Methods of feasible directions.* Elsevier, New York

Chapter 7
Methods for Nonlinearly Constrained Problems

As opposed to the previous chapter, the methods presented in this chapter allow the given constraints to be nonlinear. In the first section of this chapter, we consider problems with convex constraints, starting with the cutting plane method for nonlinear programming due to Kelley (1960). Although Kelley's method suffers from a numerically slow convergence in comparison with other methods (except possibly for highly nonlinear constraints), we cover it here because of the considerable theoretical interest of the cutting plane principle. We continue with the generalized reduced gradient (*GRG*) method, which may be regarded as a standard technique for convex differentiable programming. Techniques for handling nondifferentiable functions using subgradients as well as methods for concave objective functions are then described.

The last section of this chapter briefly discusses difference-convex or so-called *D.C. programming*, which involves the difference of two convex functions, as well as semidefinite and conical programming.

7.1 Problems with Linear Objective Function

Cutting plane methods were first introduced by Gomory (1963) for solving integer programming problems (see, e.g., Eiselt and Sandblom 2000). The cutting plane approach was then developed by Kelley (1960) to handle nonlinear programming problems. The basic idea is as follows: Suppose that the given problem has a linear objective function as well as a bounded feasible region defined by linear and/or nonlinear convex constraints. Then one could circumscribe the feasible region by a linear polytope which can be seen as an outer approximation of the feasible set; see Fig. 7.1.

The resulting auxiliary linear programming problem with this polytope as feasible region can be solved using any suitable linear programming method. If the optimal

© Springer Nature Switzerland AG 2019

H. A. Eiselt, C.-L. Sandblom, *Nonlinear Optimization*, International Series in Operations Research & Management Science 282,
https://doi.org/10.1007/978-3-030-19462-8_7

Fig. 7.1 Optimization
problem with nonlinearly
bounded feasible set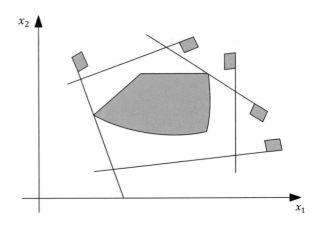

solution of the auxiliary linear programming problem is also feasible, or sufficiently close to feasibility, for the original problem, then it is also optimal or near-optimal for the original problem and the procedure terminates. Otherwise, we create a cutting plane, i.e. a linear constraint that cuts off the optimal solution point of the previous linear programming approximation but no point of the feasible region. This procedure is repeated making the polytope locally approximate the feasible region of the original problem tighter and tighter in the vicinity of the optimal solution of the original problem.

Kelley's cutting plane algorithm is applicable to problems of the type

$$P: \text{Min } z = \mathbf{c}\mathbf{x}$$
$$\text{s.t. } \mathbf{g}(\mathbf{x}) \leq \mathbf{0}$$
$$\mathbf{x} \in \mathbb{R}^n,$$

where $\mathbf{g} : \mathbb{R}^n \to \mathbb{R}^m$ is a convex differentiable function. In other words, P is a convex problem with linear objective function and linear or nonlinear constraints. Although it appears that this formulation is less general than the case with nonlinear objective function as well as nonlinear constraints, this is not so. To see this, let us show that the problem

$$P': \text{Min } z' = f(\mathbf{x})$$
$$\text{s.t. } g_i(\mathbf{x}) \leq 0 \ \forall i = 1, \ldots, m$$
$$x_j \in \mathbb{R} \ \forall j = 1, \ldots, n$$

can be written as a problem with a linear objective function. Introducing an additional variable x_{n+1}, we can write P′ in the equivalent form

$$P'' : \text{Min } z'' = x_{n+1}$$
$$\text{s.t.} g_i(\mathbf{x}) \leq 0 \ \forall i = 1, \ldots, m$$
$$f(\mathbf{x}) - x_{n+1} \leq 0$$
$$x_j \in \mathbb{R} \ \forall j = 1, \ldots, n+1,$$

i.e., a problem in $(n + 1)$ variables, with a linear objective function and $(m + 1)$ linear and/or nonlinear constraints.

Returning to our formulation of the problem P with a linear objective function, suppose now that in solving P the points $\mathbf{x}^1, \mathbf{x}^2, \ldots, \mathbf{x}^k$ have been generated. In the next iteration of the cutting plane algorithm the following auxiliary linear programming problem is solved.

$$P^k: \text{Min } z^k = \mathbf{cx}$$
$$\text{s.t. } g_{i_\ell}\left(\mathbf{x}^\ell\right) + \left(\mathbf{x} - \mathbf{x}^\ell\right)^T \nabla g_{i_\ell}\left(\mathbf{x}^\ell\right) \leq 0, \ \ell = 1, \ldots, k$$
$$-\mathbf{e}M \leq \mathbf{x} \leq \mathbf{e}M$$
$$\mathbf{x} \in \mathbb{R}^n,$$

where M is a preselected number, sufficiently large, so that each feasible point satisfies the last constraint, and where $\mathbf{e}$ denotes the $[n \times 1]$-dimensional (summation) vector of ones. An optimal solution to the problem P^k is denoted by $\mathbf{x}^{k+1}$.

Because of the convexity of the functions $g_i(\mathbf{x})$, Theorem 1.34 implies that

$$g_{i_\ell}(\mathbf{x}^\ell) + (\mathbf{x} - \mathbf{x}^\ell)^T \nabla g_{i_\ell}(\mathbf{x}^\ell) \leq g_{i_\ell}(\mathbf{x}) \ \forall \mathbf{x} \in \mathbb{R}^n, \ \ell = 1, \ldots, k,$$

and therefore the feasible set of P^k contains the feasible set of the original problem P. It follows that if $\mathbf{g}(\mathbf{x}^{k+1}) \leq \mathbf{0}$, i.e., if $\mathbf{x}^{k+1}$ is feasible for P, then $\mathbf{x}^{k+1}$ is also an optimal solution for P. If, on the other hand, $\mathbf{g}(\mathbf{x}^{k+1}) \not\leq \mathbf{0}$, then for some $i = i_{k+1}$ we must have $g_{i_{k+1}}\left(\mathbf{x}^{k+1}\right) > 0$. The constraint

$$g_{i_{k+1}}\left(\mathbf{x}^{k+1}\right) + (\mathbf{x} - \mathbf{x}^{k+1})^T \nabla g_{i_{k+1}}(\mathbf{x}^{k+1}) \leq 0$$

is then added to problem P^k so that the next problem P^{k+1} results.

The current polytope is $ABCDE$ where $B = \mathbf{x}^{k+1}$ is the optimal point for problem P^k. The constraint $g_{i_{k+1}}(\mathbf{x}) \leq 0$ is one of the constraints violated at $\mathbf{x}^{k+1}$ and the cutting plane indicated in Fig. 7.2 is introduced, so that the next polytope for P^{k+1} is $AFGDE$.

Recall that problem P is defined as

$$P : \text{Min } z = \mathbf{cx}$$
$$\text{s.t. } \mathbf{g}(\mathbf{x}) \leq \mathbf{0}$$
$$\mathbf{x} \in \mathbb{R}^n.$$

Fig. 7.2 Cutting planes

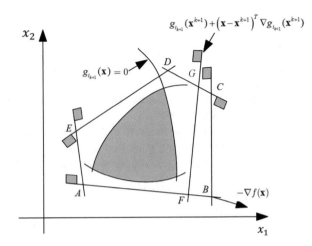

The cutting plane method can now formally be described as follows. Note that there is no Step 0 for solving P without any constraints, since it would have no finite optimal solution, barring the trivial case with $\mathbf{c} = \mathbf{0}$.

The Cutting Plane Method

Step 1: Define an initial polytope, e.g., $\{\mathbf{x}: -\mathbf{e}M \leq \mathbf{x} \leq \mathbf{e}M\}$ with a sufficiently large number M, so as to contain the feasible set of problem P. Set $k := 0$.

Step 2: Solve the auxiliary linear programming problem

$$\mathrm{P}^k: \mathrm{Min} z^k = \mathbf{c}\mathbf{x}$$
$$\text{s.t. } g_{i_\ell}\left(\mathbf{x}^\ell\right) + \left(\mathbf{x} - \mathbf{x}^\ell\right)^T \nabla g_{i_\ell}\left(\mathbf{x}^\ell\right) \leq 0, \ \ell = 1, \ldots, k$$
$$-\mathbf{e}M \leq \mathbf{x} \leq \mathbf{e}M$$
$$\mathbf{x} \in \mathbb{R}^n,$$

and denote an optimal solution by $\mathbf{x}^{k+1}$. If P^k has no feasible solution, stop; problem P has then no feasible solution.

Step 3: Is $g_i(\mathbf{x}^{k+1}) \leq 0 \ \forall \ i = 1, \ldots, m$?

If yes: Stop $\mathbf{x}^{k+1}$ is an optimal solution to problem P.
If no: Go to step 4.

Step 4: Select $i = i_{k+1}$, such that $g_{i_{k+1}}\left(\mathbf{x}^{k+1}\right) = \max_i \left\{g_i\left(\mathbf{x}^{k+1}\right)\right\}$. Is $\nabla g_{i_{k+1}}\left(\mathbf{x}^{k+1}\right) = \mathbf{0}$?

If yes: Stop, P has no feasible solution.
If no: Set $k := k+1$ and go to Step 2.

A few comments about the algorithm are in order. First, no initial feasible point is needed to start the algorithm. If a "better" approximation of the feasible set than the polytope $-\mathbf{e}M \leq \mathbf{x} \leq \mathbf{e}M$ is initially available, it should preferably be used. The structural constraints of P^k are linearizations of the constraints of P. If some of the constraints of problem P are linear, they can be treated specially, by always including them in the linear programming problems P^k. It may then be that some or all of the upper and lower bound constraints $-eM \leq x \leq eM$ can be omitted. During the first iteration ($k = 0$), no cutting planes in Step 2 are yet generated. In Step 4, we can actually select any i such that $g_i(x^{k+1}) > 0$ Also, if $\nabla g_{i_{k+1}}(x^{k+1}) = 0$, then x^{k+1} minimizes $g_{i_{k+1}}(x)$, which is convex. Consequently, $g_{i_{k+1}}(x) \geq g_{i_{k+1}}(x^{k+1}) > 0 \, \forall \, x \in \mathbb{R}^n$, which clearly indicates that P has no feasible solution. Finally, it is advantageous to design "deep cuts" which cut off large portions of the polytope for P^k. The deepest cuts are those whose hyperplanes are tangents to the feasible set of P.

Example   The cutting plane method will be illustrated by carrying out three complete iterations for our usual problem.

$$P: \text{Min } z = 3x_1^2 - 2x_1x_2 + 2x_2^2 - 26x_1 - 8x_2$$
$$\text{s.t. } x_1 + 2x_2 \leq 6$$
$$x_1 - x_2 \geq 1$$
$$x_1, x_2 \geq 0.$$

First problem P is transformed into an equivalent problem with a linear objective function, i.e.,

$$
\begin{aligned}
P' : \text{Min } z' = \quad & & x_3 & \\
\text{s.t. } g_1(x) = \quad & & x_1 + 2x_2 - 6 & \quad \leq 0 \\
g_2(x) = \quad & & -x_1 + x_2 + 1 & \quad \leq 0 \\
g_3(x) = \quad & & -x_1 & \quad \leq 0 \\
g_4(x) = \quad & & -x_2 & \quad \leq 0 \\
g_5(x) = \quad & 3x_1^2 - 2x_1x_2 + 2x_2^2 - 26x_1 - 8x_2 - x_3 & & \quad \leq 0 \\
& x \in \mathbb{R}^3. & &
\end{aligned}
$$

As $g_1(x) \leq 0$ for $i = 1, 2, 3, 4$, are linear constraints, they will be included in all problems P^k. We select $M = 100$, judging this number to be sufficiently large. For $k = 0$, the linear programming problem P^1 is

$$P^1: \text{Min } z^1 = x_3$$
$$\text{s.t. } x_1 + 2x_2 \leq 6$$
$$x_1 - x_2 \geq 1$$
$$x_1, x_2 \geq 1$$
$$-100 \leq x_3 \leq 100.$$

One of its optimal solutions is $x_1^1 = 3$, $x_2^1 = 1$, $x_3^1 = -100$ with $z^1 = -100$, where x_1^1 and x_2^1 are selected to facilitate a comparison with the methods covered in the previous two chapters.

We find that $g_5(\mathbf{x}^1) = 37 > 0$ and as $\nabla g_5(\mathbf{x}^1) = \begin{bmatrix} -10 \\ -10 \\ -1 \end{bmatrix} \neq \begin{bmatrix} 0 \\ 0 \\ 0 \end{bmatrix}$, we set $k := 1$

and proceed to the next iteration. With $k = 1$ and $g_5(\mathbf{x}^1) + (\mathbf{x} - \mathbf{x}^1)^T \nabla g_5(\mathbf{x}^1) =$

$37 + [x_1 - 3, x_2 - 1, x_3 + 100] \begin{bmatrix} -10 \\ -10 \\ -1 \end{bmatrix} = -10x_1 - 10x_2 - x_3 - 23$, the auxiliary

linear programming problem P^2 becomes

$$P^2: \text{Min } z^2 = x_3$$
$$\text{s.t. } x_1 + 2x_2 \leq 6$$
$$x_1 - x_2 \geq 1$$
$$-10x_1 - 10x_2 - x_3 \leq 23$$
$$x_1, x_2 \geq 0$$
$$x_3 \in [-100, 100].$$

The unique optimal solution of P^2 is $x_1^2 = 6$, $x_2^2 = 0$, $x_3^2 = -83$ with $z^2 = -83$.

We now find that $g_5(\mathbf{x}^2) = 35 > 0$, and as $\nabla g_5(\mathbf{x}^2) = \begin{bmatrix} 10 \\ -20 \\ -1 \end{bmatrix} \neq \begin{bmatrix} 0 \\ 0 \\ 0 \end{bmatrix}$, we set $k := 2$

and go to the next iteration.

With $k = 2$ and $g_5(\mathbf{x}^2) + (\mathbf{x} - \mathbf{x}^2)\nabla g_5(\mathbf{x}^2) = 35 + [x_1 - 6, x_2, x_3 + 83] \begin{bmatrix} 10 \\ -20 \\ -1 \end{bmatrix}$

$= 10x_1 - 20x_2 - x_3 - 108$, the auxiliary linear programming problem P^3 becomes

$$P^3: \text{Min } z^3 = x_3$$
$$\text{s.t. } x_1 + 2x_2 \leq 6$$
$$x_1 - x_2 \geq 1$$
$$-10x_1 - 10x_2 - x_3 \leq 23$$
$$10x_1 - 20x_2 - x_3 \leq 108$$
$$x_1, x_2 \geq 0$$
$$x_3 \in [-100, 100].$$

The unique optimal solution of P^3 is $x_1^3 = 4\frac{3}{5}, x_2^3 = \frac{7}{10}, x_3^3 = -76$ with $z^3 = -76$.

We find that $g_5(\mathbf{x}^3) = 8\frac{41}{50}$ and as $\nabla g_5(\mathbf{x}^3) = \begin{bmatrix} \frac{1}{5} \\ -14\frac{2}{5} \\ -1 \end{bmatrix} \neq \begin{bmatrix} 0 \\ 0 \\ 0 \end{bmatrix}$, we should proceed

to the fourth iteration. Having decided to carry out three complete iterations, we terminate the process at this point. Since the addition of the variable x_3 causes the feasible region of the auxiliary problems P^k to be three-dimensional, graphical illustrations of our example become awkward. However, the progression of the points $\mathbf{x}^k = \left(x_1^k, x_2^k\right)$ is shown in the two-dimensional (x_1, x_2)-space in Fig. 7.3.

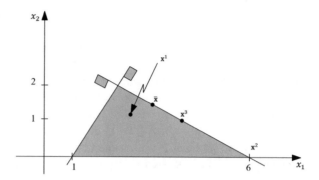

Fig. 7.3 Iterative solutions with the cutting plane method

Note that the performance of the cutting plane method in the above numerical example is much poorer than for the other methods considered so far. This is no coincidence; the method is preferable only when the nonlinear constraints are highly nonlinear, and when the polytope which is the outer approximation of the feasible set is "not much larger" than the feasible set itself, see Neumann (1975). For discussions of the convergence properties of the cutting plane algorithm, the reader is referred to Luenberger and Ye (2008), see also Bertsekas (2016) and Wolfe (1970).

7.2 Problems with Convex Objective

7.2.1 Problems with Differentiable Objective

The gradient projection method which we described in Sect. 6.2.2 belongs to the class of feasible direction methods. Another such method is the *reduced gradient method* due to Wolfe (1963). As for the gradient projection method it was first developed for linear constraints and later generalized to admit nonlinear constraints; thus the *generalized reduced gradient method, GRG* was created. It is credited to Abadie and Carpentier (1969) and is currently one of the most widely used constrained optimization techniques. In order to describe the *GRG* method, we will make frequent reference to the material on Karush-Kuhn-Tucker optimality conditions from Chap. 4.

The problem under consideration is assumed to be

$$P : \text{Min } z = f(\mathbf{t})$$
$$\text{s.t.} \quad \mathbf{g(t)} = \mathbf{0}$$
$$\mathbf{a} \leq \mathbf{t} \leq \mathbf{b},$$

where $f: \mathbb{R}^{m+n} \to \mathbb{R}$ and $\mathbf{g}: \mathbb{R}^{m+n} \to \mathbb{R}^m$ are convex, continuously differentiable functions and $\mathbf{t}$ is an $[m + n]$-dimensional vector. Any nonlinear programming problem can actually be written in this form by adding slack variables to all inequality constraints and by letting components of $\mathbf{a}$ and $\mathbf{b}$ be $-M \ll 0$ and

$M \gg 0$, respectively, whenever necessary. It should be noted that if the vector-valued function $\mathbf{g}(\mathbf{t})$ is nonlinear but convex, then the set $\{\mathbf{t}: \mathbf{g}(\mathbf{t}) = \mathbf{0}\}$ is never convex unless it consists of only one point. For example, $g(x_1, x_2) = x_1^2 + x_2 - 1$ is a convex function, yet the set $X = \{(x_1, x_2) : x_1^2 + x_2 - 1 = 0\}$ is not convex. This can be seen from the fact that the points $(1, 0)$ and $(-1, 0)$ both belong to X, whereas $(0, 0) = \frac{1}{2}(1, 0) + \frac{1}{2}(-1, 0)$ does not. Thus even if $\{x_1 : x_1^2 - 1 \le 0\}$ defines a convex set in $\mathbb{R}$, adding a nonnegative slack variable x_2 makes the set $\{(x_1, x_2) : x_1^2 - 1 + x_2 = 0, x_2 \ge 0\} \subseteq \mathbb{R}^2$ nonconvex.

In the above problem P we assume that $\mathbf{a} < \mathbf{b}$; otherwise one or more of the components of $\mathbf{t}$ will be fixed at the level of the corresponding lower and upper bound. A partition of $\mathbf{t}$ into two components $\mathbf{t} = \begin{bmatrix} \mathbf{x} \\ \mathbf{y} \end{bmatrix}$, where $\mathbf{x} \in \mathbb{R}^n$, and $\mathbf{y} \in \mathbb{R}^m$, will be made in a way and for reasons to become clear below. The bounds $\mathbf{a}$ and $\mathbf{b}$ are partitioned accordingly, so that $\mathbf{a} = \begin{bmatrix} \boldsymbol{\alpha} \\ \boldsymbol{\alpha}' \end{bmatrix}$ and $\mathbf{b} = \begin{bmatrix} \boldsymbol{\beta} \\ \boldsymbol{\beta}' \end{bmatrix}$ with $\boldsymbol{\alpha}, \boldsymbol{\beta} \in \mathbb{R}^n$ and $\boldsymbol{\alpha}'$, $\boldsymbol{\beta}' \in \mathbb{R}^m$. The problem can then be rewritten as

$$P': \text{Min } z = f(\mathbf{x}, \mathbf{y})$$
$$\text{s.t. } \mathbf{g}(\mathbf{x}, \mathbf{y}) = \mathbf{0}$$
$$\boldsymbol{\alpha} \le \mathbf{x} \le \boldsymbol{\beta}$$
$$\boldsymbol{\alpha}' \le \mathbf{y} \le \boldsymbol{\beta}',$$

where $f : \mathbb{R}^{m+n} \to \mathbb{R}$ and $\mathbf{g} : \mathbb{R}^{n+m} \to \mathbb{R}^m$. A nondegeneracy assumption will be needed, namely that every feasible $\tilde{\mathbf{t}}$ can be partitioned as $\tilde{\mathbf{t}} = \begin{bmatrix} \tilde{\mathbf{x}} \\ \tilde{\mathbf{y}} \end{bmatrix}$ with $\tilde{\mathbf{x}} \in \mathbb{R}^n$, $\tilde{\mathbf{y}} \in \mathbb{R}^m$, such that

(i) $\boldsymbol{\alpha}' < \tilde{\mathbf{y}} < \boldsymbol{\beta}'$

(ii) the $[m \times m]$-dimensional matrix $\left(\dfrac{\partial \mathbf{g}}{\partial \mathbf{y}} \right)_{\mathbf{t} = \tilde{\mathbf{t}}}$ with rows $\nabla_{\mathbf{y}}^T g_i(\mathbf{t})$ is nonsingular. If

singularity occurs, a small perturbation of $\tilde{\mathbf{t}}$ may remedy the situation.

In the *GRG* method a sequence of points is generated using an approach resembling that of the simplex methods, moving from one so-called basic solution to another. In particular, the vector $\tilde{\mathbf{t}}$ is first partitioned into (basic) dependent variables $\tilde{\mathbf{y}}$ and (nonbasic) independent variables $\tilde{\mathbf{x}}$. Because of the nondegeneracy assumption, the relationship $\mathbf{g}(\mathbf{x}, \mathbf{y}) = \mathbf{0}$ will define $\mathbf{y}$ as a function of $\mathbf{x}$, i.e., $\mathbf{y} = \mathbf{y}(\mathbf{x})$, and $\mathbf{y}$ is said to be an *implicit function* of $\mathbf{x}$ (see the implicit function theorem in for instance Nocedal and Wright (2006), Bertsekas (2016), or Cottle and Thapa (2017). As an example, $y^3 - x - 7 = 0$ defines $y = \sqrt[3]{x+7}$ as a function of x. With $\mathbf{y} = \mathbf{y}(\mathbf{x})$ we have $\mathbf{g}(\mathbf{x}, \mathbf{y}(\mathbf{x})) = \mathbf{0}$ identically and the objective function f can be regarded as a function $\hat{f}$ of only $\mathbf{x}$, i.e., $\hat{f}(\mathbf{x}) = f(\mathbf{x}, \mathbf{y}(\mathbf{x}))$.

In Fig. 7.4 the case is illustrated where $g(x, y) = 0$ is a plane and the curve $\{(x, y, z) : z = f(x, y), g(x, y) = 0\}$ is drawn in the plane $g(x, y) = 0$. Its projection in this plane would be $z = \hat{f}(x)$.

Fig. 7.4 3D-plot and projection in the *GRG* method

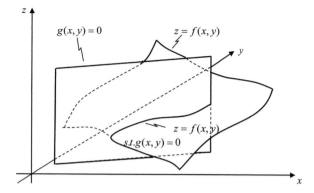

The gradient of $\hat{f}$, called the *reduced gradient,* is then computed at $\tilde{\mathbf{x}}$, i.e., $\nabla_{\mathbf{x}}\hat{f}\left(\tilde{\mathbf{x}}\right) = \left(\nabla_{\mathbf{x}}\hat{f}(\mathbf{x})\right)_{\mathbf{x}=\tilde{\mathbf{x}}}$, and $\tilde{\mathbf{x}}$ is updated by taking a step in the direction of the negative reduced gradient, neglecting components of this direction that correspond to infeasible changes. Finally, a corresponding updating of $\tilde{\mathbf{y}}$ to maintain feasibility is carried out. Details of the updating procedure are outlined below.

The Lagrangean function for problem P′ is

$$L(\mathbf{x}, \mathbf{y}, \mathbf{u}, \mathbf{v}^+, \mathbf{v}^-, \mathbf{w}^+, \mathbf{w}^-) = f(\mathbf{x}, \mathbf{y}) + \sum_{i=1}^{m} u_i g_i(\mathbf{x}, \mathbf{y}) + \sum_{j=1}^{n} v_j^+ \left(\alpha_j - x_j\right)$$

$$+ \sum_{j=1}^{n} v_j^- \left(x_j - \beta_j\right) + \sum_{i=1}^{m} w_i^+ \left(\alpha_i' - y_i\right)$$

$$+ \sum_{i=1}^{m} w_i^- \left(y_i - \beta_i'\right).$$

From the results of Chap. 4, the Karush-Kuhn-Tucker conditions for optimality at $\left(\tilde{\mathbf{x}}, \tilde{\mathbf{y}}\right)$ are

$$\nabla_{\mathbf{x}}^T f\left(\tilde{\mathbf{x}}, \tilde{\mathbf{y}}\right) + \sum_{i=1}^{m} u_i \nabla_{\mathbf{x}}^T g_i\left(\tilde{\mathbf{x}}, \tilde{\mathbf{y}}\right) = \mathbf{v}^+ - \mathbf{v}^- \qquad (7.1)$$

$$\nabla_{\mathbf{y}}^T f\left(\tilde{\mathbf{x}}, \tilde{\mathbf{y}}\right) + \sum_{i=1}^{m} u_i \nabla_{\mathbf{y}}^T g_i\left(\tilde{\mathbf{x}}, \tilde{\mathbf{y}}\right) = \mathbf{w}^+ - \mathbf{w}^- \qquad (7.2)$$

$$\mathbf{v}^+ \left(\boldsymbol{\alpha} - \tilde{\mathbf{x}}\right) = 0 \qquad (7.3)$$

$$\mathbf{v}^- \left(\tilde{\mathbf{x}} - \boldsymbol{\beta}\right) = 0 \qquad (7.4)$$

$$\mathbf{w}^+ \left(\boldsymbol{\alpha}' - \tilde{\mathbf{y}}\right) = 0 \qquad (7.5)$$

$$\mathbf{w}^- \left(\tilde{\mathbf{y}} - \boldsymbol{\beta}' \right) = 0 \tag{7.6}$$

$$\mathbf{u} \in \mathbb{R}^m; \, \mathbf{v}^+, \mathbf{v}^- \in \mathbb{R}^n_+; \, \mathbf{w}^+, \mathbf{w}^- \in \mathbb{R}^m_+. \tag{7.7}$$

Taking the first part of the nondegeneracy assumption into account, the conditions (7.5) and (7.6) imply that $\mathbf{w}^+ = \mathbf{0} = \mathbf{w}^-$. By defining $\mathbf{v} := \mathbf{v}^+ - \mathbf{v}^-$, the conditions (7.3) and (7.4) can be written as

$$v_j = \begin{cases} = 0, & \text{if } \alpha_j < \tilde{x}_j < \beta_j \\ \geq 0, & \text{if } \quad \tilde{x}_j = \alpha_j \\ \leq 0, & \text{if } \quad \tilde{x}_j = \beta_j \end{cases}. \tag{7.8}$$

Writing $\dfrac{\partial \mathbf{g}}{\partial \tilde{\mathbf{x}}}$ for $\left(\dfrac{\partial \mathbf{g}}{\partial \mathbf{x}} \right)_{\mathbf{x} = \tilde{\mathbf{x}}}$ and $\nabla_{\tilde{\mathbf{x}}} f$ for $\nabla_{\mathbf{x}} f \left(\tilde{\mathbf{x}}, \tilde{\mathbf{y}} \right)$, etc., the conditions (7.1) and (7.2) reduce to

$$\nabla_{\tilde{\mathbf{x}}}^T f + \mathbf{u} \, \frac{\partial \mathbf{g}}{\partial \tilde{\mathbf{x}}} = \mathbf{v} \tag{7.9}$$

$$\nabla_{\tilde{\mathbf{y}}}^T f + \mathbf{u} \, \frac{\partial \mathbf{g}}{\partial \tilde{\mathbf{y}}} = \mathbf{0}. \tag{7.10}$$

Given the nonsingularity of $\dfrac{\partial \mathbf{g}}{\partial \tilde{\mathbf{y}}}$, relation (7.10) can be solved in terms of $\mathbf{u}$, resulting in

$$\mathbf{u} = -\nabla_{\tilde{\mathbf{y}}}^T f \left(\frac{\partial \mathbf{g}}{\partial \tilde{\mathbf{y}}} \right)^{-1}.$$

Inserting this result into (7.9) yields the reduced gradient

$$\mathbf{v} = \nabla_{\tilde{\mathbf{x}}}^T f - \nabla_{\tilde{\mathbf{y}}}^T f \left(\frac{\partial \mathbf{g}}{\partial \tilde{\mathbf{y}}} \right)^{-1} \frac{\partial \mathbf{g}}{\partial \tilde{\mathbf{x}}}.$$

If $\mathbf{v}$ according to this equation satisfies the relations (7.8), then $\tilde{\mathbf{t}} = \begin{bmatrix} \tilde{\mathbf{x}} \\ \tilde{\mathbf{y}} \end{bmatrix}$ might represent a solution to the original problem. If not, a step from $\tilde{\mathbf{x}}$ would be taken during a *GRG* iteration in the direction $\mathbf{s} \in \mathbb{R}^n$ defined as

$$s_j := \begin{cases} 0, & \text{if } \tilde{x}_j = \alpha_j \text{ and } v_j > 0 \\ 0, & \text{if } \tilde{x}_j = \beta_j \text{ and } v_j < 0 \\ v_j, & \text{otherwise} \end{cases}.$$

One can show that $\mathbf{s} \neq \mathbf{0}$. As $\dfrac{\partial \mathbf{g}}{\partial \tilde{\mathbf{y}}}$ is non-singular by assumption, the equation $\mathbf{g}(\mathbf{x}, \mathbf{y}) = \mathbf{0}$ will define $\mathbf{y}$ as a continuous function of $\mathbf{x}$ in a vicinity of $\mathbf{x}$ by the implicit function theorem. This function will therefore define a continuous curve Γ by the equations $\mathbf{x} = \tilde{\mathbf{x}} - \lambda \mathbf{s}$ and $\mathbf{g}\left(\tilde{\mathbf{x}} - \lambda \mathbf{s}, \mathbf{y}(\lambda)\right) = \mathbf{0}, \lambda \in \mathbb{R}_+$. Clearly, $\mathbf{x}(0) = \tilde{\mathbf{x}}$ and $\mathbf{y}(0) = \tilde{\mathbf{y}}$. Moreover, note that the tangent to Γ at the point $\left(\tilde{\mathbf{x}}, \tilde{\mathbf{y}}\right)$ is simply they ray $\mathbf{x} = \tilde{\mathbf{x}} - \lambda \mathbf{s}$ and $\mathbf{y} = \tilde{\mathbf{y}} - \lambda \mathbf{r}, \lambda \in \mathbb{R}_+$, where $\mathbf{r} = \left(\dfrac{d\mathbf{y}}{d\lambda}\right)_{\lambda=0}$ is defined by the equation $\dfrac{\partial \mathbf{g}}{\partial \mathbf{x}} \mathbf{s} + \dfrac{\partial \mathbf{g}}{\partial \mathbf{y}} \mathbf{r} = \mathbf{0}$. Keeping in mind that from the requirement $\mathbf{a} \leq \mathbf{t} \leq \mathbf{b}$ there are bounds on the feasible values of λ, the minimization of the function f along the tangent above yields an optimal $\bar{\lambda}$ and we set $\tilde{\mathbf{x}}' := \tilde{\mathbf{x}} - \bar{\lambda} \mathbf{s}$, $\tilde{\mathbf{y}}' := \tilde{\mathbf{y}} - \bar{\lambda} \mathbf{r}$. However, $\tilde{\mathbf{t}}' = \begin{bmatrix} \tilde{\mathbf{x}}' \\ \tilde{\mathbf{y}}' \end{bmatrix}$ is in general not feasible, but from $\tilde{\mathbf{y}}'$ one can proceed to find $\tilde{\mathbf{y}}''$, such that $\tilde{\mathbf{t}}'' = \begin{bmatrix} \tilde{\mathbf{x}}' \\ \tilde{\mathbf{y}}'' \end{bmatrix}$ is feasible. This is accomplished by solving the equations $\mathbf{g}\left(\tilde{\mathbf{x}}', \mathbf{y}\right) = \mathbf{0}$ (e.g., with the Newton-Raphson method of Chap. 2), starting specifically with $\tilde{\mathbf{y}}'$ as a first approximation. The iterations of the Newton-Raphson method would proceed according to $\tilde{\mathbf{y}}^{\ell+1} = \tilde{\mathbf{y}}^{\ell} - \left(\dfrac{\partial \mathbf{g}}{\partial \tilde{\mathbf{y}}^{\ell}}\right)^{-1} \mathbf{g}(\tilde{\mathbf{x}}', \mathbf{y}^{\ell}), \ell = 1, 2, \ldots$ starting with $\tilde{\mathbf{y}}^1 = \mathbf{y}'$ and terminating with the result $\tilde{\mathbf{y}}''$. If $\tilde{\mathbf{y}}''$ is such that $\boldsymbol{\alpha}' < \tilde{\mathbf{y}}'' < \boldsymbol{\beta}'$, we can then again start with $\tilde{\mathbf{x}} := \tilde{\mathbf{x}}'$, $\tilde{\mathbf{y}} := \tilde{\mathbf{y}}''$, so that $\tilde{\mathbf{t}} := \tilde{\mathbf{t}}''$. If, on the other hand, for some i we have $\tilde{y}''_i = \alpha'_i$ or $\tilde{y}''_i = \beta'_i$, then this variable y_i will have to be replaced by a variable x_j in a so-called change of basis.

We are now ready to state the generalized reduced gradient method in a formalized procedure. Recall that the problem under consideration is

$$P': \text{Min } z = f(\mathbf{x}, \mathbf{y})$$
$$\text{s.t. } \mathbf{g}(\mathbf{x}, \mathbf{y}) = \mathbf{0}$$
$$\boldsymbol{\alpha} \leq \mathbf{x} \leq \boldsymbol{\beta}$$
$$\boldsymbol{\alpha}' \leq \mathbf{y} \leq \boldsymbol{\beta}'.$$

Note that in the algorithm described below the variables $\mathbf{x}^k, \mathbf{y}^k, \mathbf{x}^{k+1}, \tilde{\mathbf{y}}^{k+1}$ and $\mathbf{y}^{k+1}$ replace $\tilde{\mathbf{x}}, \tilde{\mathbf{y}}, \tilde{\mathbf{x}}', \tilde{\mathbf{y}}$ and $\tilde{\mathbf{y}}''$, respectively, from our previous discussion.

The Generalized Reduced Gradient Method (GRG)

Step 0: Disregarding the constraints, find a minimum for the unconstrained problem. Is this solution feasible?

If yes: Stop, an optimal solution has been found.
If no: Go to Step 1.

Step 1: Start with an initial feasible solution $\mathbf{t}^1 = \begin{bmatrix} \mathbf{x}^1 \\ \mathbf{y}^1 \end{bmatrix}$ satisfying the nondegeneracy assumption. Set $k := 1$ and go to Step 2.

Step 2: Compute the gradients $\nabla_{\mathbf{x}^k} f$, $\nabla_{\mathbf{y}^k} f$, and the Jacobian matrices $\dfrac{\partial \mathbf{g}}{\partial \mathbf{x}^k}, \dfrac{\partial \mathbf{g}}{\partial \mathbf{y}^k}$. Determine the reduced gradient $\mathbf{v} := \nabla_{\mathbf{x}^k}^T f - \nabla_{\mathbf{y}^k}^T f \left(\dfrac{\partial \mathbf{g}}{\partial \mathbf{y}^k}\right)^{-1} \dfrac{\partial \mathbf{g}}{\partial \mathbf{x}^k}$. Are the following relations satisfied?

$$v_j \begin{cases} = 0, \text{if } \alpha_j < x_j^k < \beta_j \\ \geq 0, \text{if } x_j^k = \alpha_j \quad \forall j = 1, \ldots, n. \\ \leq 0, \text{if } x_j^k = \beta_j \end{cases}$$

If yes: Stop, an optimal solution has been found.
If no: Go to Step 3.

Step 3: Define $\mathbf{s} \in \mathbb{R}^n$ by $s_j := 0$, if $x_j^k = \alpha_j$ and $v_j > 0$, or if $x_j^k = \beta_j$ and $v_j < 0$.

Otherwise, let $s_j := v_j$. Then determine $\mathbf{r} \in \mathbb{R}^m$ from $\mathbf{r} = -\left(\dfrac{\partial \mathbf{g}}{\partial \mathbf{y}^k}\right)^{-1} \left(\dfrac{\partial \mathbf{g}}{\partial \mathbf{x}^k}\right) \mathbf{s}$.

Step 4: Minimize with respect to $\lambda \in \mathbb{R}_+$ the function $f(\mathbf{x}^k - \lambda \mathbf{s}, \mathbf{y}^k - \lambda \mathbf{r})$ subject to the restrictions $\boldsymbol{\alpha} \leq \mathbf{x}^k - \lambda \mathbf{r} \leq \boldsymbol{\beta}$ and $\boldsymbol{\alpha}' \leq \mathbf{y}^k - \lambda \mathbf{s} \leq \boldsymbol{\beta}'$. Denote the optimal value of λ by $\bar{\lambda}$ and let $\mathbf{x}^{k+1} := \mathbf{x}^k - \bar{\lambda}\mathbf{r}$ and $\tilde{\mathbf{y}}^{k+1} := \mathbf{y}^k - \bar{\lambda}\mathbf{s}$.

Step 5: Is $\left(\mathbf{x}^{k+1}, \tilde{\mathbf{y}}^{k+1}\right)$ feasible?

If yes: Set $\mathbf{y}^{k+1} := \tilde{\mathbf{y}}^{k+1}$, $k := k+1$ and go to Step 7.
If no: Go to Step 6.

Step 6: Beginning with $\tilde{\mathbf{y}}^{k+1}$, find $\mathbf{y}$ such that $\mathbf{g}(\mathbf{x}^{k+1}, \mathbf{y}) = \mathbf{0}$, e.g., by means of the Newton-Raphson technique. Denote $\mathbf{y}$ by $\mathbf{y}^{k+1}$.

Step 7: Is $\boldsymbol{\alpha}' < \mathbf{y}^{k+1} < \boldsymbol{\beta}'$?

If yes: Set $k := k+1$ and go to Step 8.
If no: Go to Step 8.

Step 8: Replace all variables y_i for which $y_i^{k+1} = \alpha_i'$ or $y_i^{k+1} = \beta_i'$ by variables x_j, selecting the largest variable x_j which is still below its respective upper bound. Set $k := k+1$ and go to Step 2.

Some comments about the above algorithm are in order. In Step 2 special updating schemes may be used to calculate $\left(\dfrac{\partial \mathbf{g}}{\partial \mathbf{y}^k}\right)^{-1}$. Also, there are alternate ways of computing $\mathbf{s}$ in Step 3. Modifications in the algorithm are necessary if the Newton-Raphson technique in Step 6 finds a point $\mathbf{y}$ outside the feasible region, if it produces a poorer solution than in the previous iteration or if it fails to converge. For further details see Abadie and Carpentier (1969), Sun and Yuan (2006), or Luenberger and Ye (2008). The required initial feasible solution $\mathbf{t}^1 = \begin{bmatrix} \mathbf{x}^1 \\ \mathbf{y}^1 \end{bmatrix}$ can be obtained by using a phase one technique corresponding to that of the primal simplex method for linear programming. As an alternative, one can also proceed with the phase one procedure described at the beginning of Chap. 5.

Example We now solve the following quadratic programming problem by the *GRG* method:

$$P: \text{Min } z = 3x_1^2 - 2x_1x_2 + 2x_2^2 - 26x_1 - 8x_2$$
$$\text{s.t. } x_1 + 2x_2 \leq 6$$
$$x_1 - x_2 \leq 1$$
$$x_1, x_2 \geq 0.$$

This is the same problem we have used as a numerical example for previous methods. First the problem must be brought into the desired form

$$P': \text{Min } z = f(\mathbf{x}, \mathbf{y})$$
$$\text{s.t. } \mathbf{g}(\mathbf{x}, \mathbf{y}) = \mathbf{0}$$
$$\boldsymbol{\alpha} \leq \mathbf{x} \leq \boldsymbol{\beta}$$
$$\boldsymbol{\alpha}' \leq \mathbf{y} \leq \boldsymbol{\beta}'.$$

This is done be introducing nonnegative slack variables $\mathbf{y}$ and sufficiently large upper bounds $\boldsymbol{\beta}$ and $\boldsymbol{\beta}'$, say 100, resulting in

$$P': \text{Min } z = 3x_1^2 - 2x_1x_2 + 2x_2^2 - 26x_1 - 8x_2$$
$$\text{s.t. } x_1 + 2x_2 + y_3 - 6 = 0$$
$$-x_1 + x_2 + y_4 + 1 = 0$$
$$0 \leq x_i \leq 100, i = 1,2$$
$$0 \leq y_i \leq 100, i = 3,4.$$

Here we have $\mathbf{x} = [x_1, x_2]^T$ and $\mathbf{y} = [y_3, y_4]^T$. Obviously tighter upper bounds such as 6 and 3 could have been chosen for x_1 and x_2, respectively, as can be deduced from the first constraint. With this reformulation at hand, $\dfrac{\partial \mathbf{g}}{\partial \mathbf{y}} = \begin{bmatrix} 1 & 0 \\ 0 & 1 \end{bmatrix}$, so that, with $[x_1,$ $x_2, y_3, y_4]^T = [3, 1, 1, 1]^T$ and $\dfrac{\partial \mathbf{g}}{\partial \mathbf{y}} = \begin{bmatrix} 1 & 0 \\ 0 & 1 \end{bmatrix}$ being non-singular, the nondegeneracy assumption is satisfied. Furthermore, with $\nabla_{\mathbf{y}} f = \mathbf{0}$, we obtain $\mathbf{v}^T = \nabla_{\mathbf{x}} f = \begin{pmatrix} 6x_1 - 2x_2 - 26 \\ -2x_1 + 4x_2 - 8 \end{pmatrix}$ and $\left(\dfrac{\partial \mathbf{g}}{\partial \mathbf{y}} \right)^{-1} \dfrac{\partial \mathbf{g}}{\partial \mathbf{x}} = \begin{bmatrix} 1 & 0 \\ 0 & 1 \end{bmatrix} \begin{bmatrix} 1 & 2 \\ -1 & 1 \end{bmatrix} = \begin{bmatrix} 1 & 2 \\ -1 & 1 \end{bmatrix}$.
The procedure can now be initiated.

Since the minimum for the unconstrained problem is attained at $[\hat{x}_1, \hat{x}_2]^T = [6, 5]^T$, which is infeasible, we arbitrarily choose $\mathbf{t}^1 = [x_1^1, x_2^1, y_3^1, y_4^1]^T = [3, 1, 1, 1]^T$, have the same $\mathbf{x}$-coordinates as the starting point in previous methods, and proceed to Step 2. We find $\mathbf{v}^T = \begin{bmatrix} 6x_1^1 - 2x_2^1 - 26 \\ -2x_1^1 + 4x_2^1 - 8 \end{bmatrix}$ $= \begin{bmatrix} -10 \\ -10 \end{bmatrix}$ and notice that the optimality condition is not satisfied. With $\mathbf{s} = \mathbf{v}^T$ and $\mathbf{r} = -\begin{bmatrix} 1 & 2 \\ -1 & 1 \end{bmatrix} \begin{bmatrix} -10 \\ -10 \end{bmatrix} = \begin{bmatrix} 30 \\ 0 \end{bmatrix}$ the function $f(\mathbf{x}^1 - \lambda \mathbf{s}, \mathbf{y}^1 - \lambda \mathbf{r}) = f(3 + 10\lambda,$ $1 + 10\lambda, 1 - 30\lambda, 1)$ is minimized for $\lambda \geq 0$ and subject to the upper and lower bound constraints on the variables $\mathbf{x}$ and $\mathbf{y}$. However, as f does not depend on $\mathbf{y}$, the minimization takes place with respect to the nonbasic variables $\mathbf{x}$ only. From

$$\frac{df}{d\lambda} = 10(6x_1 - 2x_2 - 26) + 10(-2x_1 + 4x_2 - 8) = 40x_1 + 20x_2 - 340 = 600\lambda - 200,$$

we find that $\bar{\lambda} = \min\left\{\frac{1}{3}, \frac{1}{30}\right\} = \frac{1}{30}$. Hence, $x_1^2 = 3 + 10\bar{\lambda} = \frac{10}{3}, x_2^2 = 1 + 10\bar{\lambda} = \frac{4}{3}$, $\tilde{y}_3^2 = 1 - 30\bar{\lambda} = 0$, and $\tilde{y}_4^2 = 1$. In Step 5, it turns out that $\left[x_1^2, x_2^2, \tilde{y}_3^2, \tilde{y}_4^2\right]^T$ is feasible so that $\mathbf{y}^2 = \tilde{\mathbf{y}}^2$; therefore $\left[x_1^2, x_2^2, y_3^2, y_4^2\right]^T := \left[\frac{10}{3}, \frac{4}{3}, 0, 1\right]^T$. Since y_3^2 is at its lower bound, y_3 must become a nonbasic variable denoted by x_3, whereas one of x_1 and x_2 must become a basic variable denoted by y_1 or y_2, respectively. As $x_1^2 = \frac{10}{3}$ is the largest of the two nonbasic variables, it is selected to enter the basis, so that $\left[y_1^2, \ x_2^2, \ x_3^2, \ y_4^2\right]^T := \left[\frac{10}{3}, \frac{4}{3}, 0, 1\right]^T$ and we proceed to the next iteration.

Now $\nabla_{\mathbf{x}} f = \begin{bmatrix} -2y_1 + 4x_2 - 8 \\ 0 \end{bmatrix}$, $\nabla_{\mathbf{y}} f = \begin{bmatrix} 6y_1 - 2x_2 - 26 \\ 0 \end{bmatrix}$, $\frac{\partial \mathbf{g}}{\partial \mathbf{x}} = \begin{bmatrix} 2 & 1 \\ 1 & 0 \end{bmatrix}$,

and $\frac{\partial \mathbf{g}}{\partial \mathbf{y}} = \begin{bmatrix} 1 & 0 \\ -1 & 1 \end{bmatrix}$, so that $\left(\frac{\partial \mathbf{g}}{\partial \mathbf{y}}\right)^{-1} \frac{\partial \mathbf{g}}{\partial \mathbf{x}} = \begin{bmatrix} 1 & 0 \\ 1 & 1 \end{bmatrix} \begin{bmatrix} 2 & 1 \\ 1 & 0 \end{bmatrix} = \begin{bmatrix} 2 & 1 \\ 3 & 1 \end{bmatrix}$ and $\mathbf{v}$

$= [-2y_1 + 4x_2 - 8, 0] - [6y_1 - 2x_2 - 26, 0] \begin{bmatrix} 2 & 1 \\ 3 & 1 \end{bmatrix} = [-14y_1 + 8x_2 + 44,$

$-6y_1 + 2x_2 + 26]$. Inserting $y_1 = y_1^2 = \frac{10}{3}$ and $x_2 = x_2^2 = \frac{4}{3}$ into this expression, we obtain $\mathbf{v} = [v_2, \ v_3] = \left[8, \frac{26}{3}\right]$. The optimality condition is not yet satisfied, in particular $v_2 = 8 \neq 0$ although $x_2^2 = \frac{4}{3}$ is at neither of its bounds. Then $\mathbf{s} = \begin{bmatrix} s_2 \\ s_3 \end{bmatrix}$ is determined as follows. We have $s_2 = v_2 = 8$, and $s_3 = 0$, since $x_3^2 = 0$ and $v_3 = \frac{26}{3} > 0$. Furthermore, $\mathbf{r} = \begin{bmatrix} r_1 \\ r_4 \end{bmatrix} = -\left(\frac{\partial \mathbf{g}}{\partial \mathbf{y}^1}\right)^{-1} \left(\frac{\partial \mathbf{g}}{\partial \mathbf{x}^1}\right) \mathbf{s} = -\begin{bmatrix} 2 & 1 \\ 3 & 1 \end{bmatrix} \begin{bmatrix} 8 \\ 0 \end{bmatrix} = \begin{bmatrix} -16 \\ -24 \end{bmatrix}$. The function $f(\mathbf{x}^2 - \lambda \mathbf{s}, \mathbf{y}^2 - \lambda \mathbf{r}) = f\left(y_1^2 - \lambda r_1, x_2^2 - \lambda s_2, x_3^2 - \lambda s_3, y_4^2 - \lambda r_4\right) = f\left(\frac{10}{3} + 16\lambda, \frac{4}{3} - 8\lambda, 0, 1 + 24\lambda\right)$ will now be minimized with respect to $\lambda \geq 0$ subject to the upper and lower bound constraints on the variables. These upper and lower bounds provide $0 \leq \lambda \leq \frac{1}{6}$. The calculation of $\frac{df}{d\lambda} = 16(6y_1 - 2$ $x_2 - 26) + (-8)(-2y_1 + 4x_2 - 8) = 16(7y_1 - 4x_2 - 22) = 64(36\lambda - 1)$ yields $\bar{\lambda} = \frac{1}{36}$ and therefore $\tilde{y}_1^3 = \frac{10}{3} + 16\bar{\lambda} = \frac{34}{9}, x_2^3 = \frac{4}{3} - 8\bar{\lambda} = \frac{10}{9}, x_3^3 = 0$ and $\tilde{y}_4^3 = 1 + 24\bar{\lambda} = \frac{5}{3}$.

Now $\left[\tilde{y}_1^3, x_2^3, x_3^3, \tilde{y}_4^3\right]^T$ is feasible, so that $\mathbf{y}^3 = \tilde{\mathbf{y}}^3$ and $\left[y_1^3, x_2^3, x_3^3, y_4^3\right]^T = \left[\frac{34}{9}, \frac{10}{9}, 0, \frac{5}{3}\right]^T$. Since $\mathbf{y}^2$ is strictly within bounds, we go to the third iteration. Now $\mathbf{r} = [r_2, r_3] = \left[-14y_1^3 + 8x_2^3 + 44, -6y_1^3 + 2x_2^3 + 26\right] = \left[0, \frac{50}{9}\right]$. At this point the optimality condition is satisfied since $r_2 = 0$ and x_2^3 is strictly within bounds, $r_3 = \frac{50}{9} \geq 0$ and x_3^3 is at lower bound; we conclude that an optimal solution has been reached. The algorithm terminates with the solution $\left[y_1^3, x_2^3, x_3^3, y_4^3\right]^T = \left[\frac{34}{9}, \frac{10}{9}, 0, \frac{5}{3}\right]^T$. In terms of the original problem, $\bar{x}_1 = 3\frac{7}{9} \approx 3.78$ and $\bar{x}_2 = 1\frac{1}{9} \approx 1.11$, which is the same as the result obtained previously.

We note that the points obtained during the iterations of the *GRG* method correspond exactly to those obtained by the gradient projection method; Fig. 6.8 in Sect. 6.2 will therefore provide a graphical representation of the points generated by the *GRG* algorithm as well. Since the problem had only linear constrains, the *GRG* algorithm actually took the form of the Wolfe reduced gradient method in this particular example. One can show that the gradient projection method and the reduced gradient method do not necessarily produce the same sequence of points. This could be seen in the above example by starting at a corner point instead of at an interior point; see also the numerical example in Bazaraa et al. (2013).

7.2.2 Problems with Nondifferentiable Objective

In many optimization models arising from real-world situations, the objective function is not necessarily smooth. For the purpose of this book, we use the term "smooth" synonymously with "differentiable," following a convention accepted by many authors, e.g., Nocedal and Wright (2006) or Sun and Yuan (2006). Another convention, adopted by, e.g., Cottle and Thapa (2017), equates "smooth" with "twice continuously differentiable." In our notation, "nonsmooth" is then synonymous with "nondifferentiable." Cases with nonsmooth objective function include instances, in which a process or a machine switches between different modes of operation. Other examples that involve nondifferentiable functions are minimax and multi-stage problems.

The study of algorithms for minimizing convex but not necessarily everywhere differentiable (smooth) functions originated with Shor (1964) and Polyak (1967, 1969), but it was not until the mid-1970s that their results became widely known outside the USSR. Since then, great interest has been attracted by this field of optimization, witnessed by an increasing number of publications written on the subject. Useful references are Bertsekas (2015, 2016) and the references mentioned above.

Before considering algorithms specifically designed for nondifferentiable functions, it should be mentioned that a pragmatic way to attack a nondifferentiable optimization problem is to simply disregard the lack of differentiability. If a method designed for smooth functions does not require analytical calculations of derivatives, such as the *SUMT* method, which we covered in Chap. 5, this can easily be done. Although differentiability was invoked to prove the convergence properties of the *SUMT* method, the algorithm itself does not require differentiability. In other cases, it may be that subgradients are readily available and could replace any gradients required in the algorithm. The cutting plane method of Kelley is an example of this. Wolfe (1970) has modified the cutting plane method to use subgradients instead of gradients. Computational evidence seems to indicate that algorithms for smooth objective functions are quite efficient even for nonsmooth functions, although convergence may then be difficult or impossible to prove.

Assume for now that all functions involved are continuous are convex but not necessarily everywhere differentiable.

Definition 7.1 A function $f: M \to \mathbb{R}$ where $M \subset \mathbb{R}^n$ is said to be *subdifferentiable* at the point $\tilde{\mathbf{x}} \in M$ with *subgradient* $\mathbf{y} \in \mathbb{R}^n$ of f at $\tilde{\mathbf{x}}$, if

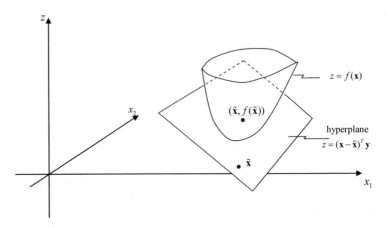

Fig. 7.5 Tangent hyperplane to a convex function

$f(\mathbf{x}) \ge f(\tilde{\mathbf{x}}) + (\mathbf{x} - \tilde{\mathbf{x}})^T \mathbf{y} \ \forall \ \mathbf{x} \in M.$

Geometrically speaking $z = (\mathbf{x} - \tilde{\mathbf{x}})^T \mathbf{y}$ is the hyperplane, tangent to the function $z = f(\mathbf{x})$ at the point $(\tilde{\mathbf{x}}, f(\tilde{\mathbf{x}}))$ as shown in Fig. 7.5.

A convex function which only takes finite values can be shown to be subdifferentiable on the interior of the set where it is defined; see Bertsekas (2015) and Sun and Yuan (2006). Any subgradient of f at $\tilde{\mathbf{x}}$ will be denoted by $\nabla f(\tilde{\mathbf{x}})$; a full treatment of subdifferentiability and subgradients can be found in the aforementioned references. Polyak (1967, 1969) has outlined a general theory for subgradient minimization algorithms based on previous work by Shor (1964) and has given results on their convergence properties. The Shor-Polyak subgradient algorithm will now be described after first introducing the notion of projection of a point onto a set. For the sake of simplicity, projection onto compact convex (i.e., closed and bounded) sets is defined; the extension to the case with closed convex, but not necessarily bounded, sets is straightforward.

Definition 7.2 Given a point $\mathbf{x} \in \mathbb{R}^n$ and a compact convex set $M \subset \mathbb{R}^n$, the *projection* of $\mathbf{x}$ onto M, denoted by $P_M(\mathbf{x})$, is defined as a solution to the equation

$$\mathbf{x} - P_M(\mathbf{x}) = \inf_{\mathbf{y} \in M} \|\mathbf{x} - \mathbf{y}\|.$$

The concept of projection is illustrated in Fig. 7.6.

Since the distance from $\mathbf{y}$ to $\mathbf{x}$ is a continuous function of $\mathbf{y}$ and the set M is compact, the infimum is attained at a point in M. If $\mathbf{x} \in \mathbb{R}^n$, obviously $P_M(\mathbf{x}) = \mathbf{x}$. Having proved the existence of $P_M(\mathbf{x})$, we will now consider its uniqueness.

Theorem 7.3 For any $\tilde{\mathbf{x}} \in \mathbb{R}^n$ and for any compact convex set $M \subset \mathbb{R}^n$, the projection $P_M(\tilde{\mathbf{x}})$ of $\tilde{\mathbf{x}}$ onto M is unique.

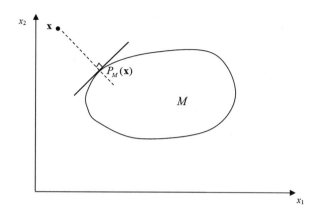

Fig. 7.6 Projection of a point onto a convex set

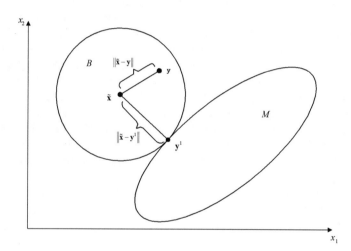

Fig. 7.7 Illustration to the proof of uniqueness of the projection point

Proof Having seen that the case in which $\tilde{\mathbf{x}} \in M$ is trivial, we now assume that $\tilde{\mathbf{x}} \notin M$ and will prove uniqueness by contradiction. Suppose therefore that $\left\| \tilde{\mathbf{x}} - \mathbf{y}^1 \right\| = \inf\limits_{\mathbf{y} \in M}$ $\left\| \tilde{\mathbf{x}} - \mathbf{y} \right\| = \left\| \tilde{\mathbf{x}} - \mathbf{y}^2 \right\|$ with $\mathbf{y}^1 \neq \mathbf{y}^2$ and $\mathbf{y}^1, \mathbf{y}^2 \in M$. Then define $B := \left\{ \mathbf{y} \in \mathbb{R}^n : \left\| \tilde{\mathbf{x}} - \mathbf{y} \right\| \leq \left\| \tilde{\mathbf{x}} - \mathbf{y}^1 \right\| \right\}$, which is a closed hypersphere, i.e., a ball with center $\tilde{\mathbf{x}}$ and radius $\left\| \tilde{\mathbf{x}} - \mathbf{y}^1 \right\|$, see Fig. 7.7. By definition, $\mathbf{y}^1, \mathbf{y}^2 \in B \cap M$ and since both B and M are convex, it follows that $\tilde{\mathbf{y}} = \frac{1}{2}\mathbf{y}^1 + \frac{1}{2}\mathbf{y}^2 \in B \cap M$. Furthermore, $\tilde{\mathbf{y}}$ is in the interior of B, since B is strictly convex. Thus $\left\| \tilde{\mathbf{x}} - \tilde{\mathbf{y}} \right\| < \left\| \tilde{\mathbf{x}} - \mathbf{y}^1 \right\|$ $= \inf\limits_{\mathbf{y} \in M} \left\| \tilde{\mathbf{x}} - \mathbf{y} \right\|$ which is a contradiction and consequently $P_M(\tilde{\mathbf{x}})$ must be unique. $\square$

Consider now the problem

$$P: \text{Min } z = f(\mathbf{x})$$
$$\text{s.t.} \mathbf{g}(\mathbf{x}) \leq \mathbf{0}$$
$$\mathbf{x} \in Q_2,$$

where $f: \mathbb{R}^n \to \mathbb{R}$ and $\mathbf{g}: \mathbb{R}^n \to \mathbb{R}^m$ are continuous convex functions and $Q_2 \in \mathbb{R}^n$ is a given closed convex set. Defining $Q_1 = \{\mathbf{x} : \mathbf{g}(\mathbf{x}) \leq \mathbf{0}\}$, problem P can be written as

$$P': \underset{\mathbf{x} \in Q_1 \cap Q_2}{\text{Min}} \ z = f(\mathbf{x}).$$

In the following, we need the nondegeneracy assumption that the set $\{\mathbf{x} : \mathbf{g}(\mathbf{x}) < \mathbf{0}\} \cap Q_2$ is nonempty. For the slightly weaker condition that the intersection of Q_2 with the interior of Q_1 should be nonempty, see Sandblom (1980). Note that with $Q_2 = \mathbb{R}^n$, the above problem description turns into a standard formulation. We will now describe an iteration of a very general subgradient algorithm first considered by Shor (1964) and Polyak (1967). Assume that a point $\mathbf{x}^k \in Q_2$, but not necessarily in Q_1, has been generated. If $\mathbf{x}^k$ belongs to Q_1, then it is feasible for P and a step along a negative subgradient of f at $\mathbf{x}^k$ is taken. If $\mathbf{x}^k$ does not belong to Q_1, a recovery step is taken along a negative subgradient of the "most violated" constraint at $\mathbf{x}^k$.

Using Definition 7.2, the resulting point is then projected onto the set Q_2 and we go to the next iteration. This procedure is illustrated in Fig. 7.8, where $\mathbf{x}^k$ is not feasible, so that some constraint $g_i(\mathbf{x}) \leq 0$ is violated. In Fig. 7.8, the set Q_2 is arbitrarily taken to be a simple axis-parallel polytope, making projection an easy operation.

The above algorithm can now be summarized as follows. Recall that the problem under consideration is

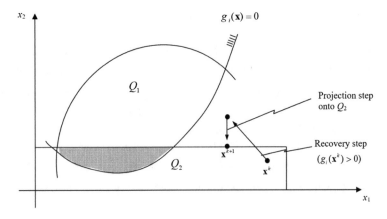

Fig. 7.8 Projection and recovery steps

$$P: \text{Min } z = f(\mathbf{x})$$
$$\text{s.t. } \mathbf{x} \in Q_1 = \{\mathbf{x} : \mathbf{g}(\mathbf{x}) \leq \mathbf{0}\}$$
$$\mathbf{x} \in Q_2,$$

where $f : \mathbb{R}^n \to \mathbb{R}$ and $\mathbf{g} : \mathbb{R}^n \to \mathbb{R}^m$ are convex functions and $Q_2 \subset \mathbb{R}^n$ is a closed convex set. Furthermore we assume that $\{\mathbf{x} : \mathbf{g}(\mathbf{x}) < \mathbf{0}\} \cap Q_2 \neq \varnothing$.

The Shor-Polyak Subgradient Algorithm

Step 0: Disregarding the constraints, find a minimal point for the unconstrained problem. Is this point feasible?

If yes: Stop, an optimal solution for P has been found.
If no: Go to Step 1.

Step 1: Select any $\mathbf{x}^0 \in Q_2$ and set $k := 0$.
Step 2: Is $\mathbf{x}^k \in Q_1$?

If yes: Go to Step 3.
If no: Go to Step 4.

Step 3: Select some subgradient $\mathbf{s}^k$ of f at $\mathbf{x}^k$. Is $\mathbf{s}^k = \mathbf{0}$?

If yes: Stop, $\mathbf{x}^k$ is an optimal solution for P.
If no: Go to Step 5.

Step 4: Determine i' such that $g_{i'}(\mathbf{x}^k) = \max\limits_{1 \leq i \leq m} \{g_i(\mathbf{x}^k)\}$ and select some subgradient $\mathbf{s}^k \in \nabla g_{i'}(\mathbf{x}^k)$.

Step 5: Compute $\lambda_k := \dfrac{1}{k+1} \dfrac{1}{\|\mathbf{s}^k\|}$ and the projection of $\mathbf{x}^k - \lambda_k \mathbf{s}^k$ onto the set Q_2. This new point is denoted by $\mathbf{x}^{k+1}$ and it is the solution of $\|\mathbf{x}^k - \lambda_k \mathbf{s}^k - \mathbf{x}^{k+1}\| = \inf\limits_{\mathbf{x} \in Q_2} \|\mathbf{x}^k - \lambda_k \mathbf{s}^k - \mathbf{x}\|$. Set $k := k+1$ and go to Step 2.

In Step 0 of this algorithm, the choice of a suitable unconstrained minimization technique will obviously depend on whether the objective function $f(\mathbf{x})$ is differentiable or not. In Steps 3 and 4 the issue of how to select the subgradients has not been addressed here, this question has been dealt with in a number of studies; see, e.g., Lemarechal (1975), Wolfe (1975), Goffin (1977), Mifflin (1982), Kiwiel (1983) and Fukushima (1984) as well as Bertsekas (2015) and Sun and Yuan (2006). In Step 3, one might also stop if $\|\mathbf{s}^k\| \leq \varepsilon$ with some small predetermined $\varepsilon > 0$. The choice of step length in Step 5 has been made for its simplicity. Polyak (1967) showed (see also Sandblom 1980) that in general the sequence of function values $f(\mathbf{x}^k)$ contains a subsequence converging to the optimal function value $f(\bar{\mathbf{x}})$ provided that $\lim\limits_{k \to \infty} \|\lambda_k \mathbf{s}^k\| = 0$ and that $\sum\limits_{k=0}^{\infty} \|\lambda_k \mathbf{s}^k\|$ is divergent. In other words, the step lengths should be neither "too long" nor "too short". Alternatively, if the optimal objective function value $f(\bar{\mathbf{x}})$ is known, the step length may be related to the difference

$f(\bar{\mathbf{x}}) - f(\mathbf{x}^k)$ or to $\tilde{z} - f(\mathbf{x}^k)$ with given $\tilde{z} > f(\bar{\mathbf{x}})$, see Polyak (1969) who provides a detailed discussion of the convergence properties of the method. Finding $P_{Q_2}(\mathbf{x}^k - \lambda_k \mathbf{s}^k)$ in Step 5 is equivalent to solving the problem of minimizing, for all possible $\mathbf{x} \in Q_2$, the expression $\|\mathbf{x}^k - \lambda_k \mathbf{s}^k - \mathbf{x}\|$, according to our Definition 7.2 of projection. If the variables have upper and lower bounds, it will be advantageous to associate Q_2 with these bounds; the projection will then consist of simple move (s) parallel to the coordinate axes.

Example The above subgradient algorithm will be used to solve our usual problem

$$P: \text{Min } z = 3x_1^2 - 2x_1 x_2 + 2x_2^2 - 26x_1 - 8x_2$$
$$\text{s.t. } x_1 + x_2 \leq 6$$
$$x_1 - x_2 \geq 1$$
$$x_1, x_2 \geq 0$$

by carrying out four iterations of the method.

We let Q_2 be the nonnegative orthant $\mathbb{R}_+^2$ and the subgradients in Step 3 are simply taken to be the gradients of f. Furthermore, as feasibility is not required to start the method, we select $\mathbf{x}^0 = [x_1^0, x_2^0]^T = [0,0]^T$. The unconstrained minimal point $[\hat{x}_1, \hat{x}_2]^T = [6,5]^T$ is infeasible. Start with the arbitrarily chosen point $[x_1^0, x_2^0]^T = [0,0]^T \in Q_2$; set $k := 0$ and proceed to Step 2.

Since $\mathbf{x}^0$ violates the constraint $g_2(\mathbf{x}) = -x_1 + x_2 + 1 \leq 0$, it follows that $\mathbf{x}^0 \notin Q_1$ and we go to Step 4. With $\mathbf{s}^0 = \nabla g_2(\mathbf{x}^0) = \begin{bmatrix} -1 \\ 1 \end{bmatrix}$, we compute

$$\lambda_0 = \frac{1}{0+1} \frac{1}{\sqrt{(-1)^2 + 1^2}} = \frac{1}{\sqrt{2}} \quad \text{and} \quad \mathbf{x}^1 = P_{Q_2}(\mathbf{x}^0 - \lambda_0 \mathbf{s}^0) = P_{Q_2}\left(\begin{bmatrix} 0 \\ 0 \end{bmatrix} - \frac{1}{\sqrt{2}} \begin{bmatrix} -1 \\ 1 \end{bmatrix} \right)$$

$$= P_{Q_2}\left(\begin{bmatrix} 1/\sqrt{2} \\ -1/\sqrt{2} \end{bmatrix} \right) = \begin{bmatrix} 1/\sqrt{2} \\ 0 \end{bmatrix} \approx \begin{bmatrix} 0.707 \\ 0 \end{bmatrix}. \text{ Set } k := 1 \text{ and go to Step 2 of the next iteration.}$$

Since $\mathbf{x}^1 \notin Q_1$, we go to Step 4. With $\mathbf{s}^1 = \nabla g_2(\mathbf{x}^1) = \begin{bmatrix} -1 \\ 1 \end{bmatrix}$ we obtain

$$\lambda^1 = \frac{1}{1+1} \frac{1}{\sqrt{2}} = \frac{1}{2\sqrt{2}} \quad \text{and} \quad \mathbf{x}^2 = P_{Q_2}(\mathbf{x}^1 - \lambda_1 \mathbf{s}^1) = P_{Q_2}\left(\begin{bmatrix} 1/\sqrt{2} \\ 0 \end{bmatrix} - \frac{1}{2\sqrt{2}} \begin{bmatrix} -1 \\ 1 \end{bmatrix} \right) =$$

$$P_{Q_2}\left(\begin{bmatrix} \frac{3}{2\sqrt{2}} \\ -\frac{1}{2\sqrt{2}} \end{bmatrix} \right) = \begin{bmatrix} \frac{3}{2\sqrt{2}} \\ 0 \end{bmatrix} \approx \begin{bmatrix} 1.061 \\ 0 \end{bmatrix}. \text{ Set } k := 2 \text{ and go to Step 2 of the third iteration,}$$

and since $\mathbf{x}^2 \in Q_1$, we move on to Step 3.

$$\text{With} \quad \mathbf{s}^2 = \nabla f(\mathbf{x}^2) = \begin{bmatrix} \dfrac{9}{\sqrt{2}} - 26 \\ -\dfrac{3}{\sqrt{2}} - 8 \end{bmatrix} \approx \begin{bmatrix} -19.636 \\ -10.121 \end{bmatrix} \neq \begin{bmatrix} 0 \\ 0 \end{bmatrix} \quad \text{we calculate}$$

$$\lambda^2 \approx \frac{1}{2+1} \frac{1}{\sqrt{(-19.636)^2 + (-10.121)^2}} \approx 0.01509, \quad \text{and} \quad \mathbf{x}^3 =$$

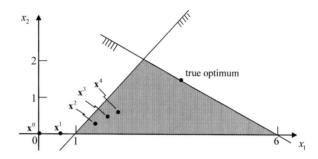

Fig. 7.9 Iterative solutions with the Shor-Polyak subgradient algorithm

$$P_{Q_2}\left(\begin{bmatrix}1.061\\0\end{bmatrix}-0.01509\begin{bmatrix}-19.636\\-10.121\end{bmatrix}\right)\approx P_{Q_2}\left(\begin{bmatrix}1.357\\0.1527\end{bmatrix}\right)=\begin{bmatrix}1.357\\0.1527\end{bmatrix}.\text{ Now we set}$$

$k := 3$ and proceed to Step 2 of the fourth iteration.

As $\mathbf{x}^3 \in Q_1$, we go to Step 3. With $\mathbf{s}^3 = \nabla f(\mathbf{x}^3) \approx \begin{bmatrix}-18.164\\-10.103\end{bmatrix} \neq \begin{bmatrix}0\\0\end{bmatrix}$, we

obtain $\qquad \lambda^3 \approx \dfrac{1}{3+1}\dfrac{1}{\sqrt{(-18.164)^2+(-10.103)^2}} \approx 0.01203 \qquad$ and consequently

$$\mathbf{x}^4 = P_{Q_2}(\mathbf{x}^3 - \lambda_3 \mathbf{s}^3) \approx P_{Q_2}\left(\begin{bmatrix}1.357\\0.1527\end{bmatrix}-0.01203\begin{bmatrix}-18.164\\-10.103\end{bmatrix}\right)\approx P_{Q_2}\left(\begin{bmatrix}1.575\\0.2742\end{bmatrix}\right)$$

$= \begin{bmatrix}1.575\\0.2742\end{bmatrix}$. Having carried out four iterations, we terminate the procedure. The solution

points generated above are shown in Fig. 7.9.

As we can see in Fig. 7.9, the restricted step length makes the convergence quite slow in comparison with the other methods we have used to solve the same problem. It should be noted, though, that this method was started from the origin rather than from $[3, 1]^T$ as were most of the other methods. The points $\mathbf{x}^k$ will continue to move inside the feasible region until they cross the hyperplane $x_1 + 2x_2 = 6$, when a recovery step back into the feasible region needs to be taken.

Nondifferentiable problems arise in many different contexts. For example, consider the problem of finding a solution $\mathbf{x}$ to the system of linear inequalities $\mathbf{Ax} \leq \mathbf{b}$, which can be written as $\mathbf{a}_i\mathbf{.x} \leq b_i, i = 1, \ldots, m$. This problem can be solved by minimizing the function $f(\mathbf{x}) = \max_{1 \leq i \leq m}\{\mathbf{a}_i\mathbf{.x} - b_i\}$, a so-called *minimax problem*, see, e.g., Bertsekas (2016) or Eiselt and Sandblom (2007). Note that the function $f(\mathbf{x})$ has a minimal point whether or not the inequality system $\mathbf{Ax} \leq \mathbf{b}$ is consistent.

A one-dimensional example of the function $z = f(\mathbf{x})$ is shown in Fig. 7.10, where the bold line is an upper envelope function of the constraint lines (hyperplanes). One can easily prove that $f(\mathbf{x})$ is a convex function which in general is not differentiable. If $\tilde{\mathbf{x}}$ can be found such that $f(\tilde{\mathbf{x}}) \leq 0$, then $\tilde{\mathbf{x}}$ solves the linear inequality system $\mathbf{Ax} \leq \mathbf{b}$, but if no such $\tilde{\mathbf{x}}$ can be found, then the system $\mathbf{Ax} \leq \mathbf{b}$ has no solution. This follows because $0 \geq f(\tilde{\mathbf{x}}) \geq \mathbf{a}_i\mathbf{.}\tilde{\mathbf{x}} - b_i, i = 1, \ldots, m$, and consequently $\mathbf{0} \geq \mathbf{A}\tilde{\mathbf{x}} - \mathbf{b}$. Conversely, if $f(\tilde{\mathbf{x}}) > 0 \,\forall\, \tilde{\mathbf{x}} \in \mathbb{R}^n$, then $\mathbf{a}_i\mathbf{.x} - b_i$ for at least one $i \in [1, m]$, and consequently no $\tilde{\mathbf{x}}$ would exist such that $\mathbf{A}\tilde{\mathbf{x}} \leq \mathbf{b}$. Thus, instead of using a phase one simplex procedure, one might try to minimize $f(\mathbf{x})$ using a subgradient procedure. This approach turns out to be closely related to the *relaxation method* of Agmon

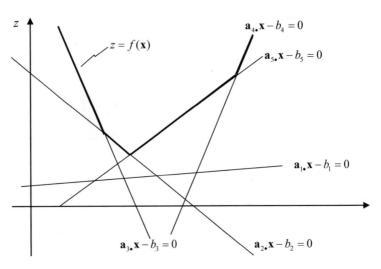

Fig. 7.10 Nondifferentiable minimax problem

(1954) and Motzkin and Schönberg (1954); see Eiselt et al. (1987) and also Held et al. (1974).

7.3 Problems with Concave Objective

In concave programming, the general setting involves minimizing a concave function over a convex set. As usual, we write the problem as

$$P : \text{Min } z = f(\mathbf{x})$$
$$\text{s.t.} g_i(\mathbf{x}) \leq 0, i = 1, \ldots, m$$
$$\mathbf{x} \in \mathbb{R}^n.$$

Definition 7.4 If the function f is concave and the functions $g_i, i = 1, \ldots, m$ are convex, we say that problem P above is a *concave programming problem*.

The monopolistic model in Sect. 6.3 with positively sloping demand functions provides an example of a concave quadratic programming problem. Other examples are based on economies of scale. Often, production costs increase sublinearly, resulting in a concave cost objective function. Fixed charge problems also tend to fall into this category. Consider the rental of a vehicle without free mileage allowance. The fixed charge are the daily costs of the rental, and once the vehicle is rented, each mile costs a certain amount, so that the cost function for all positive mileages is linear. Given a free mileage allowance, though, destroys the concavity of the

Fig. 7.11 Cost functions
for different technologies

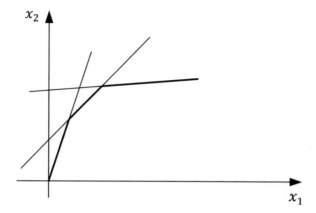

function as the costs jump from zero to the rental charge, stay constant until the mileage allowance is reached, and then increase linearly.

A piecewise linear version of economies of scale exist in the presence of price breaks and quantity discounts. As a simple example, consider the possibility of manufacturing a product on different machines, some with higher and some with lower technology. Typically, lower tech machines have lower fixed costs and higher variable costs, making them useful for smaller production quantities. The opposite holds for hi-tech machines, which usual have higher fixed cost and lower variable costs, making them a good choice for large production quantities. A typical function with three such machines is shown in Fig. 7.11.

In all of these cases, $f(\mathbf{x})$ is a concave function (in the fixed-cost case discontinuous, but still concave), so that minimizing the cost $f(\mathbf{x})$ over some convex set would constitute a concave programming problem as defined above.

As another example, consider the *nonlinear complementarity problem*, in which two vector-valued functions $\mathbf{g}$ and $\mathbf{h}$ are given, such that $\mathbb{R}^n \ni \mathbf{x} \to \mathbf{g}(\mathbf{x}), \mathbf{h}(\mathbf{x}) \in \mathbb{R}^m$, and we wish to find a solution $\mathbf{x}$ to the following system:

$$\text{NCP: } \mathbf{g}(\mathbf{x}) \leq \mathbf{0}$$
$$\mathbf{h}(\mathbf{x}) \leq \mathbf{0}$$
$$[\mathbf{g}(\mathbf{x})]^T \mathbf{h}(\mathbf{x}) = 0.$$

We have previously considered the linear complementarity problem in Sect. 6.1.2, for which $g(\mathbf{x})$ is linear and $h(\mathbf{x}) = \mathbf{x}$.

We will now show that the complementarity problem can be rewritten as the optimization problem

$$\text{NCP}': \text{Min } z = \sum_{i=1}^{m} \min\{-g_i(\mathbf{x}), -h_i(\mathbf{x})\}$$
$$\text{s.t. } \mathbf{g}(\mathbf{x}) \leq \mathbf{0}$$
$$\mathbf{h}(\mathbf{x}) \leq \mathbf{0}.$$

Assuming g_i and h_i are convex, $-g_i$ and $-h_i$, and therefore also $\min\{-g_i, -h_i\}$ are concave. Therefore, NCP' is a concave programming problem according to Definition 7.4. Since $\min\{-g_i(\mathbf{x}), -h_i(\mathbf{x})\} \geq 0$ for feasible $\mathbf{x}$, we must have $z \geq 0$. If for some point $\bar{\mathbf{x}}$ it turns out that $\bar{z} = 0$, we have found a solution to the original problem NCP; if, on the other hand, $\bar{z} > 0$, NCP has no solution. Complementarity problems occur in fixed point theory and have formal applications in engineering, economics, and operations research; for references see, e.g., Pang (1995) as well as Cottle et al. (2009).

Returning to the concave programming problem, a fundamental property is given by the following theorem, a proof of which may be found in, e.g., Horst et al. (2000).

Theorem 7.5 If the feasible region for the concave programming problem P is compact, then its global minimum is attained at an extreme point.

If the feasible region of P happens to be a polyhedron, one could therefore solve P by determining all extreme points of the polyhedron and pick those for which $f(\mathbf{x})$ is minimal. Although there are algorithms for determining all extreme points of a polyhedral set, this is in general not a good idea since the number of such points may be extremely large.

On the other hand, since the concave programming problem P is **NP**-hard even for the simple special case with f concave quadratic and g_i all linear, we will here describe a branch and bound method for solving P based on partitioning the feasible region into simplices. An initial step of this algorithm would be to construct a simplex enclosing the feasible region of P as tightly as possible. This can be accomplished as follows, assuming that the feasible region $\{\mathbf{x} \in \mathbb{R}^n : g_i(\mathbf{x}) \leq 0, i = 1, \ldots, m\}$ is bounded.

First solve the following problems $P_j, j = 1, \ldots, n$ and P_0:

$$\begin{array}{ll} P_j : \text{Min } z_j = x_j \\ \quad\;\; \text{s.t.} \mathbf{g}(\mathbf{x}) \leq \mathbf{0} \end{array} \quad \text{and} \quad \begin{array}{l} P_0 : \text{Max } z_0 = \sum_{j=1}^{n} x_j \\ \quad\;\; \text{s.t.} \mathbf{g}(\mathbf{x}) \leq \mathbf{0}, \end{array}$$

which are $n+1$ optimization problems with linear objective functions and convex feasible region. This is the format employed in Kelley's cutting plane method, discussed in Sect. 7.1, but any algorithm for convex programming could be used. Let $\bar{x}_j$ be the x_j values in the optimal solutions to the problems P_j and denote the optimal objective function values of the problems P_j by $\bar{z}_j, j = 1, \ldots, n$, and that of the problem P_0 by $\bar{z}_0$. Next, we perform a variable transformation in which x_j is replaced by $x'_j - \bar{z}_j$. Similarly, we replace $\bar{z}_0$ by $\bar{z}_0' - \sum_{j=1}^{n} \bar{z}_j$. In the transformed space, the

set $S^0 := \left\{ \mathbf{x} \geq \mathbf{0} : \sum_{j=1}^{n} x_j \leq \bar{z}_0 \right\}$ will then be an $(n+1)$-simplex, with one vertex at the origin, and the other vertices on the coordinate axes, all at the same distance of $\bar{z}_0$ from the origin. The simplex S^o contains the feasible region for P with a tightest fit,

Fig. 7.12 The simplex S^0

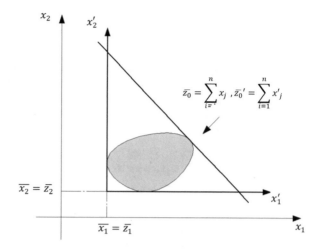

$$\bar{z}_0 = \sum_{i=1}^{n} x_j \ , \bar{z}_0' = \sum_{i=1}^{n} x'_j$$

in the sense that each of its $n + 1$ faces touches the feasible region. This is shown in Fig. 7.12.

If lower and upper bounds are readily available for each x_j, e.g., $m_j \le x_j \le M_j, j = 1, \ldots, n$, another and easier way to determine a simplex S^0 containing the feasible region would be to use the lower bounds m_j instead of the values $\bar{x}_j$ in the above construction of S^0, as well as $\sum_{j=1}^{n} M_j$ instead of $\bar{z}_0$. This might however produce an S^0 with a much looser fit.

The procedure then continues by partitioning the simplex S^0 into two subsimplices by way of what is called a *bisection* of S^0 (see, e.g., Horst et al. 2000). We accomplish this by selecting any two vertices of the simplex that are the furthest apart from each other (this involves making $\frac{1}{2}n(n + 1)$ comparisons); call them $\mathbf{v}^1$ and $\mathbf{v}^2$. In the case of S^0, all the vertices are equally far apart except the origin, which is closer. Form the midpoint on the edge between $\mathbf{v}^1$ and $\mathbf{v}^2$, i.e., $\frac{1}{2}(\mathbf{v}^1 + \mathbf{v}^2)$. Two subsimplices of S^0 are now created as follows. One of them will have all vertices of S^0, except $\mathbf{v}^1$, which is replaced by $\frac{1}{2}(\mathbf{v}^1 + \mathbf{v}^2)$, the other one will have all vertices of S^0 except $\mathbf{v}^2$, which is also replaced by $\frac{1}{2}(\mathbf{v}^1 + \mathbf{v}^2)$. Remember that in a simplex every vertex is adjacent to every other vertex. The two new simplices each have a volume which is half of the original simplex S^0, they have no interior point in common, and their union equals S^0.

These two subsimplices can, in turn, each be bisected in the same manner. Continuing this procedure, we obtain a hierarchy of ever smaller, included subsimplices. Horst et al. (2000) prove the intuitively clear result that any infinite sequence of bisected simplices, each one included in the previous one, will converge to a single point. Using this idea, the following branch and bound method for solving the concave programming problem P can now be described. Formally, the problem under investigation is

$$P : \text{Min } z = f(\mathbf{x})$$
$$\text{s.t.} g_i(\mathbf{x}) \leq 0 \ \forall \ i = 1, \ldots, m,$$

where $\mathbf{x} \in \mathbb{R}^n$, the functions $-f$ and g_i, $i = 1, \ldots, m$ are convex, and the feasible region of P is compact. The problem is initialized by constructing the initial simplex S^0, containing the feasible region of P. Note that there is no Step 0 for solving P without any constraints, since it would have no finite optimal solution, barring the trivial case with constant f.

A Simplicial Branch-and-Bound Method for Concave Minimization

Step 1: Using simplicial bisection, partition the initial simplex S^0 repeatedly, until a feasible vertex is found. Denote the best feasible vertex by $\hat{\mathbf{x}}$ and set the upper bound $UB := f(\hat{\mathbf{x}})$. Evaluate f at all vertices of S^0 and set the smallest of the vertex objective function values equal to the lower bound LB.

Step 2: Is $LB = UB$?

If yes: Stop, $\hat{\mathbf{x}}$ is an optimal solution.
If no: Go to Step 3.

Step 3: Discard any simplex whose vertices are all infeasible. Let LB be the smallest objective function value at any vertex of all remaining simplices.

Step 4: Eliminate any simplex whose vertices are all feasible, as well as any simplex whose best vertex is no better than $\hat{\mathbf{x}}$.

Step 5: Perform a bisection of each of the remaining simplices. Update $UB := f(\hat{\mathbf{x}})$ to be the function value of the best vertex $\hat{\mathbf{x}}$ found so far. Go to Step 2.

In the above algorithm it is clear that all simplices that fall completely outside the feasible region can be eliminated, as in Step 3. Since we know that the optimal points must be located on the boundary of the feasible region (apart from the trivial case with constant objective function f), Step 4 will eliminate all simplices which are completely inside the feasible region. It will also discard "dominated" simplices. Those simplices that survive to Step 5 will therefore have at least one vertex which is feasible and one which is infeasible, i.e., they will straddle the boundary of the feasible region and would be of interest for further bisection. These simplices are called *live* or *active*. It is also clear that the algorithm need not have finite convergence, so that some stopping criterion, based for instance on the bound gap $UB - LB$, on the size of the largest remaining simplex in Step 5, or on the number of iterations may have to be used.

At some additional computational expense, the lower bounds LB could be sharpened in the following way. Horst et al. (2000) have shown that the largest convex underestimate f_{conv} of a concave function over a simplex $S \subseteq \mathbb{R}^n$ with vertices $\mathbf{v}^1, \ldots, \mathbf{v}^{n+1}$ is the unique linear function f_{conv} which coincides with f at the vertices. Specifically, let $\mathbf{x} \in S$ be expressed as the linear convex combination $\mathbf{x} = \sum_{k=1}^{n+1} \alpha_k \mathbf{v}^k$ where $\alpha_k \geq 0$, $k = 1, \ldots, n + 1$ and $\sum_{k=1}^{n+1} \alpha_k = 1$, then $f_{\text{conv}}(\mathbf{x}) = f_{\text{conv}}$

$\left(\sum\limits_{k=1}^{n+1} \alpha_k \mathbf{v}^k \right) := \sum\limits_{k=1}^{n+1} \alpha_k f\left(\mathbf{v}^k\right)$. A sharper bound LB than in Steps 1 and 3 above can then be obtained by solving the problem

$$P_{LB}: \text{Min } z_{LB} = \sum_{k=1}^{n+1} \alpha_k f\left(\mathbf{v}^k\right)$$

$$\text{s.t. } \mathbf{g}\left(\sum_{k=1}^{n+1} \alpha_k \mathbf{v}^k \right) \leq \mathbf{0}$$

$$\sum_{k=1}^{n+1} \alpha_k = 1$$

$$\alpha_k \geq 0, k = 1, \dots, n+1.$$

We can see that P_{LB} is a convex programming problem with a linear objective function. Denoting the optimal objective function value of P_{LB} by $\bar{z}_{LB}$, we then set $LB := \bar{z}_{LB}$. Alternative approaches to the above partitioning of the feasible region using simplices have also been developed, using cones or rectangles, each of them with their advantages and drawbacks compared with the simplicial procedure, see, e.g., Horst et al. (2000).

Example As a numerical illustration, we will use the simplicial branch and bound algorithm to solve the following concave programming problem. Four iterations of the method will be carried out.

$$P: \text{Min } z = -x_1^2$$
$$\text{s.t.} (x_1 - 1)^2 + (x_2 - 1)^2 \leq 1.$$

We can see that the feasible region is the circular disk in $\mathbb{R}^2$ with radius 1 that is centered at $(x_1, x_2) = (1, 1)$. This feasible region is shown as the shaded area in Fig. 7.13. We also find that the initial simplex S^0 becomes the triangle with corners at $\mathbf{v}^1 = (0, 0)$, $\mathbf{v}^2 = (2 + \sqrt{2}, 0)$, and $\mathbf{v}^3 = (0, 2 + \sqrt{2})$.

In Step 1, we see that $\mathbf{v}^2$ and $\mathbf{v}^3$ are furthest apart, so that we calculate $\mathbf{v}^4 := \frac{1}{2}(\mathbf{v}^2 + \mathbf{v}^3) = \left(1 + \dfrac{1}{\sqrt{2}}, 1 + \dfrac{1}{\sqrt{2}} \right) \approx (1.707, 1.707)$, which happens to be feasible (on the boundary of the feasible region, to be precise). So far, $\mathbf{v}^4$ is the only feasible vertex found in the process. Therefore, $\hat{\mathbf{x}} := \mathbf{v}^4$, so that the upper bound is $UB := f(\hat{\mathbf{x}}) = -\left(1 + \dfrac{1}{\sqrt{2}} \right)^2 = -\dfrac{3}{2} - \sqrt{2} \approx -2.914$. Furthermore, $f(\mathbf{v}^1) = 0, f(\mathbf{v}^2) = -6 - 4\sqrt{2} \approx -11.657$ and $f(\mathbf{v}^3) = 0$, so that $LB := f(\mathbf{v}^2) \approx -11.657$. Step 4 eliminates the simplex with corners $(\mathbf{v}^1, \mathbf{v}^4, \mathbf{v}^3)$ since its best vertex is $\mathbf{v}^4 = \hat{\mathbf{x}}$ and therefore only the simplex with extreme points $(\mathbf{v}^1, \mathbf{v}^2, \mathbf{v}^4)$ remains active. In Step 5, the longest edge of the simplex $(\mathbf{v}^1, \mathbf{v}^2, \mathbf{v}^4)$ is between $\mathbf{v}^1$ and $\mathbf{v}^2$, so that $\mathbf{v}^5 := \frac{1}{2}(\mathbf{v}^1 + \mathbf{v}^2) = \left(1 + \dfrac{1}{\sqrt{2}}, 0 \right) \approx (1.707, 0)$, which is not feasible. Then we proceed to Step 2 of the second iteration,

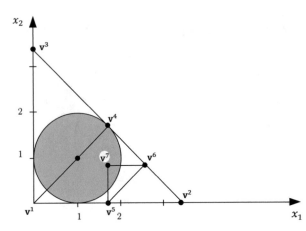

Fig. 7.13 Iterative solutions with the simplicial branch-and-bound method

where we find that $LB < UB$. No simplices are eliminated and $LB = f(\mathbf{v}^2)$ is not updated.

The remaining simplices are now $(\mathbf{v}^1, \mathbf{v}^5, \mathbf{v}^4)$ and $(\mathbf{v}^5, \mathbf{v}^2, \mathbf{v}^4)$; the upper bound $UB = f(\mathbf{v}^4)$ is not updated and we proceed to the third iteration. Step 4 determines that $f(\mathbf{v}^5) = -\dfrac{3}{2} - \sqrt{2} = UB$, so that the simplex $(\mathbf{v}^1, \mathbf{v}^5, \mathbf{v}^4)$ has no vertex better than $\hat{\mathbf{x}} = \mathbf{v}^4$, it can therefore be eliminated, leaving $(\mathbf{v}^5, \mathbf{v}^2, \mathbf{v}^4)$ as the only active simplex. The vertices $\mathbf{v}^2$ and $\mathbf{v}^4$ are furthest apart, so that the bisection point in Step 5 is $\mathbf{v}^6 :=$ $\frac{1}{2}(\mathbf{v}^2 + \mathbf{v}^4) = \left(\dfrac{3}{2} + \dfrac{3\sqrt{2}}{4}, 1/2 + \dfrac{\sqrt{2}}{4}\right) \approx (2.561, 0.854)$, which is infeasible with f $(\mathbf{v}^6) = -\dfrac{9}{8}\left(3 + 2\sqrt{2}\right) \approx -6.557$. Remaining simplices are $(\mathbf{v}^5, \mathbf{v}^6, \mathbf{v}^4)$ and $(\mathbf{v}^5, \mathbf{v}^2, \mathbf{v}^6)$. The upper bound UB is not updated and we continue with the fourth iteration.

In Step 3 we eliminate $(\mathbf{v}^5, \mathbf{v}^2, \mathbf{v}^6)$ with all vertices infeasible, updating the lower bound $LB := f(\mathbf{v}^6) \approx -6.557$. Step 5 computes $\mathbf{v}^7 := \frac{1}{2}(\mathbf{v}^5 + \mathbf{v}^4) = \left(1 + \dfrac{1}{\sqrt{2}}, 1/2 + \dfrac{1}{2\sqrt{2}}\right) \approx (1.707, 0.854)$ with $f(\mathbf{v}^7) = f(\mathbf{v}^4) = -\dfrac{3}{2} - \sqrt{2} = UB \approx$ -2.914. Having carried out four iterations of the method, we stop here with $LB \approx -6.557$. The vertex $\mathbf{v}^7$ is feasible, so that $UB = f(\mathbf{v}^7) \approx f(1.707, .854) \approx -2.914$, see Fig. 7.13. Continuing in this fashion, the algorithm will converge to the unique optimal point which is $\bar{\mathbf{x}} = (2, 1)$ with objective value $\bar{z} = f(\bar{\mathbf{x}}) = -4$.

In closing, we should note that the feasible region could be partitioned into other geometric shapes than simplices. Hendrix and G.-Tóth (2010) describe an algorithm that uses axis-parallel hyperrectangles to partition the feasible region. This approach, although geometrically appealing, suffers from the "curse of dimensionality" in that the number of corner points in the subdivided regions will grow exponentially with the number of variables of the problem. This is not much of a drawback here, since most practical problems in nonlinear programming are of relatively low dimension as compared to typical linear programming problems.

7.4 D.C., Conic, and Semidefinite Programming

7.4.1 D.C. Programming

In previous sections we have dealt with problems involving convex and also concave functions. In this section, we extend our discussion to a rather large class of functions, namely those that can be expressed as the difference between two convex functions. Such functions are called *difference-convex*, or more commonly simply *d. c. functions*. For example, if $f(x) = (x - 3)^2 + 1$ and $g(x) = \dfrac{4}{x} - 2$, both of which are convex functions for $x > 0$, then the function $f(x) - g(x) = (x - 3)^2 - \frac{4}{x} + 3$ is a *d.c.* function. For a graphical depiction of the functions $f(x)$, $g(x)$, and $f(x) - g(x)$, see Fig. 7.14.

Note that the representation of a *d.c.* function as $f(\mathbf{x}) - g(\mathbf{x})$, where f and g are convex functions, is not unique. This is clear since for any nontrivial convex function $h(\mathbf{x})$, we can write $f(\mathbf{x}) - g(\mathbf{x}) = (f(\mathbf{x}) + h(\mathbf{x})) - (g(\mathbf{x}) + h(\mathbf{x}))$, thus obtaining two different representations of the same *d.c.* function. Furthermore, we will refer to a problem P of the type

$$\text{P: Min } z = f(\mathbf{x})$$
$$\text{s.t. } g_i(\mathbf{x}) \leq 0, i = 1, \ldots, m$$
$$\mathbf{x} \in \mathbb{R}^n,$$

where $f(\mathbf{x})$, $g_1(\mathbf{x})$, $g_2(\mathbf{x})$, ..., $g_m(\mathbf{x})$ are *d.c.* functions, as a *d.c. minimization problem*.

The class of *d.c.* functions is quite large. To begin with, it includes all convex and all concave functions. To see this, let $f(\mathbf{x})$ be any convex and $g(\mathbf{x})$ be any concave function. Then $h_1(\mathbf{x}) := f(\mathbf{x}) - 0 = f(\mathbf{x})$, and $h_2(\mathbf{x}) := 0 - (-g(\mathbf{x})) = g(\mathbf{x})$ are both

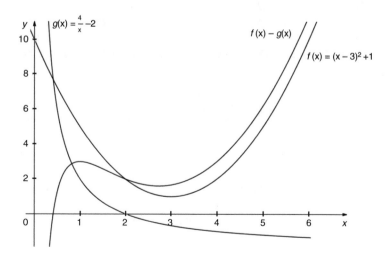

Fig. 7.14 A difference-convex (d.c.) function

expressed as the difference between two convex functions and are therefore *d.c.*
Furthermore, any linear combination, product, absolute value of *d.c.* functions is also
d.c., as are the upper and lower envelopes of any collection of *d.c.* functions. For a
proof of this statement as well as many others in this section, readers are referred to
Horst et al. (2000), see also Hendrix and G-Tóth (2010). Two main references are
Tuy (1995, 1998), one of the pioneers of *d.c.* programming.

When dealing with *d.c.* optimization, a so-called *reverse convex constraint* is of
particular interest. Given some convex function $g(\mathbf{x})$ it is a constraint of the type $g(\mathbf{x})$
≥ 0. Writing it as $(0 - g(\mathbf{x})) \leq 0$ shows that it is indeed a proper *d.c.* constraint.
Furthermore, it is the complement of the open convex set defined by $g(\mathbf{x}) < 0$. An
obvious use of reverse convex constraints is in location problems. We may refer to
the situation described in Sect. 3.8 of this volume, where an obnoxious facility is to
be located in some set $S \subseteq \mathbb{R}^2$. Suppose now that S includes all points that are
at least a given distance d_i from the i-th population center with coordinates
(a_i, b_i), $i = 1, \ldots, m$, then we find that the set S can be described as
$S := \left\{ (x,y) : (a_i - x)^2 + (b_i - y)^2 \geq d_i^2, i = 1, \ldots, m \right\}$, which involves m reverse
convex constraints. Other classical problems in location analysis such as the
Weber problem and the Steiner tree problem have also been formulated and studied
as *d.c.* programming problems, see, e.g., Horst et al. (2000).

Another simple application of *d.c.* programming is the following modification of
Hitchcock's classical transportation problem (Hitchcock 1941), in which a commodity is being produced in quantities s_i, $i = 1, \ldots, m$ at m specific origins and at
costs described by increasing concave functions $g_i(s_i)$. Concavity of the functions
$g_i(s_i)$ is a natural assumption in the presence of economies of scale. The commodity
produced at the origins will be sent directly from each origin via direct link to each of
n destinations, which require at least given quantities $d_j, j = 1, \ldots, n$. The transportation costs from origin i to destination j are given by the increasing convex functions
$c_{ij}(x_{ij})$, where x_{ij} is the quantity shipped from origin i to destination j; $i = 1, \ldots, m$;
$j = 1, \ldots, n$. The convexity of the functions $c_{ij}(x_{ij})$ may be due to congestion effects
in the transportation links. The problem of finding a production—transportation plan
that minimizes the total costs of production and transportation can then be formulated as

$$\text{P: } \min_{x_{ij}, s_i} z = \sum_{i=1}^{m} \sum_{j=1}^{n} c_{ij}(x_{ij}) + \sum_{i=1}^{m} g_i(s_i)$$

$$\text{s.t. } \sum_{j=1}^{n} x_{ij} = s_i, i = 1, \ldots, m$$

$$\sum_{i=1}^{m} x_{ij} \geq d_j, j = 1, \ldots, n$$

$$x_{ij} \geq 0; i = 1, \ldots, m; j = 1, \ldots, n$$

$$s_i \geq 0, i = 1, \ldots, m.$$

Since the objective function is the sum of convex and concave functions, problem P is a *d.c.* programming problem. For a description of several other applications of *d. c.* optimization models, readers are referred to Tuy (1995, 1998) and Horst et al. (2000).

A number of algorithms is available for the solution of *d.c.* optimization problems. For example, the simplicial branch-and-bound method for concave minimization, described in Sect. 7.3 can be adapted to solve *d.c.* minimization problems by using appropriate modifications. For further details, see, e.g., Horst et al. (2000). As far as extensions go, there is an even more general class of functions, called *difference-monotonic*, or *d.m. functions*. Functions in this class can be expressed as the difference between two monotonic functions. Theory and optimization techniques for problems in this class have also been developed, see, e.g., Tuy (2005) and Tuy et al. (2005), who also treat the question of how a given *d.c.* function can be decomposed into the difference between two convex functions. The decomposition of a given *d.m.* function is similarly discussed.

7.4.2   Conic and Semidefinite Programming

In conic optimization we restrict the movement of variables by requiring that they satisfy conic constraints, i.e., that they belong to some cone C (for the definition and a description of a cone, see Appendix 9C). To formalize, consider the problem

$$\text{P: } \min_{\mathbf{x}} \; z = \mathbf{cx}$$
$$\text{s.t. } \mathbf{Ax} \geq \mathbf{b}$$
$$\mathbf{x} \in C,$$

where $\mathbf{x} \in \mathbb{R}^n$ is an n-dimensional column vector, $\mathbf{c}$ is a n-dimensional row vector of coefficients, $\mathbf{A}$ is an $[m \times n]$-dimensional matrix, $\mathbf{b}$ is an m-dimensional column vector, and C is some given cone. If C is the nonnegative orthant, i.e., $C \in \mathbb{R}_+^n$, then P is a linear programming problem in canonical form.

An interesting case with important applications is the *second-order cone*, also referred to as the *ice cream cone* or the *Lorentz cone* C_q, where "q" symbolizes "quadratic." (Hendrik Lorentz (1853–1928) was a Dutch physicist, who is remembered today mostly via his Lorentz transformation equations in Einstein's relativity theory in physics.) Such a cone is defined as

$$C_q = \left\{ \mathbf{x} \in \mathbb{R}^n : x_n \geq \sqrt{x_1^2 + x_2^2 + \ldots + x_{n-1}^2} \right\}.$$

Figure 7.15 shows the cone C_q in three dimensions, and it is apparent that C_q looks like a wide, infinite ice cream cone. It turns out that any convex quadratic

Fig. 7.15 A Lorentz cone

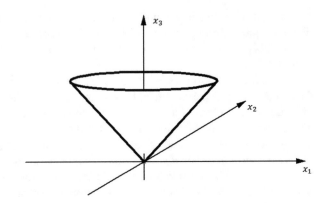

constraint can be expressed using the ice cream cone (maybe rotated), supplemented by linear constraints.

Another important cone is the set of symmetric positive semidefinite matrices C_s (where the subscript "s" stands for semidefinite) defined as

$$
C_s = \left\{ \mathbf{A} = \begin{bmatrix} a_{11} & a_{12} & \cdots & a_{1n} \\ a_{21} & a_{22} & \cdots & a_{2n} \\ \vdots & \vdots & \ddots & \vdots \\ a_{n1} & a_{n2} & \cdots & a_{nn} \end{bmatrix} \in \mathbb{R}^{n \times n} : \mathbf{A} = \mathbf{A}^T, \mathbf{A} \text{ positive semidefinite} \right\}.
$$

A duality theory, involving weak and strong duality of conic programming as well as optimality conditions has been developed for both general and special types of cones, see, e.g., Bertekas (2016). The interest in conic programming has also been furthered by research showing that so-called interior point methods for linear programming can be adapted to conic optimization, see, e.g., Alizadeh and Goldfarb (2003), Boyd and Vandenberghe (2004), Nemirovski (2007), and Anjos and Lasserre (2012). Although this has been known for some time, the development of numerically efficient methods for interior point algorithms has increased the attractiveness of conic optimization and its applications. For an introduction to interior point methods, readers are referred to Eiselt and Sandblom (2007).

Returning to the case with second-order cones, optimization models involving such cones have found widespread applications in finance, concerning financial risk, volatility and covariance estimation (see, e.g., Cornuejols and Tütüncü, 2007).

Let us now consider semidefinite programming. Specifically, we define the following problem:

$$
P_{SD}: \underset{\mathbf{x} \in \mathbb{R}^n}{\text{Min }} z_{SD} = \mathbf{cx}
$$
$$
\mathbf{A(x)} \succeq = \mathbf{0},
$$

where $\mathbf{A} := \mathbf{A}_0 + \sum_{j=1}^{n} \mathbf{A}_j x_j$ and $\mathbf{A}, \mathbf{A}_j, j = 0, 1, \ldots, n$ are $[m \times m]$-dimensional matrices and "$\succ = 0$" indicates positive semidefiniteness. We say that P_{SD} is a *linear semidefinite programming problem* and that $\mathbf{A}(\mathbf{x}) \succ = \mathbf{0}$ is a *linear matrix inequality*. The feasible region of problem P_{SD} is convex, since if some $\mathbf{x}^1$ and $\mathbf{x}^2$ are both feasible, i.e., $\mathbf{A}(\mathbf{x}^1) \succ = \mathbf{0}$ and $\mathbf{A}(\mathbf{x}^2) \succ = \mathbf{0}$, then for any $\lambda \in [0, 1]$, we must have $\mathbf{A}$ $(\lambda \mathbf{x}^1 + (1 - \lambda)\mathbf{x}^2) = \lambda \mathbf{A}(\mathbf{x}^1) + (1 - \lambda)\mathbf{A}(\mathbf{x}^2)$, which is positive semidefinite, being the linear convex combination of two positive semidefinite matrices.

The class of linear semidefinite programming problems includes the class of ordinary linear programming problems. To see this, consider the linear programming problem

$$P_{LP}: \min_{\mathbf{x} \in \mathbb{R}^n} z_{LP} = \mathbf{cx}$$
$$\text{s.t. } \mathbf{A}^{LP}\mathbf{x} \geq \mathbf{b}.$$

We now define the diagonal matrix $\mathbf{A}^{SD}(\mathbf{x}) := \mathrm{diag}\,(\mathbf{A}^{LP}\mathbf{x} - \mathbf{b})$, i.e., the diagonal matrix $\mathbf{A}^{SD}$, which has its diagonal elements equal to the corresponding components of the column matrix $\mathbf{A}^{LP}\mathbf{x} - \mathbf{b}$. Since a diagonal matrix is positive semidefinite if and only if all of its diagonal elements are nonnegative, the linear programming problem P_{LP} is equivalent to the linear semidefinite programming problem P_{SD} given as

$$P_{SD}: \min_{\mathbf{x} \in \mathbb{R}^n} z_{SD} = \mathbf{cx}$$
$$\mathbf{A}^{SD}(\mathbf{x}) \succ = \mathbf{0}.$$

There are also nonlinear programming problems that can be formulated as linear semidefinite problems. For example, consider the following problem (see Vandenberghe and Boyd 1996):

$$P: \min_{\mathbf{x}} z = \frac{(\mathbf{cx})^2}{\mathbf{dx}}$$
$$\text{s.t. } \mathbf{Ax} \geq \mathbf{b},$$

where we assume that $\mathbf{Ax} \geq \mathbf{b}$ implies that $\mathbf{dx} > 0$, so that the denominator in the objective function will not cause any difficulties. Now we introduce an auxiliary variable y as an upper bound on the objective function value, i.e., $y \geq \frac{(\mathbf{cx})^2}{\mathbf{dx}}$, so that problem P can be written as

$$P_1: \min_{y, \mathbf{x}} z_1 = y$$
$$\text{s.t } \mathbf{Ax} \geq \mathbf{b}$$
$$y \geq \frac{(\mathbf{cx})^2}{\mathbf{dx}},$$

where obviously $y \geq 0$ for any feasible solution y. Next, we formulate problem P_{SD} as

$$P_{SD}: \underset{y, \mathbf{x}}{\text{Min}} \ z_{SD} = y$$

$$\text{s.t.} \begin{bmatrix} diag(\mathbf{Ax} - \mathbf{b}) & 0 & 0 \\ 0 & y & \mathbf{cx} \\ 0 & \mathbf{cx} & \mathbf{dx} \end{bmatrix} \succ = \mathbf{0}.$$

The constraint matrix of P_{SD} is "almost" diagonal, with diagonal elements required to be nonnegative, except for the bottom right $[2 \times 2]$-dimensional submatrix $\begin{bmatrix} y & \mathbf{cx} \\ \mathbf{cx} & \mathbf{dx} \end{bmatrix}$. This submatrix is positive semidefinite if and only if $y\mathbf{dx} - (\mathbf{cx})^2 \geq 0$ and $\mathbf{dx} \geq 0$. Now, since $\mathbf{Ax} \geq \mathbf{b}$, $\mathbf{dx} > 0$, so that we can write $y - \dfrac{(\mathbf{cx})^2}{\mathbf{dx}} \geq 0$ or simply $\dfrac{(\mathbf{cx})^2}{\mathbf{dx}} \leq y$. Therefore, the linear semidefinite programming problem P_{SD} is equivalent to the nonlinear programming problem P.

We will now show that any quadratic inequality constraint in an optimization problem can be converted to a corresponding linear semidefinite matrix inequality. To accomplish this, we need the following

Lemma 7.6 Given an $[n \times n]$-dimensional positive definite matrix $\mathbf{A}$, an $[n \times m]$-dimensional matrix $\mathbf{B}$, and a symmetric $[m \times m]$-dimensional matrix $\mathbf{C}$. Then the $[(n + m) \times (n + m)]$-dimensional matrix $\begin{bmatrix} \mathbf{A} & \mathbf{B} \\ \mathbf{B}^T & \mathbf{C} \end{bmatrix}$ is positive semidefinite, if and only if its *Schur complement* of $\mathbf{A}$, defined by $\mathbf{C} - \mathbf{B}^T \mathbf{A}^{-1} \mathbf{B}$ is also positive semidefinite.

Proof (see, e.g., Jahn 2007, or Luenberger and Ye 2008): For all vectors $\mathbf{x}$ and $\mathbf{y}$ of appropriate dimension, we have $\begin{bmatrix} \mathbf{x} \\ \mathbf{y} \end{bmatrix}^T \begin{bmatrix} \mathbf{A} & \mathbf{B} \\ \mathbf{B}^T & \mathbf{C} \end{bmatrix} \begin{bmatrix} \mathbf{x} \\ \mathbf{y} \end{bmatrix} = \mathbf{x}^T \mathbf{Ax} + 2\mathbf{y}^T \mathbf{B}^T \mathbf{x} + \mathbf{y}^T \mathbf{Cy}$, and minimizing this expression with respect to $\mathbf{x}$, while regarding $\mathbf{y}$ as fixed, we obtain $\bar{\mathbf{x}} = -\mathbf{A}^{-1} \mathbf{B}^T \mathbf{y}$. Inserting this value of $\bar{\mathbf{x}}$, we find after some calculations $\begin{bmatrix} \bar{\mathbf{x}} \\ \mathbf{y} \end{bmatrix}^T \begin{bmatrix} \mathbf{A} & \mathbf{B} \\ \mathbf{B}^T & \mathbf{C} \end{bmatrix} \begin{bmatrix} \bar{\mathbf{x}} \\ \mathbf{y} \end{bmatrix} = \mathbf{y}^T (\mathbf{C} - \mathbf{B}^T \mathbf{A}^{-1} \mathbf{B}) \mathbf{y}$. Therefore, the matrix $\begin{bmatrix} \mathbf{A} & \mathbf{B} \\ \mathbf{B}^T & \mathbf{C} \end{bmatrix}$ is positive semidefinite if and only if $(\mathbf{C} - \mathbf{B}^T \mathbf{A}^{-1} \mathbf{B})$ is positive semidefinite. □

The result of the above lemma will now be used in the following way. Consider a quadratic constraint given by $-\mathbf{x}^T \mathbf{B}^T \mathbf{Bx} + \mathbf{cx} + d \geq 0$ and construct the matrix $\begin{bmatrix} \mathbf{I} & \mathbf{Bx} \\ \mathbf{x}^T \mathbf{B} & \mathbf{cx} + d \end{bmatrix}$. Then we find that the Schur complement of $\mathbf{I}$ in this matrix is $\mathbf{cx} + d - (\mathbf{Bx})^T \mathbf{I}^{-1} \mathbf{Bx} = \mathbf{cx} + d - \mathbf{x}^T \mathbf{B}^T \mathbf{Bx}$, which by virtue of Lemma 7.6 is nonnegative (which is equivalent to positive semidefinite, as this is a $[1 \times 1]$-dimensional matrix, i.e., a real number), if and only if $\begin{bmatrix} \mathbf{I} & \mathbf{Bx} \\ \mathbf{x}^T \mathbf{B} & \mathbf{cx} + d \end{bmatrix}$ is positive semidefinite. Since **x** appears only in linear form in the larger matrix, this shows that quadratic constraints can be handled in a linear semidefinite programming framework. As mentioned above, interior point methods for linear programming have been adapted to solve linear semidefinite optimization problems efficiently, making this framework important in practical applications.

References

Abadie J, Carpentier J (1969) Generalization of the Wolfe reduced gradient method to the case of nonlinear constraints. In: Fletcher R (ed.) *Optimization*. Academic Press, New York, pp. 37-47

Agmon S (1954) The relaxation method for linear inequalities. *Canadian Journal of Mathematics* **6**: 382-392

Alizadeh F, Goldfarb D (2003) Second-order cone programming. *Mathematical Programming* **95**: 3-51

Anjos MF, Lasserre JB (2012) Introduction to semidefinite, conic, and polynomial optimziation. pp 1-22 in Anjos MF, Lasserre JB (eds.) *Handbook on semidefinite, conic, and polynomial optimization*. Springer, New York

Bazaraa MS, Sherali HD, Shetty CM (2013) *Nonlinear programming: theory and algorithms.* (3rd ed.) Wiley, New York

Bertsekas DP (2015) *Convex optimization algorithms*. Athena Scientific, Belmont, MA

Bertsekas DP (2016) *Nonlinear programming*. (3rd ed.) Athena Scientific, Belmont, MA

Boyd S, Vandenberghe L, (2004) *Convex optimization*. Cambridge University Press, Cambridge, UK

Cornuejols G, Tütüncü R (2007) *Optimization methods in finance*. Cambridge University Press, Cambridge, UK

Cottle RW, Pang JS, Stone RE (2009) *The linear complementarity problem*. Classics in Applied Mathematics, SIAM, Philadelphia

Cottle RW, Thapa MN (2017) *Linear and nonlinear optimization*. Springer-Verlag, Berlin-Heidelberg-New York

Eiselt HA, Pederzoli G, Sandblom C-L (1987) *Continuous optimization models*. De Gruyter, Berlin-New York

Eiselt HA, Sandblom C-L (2000) *Integer programming and network models*. Springer-Verlag, Berlin-Heidelberg-New York

Eiselt HA, Sandblom C-L (2007) *Linear programming and its applications*. Springer-Verlag, Berlin-Heidelberg

Fukushima M (1984) A descent algorithm for nonsmooth convex optimization. *Mathematical Programming* **30**: 163-175

Gomory RE (1963) An algorithm for integer solutions to linear programs. In Graves RL, Wolfe P (eds.) *Recent advances in mathematical programming*: pp. 269-302. McGraw-Hill, New York

Goffin JL (1977) On the convergence rates of subgradient optimization methods. *Mathematical Programming* **13**: 329-347

Held M, Wolfe P, Crowder HP (1974) Validation of subgradient optimization. *Mathematical Programming* **6**: 62-88

Hendrix EMT, G.-Tóth B (2010) *Introduction to nonlinear and global optimization*. Springer-Verlag, Gerlin-Heidelberg-New York

Hitchcock FL (1941) The distribution of a product from several sources to numerous localities. *Journal of Mathematical Physics* **20**: 224-230

Horst R, Pardalos PM, Thoai NV (2000) *Introduction to global optimization*, vol. 2. Kluwer, Dordrecht, The Netherlands

Jahn J (2007) *Introduction to the theory of nonlinear optimization*. (3rd ed.), Springer-Verlag, Berlin-Heidelberg-New York

Kelley JE (1960) The cutting-plane method for solving convex programs. *SIAM Journal* **8**: 703-712

Kiwiel, KC (1983) An aggregate subgradient method for nonsmooth convex minimization. *Mathematical Programming* **27**: 320-341

Lemarechal C (1975) An extension of Davidon methods to nondifferentiable problems. *Mathematical Programming Study* **3**: 95-109

Luenberger DL, Ye Y (2008) *Linear and nonlinear programming*. (3rd ed.) Springer-Verlag, Berlin-Heidelberg-New York

Mifflin R (1982) A modification and an extension of Lemarechal's algorithm for nonsmooth minimization. *Mathematical Programming Study* **17**: 77-90

Motzkin TS, Schönberg IJ (1954) The relaxation method for linear inequalities. *Canadian Journal of Mathematics* **6**: 393-404

Nemirovski A (2007) Advances in convex optimization: conic programming. *Proceedings of the International Congress of Mathematics*, Madrid 2006. European Mathematical Society: 413-444

Neumann K (1975) *Operations Research Verfahren*, Band 1. C. Hanser Verlag, München-Wien

Nocedal J, Wright SJ (2006) *Numerical optimization* (2nd ed.). Springer-Verlag, Berlin-Heidelberg-New York

Pang JS (1995) Complementarity problems. In Horst R, Pardalos PM (eds.) *Handbook of global optimization. Nonconvex optimization and its applications, vol 2*. Springer, Boston, MA, pp. 271-338

Polyak BT (1967) A general method of solving extremum problems. *Soviet Mathematics Doklady* **8**: 593-597 (translated from the Russian)

Polyak BT (1969) Minimization of unsmooth functionals. *USSR Computational Mathematics and Mathematical Physics* **9**: 509-521 (translated from the Russian)

Sandblom C-L (1980) Two approaches to nonlinear decomposition. *Mathematische Operationsforschung und Statistik, Series Optimization* **11**: 273-285

Shor Z (1964) *On the structure of algorithms for the numerical solution of optimal planning and design problems*. Dissertation, Cybernetics Institute, Kiev, Ukraine

Sun W, Yuan Y (2006) *Optimization theory and methods. Nonlinear programming*. Springer-Verlag, Berlin-Heidelberg-New York

Tuy H (1995) *D.C. optimization theory, methods and algorithms*. In Horst R, Pardalos P (eds.) *Handbook of global optimization*. Kluwer, Boston, MA

Tuy H (1998) *Convex analysis and global optimization*. Kluwer, Boston, MA

Tuy H (2005) Polynomial optimization: a robust approach. *Pacific Journal of Optimization* **1**: 257-274

Tuy H, Al-Khayyal F, Thach PT (2005) Monotonic optimization: branch-and-cut methods. In: Audet C, Hansen P, Savard G (eds.) (2005) *Surveys in global optimization*. Springer, Berlin-Heidelberg-New York

Vandenberghe L, Boyd S (1996) Semidefinite programming. *SIAM Review* **38**: 49-95

Wolfe P (1963) Methods of nonlinear programming. pp. 67-86 in Graves RL, Wolfe P (eds.) *Recent advances in mathematical programming*. McGraw Hill, New York

Wolfe P (1970) Convergence theory in nonlinear programming. pp. 1-36 in Abadie J (ed.) *Integer and nonlinear programming*. North Holland, Amsterdam

Wolfe P (1975) A method of conjugate subgradients for minimizing nondifferentiable functions. *Mathematical Programming Study* **3**: 145-173

Chapter 8
Geometric Programming

This chapter is devoted to a branch of optimization called *geometric programming*. It originated in the 1960s and early references are Zener (1961) and Duffin (1962). The term "geometric programming" is actually a misnomer as explained below, but it has stuck. A better term would be "posynomial programming," since the problems under investigation involve posynomial functions, which we will define below. Our discussion commences with unconstrained geometric programming. Readers may wonder why this was not covered in Chap. 2. As we develop the theory of geometric programming below, we will have to resort to results from duality for nonlinear programming, an issue not covered until Chap. 4. General references are Beightler and Phillips (1976), Eiselt et al. (1987), Avriel (2013), and Bazaraa et al. (2013). An entertaining account can be found in Woolsey and Swanson (1975).

After introducing concepts and the philosophy of geometric programming, we will introduce duality considerations and discuss how to handle constrained problems. Subsequently, solution methods for geometric programming problems will be described. The chapter ends with a section on how various problems can be transformed into geometrical programming form, followed by a few illustrations of geometric programming applications.

8.1 Fundamental Concepts of Geometric Programming

As a motivation for the following discussion, the ideas underlying the geometric programming approach are best introduced by two examples. To begin, consider the basic inventory management problem already discussed in Sect. 3.9, in which an item of a homogeneous commodity is kept in stock at a carrying cost of c_h monetary units per year and per quantity unit of the commodity. The costs associated with placing an order are c_0 regardless of the order size, and the total annual demand is D units. Shortages will not be allowed and the demand rate is assumed to be

© Springer Nature Switzerland AG 2019
H. A. Eiselt, C.-L. Sandblom, *Nonlinear Optimization*, International Series in
Operations Research & Management Science 282,
https://doi.org/10.1007/978-3-030-19462-8_8

constant. Furthermore, suppose that each time an order is placed, the same amount will be ordered.

Denoting this order quantity by Q, the annual costs TC are given by

$$TC = c_h \frac{Q}{2} + c_0 \frac{D}{Q},$$

where the first term represents the annual carrying cost and the second the annual ordering cost. Considering the order quantity Q as the variable, the problem of minimizing total annual inventory cost can then be stated as

$$\text{P: } \underset{x}{\text{Min}}\, z = c_1 x + c_2 x^{-1},$$

where $z := TC$, $x := Q$, $c_1 := \frac{1}{2}c_h$, and $c_2 := c_0 D$. This is the basic version of the economic order quantity (EOQ) model, which is an unconstrained optimization problem. The minimal point of the convex objective function is determined by differentiating with respect to x, i.e., calculating $\frac{dz}{dx} = c_1 - c_2 x^{-2}$ and solving the nonlinear equation $\frac{dz}{dx} = 0$. The optimal order quantity $\bar{x} = \sqrt{c_2/c_1}$ can then be substituted into the cost function to determine the minimal cost $\bar{z} = 2\sqrt{c_1 c_2}$. Instead of following this "classical" optimization approach of first determining the optimal $\bar{x}$ and then finding the optimal $\bar{z}$ by substitution, the problem will now be turned around, so that rather than first determining $\bar{x}$ and then $\bar{z}$, we will first find the relative sizes of the two terms $c_1 x$ and $c_2 x^{-1}$ in the objective function when optimality occurs, i.e., values of $\bar{\lambda}_1$ and $\bar{\lambda}_2$ such that $\bar{\lambda}_1 = \dfrac{c_1 \bar{x}}{\bar{z}}$ $\left(\text{or } \bar{z} = \dfrac{c_1 \bar{x}}{\bar{\lambda}_1}\right)$ and $\bar{\lambda}_2 = \dfrac{c_2 \bar{x}^{-1}}{\bar{z}}$ $\left(\text{or } \bar{z} = \dfrac{c_2 \bar{x}^{-1}}{\bar{\lambda}_2}\right)$. By construction, $\bar{\lambda}_1 + \bar{\lambda}_2 = 1$, $\bar{\lambda}_1 > 0$, $\bar{\lambda}_2 > 0$, so that $\bar{\lambda}_1$ and $\bar{\lambda}_2$ can be regarded as weights indicating the contribution of each of the respective terms $c_1 \bar{x}$ and $c_2 \bar{x}^{-1}$ to the optimal function value

$$\bar{z} = c_1 \bar{x} + c_2 \bar{x}^{-1} = \frac{c_1 \bar{x}}{\bar{z}}\bar{z} + \frac{c_2 \bar{x}^{-1}}{\bar{z}}\bar{z} = \bar{\lambda}_1 \bar{z} + \bar{\lambda}_2 \bar{z}.$$

One can actually say something more about the weights. The optimality condition can be written as $c_1 - c_2 \bar{x}^{-2} = 0$; multiplication with $\bar{x}$ yields $c_1 \bar{x} - c_2 \bar{x}^{-1} = 0$ and division by $\bar{z} > 0$ results in $\dfrac{c_1 \bar{x}}{\bar{z}} - \dfrac{c_2 \bar{x}^{-1}}{\bar{z}} = 0$ which is equivalent to $\bar{\lambda}_1 - \bar{\lambda}_2 = 0$. It turns out that the optimal objective function value $\bar{z}$ can be expressed in terms of the problem coefficients and the optimal weights. For reasons to become clear later, this is referred to as the geometric programming approach. Specifically,

$$\bar{z} = \bar{z}^1 = \bar{z}^{\bar{\lambda}_1 + \bar{\lambda}_2} \quad \left(\text{because } \bar{\lambda}_1 + \bar{\lambda}_2 = 1\right)$$

$$= \left(\frac{c_1 \bar{x}}{\bar{\lambda}_1}\right)^{\bar{\lambda}_1} \left(\frac{c_2 \bar{x}^{-1}}{\bar{\lambda}_2}\right)^{\bar{\lambda}_2} \quad \text{(using the definitions of } \bar{\lambda}_1 \text{ and } \bar{\lambda}_2)$$

$$= \left(\frac{c_1}{\bar{\lambda}_1}\right)^{\bar{\lambda}_1} \left(\frac{c_2}{\bar{\lambda}_2}\right)^{\bar{\lambda}_2} \quad \underbrace{(\bar{x})^{\bar{\lambda}_1} (\bar{x}^{-1})^{\bar{\lambda}_2}}_{},$$

$$= \bar{x}^{\bar{\lambda}_1 - \bar{\lambda}_2} = \bar{x}^0 = 1 \quad \left(\text{since } \bar{\lambda}_1 - \bar{\lambda}_2 = 0\right)$$

and therefore $\bar{z} = \left(\dfrac{c_1}{\bar{\lambda}_1}\right)^{\bar{\lambda}_1} \left(\dfrac{c_2}{\bar{\lambda}_2}\right)^{\bar{\lambda}_2}$, which does not depend on $\bar{x}$. It follows that if one could find the optimal weights $\bar{\lambda}_1$ and $\bar{\lambda}_2$, then it would be possible to compute the optimal value $\bar{z}$ of the objective function without knowing the value of $\bar{x}$. As $\bar{\lambda}_1 + \bar{\lambda}_2 = 1$ and $\bar{\lambda}_1 - \bar{\lambda}_2 = 0$ with $\bar{\lambda}_1 > 0$, $\bar{\lambda}_2 > 0$, we obtain $\bar{\lambda}_1 = \bar{\lambda}_2 = \frac{1}{2}$ and therefore $\bar{z} = \sqrt{2c_1}\sqrt{2c_2} = 2\sqrt{c_1 c_2}$. Finally, since $\bar{\lambda}_1 = \dfrac{c_1 \bar{x}}{\bar{z}}$, one can write $\bar{x} = \bar{\lambda}_1 \dfrac{\bar{z}}{c_1} = \frac{1}{2} \dfrac{2\sqrt{c_1 c_2}}{c_1} = \sqrt{\dfrac{c_2}{c_1}}$. These two expressions for $\bar{x}$ and $\bar{z}$ provide the unique solution to the original inventory problem. In terms of the original notation, the optimal solution is $\bar{Q} = \sqrt{\dfrac{c_2}{c_1}} = \sqrt{\dfrac{2Dc_0}{c_h}}$ which is the well known economic order quantity (*EOQ*) formula in inventory theory.

Comparing the above two approaches for solving problem P, we realize that while the classical method requires the solution of a nonlinear equation, the geometric programming technique consists of solving a system of linear equations. Given the nature of this illustration the linearity feature does not seem all that worthwhile, but for more complicated problems the situation may be quite different as the next example will show. Consider the unconstrained minimization problem

$$\text{P: } \underset{x_1, x_2 > 0}{\text{Min}} \; z = c_1 x_1 x_2^{-3} + c_2 x_1^2 x_2^2 + c_3 x_1^{-1} x_2.$$

Proceeding according to classical methods of differential calculus, we would take the first partial derivatives with respect to x_1 and x_2 obtaining

$$\frac{\partial z}{\partial x_1} = c_1 x_2^{-3} + 2c_2 x_1 x_2^2 - c_3 x_1^{-2} x_2$$

$$\frac{\partial z}{\partial x_2} = -3c_1 x_1 x_2^{-4} + 2c_2 x_1^2 \bar{x}_2 + c_3 x_1^{-1}.$$

Finding a minimal point $(\bar{x}_1, \bar{x}_2)$ for P requires solving the following system of nonlinear equations, which constitutes the optimality conditions:

$$c_1 \bar{x}_2^{-3} + 2c_2 \bar{x}_1 \bar{x}_2^2 - c_3 \bar{x}_1^{-2} \bar{x}_2 = 0$$

$$-3c_1 \bar{x}_1 \bar{x}_2^{-4} + 2c_2 \bar{x}_1^2 \bar{x}_2 + c_3 \bar{x}_1^{-1} = 0.$$

This is a rather difficult task even though the example is fairly simple. In contrast, in order to solve this problem by the geometric programming approach we would define

$$\bar{\lambda}_1 := \frac{c_1 \bar{x}_1 \bar{x}_2^{-3}}{\bar{z}}, \quad \bar{\lambda}_2 := \frac{c_2 \bar{x}_1^2 \bar{x}_2^2}{\bar{z}}, \quad \bar{\lambda}_3 := \frac{c_3 \bar{x}_1^{-1} \bar{x}_2}{\bar{z}},$$

so that $\bar{\lambda}_1 + \bar{\lambda}_2 + \bar{\lambda}_3 = 1$ with $\bar{\lambda}_1$, $\bar{\lambda}_2$, $\bar{\lambda}_3 > 0$. Imitating the development carried out in the previous example, we divide both optimality conditions by $\bar{z}$ and multiply them by $\bar{x}_1$ and $\bar{x}_2$ respectively. This results in

$$\frac{c_1 \bar{x}_1 \bar{x}_2^{-3}}{\bar{z}} + 2\frac{c_2 \bar{x}_1^2 \bar{x}_2^2}{\bar{z}} - \frac{c_3 \bar{x}_1^{-1} \bar{x}_2}{\bar{z}} = 0$$

$$-3\frac{c_1 \bar{x}_1 \bar{x}_2^{-3}}{\bar{z}} + 2\frac{c_2 \bar{x}_1^2 \bar{x}_2^2}{\bar{z}} + \frac{c_3 \bar{x}_1^{-1} \bar{x}_2}{\bar{z}} = 0,$$

which, in terms of $\bar{\lambda}_1, \bar{\lambda}_2$ and $\bar{\lambda}_3$ turns out to be

$$\bar{\lambda}_1 + 2\bar{\lambda}_2 - \bar{\lambda}_3 = 0$$

$$-3\bar{\lambda}_1 + 2\bar{\lambda}_2 + \bar{\lambda}_3 = 0.$$

There are now three linear equations (the two immediately above and the requirement that the weights sum up to one) in three unknowns. The unique solution to this system is $\bar{\lambda}_1 = \frac{2}{7}$, $\bar{\lambda}_2 = \frac{1}{7}$, $\bar{\lambda}_3 = \frac{4}{7}$. Using the definitions of $\bar{\lambda}_1$, $\bar{\lambda}_2$, $\bar{\lambda}_3$, we obtain $\bar{z} = \bar{z}^1 = \bar{z}^{\bar{\lambda}_1 + \bar{\lambda}_2 + \bar{\lambda}_3} = \bar{z}^{\bar{\lambda}_1} \bar{z}^{\bar{\lambda}_2} \bar{z}^{\bar{\lambda}_3}$

$$= \left(\frac{c_1 \bar{x}_1 \bar{x}_2^{-3}}{\bar{\lambda}_1}\right)^{\bar{\lambda}_1} \left(\frac{c_2 \bar{x}_1^2 \bar{x}_2^2}{\bar{\lambda}_2}\right)^{\bar{\lambda}_2} \left(\frac{c_3 \bar{x}_1^{-1} \bar{x}_2}{\bar{\lambda}_3}\right)^{\bar{\lambda}_3} =$$

$$= \left(\frac{c_1}{\bar{\lambda}_1}\right)^{\bar{\lambda}_1} \left(\frac{c_2}{\bar{\lambda}_2}\right)^{\bar{\lambda}_2} \left(\frac{c_3}{\bar{\lambda}_3}\right)^{\bar{\lambda}_3} (x_1)^{\bar{\lambda}_1 + 2\bar{\lambda}_2 - \bar{\lambda}_3} (x_2)^{-3\bar{\lambda}_1 + 2\bar{\lambda}_2 + \bar{\lambda}_3}$$

$$= \left(\frac{c_1}{\bar{\lambda}_1}\right)^{\bar{\lambda}_1} \left(\frac{c_2}{\bar{\lambda}_2}\right)^{\bar{\lambda}_2} \left(\frac{c_3}{\bar{\lambda}_3}\right)^{\bar{\lambda}_3}.$$

Substituting the values of $\bar{\lambda}_1$, $\bar{\lambda}_2$, $\bar{\lambda}_3$ into this expression yields the optimal value $\bar{z}$ of the objective function. Finally, it is then a simple matter to solve for $\bar{x}_1$ and $\bar{x}_2$ by using the definitions of $\bar{\lambda}_1$, $\bar{\lambda}_2$ and $\bar{\lambda}_3$. For instance

$$\bar{\lambda}_1\bar{\lambda}_3 = \frac{c_1\bar{x}_1\bar{x}_2^{-3}}{\bar{z}}\frac{c_3\bar{x}_1^{-1}\bar{x}_2}{\bar{z}} = \frac{c_1c_3}{\bar{z}^2}\bar{x}_2^{-2},$$

which can be used obtain $\bar{x}_2$. Similarly one finds that

$$\bar{\lambda}_1\ \bar{\lambda}_2\ \bar{\lambda}_3 = \frac{c_1c_2c_3}{\bar{z}^3}\bar{x}_1^2,$$

which can be solved for $\bar{x}_1$. Here, as in the inventory problem, it turns out that the optimal weights $\bar{\lambda}_j$ do not depend on the objective function coefficients c_j. The optimal objective function value $\bar{z}$ is expressed in terms of the coefficients c_j and the optimal weights $\bar{\lambda}_j$ only. Again, instead of having to solve a system of nonlinear equations, only a system of three simultaneous linear equations has to be solved with the geometric programming approach. In other words, without this change of variables we would have been faced with the not so easy task of solving a system of two simultaneous nonlinear equations. Furthermore, solution techniques for solving this system could fail to find the proper roots even if they converge.

Summarizing, in the above two examples we were able to solve for the optimal weights $\bar{\lambda}_j$ using systems of linear equations and then finding the optimal solution points and values of the objective function, specified in terms of the optimal weights. This feature may intuitively appear surprising and perhaps applicable only in very special cases, but as the development below will demonstrate, this can systematically be exploited to solve a particular class of problems. We now introduce some basic terminology and notation.

Definition 8.1   A function $f(\mathbf{x})$ of the vector $\mathbf{x} = [x_1, x_2, \ldots, x_n]^T \in \mathbb{R}^n$ is a *polynomial* if it can be written as

$$z = f(\mathbf{x}) = \sum_{j=1}^{T} c_j p_j(\mathbf{x}),$$

where the numbers c_j are given nonzero constants, and the terms $p_j(\mathbf{x})$ are of the form $p_j(\mathbf{x}) = \prod_{i=1}^{n} x_i^{a_{ij}}$ with nonnegative integer coefficients a_{ij}. Such a polynomial is said to consist of n *variables*, T *terms*, and to be of *degree* $\max_{j}\left\{\sum_{i=1}^{n} a_{ij}\right\}$.

For instance, the function $f(x_1, x_2, x_3) = x_1x_2^3x_3^2 - \sqrt{2}x_2^4 + \frac{5}{7}x_1^5x_2^3 + x_2$ is a polynomial in three variables with four terms and of degree eight. The numbers c_j are referred to as the *economic coefficients* and the exponents a_{ij} are called the *technological coefficients*. Note that in keeping with accepted terminology, we use the letter T to indicate the number of terms, even though this might occasionally be confusing, since we also use the letter T to indicate the transpose of a matrix.

A *posynomial function* is similar to a polynomial except that the exponents, i.e., the technological coefficients a_{ij} can be any real numbers and that the economic coefficients c_j must be positive numbers. Thus the above polynomial is not a posynomial, because of the negative term $-\sqrt{2}x_2^4$. On the other hand, the function $f(x_1, x_2) = 2.31x_1x_2^2 + x_1^{-1}x_2^{0.5} + x_1^{0.19}x_2^3$ is a posynomial but not a polynomial since some of the exponents are noninteger numbers. Finally, $f(x) = \pi x^5$ is a polynomial as well as a posynomial, whereas the function $f(x) = -x^{0.5}$ is neither a polynomial nor a posynomial. It does however belong to the class of *signomial functions* where the a_{ij} are any real numbers and the c_j are nonzero real numbers. A signomial (posynomial, polynomial) function is a *monomial*, if it has only one term, i.e., $T = 1$. The relation between the set of polynomials and the set of posynomials as subsets of signomials can be visualized in Fig. 8.1.

Consider now the problem of minimizing a posynomial function without any constraints, assuming only that the optimal solution $\bar{\mathbf{x}} = \left[\bar{x}_1, \bar{x}_2, \ldots, \bar{x}_n\right]^T$ will be in the positive orthant, i.e., $\bar{x}_i > 0$ for $i = 1, 2, \ldots, n$. The problem is then

$$\text{P}: \underset{\mathbf{x} > \mathbf{0}}{\text{Min}}\, z = f(\mathbf{x}) = \sum_{j=1}^{T} c_j p_j(\mathbf{x}) = \sum_{j=1}^{T} c_j \prod_{i=1}^{n} x_i^{a_{ij}}.$$

At a minimal point $\bar{\mathbf{x}}$ we must have $\nabla f(\bar{\mathbf{x}}) = \mathbf{0}$, i.e.,

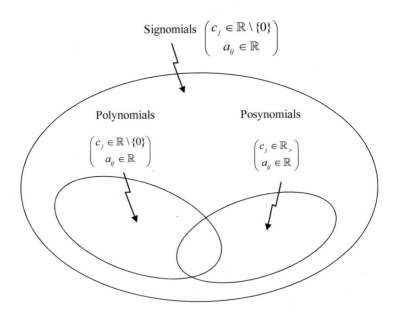

Fig. 8.1 Relations between polynomials, posynomials, and signomials

$$\frac{\partial z}{\partial x_i} = \sum_{j=1}^{T} c_j a_{ij} \bar{x}_1^{-a_{1j}} \bar{x}_2^{-a_{2j}} \cdot \ldots \cdot \bar{x}_{i-1}^{-a_{i-1,j}} \bar{x}_i^{-a_{ij}-1} \bar{x}_{i+1}^{-a_{i+1,j}} \cdot \ldots \cdot \bar{x}_n^{-a_{nj}} = 0, \quad i = 1, 2, \ldots, n.$$

Since $\bar{x}_i > 0$ is required, each of the above equations can be multiplied by $\bar{x}_i$ to obtain $\sum_{j=1}^{T} c_j a_{ij} p_j(\bar{\mathbf{x}}) = 0$, $i = 1, 2, \ldots, n$, which is a system of nonlinear equations. Again, instead of solving the nonlinear system for an optimal point $\bar{\mathbf{x}}$, we define weights $\bar{\lambda}_j$, $j = 1, 2, \ldots, T$ by $\bar{\lambda}_j := \dfrac{c_j p_j(\bar{\mathbf{x}})}{\bar{z}}$ where $\bar{z}$ stands for the optimal function value, i.e.,

$$\bar{z} = f(\bar{\mathbf{x}}) = \sum_{j=1}^{T} c_j p_j(\bar{\mathbf{x}}) = \sum_{j=1}^{T} c_j \prod_{i=1}^{n} \bar{x}_i^{a_{ij}}.$$

Since c_j, $p_j(\bar{\mathbf{x}})$ and $\bar{z}$ are all positive, it follows that $\bar{\lambda}_j > 0$, $j = 1, 2, \ldots, T$. Furthermore, the weights must sum up to one, i.e., $\sum_{j=1}^{T} \bar{\lambda}_j = 1$. This requirement is called the *normality condition*. Dividing the nonlinear equations above by the positive number $\bar{z}$, the so-called *orthogonality conditions* $\sum_{j=1}^{T} a_{ij} \bar{\lambda}_j = 0$, $i = 1, 2, \ldots, n$ are obtained. Geometrically this means that the vector $[\bar{\lambda}_1, \bar{\lambda}_2, \ldots, \bar{\lambda}_T]^T$ is orthogonal to each of the vectors $\mathbf{a}_{i\bullet} = [a_{i1}, a_{i2}, \ldots, a_{iT}]^T$, $i = 1, \ldots, n$; hence the name orthogonality conditions. Next, along the lines of the discussion concerning the two examples above we write

$$\bar{z} = (\bar{z})^{\sum_{j=1}^{T} \bar{\lambda}_j} = \prod_{j=1}^{T} (\bar{z})^{\bar{\lambda}_j} = \prod_{j=1}^{T} \left(\frac{c_j p_j(\bar{\mathbf{x}})}{\bar{\lambda}_j} \right)^{\bar{\lambda}_j} = \left[\prod_{j=1}^{T} \left(\frac{c_j}{\bar{\lambda}_j} \right)^{\bar{\lambda}_j} \right] \left[\prod_{j=1}^{T} (p_j(\bar{\mathbf{x}}))^{\bar{\lambda}_j} \right].$$

However, $\prod_{j=1}^{T} (p_j(\bar{\mathbf{x}}))^{\bar{\lambda}_j} = \prod_{j=1}^{T} \left(\prod_{i=1}^{n} \bar{x}_i^{a_{ij}} \right)^{\bar{\lambda}_j} = \prod_{i=1}^{n} (\bar{x}_i)^{\sum_{j=1}^{T} a_{ij} \bar{\lambda}_j} = \prod_{i=1}^{n} (\bar{x}_i)^0 = 1$ and therefore

$$\bar{z} = \prod_{j=1}^{T} \left(\frac{c_j}{\bar{\lambda}_j} \right)^{\bar{\lambda}_j}.$$

One can easily verify that the above relations also hold without necessarily requiring the weights λ_j to be optimal as long as they satisfy the normality and orthogonality conditions. Thus one can state

Theorem 8.2 Let $\lambda_j > 0$, $j = 1,\ldots, T$ be a set of positive weights satisfying the normality condition $\sum_{j=1}^{T} \lambda_j = 1$ and the orthogonality conditions $\sum_{j=1}^{T} a_{ij}\lambda_j = 0$, $i = 1, 2, \ldots, n$. Then

$$z = \prod_{j=1}^{T} \left(\frac{c_j p_j}{\lambda_j} \right)^{\lambda_j} = \prod_{j=1}^{T} \left(\frac{c_j}{\lambda_j} \right)^{\lambda_j}.$$

Definition 8.3 The expression $\prod_{j=1}^{T} \left(\frac{c_j p_j}{\lambda_j} \right)^{\lambda_j}$ is called the *predual function* and the expression $\prod_{j=1}^{T} \left(\frac{c_j}{\lambda_j} \right)^{\lambda_j}$ is referred to as the *dual function* of problem P.

Back now to the case with optimal weights $\bar{\lambda}_j$. If optimal $\bar{\lambda}_j$ values are known, then the optimal objective function value $\bar{z}$ can be calculated by using the result of Theorem 8.2. In case both $\bar{z}$ and the values $\bar{\lambda}_j$ are known, then the values for $\bar{x}_i$ can be computed in the following way. Given that $\bar{\lambda}_j\bar{z} = c_j p_j(\bar{\mathbf{x}}) = c_j \prod_{i=1}^{n} \bar{x}_i^{a_{ij}}, j = 1, 2, \ldots, T$, one can take logarithms resulting in

$$\ln \left(\bar{\lambda}_j \bar{z} \right) = \ln c_j + \sum_{i=1}^{n} a_{ij} \ln \bar{x}_i, \; j = 1, \; 2, \; \ldots, \; T,$$

which is now a system of linear equations in $\ln \bar{x}_i$. If there are enough equations so that $\ln \bar{x}_i$ can be found, then it is easy to calculate the values of $\bar{x}_i$ for $i = 1, 2,\ldots n$. A unique solution for the T unknowns $\bar{\lambda}_j, j = 1, 2,\ldots, T$ is of course available only if there are at least T equations to be solved for them. Here one normality condition and n orthogonality conditions are given; so that uniqueness for the $\bar{\lambda}_j$ values requires that $T \leq n + 1$. If T is greater than $n + 1$, the optimal weights cannot be determined uniquely. The number $T - (n + 1)$ is called the *degree of difficulty*. The larger this number, the more difficult it is to solve the problem. With a zero degree of difficulty, then not only can a unique solution for the weights be found but there are also enough equations to obtain the optimal $\bar{x}_i$ values. Upon checking, we find that both of the examples we solved above have a zero degree of difficulty.

8.2 Duality Theory for Geometric Programming

Having shown that a geometric programming problem with a zero degree of difficulty can be solved using the normality and the orthogonality conditions, we will now explore how proceed if the optimal weights cannot be uniquely determined by the equations of normality and orthogonality. In other words, we will consider problems with a degree of difficulty of one or more.

As a beginning consider the inequality of arithmetic and geometric means which in the simplest version states that for two given positive numbers a and b, the arithmetic mean $\frac{1}{2}(a + b)$ is greater than or equal to the geometric mean $(ab)^{1/2}$. Equality occurs if and only if $a = b$. Geometrically speaking, if a and b are the sides of a rectangle, then the geometric mean is equal to the side of a square having the same area. The inequality then implies that the square has the least perimeter among all rectangles with the same area. Generalizing, we can state

Definition 8.4 Given positive numbers α_j and $\lambda_j, j = 1, \ldots, T$, such that $\sum_{j=1}^{T} \lambda_j = 1$, then the *arithmetic mean A* of the numbers α_j with *weights* λ_j is defined by the expression $A = \sum_{j=1}^{T} \alpha_j \lambda_j$, and the *geometric mean G* is defined as $G = \prod_{j=1}^{T} \alpha_j^{\lambda_j}$.

We continue by stating and proving the following

Theorem 8.5 Let $\alpha_1, \alpha_2, \ldots, \alpha_T$ be given positive numbers and let $\lambda_1, \lambda_2, \ldots, \lambda_T$ be a given set of weights with $\lambda_j > 0, j = 1, 2, \ldots, T$ and $\sum_{j=1}^{T} \lambda_j = 1$. Then the arithmetic–geometric mean inequality can be stated as $A \geq G$ or

$$\sum_{j=1}^{T} \lambda_j \alpha_j \geq \prod_{j=1}^{T} \alpha_j^{\lambda_j},$$

with equality if and only if $\alpha_1 = \alpha_2 = \ldots = \alpha_T$.

Proof Let $f(\alpha) = \ln \alpha$ for $\alpha > 0$. Taking logarithms in the statement of the theorem, and since the logarithmic function is strictly increasing, we obtain the equivalent inequality

$$f\left(\sum_{j=1}^{T} \lambda_j \alpha_j\right) \geq f\left(\prod_{j=1}^{T} \alpha_j^{\lambda_j}\right) = \sum_{j=1}^{T} \lambda_j f(\alpha_j),$$

with equality if and only if all α_j are identical. First we prove this inequality for $T = 2$. Then we set $\lambda_1 := t$ and $\lambda_2 := 1 - t$, and for $0 \leq t \leq 1$, we define the function $g(t) = \ln [(1 - t) \alpha_1 + t\alpha_2] - (1 - t)\ln\alpha_1 - t\ln\alpha_2$. If $\alpha_1 \neq \alpha_2$, this function satisfies $g(0) = g(1) = 0$. Furthermore, differentiating $g(t)$ twice yields $g''(t) = -\dfrac{(\alpha_1 - \alpha_2)^2}{(1 - t)\alpha_1 + t\alpha_2}$, which is negative for $0 \leq t \leq 1$. We now claim that $g(t) > 0$ for $0 < t < 1$. If not, then $g(t) \leq 0$ at some interior point t of the interval $[0; 1]$ and therefore $g(t)$ would attain its minimal value at some interior point t_0 where $g'(t_0)$ is required to be zero and $g''(t_0)$ is nonnegative. This is impossible since we showed that since $g''(t) < 0$. Hence the claim of strict inequality holds true, and if we take exponentials the required result follows for the case $T = 2$. It is easily seen that for $\alpha_1 = \alpha_2$ the two sides of the inequality are $\alpha_1 = \alpha_2$.

Consider now the case $T > 2$ and proceed by induction, assuming that the desired result is true for $T - 1$. For that purpose, let β be a positive number such that $\lambda_1\alpha_1 + \lambda_2\alpha_2 + \ldots + \lambda_T\alpha_T = \lambda_1\alpha_1 + (1 - \lambda_1)\beta$, noting that by assumption, all $\lambda_j < 1$, so that $\beta = \dfrac{\lambda_2\alpha_2 + \ldots + \lambda_T\alpha_T}{1 - \lambda_1}$. Then the result for $T = 2$ implies that $f[\lambda_1\alpha_1 + (1 - \lambda_1)\beta] \geq \lambda_1 f(\alpha_1) + (1 - \lambda_1)f(\beta)$ and the induction argument gives $f(\beta) = f[(1 - \lambda_1)^{-1}(\lambda_2\alpha_2 + \ldots + \lambda_T\alpha_T)] \geq (1-\lambda_1)^{-1}\lambda_1 f(\alpha_2) + \ldots + (1 - \lambda_1)^{-1}\lambda_T f(\alpha_T)$.

By using this relation in the previous inequality the desired general result is

$$f(\lambda_1\alpha_1 + \cdots + \lambda_T\alpha_T) \geq \lambda_1 f(\alpha_1) + \cdots + \lambda_T f(\alpha_T),$$

and taking exponentials on both sides yields the desired inequality about arithmetic and geometric means. It is easy to see that again, equality occurs if and only if $\alpha_1 = \alpha_2 = \ldots = \alpha_T$ (which then equal β), which completes the proof. $\square$

Remark 1 In geometric programming the inequality of arithmetic and geometric means often appears in the form

$$\sum_{j=1}^{T} x_j \geq \prod_{j=1}^{T} \left(\frac{x_j}{\lambda_j}\right)^{\lambda_j},$$

where $x_j > 0, \lambda_j > 0$ for $j = 1, 2, \ldots, T$ and $\sum_{j=1}^{T}\lambda_j = 1$. Setting $x_j := \lambda_j\alpha_j$ we can see that this form is equivalent to the one presented before and that equality now holds if and only if $\dfrac{x_1}{\lambda_1} = \ldots = \dfrac{x_T}{\lambda_T}$. This means that for any sum we can construct a variety of related geometric means and although the latter ones are always less than or equal to the former there is always some set of values for $\lambda_1, \lambda_2, \ldots, \lambda_T$ that causes the two sides to be identical. (Actually, this is achieved by setting $\lambda_j = \dfrac{x_j}{\sum_{\ell=1}^{T} x_\ell}$).

If in the above formulation some, but not all, of the weights λ_j are equal to zero, we define the term $\left(\dfrac{x_j}{\lambda_j}\right)^{\lambda_j} := 1$ for $\lambda_j = 0$. With this definition, the previous specification can be replaced by the broader one $x_j > 0, \lambda_j \geq 0$ and $\sum_{j=1}^{T}\lambda_j = 1$, $j = 1, 2, \ldots, T$ where some weights λ_j are allowed to be zero. If $J := \{j : \lambda_j > 0\} \subsetneq \{1, 2, \ldots, T\}$, then $\sum_{j=1}^{T} x_j > \sum_{j \in J} x_j$ and the inequality in Remark 1 implies that

$$\sum_{j \in J} x_j \geq \prod_{j \in J} \left(\frac{x_j}{\lambda_j}\right)^{\lambda_j}.$$

The product over the set J may however be taken over all indices $j = 1, 2, \ldots,$ T since the additional factors equal one anyway. It follows that the last two expressions combined imply the strict inequality $\sum_{j=1}^{T} x_j > \prod_{j=1}^{T} \left(\frac{x_j}{\lambda_j}\right)^{\lambda_j}$ when at least one of the weights λ_j is zero.

Remark 2 Yet another modification of the arithmetic–geometric mean inequality is obtained if weights are allowed that do not necessarily sum to one. Suppose now that such weights are μ_j with $\mu_1 + \mu_2 + \ldots + \mu_T := \mu$, which is assumed to be positive, but not necessarily equal to one.

Defining $\lambda_j := \frac{\mu_j}{\mu}$, $j = 1, \ldots, T$, the inequality in Remark 1 above becomes

$$\sum_{j=1}^{T} \alpha_j \geq \prod_{j=1}^{T} \left(\frac{\alpha_j}{\lambda_j}\right)^{\lambda_j} = \prod_{j=1}^{T} \left(\frac{\alpha_j}{\mu_j}\right)^{\left(\frac{\mu_j}{\mu}\right)} \mu^{\left(\frac{\mu_j}{\mu}\right)} = \mu \prod_{j=1}^{T} \left(\frac{\alpha_j}{\mu_j}\right)^{\left(\frac{\mu_j}{\mu}\right)},$$

or, equivalently,

$$\left(\sum_{j=1}^{T} \alpha_j\right)^{\mu} \geq \mu^{\mu} \prod_{j=1}^{T} \left(\frac{\alpha_j}{\mu_j}\right)^{\mu_j},$$

with equality if and only if $\frac{\alpha_1}{\mu_1} = \frac{\alpha_2}{\mu_2} = \ldots \frac{\alpha_T}{\mu_T}$.

Remark 3 In addition to the arithmetic and geometric means, there is also the *harmonic mean H*. Based on the assumptions in Definition 8.4, H is defined as the inverted arithmetic mean of the inverted values of α_j, i.e., $H = \left(\sum_{j=1}^{T} \lambda_j \alpha_j^{-1}\right)^{-1}$. The name harmonic derives from the fact that each term in the divergent so-called harmonic series $\sum_{j=1}^{\infty} \frac{1}{j}$ is the harmonic mean of its immediate predecessor and successor terms. The harmonic mean is useful, since it is always smaller than or equal to the geometric mean G. This is easy to see, since $\frac{1}{H} = \sum_{j=1}^{T} \frac{\lambda_j}{\alpha_j} \geq \prod_{j=1}^{T} \left(\frac{1}{\alpha_j}\right)^{\lambda_j} = \frac{1}{\prod_{j=1}^{T} \alpha_j^{\lambda_j}} = \frac{1}{G}$, where the inequality follows from the arithmetic-geometric mean inequality for the numbers $1/\alpha_j$, $j = 1, \ldots, T$. With $1/H \geq 1/G$, we immediately obtain $H \leq G$, and invoking the arithmetic-geometric mean inequality for α_j, $j = 1, \ldots, T$, we finally obtain $H \leq G \leq A$, providing useful bounds for the three different means.

Returning to our problem P, we now consider again the posynomial function

$$z = f(\mathbf{x}) = \sum_{j=1}^{T} c_j p_j(\mathbf{x}) = \sum_{j=1}^{T} \left(c_j \prod_{i=1}^{n} x_i^{a_{ij}} \right).$$

Selecting any set of positive weights λ_j, $j = 1, \ldots, T$, we can write

$$z = \sum_{j=1}^{T} c_j p_j = \sum_{j=1}^{T} \left(\frac{c_j p_j}{\lambda_j} \right) \lambda_j \geq \prod_{j=1}^{T} \left(\frac{c_j p_j}{\lambda_j} \right)^{\lambda_j}$$

in view of Theorem 8.5 above. We recognize the right-hand side as the predual function of P. The inequality is satisfied for all $x_i > 0$, $i = 1, 2, \ldots, n$ and for all coefficients $\lambda_j \geq 0$ with $\sum_{j=1}^{T} \lambda_j = 1$. If $\bar{x}_i$ and $\bar{\lambda}_j$ are chosen, such that $\dfrac{c_j p_j(\bar{\mathbf{x}})}{\bar{\lambda}_j} = \bar{z}$, where $\bar{\mathbf{x}}$ and $\bar{z}$ denote an optimal solution point and objective function value of problem P, then equality would hold true above in accordance with the arithmetic−geometric mean inequality with equal values α_j. This proves

Theorem 8.6 Let $\bar{\mathbf{x}}$ and $\bar{z}$ denote an optimal solution point and its associated objective function value of problem P and set $\bar{\lambda}_j := \dfrac{c_j p_j(\bar{\mathbf{x}})}{\bar{z}}$, $j = 1, \ldots, T$. Then we have

$$z \geq \prod_{j=1}^{T} \left(\frac{c_j p_j}{\lambda_j} \right)^{\lambda_j}$$

for any set of weights $\lambda_j > 0$, $j = 1, \ldots, T$, satisfying the normality condition. Equality holds for $z = \bar{z}$, $p_j = \bar{p}_j = p_j(\bar{\mathbf{x}})$, and $\lambda_j = \bar{\lambda}_j$ for $j = 1, \ldots, T$.

Assume now that optimal values $\bar{\mathbf{x}}$ and $\bar{z}$ are used in the above theorem, so that

$$\bar{z} \geq \prod_{j=1}^{T} \left(\frac{c_j p_j(\bar{\mathbf{x}})}{\lambda_j} \right)^{\lambda_j}$$

for all values of $\lambda_j > 0$, $j = 1, \ldots, T$, and $\sum_{j=1}^{T} \lambda_j = 1$. The function of $\lambda_j, j = 1, 2, \ldots,$ T on the right−hand side of this expression is bounded from above by $\bar{z}$; note that such a bound is attained for $\lambda_j = \bar{\lambda}_j$ for $j = 1, 2, \ldots, T$. In order to find $\bar{z}$, one might try to maximize the right−hand side of the above inequality. Unfortunately, the formula contains the factors $p_j(\bar{\mathbf{x}}) = \bar{p}_j$ whose values are not known. To get around this difficulty, we consider instead the following constrained problem

$$P_D : \underset{\lambda}{\text{Max}}\; z_d(\lambda) = \prod_{j=1}^{T} \left(\frac{c_j}{\lambda_j}\right)^{\lambda_j}$$

$$\text{s.t.} \sum_{j=1}^{T} a_{ij}\lambda_j = 0,\; i = 1, \ldots, n$$

$$\sum_{j=1}^{T} \lambda_j = 1$$

$$\lambda_j > 0,\; j = 1,\, 2, \ldots, T.$$

This constrained optimization problem will be referred to as the *dual problem* with *dual variables* λ_j. Note that the objective function $z_d(\lambda)$ is the dual function of Definition 8.3. Since all the constraints of problem P_D are linear, the region of feasible solutions is convex. The justification for considering this problem instead of the maximization of the above lower bound expression for $\bar{z}$, i.e.,

$$\prod_{j=1}^{T} \left(\frac{c_j \bar{p}_j}{\lambda_j}\right)^{\lambda_j}$$

is that as long as all constraints of P_D are fulfilled, the values of the two objective functions are actually the same, due to Theorem 8.2. Furthermore, it can be shown that the unique solution to the dual problem is $\bar{\lambda}_j$, $j = 1, \ldots, T$:

Theorem 8.7 The unique optimal solution to the dual problem P_D is given by

$$\lambda_j = \frac{c_j p_j(\bar{\mathbf{x}})}{\bar{z}},\; j = 1, \ldots, T.$$

Proof Assume that an optimal solution λ_j^*, $j = 1, \ldots,\; T$ for the dual problem P_D has been found. From the previous discussion it is also known that $\bar{\lambda}_j = \frac{c_j p_j(\bar{\mathbf{x}})}{\bar{z}}$, $j = 1, \ldots, T$, is an optimal solution. Due to Theorem 8.2 we therefore have

$$\bar{z} = z_d(\lambda^*) = \prod_{j=1}^{T} \left(\frac{c_j}{\lambda_j^*}\right)^{\lambda_j^*} = \prod_{j=1}^{T} \left(\frac{c_j \bar{p}_j}{\lambda_j^*}\right)^{\lambda_j^*}.$$

On the other hand the arithmetic–geometric mean inequality yields

$$\bar{z} = \sum_{j=1}^{T} c_j \bar{p}_j = \sum_{j=1}^{T} \lambda_j^* \left(\frac{c_j \bar{p}_j}{\lambda_j^*} \right) \geq \prod_{j=1}^{T} \left(\frac{c_j \bar{p}_j}{\lambda_j^*} \right)^{\lambda_j^*},$$

with equality if and only if $\dfrac{c_j \bar{p}_j}{\lambda_j^*}$ are all equal, $j = 1, \ldots, T$; one can see that their

common value must then be $\bar{z}$. It follows that $\dfrac{c_j \bar{p}_j}{\lambda_j^*} = \bar{z}$, $j = 1, \ldots, T$ implies that

$\lambda_j^* = \dfrac{c_j \bar{p}_j}{\bar{z}} = \bar{\lambda}_j$, $j = 1, \ldots, T$ and the uniqueness of the optimal solution is
proved. $\square$

Once $\bar{z}$ and $\bar{\lambda}_j$, $j = 1, \ldots, T$ are computed, one can try to solve for $\bar{x}_i$, $i = 1,$
$\ldots, n$ as indicated at the end of Sect. 8.1. Rather than using the dual problem in its
current form it is more convenient to take the natural logarithm of the dual objective
function, i.e.,

$$\ln z_d(\lambda) = \ln \left[\prod_{j=1}^{T} \left(\frac{c_j}{\lambda_j} \right)^{\lambda_j} \right] = \sum_{j=1}^{T} \left[\lambda_j \ln \left(\frac{c_j}{\lambda_j} \right) \right] = - \sum_{j=1}^{T} \left[\lambda_j \ln \left(\frac{\lambda_j}{c_j} \right) \right].$$

One can now work with $\ln z_d(\lambda)$ rather than with $z_d(\lambda)$ as the objective function for
the dual problem, as the transformation is a strictly increasing function. The new
objective function $\ln z_d(\lambda)$ is concave with respect to the weights since it can be
expressed as the negative of a sum of convex functions, and the fact that

$$\frac{\mathrm{d}^2}{\mathrm{d}x^2} \left[x \ln \left(\frac{x}{c} \right) \right] = \frac{\mathrm{d}}{\mathrm{d}x} \left[\ln \left(\frac{x}{c} \right) + 1 \right] = \frac{1}{x} > 0 \quad \text{for} \quad x > 0$$

proves the convexity of $\lambda_j \ln \left(\dfrac{\lambda_j}{c_j} \right)$. It follows that the transformed dual problem P_D
consists of maximizing a concave objective function where the variables λ_j,
constrained to be positive, are subject to the normality and orthogonality conditions.
As shown above, the global maximal objective function value of the dual problem
will coincide with the global minimal objective function value of the primal problem
so that there is no duality gap. In general it is easier to solve the dual than the primal.
Furthermore, feasible solutions to the primal and dual problems, even if they are not
optimal, are useful since they generate upper and lower bounds for the true optimal
primal objective function value; this was referred to as "weak duality" in Sect. 4.3.

Having considered the primal and the dual of an unconstrained geometric programming problem, the constrained case will now be discussed. For now, we assume
that there is only one posynomial constraint with S terms. The case with several
constraints will be discussed later. The problem is then

$$\text{P: } \underset{\mathbf{x}}{\text{Min}}\, z = \sum_{j=1}^{T} c_j \prod_{i=1}^{n} x_i^{a_{ij}}$$

$$\text{s.t.} \sum_{k=1}^{S} d_k \prod_{i=1}^{n} x_i^{b_{ik}} \leq 1$$

$$x_i > 0, \ i = 1, \ldots, n.$$

Recall that $p_j(\mathbf{x}) = \prod_{i=1}^{n} x_i^{a_{ij}}$ so that the objective function can also be written as $f(\mathbf{x}) = \sum_{j=1}^{T} c_j p_j$. Similarly, we define $q_k(\mathbf{x}) = \prod_{i=1}^{n} x_i^{b_{ik}}$ so that the left−hand side of the posynomial constraint can be written as $g(\mathbf{x}) = \sum_{k=1}^{S} d_k q_k$. Using the arithmetic−geometric mean inequality in the version of Remark 2 above, with weights not necessarily summing to one, assign nonnegative weights λ_j, $j = 1$, ..., T with sum $\lambda > 0$ to the objective function $f(\mathbf{x})$ and nonnegative weights μ_k, $k = 1,\ldots, S$ with sum $\mu > 0$ to the constraint function $g(\mathbf{x})$. Then

$$[f(\mathbf{x})]^{\lambda} = \left(\sum_{j=1}^{T} c_j p_j \right)^{\lambda} \geq \lambda^{\lambda} \prod_{j=1}^{T} \left(\frac{c_j p_j}{\lambda_j} \right)^{\lambda_j}, \quad \text{and}$$

$$1 \geq [g(\mathbf{x})]^{\mu} = \left(\sum_{k=1}^{S} d_k q_k \right)^{\mu} \geq \mu^{\mu} \prod_{k=1}^{S} \left(\frac{d_k q_k}{\mu_k} \right)^{\mu_k}.$$

The product of the extreme left−hand sides of these inequalities must be greater than or equal to the product of the extreme right−hand sides, i.e.,

$$[f(\mathbf{x})]^{\lambda} \geq \lambda^{\lambda} \mu^{\mu} \left[\prod_{j=1}^{T} \left(\frac{c_j p_j}{\lambda_j} \right)^{\lambda_j} \right] \left[\prod_{k=1}^{S} \left(\frac{d_k q_k}{\mu_k} \right)^{\mu_k} \right].$$

In this inequality we now add the requirement that the normality and orthogonality conditions hold for the weights λ_j as in the unconstrained case, so that their sum equals one, whereas the weights μ_k are only required to satisfy orthogonality, but not normality. The reason for requiring normality for the weights λ_j is to retain their interpretation as weights of the terms in the objective function; this does not apply to the constraint. The application of Theorem 8.2 modified for the fact that the variables μ_k need not be normalized then results in

Theorem 8.8 Assume that in the inequality constrained geometric programming problem P the following normality and orthogonality conditions are satisfied:

$$\sum_{j=1}^{T} \lambda_j = 1, \quad \sum_{j=1}^{T} a_{ij}\lambda_j = 0, \ i = 1, \ldots, n, \text{ and } \lambda_j \geq 0 \text{ for } j = 1, \ldots, T,$$

as well as

$$\sum_{k=1}^{S} \mu_k = \mu > 0, \sum_{k=1}^{S} b_{ik}\mu_k = 0, \ i = 1, \ldots, \ n, \quad \mu_k \geq 0, \ k = 1, \ldots, S.$$

Then $f(\mathbf{x}) \geq \mu^{\mu} \left[\prod_{j=1}^{T} \left(\frac{c_j}{\lambda_j} \right)^{\lambda_j} \right] \left[\prod_{k=1}^{S} \left(\frac{d_k}{\mu_k} \right)^{\mu_k} \right]$ for any feasible $\mathbf{x}$.

The inequality in the above theorem can also be written in the equivalent form

$$f(\mathbf{x}) \geq \left[\prod_{j=1}^{T} \left(\frac{c_j}{\lambda_j} \right)^{\lambda_j} \right] \left[\prod_{k=1}^{S} \left(\frac{d_k \mu}{\mu_k} \right)^{\mu_k} \right],$$

and it is also easy to see that the $2n$ orthogonality conditions of Theorem 8.8 can be replaced by the n combined conditions

$$\sum_{j=1}^{T} a_{ij}\lambda_j + \sum_{k=1}^{S} b_{ik}\mu_k = 0, \ i = 1, \ldots, n.$$

In analogy with the unconstrained case, the dual problem P_D of the constrained primal problem P is then defined as

$$P_D : \text{Max } z_d(\boldsymbol{\lambda}, \boldsymbol{\mu}) = \left[\prod_{j=1}^{T} \left(\frac{c_j}{\lambda_j} \right)^{\lambda_j} \right] \left[\prod_{k=1}^{S} \left(\frac{d_k \mu}{\mu_k} \right)^{\mu_k} \right]$$

$$\text{s.t.} \sum_{j=1}^{T} a_{ij}\lambda_j + \sum_{k=1}^{S} b_{ik}\mu_k = 0, \ i = 1, \ldots, n$$

$$\sum_{j=1}^{T} \lambda_j = 1$$

$$\sum_{k=1}^{S} \mu_k = \mu$$

$$\lambda_j > 0, \ j = 1, 2, \ldots, T$$

$$\mu_k \geq 0, \quad k = 1, 2, \ldots, S.$$

Note that we have used $\mathbf{\mu}$ to denote the vector $\mathbf{\mu} = [\mu_1, \ldots, \mu_s]$, which should not be confused with the scalar μ, which is the sum of the components of the vector $\mathbf{\mu}$. As stated earlier, the weights λ_j may be thought of as the relative contributions of the individual terms of the objective function $f(\mathbf{x})$. Luptáčik (1977, 1981a, b) has shown that μ may be interpreted as an elasticity coefficient, i.e., the relative change in the value of the primal objective function $f(\mathbf{x})$ resulting from a perturbation of the right–hand side of the primal posynomial constraint.

Consider now an optimal solution $\bar{\mathbf{x}}$ to the primal problem and an optimal solution $(\bar{\lambda}, \bar{\mu})$ to the dual problem, assuming that the primal and the dual problems both have feasible solutions. Due to Theorem 8.8, we know that $f(\bar{\mathbf{x}}) \geq z_d(\bar{\lambda}, \bar{\mu})$. In order to show that equality holds just as in the unconstrained case, so that there is no duality gap, two separate cases have to be considered.

Case 1: The constraint is tight, i.e., $g(\bar{\mathbf{x}}) = 1$.

From the discussion preceding Theorem 8.8, it follows that equality holds if the arithmetic–geometric mean inequality as well as the normality and orthogonality conditions are used.

Case 2: The constraint is loose, i.e., $g(\bar{\mathbf{x}}) < 1$.

Set $\mu_k := 0$, $k = 1, \ldots, S$ and use the definition $\lim\limits_{\mu_k \to +0} \left(\dfrac{d_k \mu}{\mu_k} \right)^{\mu_k} = 1$. The dual objective function will then be $z_d(\lambda, \mu) = z_d(\lambda, \mathbf{0}) = \prod\limits_{j=1}^{T} \left(\dfrac{c_j}{\lambda_j} \right)^{\lambda_j}$ and problem P_D will take the same form as in the unconstrained case. For the primal problem P, the loose constraint can be dropped leading back to the unconstrained situation for which it has already been proved that no duality gap exists.

We will later show that Karush-Kuhn-Tucker necessary conditions for optimality are

$$\bar{\lambda}_j \bar{z} = c_j \bar{p}_j, \quad j = 1, \ldots, T$$

$$\bar{\mu}_k = \bar{\mu} d_k \bar{q}_k, \quad k = 1, \ldots, S.$$

We will also show that if the primal constraint is tight, i.e., $g(\bar{\mathbf{x}}) = 1$, then either $\bar{\mu}_k$ are all zero or all positive. If, on the other hand, the primal constraint is loose, i.e., $g(\bar{\mathbf{x}}) < 1$, then there exists a solution with $\bar{\mu}_k = 0$, $k = 1, 2, \ldots, S$. These results will be more conveniently discussed and proved as Theorem 8.9 in Sect. 8.3.2. Along the lines of the above discussion, problem with more than one primal posynomial constraint could also be treated.

8.3   Solution Techniques for Geometric Programming Problems

8.3.1   Solution Techniques for Unconstrained Posynomial Geometric Programming Problems

In this section we will restrict ourselves to posynomial unconstrained geometric programming. Recalling our discussion in the previous sections, we consider the problem

$$P: \min_{\mathbf{x} > 0} z = f(\mathbf{x}) = \sum_{j=1}^{T} c_j p_j(\mathbf{x}) = \sum_{j=1}^{T} c_j \left(\prod_{i=1}^{n} x_i^{a_{ij}} \right),$$

where $f(\mathbf{x})$ is a posynomial function, i.e. the economic coefficients $c_j, j = 1, \ldots, T$ are positive real numbers and the technological coefficients $a_{ij}, i = 1, \ldots, n; j = 1, \ldots, T$ are any real numbers. The objective function $f(\mathbf{x})$ has T terms $c_j p_j, j = 1, \ldots, T$ and n variables $x_i, j = 1, \ldots, n$. Recalling that only the strictly positive orthant is considered, i.e., $\{\mathbf{x} | x_i > 0, i = 1, \ldots, n\} = \mathbb{R}^n_>$, it is clear that a posynomial function is always continuous. The function $f(x) = x^2$ shows that a posynomial may be convex; $f(x) = \sqrt{x}$ shows that a posynomial may be concave and $f(x_1, x_2) = x_1 x_2$ shows that a posynomial may be neither convex nor concave.

It is also important to realize that an optimal solution to problem P does not necessarily exist. The function $f(x) = x, x \in \mathbb{R}$ is an example without a solution to the problem $\underset{x>0}{\text{Min}} f(x)$, although for $x \geq 0$ an optimal solution would exist. Similarly, $f(x) = 1/x$ where $x \in \mathbb{R} \backslash \{0\}$ is an example where a finite solution to the problem $\underset{x>0}{\text{Min}} f(x)$ does not exist. Also, even if an optimal solution $x > 0$ should exist it need not be unique as the function $f(x) = x^0 = 1$ shows.

Assume now that an optimal solution does indeed exist. Introducing positive weights λ_j, then the normality and orthogonality relations must hold, as shown in the discussion preceding Theorem 8.2 in Sect. 8.1 above, i.e., $\sum_{j=1}^{T} \lambda_j = 1$ and $\sum_{j=1}^{T} a_{ij} \lambda_j = 0, i = 1, \ldots, n$. At optimality, the weights λ_j must satisfy the relations

$$\bar{z} = \frac{c_1 p_1(\bar{\mathbf{x}})}{\bar{\lambda}_1} = \frac{c_2 p_2(\bar{\mathbf{x}})}{\bar{\lambda}_2} = \cdots = \frac{c_T p_T(\bar{\mathbf{x}})}{\bar{\lambda}_T}.$$

The normality and orthogonality conditions provide $n + 1$ equations to solve for the T weights λ_j. If $n + 1 \geq T$ one might therefore hope to find a unique solution of this system of equations. As an illustration consider the second example from Sect. 8.1, which was

$$\text{P: } \underset{x_1,x_2>0}{\text{Min}} \; z = c_1 x_1 x_2^{-3} + c_2 x_1^2 x_2^2 + c_3 x_1^{-1} x_2.$$

The normality and orthogonality conditions for this problem are

$$\lambda_1 + \lambda_2 + \lambda_3 = 1$$
$$\lambda_1 + 2\lambda_2 - \lambda_3 = 0$$
$$-3\lambda_1 + 2\lambda_2 + \lambda_3 = 0.$$

Solving this system of three linear equation in the three unknowns λ_1, λ_2, and λ_3, we find the unique solution

$$\bar{\lambda}_1 = \tfrac{2}{7}, \; \bar{\lambda}_2 = \tfrac{1}{7}, \; \bar{\lambda}_3 = \tfrac{4}{7}.$$

For this problem there are $T = 3$ terms and $n = 2$ variables, so that the degree of difficulty is $T - (n + 1) = 0$; this is then the number of variables minus the number of equations in the set of normality and orthogonality equations. To analyze this example further, we arbitrarily assume that $c_1 = 5$, $c_2 = 8$, and $c_3 = 6$. According to Theorem 8.2, the optimal primal objective function value $\bar{z}$ equals that of the predual function, i.e.,

$$\bar{z} = \prod_{j=1}^{3} \left(\frac{c_j}{\bar{\lambda}_j}\right)^{\bar{\lambda}_j} = \left(\frac{5}{2/7}\right)^{2/7} \left(\frac{8}{1/7}\right)^{1/7} \left(\frac{6}{4/7}\right)^{4/7} \approx 15.4323.$$

From the relation $\bar{\lambda}_j = \dfrac{c_j p_j(\bar{\mathbf{x}})}{\bar{z}}, j = 1, 2, 3$, we then obtain

$$\bar{x}_1 = \sqrt{\frac{\bar{\lambda}_1 \bar{\lambda}_2 \bar{\lambda}_3 \bar{z}^3}{c_1 c_2 c_3}} \approx \sqrt{\frac{(2/7)(1/7)(4/7)(15.4323)^3}{(5)(8)(6)}} \approx 0.59764,$$

$$\bar{x}_2 = \sqrt{\frac{c_1 c_3}{\bar{\lambda}_1 \bar{\lambda}_3 \bar{z}^2}} \approx \sqrt{\frac{(5)(6)}{(2/7)(4/7)(15.4323)^2}} \approx 0.87838,$$

and the problem is now completely solved. Although the above calculations could have been done exactly, most numbers in this section are only approximate.

The optimal values of x_1 and x_2 could also have been calculated using the results discussed at the end of Sect. 8.1. Using the optimal values of z and λ, the optimal values of x_1 and x_2 can be found from the equations

$$\log(\bar{\lambda}_j \bar{z}) = \log c_j + \sum_{i=1}^{n} a_{ij} \log \bar{x}_i, \; j = 1, \ldots, T.$$

For the above example this becomes

$$\log\left[(2/7)15.4323\right] = \log 5 + \log\bar{x}_1 - 3\log\bar{x}_2$$
$$\log\left[(1/7)15.4323\right] = \log 8 + 2\log\bar{x}_1 + 2\log\bar{x}_2$$
$$\log\left[(4/7)15.4323\right] = \log 6 - \log\bar{x}_1 + \log\bar{x}_2,$$

or

$$0.64436 = 0.69897 + \log\bar{x}_1 - 3\log\bar{x}_2$$
$$0.34333 = 0.90309 + 2\log\bar{x}_1 + 2\log\bar{x}_2$$
$$0.94539 = 0.77815 - \log\bar{x}_1 + \log\bar{x}_2.$$

Adding the first and the third equation yields

$$\log\bar{x}_2 = \frac{0.64436 - 0.69897 + 0.94539 - 0.77815}{-2} = -0.056315.$$

Substitution in the third equation gives

$$\log\bar{x}_1 = 0.77815 - 0.94539 + \log\bar{x}_2 = -0.22356,$$

so that the optimal solution is $\bar{x}_1 = 0.59764$ and $\bar{x}_2 = 0.87839$. This is the same result as that obtained above except for the round-off error in the last digit of $\bar{x}_2$.

The question is then whether or not it is always possible to solve a posynomial geometric programming problem in this way, if the degree of difficulty is zero or less. The answer is no, since the system of equations for the weights λ_j need not always possess a unique positive solution, even with the same number of equations as unknowns, i.e., with a zero degree of difficulty. As an example, consider the function $f(x) = x + x$ which gives the following system of simultaneous linear equations:

$$\lambda_1 + \lambda_2 = 1$$
$$\lambda_1 + \lambda_2 = 0.$$

This system has no solution, reflecting the fact that f has no minimal point on the real axis. On the other hand, the objective function $f(x) = x + x^2$ gives the system

$$\lambda_1 + \lambda_2 = 1$$
$$\lambda_1 + 2\lambda_2 = 0,$$

for which $\lambda_1 = 2$, $\lambda_2 = -1$ so that no positive solution exists, which again indicates that f has no minimal point on the positive real axis. Finally, the objective function $f(\mathbf{x}) = c_1 x_1 x_2^{-2} + c_2 x_1^2 x_2^{-4} + c_3 x_1^{-1} x_2^2$ gives the system

$$\lambda_1 + \lambda_2 + \lambda_3 = 1$$
$$\lambda_1 + 2\lambda_2 - \lambda_3 = 0$$
$$-2\lambda_1 - 4\lambda_2 + 2\lambda_3 = 0.$$

Since this system is linearly dependent (the second equality equals the third one if multiplied by (-2)), an infinite number of solutions exist. One can show that positive solutions are $\lambda_1 = 2 - 3t$, $\lambda_2 = 2t - 1$ and $\lambda_3 = t$ for every $t \in \,]1/2;\,2/3[$. Furthermore, by considering the functions

$$f(\mathbf{x}) = c_1 x_1 x_2 + c_2 x_1 x_2,$$
$$f(\mathbf{x}) = c_1 x_1 x_2 + c_2 x_1^2 x_2^2, \text{ and}$$
$$f(\mathbf{x}) = c_1 x_1 x_2^{-2} x_3 + c_2 x_1^2 x_2^{-4} x_3^2 + c_3 x_1^{-1} x_2^2 x_3^{-1},$$

we find that for problems with negative degree of difficulty (which is -1 in all of the above examples), the normality and orthogonality conditions may have no solution, no positive solution, or infinitely many positive solutions, respectively. Summarizing, a posynomial geometric programming problem with nonpositive degree of difficulty need not have a unique positive solution to its normality and orthogonality conditions. It must be admitted, though, that the above examples are contrived. In practice, a problem with a zero degree of difficulty will usually possess a unique positive solution λ.

In the following, posynomial geometric programming problems with positive degree of difficulty will now be discussed. From the theory of systems of linear equations we know that the normality and orthogonality conditions will then never yield a unique solution λ. It is for such problems that solution techniques based on bounding of the λ values and duality considerations are used.

As an illustration of this approach, consider the problem

$$P: \operatorname*{Min}_{x > 0} z = c_1 x + c_2 x^{-1} + c_3 x^2.$$

First we describe the technique for bounding of the weights λ_j. The problem has $T = 3$ terms and $n = 1$ variable so that the degree of difficulty is $T - (n + 1) = 1$. Its normality condition is $\lambda_1 + \lambda_2 + \lambda_3 = 1$ and its only orthogonality condition is $\lambda_1 - \lambda_2 + 2\lambda_3 = 0$. Having only two equations one cannot solve for the three weights λ_j although bounds for them can be established, remembering that all weights must be positive. Adding the two equations yields $2\lambda_1 + 3\lambda_3 = 1$, i.e., $\lambda_1 = 1/2 - 3/2\lambda_3$. Since all weights λ_j must be positive, we conclude that $\bar{\lambda}_1 \in \,]0;\,1/2[$ at optimality. Furthermore, adding twice the normality equation to the negative orthogonality equation gives $\lambda_1 + 3\lambda_3 = 2$, or $\lambda_2 = 2/3 - 1/3\lambda_1 < 2/3$. Subtracting the orthogonality equation from the normality equation yields $2\lambda_2 - \lambda_3 = 1$, or $\lambda_2 = 1/2 + 1/2\lambda_1 > 1/2$, so that $\bar{\lambda}_2 \in \,]1/2;\,2/3[$ at optimality. Finally, the equation $2\lambda_2 + 3\lambda_3 = 1$ from above can also be written as $\lambda_3 = 1/3 - 2/3\lambda_1$. From the bounds of λ_1 one can then conclude that $\bar{\lambda}_3 \in \,]0;\,1/3[$ at optimality. Summarizing, we have found that at optimality the weights λ_j must satisfy $\bar{\lambda}_1 \in \,]0, 1/2[$, $\bar{\lambda}_2 \in \,]1/2, 2/3[$, and $\bar{\lambda}_3 \in \,]0, 1/3[$.

To analyze the above model further, we arbitrarily assume that $c_1 = 32$, $c_2 = 44$ and $c_3 = 8$. Furthermore, we tentatively set $\hat{\lambda}_2 := \frac{3}{5}$ since the narrowest range was obtained for λ_2 and it seems reasonable to fix λ_2 somewhere in the middle of this range. The normality and orthogonality conditions then imply that $\hat{\lambda}_1 = \hat{\lambda}_3 = \frac{1}{5}$. Assuming that these values of the weights λ_j are optimal or at least near-optimal, we then use the fact that $\bar{z} = \frac{c_j \bar{p}_j}{\hat{\lambda}_j}, j = 1, 2, 3$, to obtain

$$\frac{32\hat{x}}{1/5} = \frac{44\hat{x}^{-1}}{3/5} = \frac{8\hat{x}^2}{1/5}.$$

Equating the first and the last of these relations (and remembering that $x > 0$) yields $\hat{x} = 4$ and $\hat{z} = 267$. Using $\lambda_1 = \lambda_1 = \frac{1}{5}, \lambda_2 = \frac{3}{5}$, and $\lambda_3 = \frac{1}{5}$, the value of the dual function z_d is

$$z_d\left(\hat{\lambda}\right) = \prod_{j=1}^{3} \left(\frac{c_j}{\hat{\lambda}_j}\right)^{\hat{\lambda}_j} = \left(\frac{32}{1/5}\right)^{1/5} \left(\frac{44}{3/5}\right)^{3/5} \left(\frac{8}{1/5}\right)^{1/5} = 75.9306.$$

Since any primal feasible solution yields an upper bound on the optimal primal objective function value and any dual feasible solution yields a lower bound, it follows that $75.9306 \leq \bar{z} \leq 267$. These bounds are certainly not impressively tight; we repeat the procedure by using the first and the second as well as the second and the third of the above equations for finding x. The respective results are

$$\hat{x}^2 = \frac{44}{32}\left(\frac{1/5}{3/5}\right) = 0.4583, \text{ i.e., } \hat{x} = 0.6770 \text{ and thus } \hat{z} = 90.32, \text{ and}$$

$$\hat{x}^2 = \frac{44}{8}\left(\frac{1/5}{3/5}\right) = 1.8333, \text{ i.e., } \hat{x} = 1.2239 \text{ and thus } \hat{z} = 87.10.$$

Selecting the best (smallest) of the $\hat{z}$ values obtained, the above results can now be summarized as follows. At optimality, we have $\bar{\lambda}_1 \in \,]0, \frac{1}{2}[$, $\bar{\lambda}_2 \in \,]\frac{1}{2}, \frac{2}{3}[$, $\bar{\lambda}_3 \in \,]0; \frac{1}{3}[$, and $\bar{z} \in [75.93; 87.10]$. The upper bound for $\bar{z}$ is the primal objective function value for $\hat{x} = 1.2239$. (Note that the true optimal primal objective function values are $\bar{\lambda}_1 \approx 0.36731$, $\bar{\lambda}_2 \approx 0.54423$, $\bar{\lambda}_3 \approx 0.08846$, $\bar{x} \approx 0.96333$ and $\bar{z} \approx 83.9225$.) We can conclude that with a few simple calculations one can obtain useful conditions for optimality and reasonably good bounds on the optimal value of the objective function.

Another possibility to obtain an approximate solution to this problem would be to drop a term that is believed to have little effect on the solution. In our example, we might have some prior indication that the third term $8x^2$ is not going to be important at optimality. The upper bound of $\frac{1}{3}$ for λ_3 could be an indication of this, but we might also have some practical experience from the specific situation that led to this problem which supports this belief. Dropping the $8x^2$ term reduces the degree of difficulty from one to zero, and the normality and orthogonality conditions are then simply $\lambda_1 + \lambda_2 = 1$ and $\lambda_1 - \lambda_2 = 0$. Hence $\hat{\lambda}_1 = \hat{\lambda}_2 = \frac{1}{2}$, from which the relationship $\dfrac{32\hat{x}}{\hat{\lambda}_1} = \dfrac{44\hat{x}^{-1}}{\hat{\lambda}_2}$ gives $\dfrac{32\hat{x}}{\frac{1}{2}} = \dfrac{44\hat{x}^{-1}}{\frac{1}{2}}$ or $\dfrac{44}{32} = \dfrac{11}{8} = 1.375$, i.e., $\hat{x} = 1.17260$. Using the convention as in Sect. 8.2 when discussing geometric means that $\left(\dfrac{8}{0}\right)^0 := 1$, the dual function z_d of the original problem then takes the value $z_d\left(\hat{\boldsymbol{\lambda}}\right) = z_d\left(\hat{\lambda}_1, \hat{\lambda}_2, \hat{\lambda}_3\right)$

$$= z_d(\tfrac{1}{2}, \tfrac{1}{2}, 0) = \left(\frac{32}{1/2}\right)^{\frac{1}{2}} \left(\frac{44}{1/2}\right)^{\frac{1}{2}} \left(\frac{8}{0}\right)^0 = 75.05 \text{ which is now a new lower bound}$$

on the optimal $\bar{z}$-value. A new upper bound is obtained by inserting the x-value which was optimal for the simplified problem into the original primal objective function, i.e. $\hat{z} = f(\hat{x}) = f(1.17260) = 86.05$. In other words, $\bar{z} \in [75.05; \ 86.05]$ and it can be seen that the lower bound is poorer, but the upper bound is better than with the previous method.

A third possibility is to further explore the dual functions. Recall that the normality and orthogonality conditions for the above example are

$$\lambda_1 + \lambda_2 + \lambda_3 = 1$$
$$\lambda_1 - \lambda_2 + 2\lambda_3 = 0.$$

Solving for λ_1 and λ_2 in terms of λ_3 yields

$$\lambda_1 = \tfrac{1}{2} - \tfrac{3}{2}\lambda_3$$
$$\lambda_2 = \tfrac{1}{2} + \tfrac{1}{2}\lambda_3.$$

Substituting these expressions in the dual objective function z_d results in

$$z_d(\boldsymbol{\lambda}) = \prod_{j=1}^{3} \left(\frac{c_j}{\lambda_j}\right)^{\lambda_j} = \left(\frac{32}{\lambda_1}\right)^{\lambda_1} \left(\frac{44}{\lambda_1}\right)^{\lambda_2} \left(\frac{8}{\lambda_3}\right)^{\lambda_3} =$$

$$= \left(\frac{32}{\frac{1}{2}(1 - 3\lambda_3)}\right)^{\frac{1}{2}(1 - 3\lambda_3)} \left(\frac{44}{\frac{1}{2}(1 + \lambda_3)}\right)^{\frac{1}{2}(1 + \lambda_3)} \left(\frac{8}{\lambda_3}\right)^{\lambda_3}.$$

Denote this function $z_d(\boldsymbol{\lambda}) = z_d(\lambda_1, \lambda_2, \lambda_3) = z_d(\frac{1}{2}[1 - 3\lambda_3], \frac{1}{2}[1 + \lambda_3], \lambda_3)$ by $z_d(\lambda_3)$ and call it the *substituted dual objective function*. From Theorem 8.6 it follows that $z_d(\lambda_3)$ is maximized for $\bar{\lambda}_3$, which is the unique optimal solution. Due to the monotonicity of the logarithmic transformation one can work with $\ln z_d(\lambda_3)$ instead of $z_d(\lambda_3)$. This results in

$$\ln z_d(\lambda_3) = \frac{1}{2}(1 - 3\lambda_3) \ln\frac{64}{1 - 3\lambda_3} + \frac{1}{2}(1 + \lambda_3) \ln\frac{88}{1 + \lambda_3} + \lambda_3 \ln\frac{8}{\lambda_3} =$$

$$= -\frac{1}{2}(1 - 3\lambda_3) \ln\frac{1 - 3\lambda_3}{64} - \frac{1}{2}(1 + \lambda_3) \ln\frac{1 + \lambda_3}{88} - \lambda_3 \ln\frac{\lambda_3}{8}.$$

To maximize this concave function of the single scalar variable λ_3 we set the first derivative equal to zero, obtaining

$$\frac{3}{2}\left(1 + \ln\frac{1 - 3\bar{\lambda}_3}{64}\right) - \frac{1}{2}\left(1 + \ln\frac{1 + \bar{\lambda}_3}{88}\right) - \left(1 + \ln\frac{\bar{\lambda}_3}{8}\right) = 0.$$

Temporarily returning to the general case, in this approach the weights λ_j are solved for with respect to some selected λ_i, i.e., $\lambda_j = \lambda_j(\lambda_i)$, $j = 1,\ldots, T$, $i \neq j$. Then one can write

$$\frac{d}{d\lambda_i}(\ln d(\lambda_i)) = \frac{d}{d\lambda_i}\left(-\sum_{j=1}^{T}\lambda_j \ln\frac{\lambda_j}{c_j}\right)$$

$$= -\sum_{j=1}^{T}\left(\frac{d\lambda_j}{d\lambda i} \ln\left(\frac{\lambda_j}{c_j}\right) + \lambda_j\frac{1}{\lambda_j}\frac{d\lambda_j}{d\lambda_i}\right)$$

$$= -\sum_{j=1}^{T}\left[1 + \ln\frac{\lambda_j}{c_j}\right]\frac{d\lambda_j}{d\lambda_i} = -\sum_{j=1}^{T}\frac{d\lambda_j}{d\lambda_i} - \sum_{j=1}^{T}\frac{d\lambda_j}{d\lambda_i} \ln\frac{\lambda_j}{c_j}.$$

However, $\sum_{j=1}^{T}\lambda_j = 1$ implies $\sum_{j=1}^{T}\lambda_j(\lambda_i) \equiv 1$ so that $\sum_{j=1}^{T}\frac{d\lambda_j}{d\lambda_i} = 0$. It follows that $\frac{d}{d\lambda_i}(\ln z_d(\lambda_i)) = -\sum_{j=1}^{T}\frac{d\lambda_j}{d\lambda_i} \ln\frac{\lambda_j}{c_j}$, which for optimality is set to zero. Returning to our specific example and using this result already obtained above, in the above example, all nonlogarithmic constants cancel, so that we find

$$0 = \frac{3}{2} \ln\frac{1 - 3\bar{\lambda}_3}{64} - \frac{1}{2} \ln\frac{1 + \bar{\lambda}_3}{88} - \ln\frac{\bar{\lambda}_3}{8} =$$

$$= \ln\left\{(1 - 3\bar{\lambda}_3)^{3/2}(1 + \bar{\lambda}_3)\hat{x}^{-1/2}\bar{\lambda}_3^{-1}\right\} - \ln\left\{64^{3/2}88^{-1/2}8^{-1}\right\} \text{ or simply}$$

$$(1 - 3\bar{\lambda}_3)^{3/2}(1 + \bar{\lambda}_3)\hat{x}^{-1/2}\bar{\lambda}_3^{-1} = 64^{3/2}88^{-1/2}8^{-1} = \frac{32}{\sqrt{22}}.$$

The solution to this equation in one variable can now be found, either by squaring it and solving the resulting fourth degree polynomial equation, or by using some iterative technique like the Newton-Raphson method. The result is $\bar{\lambda} \approx 0.08846$ from which one obtains

$$\bar{\lambda}_1 = \tfrac{1}{2} - \tfrac{3}{2}\bar{\lambda}_3 = 0.36731$$

$$\bar{\lambda}_2 = \tfrac{1}{2} + \tfrac{1}{2}\bar{\lambda}_3 = 0.54423.$$

The relationship $\dfrac{32\bar{x}}{\bar{\lambda}_1} = \dfrac{44\bar{x}^{-1}}{\bar{\lambda}_2}$ then yields $\bar{x}^2 = \dfrac{44\,\bar{\lambda}_1}{32\,\bar{\lambda}_2} = \left(\dfrac{44}{32}\right)\dfrac{0.36731}{0.54423} =$ 0.92801, and $\bar{x} = 0.96333$. The optimal objective function value can be obtained from

$$\bar{z} = \frac{32\bar{x}}{\bar{\lambda}_1} = \frac{(32)0.96333}{0.36731} = 83.9255$$

or from the substituted dual objective function

$$z_d(\boldsymbol{\lambda}) = \left(\frac{32}{\bar{\lambda}_1}\right)^{\bar{\lambda}_1}\left(\frac{44}{\bar{\lambda}_2}\right)^{\bar{\lambda}_2}\left(\frac{8}{\bar{\lambda}_3}\right)^{\bar{\lambda}_3} =$$

$$= \left(\frac{32}{0.36731}\right)^{0.36731}\left(\frac{44}{0.54423}\right)^{0.54423}\left(\frac{8}{0.08846}\right)^{0.08846} = 83.9255.$$

It was shown above that by maximizing the substituted dual objective function a geometric programming problem with a degree of difficulty of one is reduced to a single-variable maximization problem without constraints (except for lower and upper bounds). Similarly, for a problem with two degrees of difficulty the above approach leads to a two-dimensional maximization search problem with upper and lower bounds on the two scalar search variables.

The above results can now be summarized as follows. For problems with nonpositive degree of difficulty one may hope to find a unique solution to the normality and orthogonality conditions; if such a solution exists, then the weights are used for finding a solution to the original problem. For problems with a positive degree of difficulty we can try to estimate bounds on the weights and use them to bound the optimal objective function value. Alternatively, terms may be dropped so that a problem with zero degree of difficulty results. The solution to this problem can be used to establish bounds on the variables of the original problem. Finally, it is possible to solve the normality and orthogonality conditions in terms of one or more of the weights and then perform an unconstrained maximization search on the substituted dual objective function. This approach allows finding the solution to the original problem to within any required accuracy.

Our results are now summarized in algorithmic form. Recall that the following primal problem P with n variables x_i, $i = 1,.., n$ and T terms is required to be solved:

$$\text{P: } \underset{\mathbf{x}>0}{\text{Min}}\, z = \sum_{j=1}^{T} c_j p_j(\mathbf{x}) = \sum_{j=1}^{T} c_j \prod_{i=1}^{n} x_i^{a_{ij}}.$$

With each term $c_j p_j$ we associate a positive weight λ_j defined by $\lambda_j := \dfrac{c_j p_j}{z}, j = 1,..,$ T. The dual P_D of the above primal problem P is then given by

$$\text{P}_\text{D}: \underset{\lambda>0}{\text{Max}}\, z_d(\lambda) = \prod_{j=1}^{T} \left(\frac{c_j}{\lambda_j} \right)^{\lambda_j}$$

$$\text{s.t. } \sum_{j=1}^{T} \lambda_j = 1 \qquad\qquad \text{(normality)}$$

$$\sum_{j=1}^{T} a_{ij}\lambda_j = 0, \ i = 1, \ldots, n \qquad \text{(orthogonality)}$$

$$\lambda_j > 0, \ j = 1, \ldots, T.$$

For convenience, the constraints in P_D will be referred to as the system Λ of linear equations. In order to simplify matters, assume that the problem P which is to be solved possesses a unique optimal solution $\bar{\mathbf{x}}$ in the positive orthant $\mathbb{R}_>^n$.

An Unconstrained Posynomial Geometric Programming Algorithm

Step 1 Assume that the coefficient matrix of the system Λ of linear equations has full rank. Is the degree of difficulty, i.e., $T - (n + 1) > 0$?

If yes: Go to Step 3.
If no: Go to Step 2.

Step 2 Find the unique solution $\bar{\lambda}$ to the system Λ. The optimal primal objective function value is given by

$$\bar{z} = \prod_{j=1}^{T} \left(\frac{c_j}{\bar{\lambda}_j} \right)^{\bar{\lambda}_j},$$

and the value of the individual terms are $c_j p_j(\bar{x}) = \bar{\lambda}_j \bar{z}, j = 1, 2,.., T$. Then, if enough equations are available, an optimal solution $\bar{x}$ is found by solving the following system of equations, linear in $\log \bar{x}_i$

$$\sum_{i=1}^{n} a_{ij} \log \bar{x}_i = \log \frac{\bar{\lambda}_j \bar{z}}{c_j}, \ j = 1,.., T.$$

Step 3 Select one of the following three approaches:

Bounding of the weights: go to Step 4.
Dropping terms: go to Step 5.
Searching in the dual space: go to Step 6.

Step 4 (Bounding of the weights) Repeatedly manipulating the system Λ upper and lower bounds are sought for the optimal weights λ_j. Temporarily fixing some weights λ_j at values within their respective ranges, such a tentative solution $\hat{\lambda}$ is then used to find a tentative solution $\hat{x}$ via the equations $\hat{\lambda}_j \hat{z} = c_j \hat{p}_j, j = 1, 2,\ldots, T$. The optimal objective function value $\bar{z}$ is then bounded by

$$\hat{z} = \prod_{j=1}^{T} \left(\frac{c_j}{\hat{\lambda}_j}\right)^{\hat{\lambda}_j} \leq \bar{z} \leq \sum_{j=1}^{T} c_j p_j(\hat{x}).$$

If the bounds are sufficiently tight, stop; otherwise go to Step 5 or Step 6.

Step 5 (Dropping terms): The degree of difficulty may be reduced by disregarding terms $c_j p_j(\hat{x})$ that are judged to be relatively insignificant at optimality. The resulting reduced problem can be analyzed by going to Step 2 if it has zero degree of difficulty or to Step 3 otherwise.

Step 6 (Searching in the dual space) The dual problem P_D is solved to any desired degree of accuracy either using a suitable algorithm for constrained nonlinear optimization or by using the substituted dual function. The resulting tentative solution $\hat{\lambda}$ is used to find a tentative solution $\hat{x}$ by using the equations $\lambda_j \hat{z} = c_j p_j(\hat{x}), j = 1,\ldots, T$. Then

$$\hat{z} = \prod_{j=1}^{T} \left(\frac{c_j}{\hat{\lambda}_j}\right)^{\hat{\lambda}_j} \leq \bar{z} \leq \sum_{j=1}^{T} c_j p_j(\hat{x}).$$

If the bounds are sufficiently tight, stop; otherwise repeat Step 6 with an increased accuracy for solving the dual problem P_D.

Example As a numerical example, we consider an economic order quantity (*EOQ*) model, as in our first introductory example in Sect. 8.1, but here we add a term for costs due to obsolescence or deterioration. Specifically, the single-variable problem is

$$\text{P: } \underset{x>0}{\text{Min}} \ z = f(x) = c_1 x + c_2 x^{-1} + c_3 x^{1/2},$$

where the three terms express ordering, carrying, and obsolescence costs, respectively. This problem has $T = 3$ terms and $n = 1$ variable, so that the degree of freedom is $T - (n + 1) = 1$. The values of the economic coefficients are specified as

$c_1 = \frac{1}{3}$, $c_2 = 54$, and $c_3 = 2$. Since the degree of difficulty equals one, we go to Step 3. With weights defined by $\lambda_1 := \frac{c_1 p_1}{z} = \frac{c_1 x}{z}$, $\lambda_2 := \frac{c_2 x^{-1}}{z}$, and $\lambda_3 := \frac{c_3 x^{1/2}}{z}$, we obtain the system

$$\Lambda: \lambda_1 + \lambda_2 + \lambda_3 = 1 \quad \text{(normality)}$$
$$\lambda_1 - \lambda_2 + \tfrac{1}{2}\lambda_3 = 0. \quad \text{(orthogonality)}$$

Solving for λ_1 and λ_2 in terms of λ_3, we obtain

$$\lambda_1 = \tfrac{1}{2} - \tfrac{3}{4}\lambda_3$$
$$\lambda_2 = \tfrac{1}{2} - \tfrac{1}{4}\lambda_3.$$

Since $0 < \lambda_j < 1$, $j = 1, 2, 3$, is required, these relationships yield $\lambda_1 \in \,]0, \tfrac{1}{2}[$, $\lambda_2 \in \,]0, \tfrac{1}{2}[$, and $\lambda_3 \in \,]0, \tfrac{2}{3}[$, so that in Step 4 bounding of the weights is not very attractive; the bounds are quite loose. In Step 5, we have no prior information about which term might be insignificant at optimality, and therefore we proceed to Step 6 to maximize the substituted dual function z_d as a function of λ_3. We find

$$P_D: \underset{\lambda_3 > 0}{\text{Max}} \; z_d(\lambda_3) = \prod_{j=1}^{3} \left(\frac{c_j}{\lambda_j}\right)^{\lambda_j} = \left(\frac{c_1}{\tfrac{1}{2} - \tfrac{3}{4}\lambda_3}\right)^{\tfrac{1}{2} - \tfrac{3}{4}\lambda_3} \left(\frac{c_2}{\tfrac{1}{2} - \tfrac{1}{4}\lambda_3}\right)^{\tfrac{1}{2} - \tfrac{1}{4}\lambda_3} \left(\frac{c_3}{\lambda_3}\right)^{\lambda_3}.$$

Using the formula developed above for maximizing z_d, i.e.,

$$\sum_{j=1}^{T} \frac{d\lambda_j}{d\lambda_i} \ln \frac{\pi_j}{c_j} = 0,$$

we find that

$$-\tfrac{3}{4} \ln \frac{\tfrac{1}{2} - \tfrac{3}{4}\lambda_3}{c_1} - \tfrac{1}{4} \ln \frac{\tfrac{1}{2} - \tfrac{1}{4}\lambda_3}{c_2} + \ln \frac{\lambda_3}{c_3} = 0 \text{ or}$$

$$\ln \left\{ \left(\tfrac{1}{2} - \tfrac{3}{4}\lambda_3\right)^{-\tfrac{3}{4}} \left(\tfrac{1}{2} - \tfrac{1}{4}\lambda_3\right)^{-\tfrac{1}{4}} \lambda_3 \right\} = \ln \left(c_1^{-\tfrac{3}{4}} c_2^{-\tfrac{1}{4}} c_3 \right) =$$

$$= \ln \left(\left(\tfrac{1}{3}\right)^{-\tfrac{3}{4}} (54)^{-\tfrac{1}{4}} (2) \right), \text{ or}$$

$$\left(\tfrac{1}{2} - \tfrac{3}{4}\right)^{-\tfrac{3}{4}} \left(\tfrac{1}{2} - \tfrac{1}{4}\lambda_3\right)^{-\tfrac{1}{4}} \lambda_3 = \left(\tfrac{1}{3}\right)^{-\tfrac{3}{4}} (54)^{-\tfrac{1}{2}} (2) = 2^{\tfrac{3}{4}},$$

so that finally

$$(2 - 3\lambda_3)^3 (2 - \lambda_3) - 32\lambda_3^4 = 0,$$

which is a fourth-degree polynomial equation. This can be solved numerically; its solution in the interval $]0, \frac{2}{3}[$ is $\bar{\lambda}_3 = \frac{2}{5}$ as can easily be verified by simple substitution. It follows that $\bar{\lambda}_1 = \frac{1}{5}$ and $\bar{\lambda}_2 = \frac{2}{5}$.

Now $z_d(\bar{\lambda}_1, \bar{\lambda}_2, \bar{\lambda}_3) = \prod_{j=1}^{3} \left(\dfrac{c_j}{\bar{\lambda}_j} \right)^{\bar{\lambda}_j} = \left(\dfrac{\frac{1}{3}}{\frac{1}{5}} \right)^{\frac{1}{5}} \left(\dfrac{54}{\frac{2}{5}} \right)^{\frac{2}{5}} \left(\dfrac{2}{\frac{2}{5}} \right)^{\frac{2}{5}} = 15 = \bar{z}$, and with

$\bar{\lambda}_1 = \dfrac{c_1 \bar{x}}{\bar{z}}$, we find $\frac{1}{5} = \dfrac{\frac{1}{3}\bar{x}}{15}$ or $\bar{x} = 9$. Finally, inserting $\bar{x} = 9$, we obtain $f(\bar{x}) = c_1 \bar{x} + c_2 \bar{x}^{-1} + c_3 \bar{x}^{\frac{1}{2}} = (1/3)(9) + 54/9 + 2\sqrt{9} = 15$, in accordance with $\bar{z} = 15$. We should comment that the selection of the values for the economic coefficients c_j was done for pedagogical reasons, so that readers could easily check the calculations. Also, it should be noted that using the classical approach for solving P would be to differentiate $f(x)$, $f'(x) = c_1 - c_2 x^{-2} + \frac{1}{2}c_3 x^{-\frac{1}{2}}$, and setting $f'(x) = 0$ we would obtain the equation $c_1 x^2 + \frac{1}{2}c_3 x^{\frac{3}{2}} - c_2 = 0$ or $\frac{1}{3} x^2 + x^{\frac{3}{2}} - 54 = 0$, which is not an algebraic equation. This is in contrast to the geometric programming approach that leads to a fourth-degree polynomial algebraic equation, which has a closed-form solution.

8.3.2   Solution Techniques for Constrained Posynomial Geometric Programming

For ease of presentation, we start this section by studying geometric programming problems with a single posynomial constraint. Following this, problems with $m > 1$ constraints are treated.

The problem under consideration is

$$\text{P}: \underset{\mathbf{x} > 0}{\text{Min}} \; z \; = \; f(\mathbf{x}) = \sum_{j=1}^{T} c_j \prod_{i=1}^{n} x_i^{a_{ij}}$$

$$\text{s.t. } g(\mathbf{x}) = \sum_{k=1}^{S} d_k \prod_{i=1}^{n} x_i^{b_{ik}} \leq 1.$$

As in the unconstrained case, we define $p_j(\mathbf{x}) := \prod_{i=1}^{n} x_i^{a_{ij}}, j = 1, \ldots, T$ and $q_k(\mathbf{x}) :=$ $\prod_{i=1}^{n} x_i^{b_{ik}}, k = 1, \ldots, S$ so that the objective function is $f(\mathbf{x}) = \sum_{j=1}^{T} c_j p_j(\mathbf{x})$ and the constraint is $g(\mathbf{x}) = \sum_{k=1}^{S} d_k q_k(\mathbf{x}) \leq 1$. Setting the right-hand side of the constraint equal to one is done as a matter of convenience and is not restrictive; any positive number could be handled by scaling the coefficients d_k, if needed. As usual, we define weights λ_j for the terms of the objective function by setting $\lambda_j := \dfrac{c_j p_j}{z} > 0$ or

$\lambda_j z = c_j p_j, j = 1, \ldots, T$, where $\sum_{j=1}^{T} \lambda_j = 1$. Moreover, we now assign weights $\mu_k \geq 0$ to

the terms $d_k q_k$ of the constraint, defined as $\mu_k := \mu d_k q_k$, $k = 1, \ldots, S$ where $\mu := \sum_{k=1}^{S} \mu_k$.

As we can see, the weights μ_j are only defined up to a nonnegative proportionality
constant, i.e., if μ_k are weights, then so are values $\alpha\mu_k$ with any $\alpha > 0$. Then we can
state the following theorem, which was alluded to at the end of Sect. 8.2 above.

Theorem 8.9 Let λ_j and μ_j be weights for the constrained problem P. The following
expressions hold true at optimality:

$$\sum_{j=1}^{T} \bar{\lambda}_j = 1 \qquad\qquad\qquad\qquad \text{(normality)}$$

$$\sum_{j=1}^{T} a_{ij}\bar{\lambda}_j + \sum_{k=1}^{S} b_{ik}\bar{\mu}_k = 0, \ i = 1, \ldots, n \qquad \text{(orthogonality)}$$

$$\bar{\lambda}_j z = c_j \bar{p}_j \ \text{ and } \ \bar{\lambda}_j > 0, \ j = 1, \ldots, T$$

$$\bar{\mu}_k = \bar{\mu} d_k \bar{q}_k \ \text{ and } \ \bar{\mu}_k \geq 0, \ k = 1, \ldots, S$$

$$\sum_{k=1}^{S} \bar{\mu}_k = \bar{\mu},$$

$$\bar{\lambda}_j > 0, \ j = 1, \ldots, T; \ \bar{\mu}_k \geq 0, \ k = 1, \ldots, S.$$

Furthermore, if one $\bar{\mu}_k > 0$, then all $\bar{\mu}_k > 0$, $k = 1, \ldots, S$ and $g(\bar{\mathbf{x}}) = 1$. On the
other hand, if $g(\bar{\mathbf{x}}) < 1$, then we must have $\bar{\mu}_k = 0$, $k = 1, \ldots, S$.

Proof In this proof, the Karush-Kuhn-Tucker optimality conditions of Sect. 4.2 in
Chap. 4 will be used. First, problem P can be rewritten as

$$
\begin{array}{lll}
\text{P}': \underset{\mathbf{x}, \lambda, \mu'}{\text{Min}} \ z & & \text{Dual variables} \\[2pt]
 & & \text{(Lagrangean multipliers)} \\[4pt]
\text{s.t.} & z\lambda_j = c_j p_j(\mathbf{x}), j = 1, \ldots, T & u_j \\[2pt]
 & \mu'_k = d_k q_k(\mathbf{x}), \ k = 1, \ldots, S & v_k \\[4pt]
 & \displaystyle\sum_{j=1}^{T} \lambda_j = 1 & u_0 \\[6pt]
 & \displaystyle\sum_{k=1}^{S} \mu'_k \leq 1 & v_0
\end{array}
$$

Following the convention in geometric programming, the positivity constraints
for the variables x_i have not been explicitly expressed. From the constraints i
follows that the variables λ_j and μ'_k also have to be positive, so that these
requirements do not have to be stated separately. The Lagrangean function for
problem P' can be written as

$$L = z - \sum_{j=1}^{T} u_j \left[z\lambda_j - c_j p_j(\mathbf{x}) \right] - \sum_{k=1}^{S} v_k \left[\mu'_k - d_k q_k(\mathbf{x}) \right]$$

$$+ u_0 \left[\sum_{j=1}^{T} \lambda_j - 1 \right] + v_0 \left[\sum_{k=1}^{S} \mu'_k - 1 \right].$$

The Karush-Kuhn-Tucker necessary conditions for optimality are

$$0 = \frac{\partial L}{\partial z} = 1 - \sum_{j=1}^{T} \bar{u}_j \bar{\lambda}_j$$

$$0 = \frac{\partial L}{\partial x_i} = \sum_{j=1}^{T} \bar{u}_j c_j \frac{\partial p_j(\bar{\mathbf{x}})}{\partial x_i} + \sum_{k=1}^{S} \bar{v}_k d_k \frac{\partial q_k(\bar{\mathbf{x}})}{\partial x_i}, \quad i = 1, \ldots, n$$

$$0 = \frac{\partial L}{\partial \lambda_j} = -\bar{u}_j \bar{z} + \bar{u}_0, \quad j = 1, \ldots, T$$

$$0 = \frac{\partial L}{\partial \mu'_k} = -\bar{v}_k + \bar{v}_0, \quad k = 1, \ldots, S$$

$$\bar{v}_0 \geq 0; \quad \sum_{k=1}^{S} \mu'_k - 1 \leq 0; \quad \bar{v}_0 \left[\sum_{k=1}^{S} \bar{\mu}'_k - 1 \right] = 0.$$

From these optimality conditions we obtain

$$\bar{v}_k = \bar{v}_0 \geq 0, \quad k = 1, \ldots, S$$

$$\bar{u}_j = \frac{\bar{u}_0}{\bar{z}}, \quad j = 1, \ldots, T.$$

Then $1 = \sum_{j=1}^{T} \bar{u}_j \bar{\lambda}_j = \frac{\bar{u}_0}{\bar{z}} \sum_{j=1}^{T} \bar{\lambda}_j = \frac{\bar{u}_0}{\bar{z}}$ and it follows that $\bar{u}_0 = \bar{z}$ and $\bar{u}_j = 1, j = 1, \ldots,$

T. Moreover,

$$0 = \sum_{j=1}^{T} \bar{u}_j c_j \frac{\partial \bar{p}_j}{\partial x_i} + \sum_{k=1}^{S} \bar{v}_k d_k \frac{\partial \bar{q}_k}{\partial x_i} = \sum_{j=1}^{T} c_j \frac{\partial \bar{p}_j}{\partial x_i} + \bar{v}_0 \sum_{k=1}^{S} d_k \frac{\partial \bar{q}_k}{\partial x_i}, \quad i = 1, \ldots, n.$$

But as $\bar{x}_i > 0$ and $\bar{x}_i \frac{\partial \bar{p}_j}{\partial x_i} = a_{ij} \bar{p}_j$ as well as $\bar{x}_i \frac{\partial \bar{q}_j}{\partial x_i} = b_{ij} \bar{q}_j$, we obtain

$$0 = \sum_{j=1}^{T} a_{ij} c_j \bar{p}_j + \bar{v}_0 \sum_{k=1}^{S} b_{ik} d_k \bar{q}_k = \bar{z} \sum_{j=1}^{T} a_{ij} \bar{\lambda}_j + \bar{v}_0 \sum_{k=1}^{S} b_{ik} \bar{\mu}'_k, \quad i = 1, \ldots, n$$

using the definition of the weights λ_j and μ'_k. Now two cases have to be considered separately.

Case 1: $g(\bar{\mathbf{x}}) = 1$ at optimality. By assumption, $1 = g(\bar{\mathbf{x}}) = \sum\limits_{k=1}^{S} d_k \bar{q}_k = \sum\limits_{k=1}^{S} \bar{\mu}'_k$.

The complementary slackness condition $\bar{v}_0 \left[\sum\limits_{k=1}^{S} \bar{\mu}'_k - 1 \right] = 0$ is then satisfied for any

$\bar{v}_0 \geq 0$. Defining $\bar{\mu}_k := \dfrac{\bar{v}_0}{\bar{z}} \bar{\mu}'_k$ results in

$$0 = \bar{z} \sum_{j=1}^{T} a_{ij} \bar{\lambda}_j + \bar{v}_0 \sum_{k=1}^{S} b_{ik} \bar{\mu}'_k = \bar{z} \sum_{j=1}^{T} a_{ij} \bar{\lambda}_j + \bar{z} \sum_{k=1}^{S} b_{ik} \bar{\mu}_k, \quad i = 1, \ldots, n.$$

However, as $\bar{z} = \sum\limits_{j=1}^{T} c_j \bar{p}_j > 0$, we obtain the orthogonality conditions

$$\sum_{j=1}^{T} a_{ij} \bar{\lambda}_j + \sum_{k=1}^{S} b_{ik} \bar{\mu}_k = 0, \quad i = 1, \ldots, n.$$

Because $\bar{\mu}_k = \dfrac{\bar{v}_0}{\bar{z}} \bar{\mu}'_k$, and $\bar{\mu}'_k$, $\bar{z}$ are all positive, either all $\bar{\mu}_k > 0$ $\left(\text{if } \bar{v}_0 > 0 \right)$ or all

$\bar{\mu}_k = 0$ $\left(\text{if } \bar{v}_0 = 0 \right)$. Furthermore, $\bar{\mu}'_k = d_k \bar{q}_k$, $k = 1, \ldots, S$ and $\sum\limits_{k=1}^{S} \bar{\mu}'_k = 1$ imply

that

$$\bar{\mu}_k = \frac{\bar{v}_0}{\bar{z}} \bar{\mu}'_k = \frac{\bar{v}_0}{\bar{z}} d_k q_k(\bar{x}) = \left(\sum_{k=1}^{S} \frac{\bar{v}_0}{\bar{z}} \bar{\mu}'_k \right) d_k q_k(\bar{x}) =$$

$$= \left(\sum_{k=1}^{S} \bar{\mu}_k \right) d_k q_k(\bar{\mathbf{x}}) = \bar{\mu} d_k q_k(\bar{\mathbf{x}}), \quad k = 1, \ldots, S.$$

Consequently, the variables $\bar{\mu}_k$ are the "usual" constraint weights.

Case 2: $g(\bar{\mathbf{x}}) < 1$ at optimality. By assumption, $1 > g(\bar{\mathbf{x}}) = \sum\limits_{k=1}^{S} d_k \bar{q}_k = \sum\limits_{k=1}^{S} \bar{\mu}'_k$ and

the complementary slackness condition $\bar{v}_0 \left(\sum\limits_{k=1}^{S} \bar{\mu}'_k - 1 \right) = 0$ gives $\bar{v}_0 = 0$. As above,

we define the usual weights $\bar{\mu}_k = \dfrac{\bar{v}_0}{\bar{z}} \bar{\mu}'_k$, $k = 1, \ldots, S$ which, since $\bar{v}_0 = 0$, are now all zero, so that $\bar{\mu} = \sum_{k=1}^{S} \bar{\mu}_k$. Therefore $\bar{\mu}_k = \bar{\mu} d_k \bar{q}_k$, $k = 1, \ldots, S$ holds and the theorem is proved. $\square$

The following example shows that the given constraint can be tight at optimality without necessarily having positive weights:

$$\text{P: } \underset{x>0}{\text{Min}} \ z = x + x^{-1}$$

$$\text{s.t. } g(x) = \frac{1}{2}x + \frac{1}{2}x^{-1} \leq 1.$$

The normality and orthogonality conditions are $\lambda_1 + \lambda_2 = 1$, $\lambda_1 - \lambda_2 + \mu_1 - \mu_2 = 0$. A solution is $\bar{\lambda}_1 = \bar{\lambda}_2 = 1/2$, $\bar{\mu}_1 = \bar{\mu}_2 = 0$, reflecting the obvious optimal solution $\bar{x} = 1$, for which the constraint is tight (although non-binding, i.e., redundant).

By looking at the orthogonality conditions for the general case it is intuitively clear that for a tight and not redundant constraint, the weights will be positive. As opposed to the unconstrained case the degree of difficulty will now include not only the terms in the objective function, but all terms in the constraints as well, i.e., the degree of difficulty will be $T + S - (n + 1)$. Referring to the development in Sect. 8.2, recall that the dual problem of the primal problem P with one constraint is

$$\text{P}_\text{D}: \underset{\lambda>0, \mu\geq0}{\text{Max}} \ z_d(\lambda, \mu) = \left(\prod_{j=1}^{T} \left(\frac{c_j}{\lambda_j} \right)^{\lambda_j} \right) \left(\prod_{k=1}^{S} \left(\frac{d_k \mu}{\mu_k} \right)^{\mu_k} \right)$$

$$\text{s.t. } \sum_{j=1}^{T} \lambda_j = 1, \ \sum_{k=1}^{S} \mu_k = \mu,$$

$$\sum_{j=1}^{T} a_{ij} \lambda_j + \sum_{k=1}^{S} b_{ik} \mu_k = 0, \ i = 1, \ldots, n.$$

Using Theorem 8.8 and the ensuing discussion in Sect. 8.2, we now obtain

Theorem 8.10 (Weak Duality) If $\tilde{x}$ is feasible for P and if $(\tilde{\lambda}, \tilde{\mu})$ is feasible for P_D, then $z_d(\tilde{\lambda}, \tilde{\mu}) \leq \bar{z} \leq f(\tilde{x})$.

Using Karush-Kuhn-Tucker theory it is possible to further show that no duality gap exists at optimality:

Theorem 8.11 If $\bar{x}$ is optimal for P and if $(\bar{\lambda}, \bar{\mu})$ is optimal for P_D, then $z_d(\bar{\lambda}, \bar{\mu}) = \bar{z} = f(\bar{x})$.

From the weak duality Theorem 8.10 we immediately obtain the following corollary which is the converse of Theorem 8.11:

Theorem 8.12 Assume that $\bar{\mathbf{x}}$ is feasible for P and that $(\bar{\lambda}, \bar{\mu})$ is feasible for P_D. If $f(\bar{\mathbf{x}}) = \bar{z} = z_d(\bar{\lambda}, \bar{\mu})$, then $\bar{\mathbf{x}}$ is optimal for P and $(\bar{\lambda}, \bar{\mu})$ is optimal for P_D.

Consider now the general posynomial geometric programming problem with several constraints.

$$P : \underset{z>0}{\text{Min}}\, z = \sum_{j=1}^{T} c_j \prod_{i=1}^{n} x_i^{a_{ij}}$$

$$\text{s.t.} \ \sum_{k=1}^{S_\ell} d_{i\ell} \prod_{i=1}^{n} x_i^{b_{ik\ell}} \leq 1, \ell = 1, \ldots m.$$

Setting as before $p_j := \prod_{i=1}^{n} x_i^{a_{ij}}; \ j = 1, \ldots, T$, we now also define $q_{k\ell} := \prod_{i=1}^{n} x_i^{b_{ik\ell}}, \ k = 1, \ldots, S_\ell, \ \ell = 1, \ldots, m$. The objective function can then be written as $f(\mathbf{x}) = \sum_{j=1}^{T} c_j p_j$ and the constraints as $g_\ell(\mathbf{x}) = \sum_{k=1}^{S_\ell} d_{k\ell} q_{k\ell} \leq 1, \ \ell = 1, \ldots, m$. Define now positive weights λ_j and nonnegative weights $\mu_{k\ell}$ as

$$\lambda_j z := c_j p_j, \ j = 1, \ldots, T$$

$$\mu_{k\ell} := \mu_\ell d_{k\ell} q_{k\ell}, \ k = 1, \ldots, S_\ell; \ \ell = 1, \ldots, m$$

where $\mu_\ell := \sum_{k=1}^{S_\ell} \mu_{k\ell}, \ \ell = 1, \ldots, m$. Similarly to Theorem 8.9 one can prove the following theorem which gives necessary optimality conditions.

Theorem 8.13 Let λ_j and $\mu_{k\ell}$ be the weights as defined above for the problem P with several constraints. At optimality the following conditions hold true:

$$\sum_{j=1}^{T} \bar{\lambda}_j = 1 \qquad\qquad\qquad \text{(normality)}$$

$$\sum_{j=1}^{T} a_{ij} \bar{\lambda}_j + \sum_{\ell=1}^{m} \sum_{k=1}^{S_\ell} b_{ik\ell} \bar{\mu}_{k\ell} = 0, \ i = 1, \ldots, n \qquad \text{(orthogonality)}$$

$$\bar{\lambda}_j \bar{z} = c_j \bar{p}_j \ \text{ and } \ \bar{\lambda}_j > 0, \ j = 1, \ldots, T$$

$$\bar{\mu}_{k\ell} = \bar{\mu}_\ell d_{k\ell} \bar{q}_{k\ell} \ \text{ and } \ \bar{\mu}_{k\ell} \geq 0, \ k = 1, \ldots, S_\ell; \ \ell = 1, \ldots, m$$

$$\sum_{k=1}^{S_\ell} \bar{\mu}_{k\ell} = \bar{\mu}_\ell, \ \ell = 1, \ldots, m.$$

Furthermore, if any constraint term weight $\bar{\mu}_{k\ell} > 0$, then all term weights for that particular constraint are positive and the constraint is tight. Conversely, if a particular constraint is loose then we may set all the term weights for that constraint equal to zero.

The proof of this theorem follows along the lines of that of Theorem 8.9 above, we omit it here.

The dual P_D of the general posynomial problem P with several constraints is

$$P_D: \text{Max } z_d(\boldsymbol{\lambda}, \boldsymbol{\mu}_1, \ldots, \boldsymbol{\mu}_m) = \left[\prod_{j=1}^{T} \left(\frac{c_j}{\lambda_j} \right)^{\lambda_j} \right] \prod_{\ell=1}^{m} \left[\prod_{k=1}^{S_\ell} \left(\frac{d_{k\ell}\mu_\ell}{\mu_{k\ell}} \right)^{\mu_{k\ell}} \right]$$

$$\text{s.t. } \sum_{j=1}^{T} \lambda_j = 1$$

$$\sum_{j=1}^{T} a_{ij}\lambda_j + \sum_{\ell=1}^{m}\sum_{k=1}^{S_\ell} b_{ik\ell}\mu_{k\ell} = 0, \quad i = 1, \ldots, n$$

$$\sum_{k=1}^{S_\ell} \mu_{k\ell} = \mu_\ell, \quad \ell = 1, \ldots, m$$

$$\lambda_j > 0, \quad j = 1, \ldots, T$$

$$\mu_{k\ell} \geq 0, \quad k = 1, \ldots, S_\ell; \quad \ell = 1, \ldots, m.$$

Using arguments similar to those for developing Theorem 8.8, we obtain

Theorem 8.14 (Weak Duality) If $\tilde{\mathbf{x}}$ is feasible for P and if $(\tilde{\boldsymbol{\lambda}}, \tilde{\boldsymbol{\mu}}_1, \ldots, \tilde{\boldsymbol{\mu}}_m)$ is feasible for its dual P_D, then $z_d(\tilde{\boldsymbol{\lambda}}, \tilde{\boldsymbol{\mu}}_1, \ldots, \tilde{\boldsymbol{\mu}}_m) \leq \bar{z} \leq f(\tilde{\mathbf{x}})$.

Again it is possible to show that no duality gap exists at optimality:

Theorem 8.15 If $\tilde{\mathbf{x}}$ is optimal for problem P and $(\bar{\boldsymbol{\lambda}}, \bar{\boldsymbol{\mu}}_1, \ldots, \bar{\boldsymbol{\mu}}_m)$ is optimal for P_D then $z_d(\bar{\boldsymbol{\lambda}}, \boldsymbol{\mu}_1, \ldots, \bar{\boldsymbol{\mu}}_m) = \bar{z} = f(\bar{\mathbf{x}})$.

Again, as with the single-constraint problem, an immediate corollary of the Weak Duality Theorem 8.14 is the following result, which is the converse of Theorem 8.15:

Theorem 8.16 Assume that $\bar{\mathbf{x}}$ is feasible for problem P and that $(\bar{\boldsymbol{\lambda}}, \bar{\boldsymbol{\mu}}_1, \ldots, \bar{\boldsymbol{\mu}}_n)$ is feasible for P_D. If $f(\bar{\mathbf{x}}) = z_d(\bar{\boldsymbol{\lambda}}, \bar{\boldsymbol{\mu}}_1, \ldots, \bar{\boldsymbol{\mu}}_m)$, then $\bar{\mathbf{x}}$ is optimal for P and $(\bar{\boldsymbol{\lambda}}, \bar{\boldsymbol{\mu}}_1, \ldots, \bar{\boldsymbol{\mu}}_n)$ is optimal for P_D.

In analogy with the single constraint case, the degree of difficulty for the problem with several constraints will take all terms of the objective function as well as of the constraints into account i.e., the degree of difficulty is defined as

$$T + \sum_{\ell=1}^{m} S_\ell - (n + 1).$$

Just as in the unconstrained case, we can now handle constrained problems with zero degree of difficulty by solving the normality and orthogonality conditions for the weights. For problems with positive degree of difficulty, one can proceed as in the unconstrained case. First, the normality and orthogonality conditions can be repeatedly manipulated in an effort to produce upper and lower bounds on the weights; weak duality can then be invoked to bound the optimal function value. Alternatively, terms may be dropped to reduce the degree of difficulty. Thirdly, the normality and orthogonality conditions may be solved in terms of one or more of the weights, and an unconstrained maximization search be performed on the substituted dual function.

The above solution procedure is now summarized in algorithmic form. It will closely resemble the procedure in Sect. 8.3.1 above for unconstrained problems. Recall that the problem P under consideration is

$$P: \min_{\mathbf{x}>0} z = \sum_{j=1}^{T} c_j \prod_{i=1}^{n} x_i^{a_{ij}}$$

$$\text{s.t.} \sum_{k=1}^{S_\ell} d_{k\ell} \prod_{i=1}^{n} x_i^{b_{ik\ell}} \le 1, \ \ell = 1, \dots, m.$$

The normality and orthogonality conditions form the system:

$$\Lambda: \sum_{j=1}^{T} \lambda_j = 1 \qquad\qquad\qquad \text{(normality)}$$

$$\sum_{j=1}^{T} a_{ij}\lambda_j + \sum_{\ell=1}^{m}\sum_{k=1}^{S_\ell} b_{ik\ell}\mu_{k\ell} = 0, \ i = 1, \dots, n. \qquad \text{(orthogonality)}$$

For simplicity, assume that the problem P has a unique optimal solution $\bar{\mathbf{x}}$ in the positive orthant $\mathbb{R}_{>}^{n}$. Furthermore, we assume that the matrix of coefficients of the system Λ has full rank.

A Constrained Posynomial Geometric Programming Algorithm

Step 0 Disregarding the constraints, find a solution $\hat{\mathbf{x}}$ for the unconstrained problem. Is $\hat{\mathbf{x}}$ feasible?

If yes: Stop, $\hat{\mathbf{x}}$ is optimal for P.
If no: Go to Step 1.

Step 1 Is the degree of difficulty $T = \sum_{\ell=1}^{m} S_\ell - (n+1) > 0$?

If yes: Go to Step 3.
If no: Go to Step 2.

Step 2 Find the unique solution $\left(\bar{\lambda}, \bar{\mu}_1, \ldots, \bar{\mu}_m\right)$ to system Λ, i.e., the normality and orthogonality conditions. The optimal primal objective function value is then

$$\bar{z} = \left[\prod_{j=1}^{T}\left(\frac{c_j}{\bar{\lambda}_j}\right)^{\bar{\lambda}_j}\right]\prod_{\ell=1}^{m}\left[\prod_{k=1}^{S_\ell}\left(\frac{d_{k\ell}\bar{\mu}_\ell}{\bar{\mu}_{k\ell}}\right)^{\bar{\mu}_{k\ell}}\right], \text{ where } \bar{\mu}_\ell = \sum_{k=1}^{S_\ell}\bar{\mu}_{k\ell}, \ \ell = 1, \ldots, m, \text{ and the}$$

values of the individual terms are $c_j p_j(\bar{\mathbf{x}}) = \bar{\lambda}_j\bar{z}$, $j = 1, \ldots, T$. If enough equations are available, the optimal solution $\bar{\mathbf{x}}$ can be found by solving the following system of equations, linear in $\log \bar{x}_i$:

$$\sum_{i=1}^{n}a_{ij}\log \bar{x}_i = \log\frac{\bar{\lambda}_j\bar{z}}{c_j}, \ j = 1, \ldots, T,$$

Step 3 Select one of the following three approaches:

Bounding of the weights: Go to Step 4.
Dropping terms: Go to Step 5.
Searching in the dual space: Go to Step 6.

Step 4 (Bounding of the weights) By repeatedly manipulating the system Λ, lower and upper bounds are sought for the optimal weights $\bar{\lambda}_j$, $\bar{\mu}_{k1}, \ldots, \bar{\mu}_{km}$. If further progress is required, a tentative solution $\hat{\lambda}$ satisfying these bounds may be used to find a tentative solution $\hat{\mathbf{x}}$ by using the equations $\hat{\lambda}_j\hat{z} = c_j p_j(\hat{\mathbf{x}})$, $j = 1, \ldots, T$. The optimal objective function value $\bar{z}$ is then bounded by

$$\left[\prod_{j=1}^{T}\left(\frac{c_j}{\hat{\lambda}_j}\right)^{\hat{\lambda}_j}\right]\prod_{\ell=1}^{m}\left[\prod_{k=1}^{S_\ell}\left(\frac{d_{k\ell}\hat{\mu}_\ell}{\hat{\mu}_\ell}\right)^{\hat{\mu}_{k\ell}}\right] \leq \bar{z} \leq \sum_{j=1}^{T}c_j p_j(\hat{\mathbf{x}}).$$

If the bounds are sufficiently tight, stop; otherwise go to Step 5 or 6.

Step 5 (Dropping terms) The degree of difficulty may be reduced by disregarding terms $c_j p_j(\hat{\mathbf{x}})$ that are judged to be relatively insignificant at optimality. The resulting reduced problem can be analyzed by going to Step 2 if it has zero degree of difficulty or to Step 3 otherwise.

Step 6 (Searching in the dual space) The dual problem P_D is solved to any desired degree of accuracy either using a suitable algorithm for unconstrained nonlinear optimization or by using the substituted dual function. The resulting tentative solution $\hat{\lambda}$ is used to find a tentative solution $\hat{\mathbf{x}}$ by means of the equations $\hat{\lambda}_j\hat{z} = c_j p_j(\hat{\mathbf{x}}), j = 1, \ldots, T$. Bounds may be obtained by the relations in Step 4. If the bounds are sufficiently tight, stop; otherwise repeat Step 6 with an increased accuracy for solving the dual problem P_D.

Example As a numerical example, consider the single-constraint posynomial geometric programming problem

$$P: \underset{x_1, x_2 > 0}{\text{Min}} \ z = 4x_1^2 + 5x_1 x_2$$

$$\text{s.t. } 7x_1^{-2} x_2^{-1} \leq 1.$$

With $T + S = 2 + 1 = 3$ terms and $n = 2$ variables, the degree of difficulty of this problem is $T + S - (n + 1) = 2 + 1 - (2 + 1) = 0$. Since the constraint makes solutions with very small values of x_1 and x_2 fall outside the feasible region, we pass through Steps 0 and 1, and in Step 2 we obtain the system

$$
\begin{aligned}
\Lambda: \lambda_1 + \lambda_2 &= 1 & &\text{(normality)} \\
2\lambda_1 + \lambda_2 - 2\mu_1 &= 0 & &\text{(orthogonality)} \\
\lambda_2 - \mu_1 &= 0. & &\text{(orthogonality)}
\end{aligned}
$$

The unique solution to this system is $\bar{\lambda}_1 = \dfrac{1}{3}$, $\bar{\lambda}_2 = \dfrac{2}{3}$ $\bar{\mu}_1 = \dfrac{2}{3}$, and since $\bar{\mu}_1 > 0$, Theorem 8.13 tells us that the constraint must be tight, i.e., $\bar{x}_1^2 \bar{x}_2 = 7$. We find

$$\bar{z} = \left(\frac{4}{\frac{1}{3}}\right)^{\frac{1}{3}} \left(\frac{5}{\frac{2}{3}}\right)^{\frac{2}{3}} (7)^{\frac{2}{3}} = 3\sqrt[3]{35^2} \approx 32.100, \quad 4\bar{x}_1^2 = \bar{\lambda}_1 \bar{z} = \frac{1}{3}\left(3\sqrt[3]{35^2}\right) = \sqrt[3]{35^2}, \text{ so}$$

that $\bar{x}_1 = \dfrac{1}{2}\sqrt[3]{35} \approx 1.6355$. Furthermore, $5\bar{x}_1 \bar{x}_2 = \bar{\lambda}_2 \bar{z} = \left(\dfrac{2}{3}\right)\left(3\sqrt[3]{35^2}\right) = 2\sqrt[3]{35^2}$,

so that $\bar{x}_2 = \dfrac{2\sqrt[3]{35^2}}{\frac{5}{2}\sqrt[3]{35}} = \dfrac{4}{5}\sqrt[3]{35} \approx 2.6169$. We can easily check to see that the single constraint is tight at optimality. This example does not demonstrate the workings of Steps 3–6 of the algorithm, but those were already illustrated by the example for the unconstrained case in Sect. 8.3.1 above.

8.4   Signomial Geometric Programming

This section considers problems that include terms with negative economic coefficients. As we will see, such terms are easier to handle in the constraints than in the objective function. We also mention how to handle equality constraints.

The problem P under consideration is

$$P: \underset{\mathbf{x} > \mathbf{0}}{\text{Min}} \ z = \sum_{j=1}^{T} c_j \prod_{i=1}^{n} x_i^{a_{ij}}$$

$$\text{s.t. } \sum_{k=1}^{S_\ell} d_{k\ell} \prod_{i=1}^{n} x_i^{b_{ik\ell}} \leq 1, \ \ell = 1, \ldots, m.$$

As before, we define $p_j(\mathbf{x}) = \prod_{i=1}^{n} x_i^{a_{ij}}$ and $q_{k\ell}(\mathbf{x}) = \prod_{i=1}^{n} x_i^{b_{ik\ell}}$, so that the objective function can be written as $f(\mathbf{x}) = \sum_{j=1}^{T} c_j p_j$ and the m constraints as $g_\ell(\mathbf{x}) = \sum_{k=1}^{S_\ell} d_{k\ell} q_{k\ell}$, $\ell = 1, \ldots, m$. Moreover, assume for now that $f(\mathbf{x})$ is a posynomial, i.e., $c_j > 0$, $j = 1, \ldots, T$ and that $g_\ell(\mathbf{x})$, $\ell = 1, \ldots, m$ are signomials, i.e., $d_{k\ell} \in \mathbb{R}$, $k = 1, \ldots, S_\ell$; $\ell = 1, \ldots, m$. The sign of $d_{k\ell}$ may be indicated by the *signum function* $\sigma_{k\ell} := \dfrac{d_{k\ell}}{|d_{k\ell}|}$ which equals $+1$ if $d_{k\ell}$ is positive and -1 if $d_{k\ell}$ is negative. As in the posynomial case, define positive weights λ_j for the terms $c_j p_j$ of the objective function. We then have $\lambda_j z = c_j p_j$ and $\lambda_j > 0$, $j = 1, \ldots, T$, as well as $\sum_{j=1}^{T} \lambda_j = 1$.

For the signomial constraint terms we define nonnegative generalized weights $\mu_{k\ell}$, $k = 1, \ldots, S_\ell$; $\ell = 1, \ldots, m$ such that $\mu_{k\ell} = \mu_\ell |d_{k\ell}| q_{k\ell} \geq 0$, $k = 1, \ldots, S_\ell$; $\ell = 1, \ldots, m$, where

$$\mu_\ell := \sum_{k=1}^{S_\ell} \frac{d_{k\ell}}{|d_{k\ell}|} \mu_{k\ell} = \sum_{k=1}^{S_\ell} \sigma_{k\ell} \mu_{k\ell} \geq 0, \ \ell = 1, \ldots, m.$$

In analogy with the posynomial case one can show that the following are necessary conditions for optimality of the primal problem P:

$$\sum_{j=1}^{T} \lambda_j = 1 \quad \text{(normality)}$$

$$\sum_{j=1}^{T} a_{ij} \lambda_j + \sum_{\ell=1}^{m} \sum_{k=1}^{S_\ell} b_{ik\ell} \sigma_{k\ell} \mu_{k\ell} = 0, \ \ i = 1, \ldots, n \quad \text{(orthogonality)}$$

$$\lambda_j > 0, j = 1, \ldots, T; \ \mu_{k\ell} \geq 0, \ k = 1, \ldots, S_\ell; \ \ell = 1, \ldots, m.$$

In these orthogonality conditions we can see that the signum functions $\sigma_{k\ell}$ of the constraint terms enter. The dual problem P_D associated with P is then

$$P_D: \text{Optimize } z_d(\lambda, \mu_1, \ldots, \mu_m) = \left[\prod_{j=1}^{T} \left(\frac{c_j}{\lambda_j} \right)^{\lambda_j} \right] \prod_{\ell=1}^{m} \left[\prod_{k=1}^{S_\ell} \left(\frac{|d_{k\ell}| \mu_\ell}{\mu_{k\ell}} \right)^{\mu_{k\ell}} \right]$$

$$\text{s.t.} \sum_{j=1}^{T} \lambda_j = 1$$

$$\sum_{j=1}^{T} a_{ij} \lambda_j + \sum_{\ell=1}^{m} \sum_{k=1}^{S_\ell} b_{ik\ell} \sigma_{k\ell} \mu_{k\ell} = 0, \ i = 1, \ldots, n$$

$$\sum_{k=1}^{S_\ell} \sigma_{k\ell}\mu_{k\ell} = \mu_\ell, \ \ell = 1, \ldots, m$$

$$\lambda_j > 0, \ j = 1, \ldots, T$$

$$\mu_\ell \geq 0, \ \mu_{k\ell} \geq 0, \ k = 1, \ldots, S_\ell; \ \ell = 1, \ldots, m.$$

It should be pointed out that the useful weak duality result from the posynomial case is now lost, i.e., the dual objective function value need not be smaller than or equal to the primal objective function value. Also note that P_D is stated as an "optimization" problem, it turns out that either the maximal or the minimal solution will correspond to the optimal solution of P.

The procedure for solving a signomial geometric programming problem with posynomial objective function will not be stated in algorithmic form here; instead we refer to the algorithm given in the previous section for the posynomial case. The orthogonality conditions in step 1 need to be changed to include the signum functions of the constraint terms, also the weak duality bounding of the optimal $\bar{z}$-value in steps 4, 5 and 6 does no longer apply.

Suppose now that the objective function contains negative terms but that the optimal $\bar{z}$-value is still positive. Instead of minimizing the signomial objective function

$$z = \sum_{j=1}^T c_j p_j(\mathbf{x}) = \sum_{j=1}^T c_j \prod_{i=1}^n x_i^{a_{ij}}$$

subject to signomial constraints, we introduce a new variable x_0 and minimize the posynomial function $z = x_0$ with one additional constraint, namely

$$\sum_{j=1}^T c_j x_0^{-1} \prod_{i=1}^n x_i^{a_{ij}} \leq 1.$$

With this new formulation the objective function is indeed posynomial and the constraint functions are still signomial which is a case that has already been treated. A similar transformation is possible if the optimal z value is negative. A difficult question is to decide ex ante which of the two cases applies, and in practice one will have to pursue both.

Another question is how to deal with equality constraints. One obvious method is to replace each equality by two appropriate inequalities (noting that inequalities with right-hand side values of -1 can also be handled). Unfortunately, this technique would result in a substantial increase in the degree of difficulty. Another method uses the fact that an equality constraint must be tight at optimality. Consequently, one could tentatively fix the direction of the inequality for each of the given equalities and proceed with the resulting inequality constrained problem. If at optimality any of these constraints are loose, the direction of the inequality would be changed and the problem is solved again.

8.5 Transformations and Illustrations

We will now discuss how some problems whose objective functions are not in signomial form can be transformed so that geometric programming techniques can still be used. Four different transformations are described; in all these cases any number of signomial constraints may also be given. We then provide some illustrations of how geometric programming techniques can be used to model and solve practical problems.

Transformation 1 $\operatorname*{Min}_{\mathbf{x}>0} z = f(\mathbf{x}) + [t(\mathbf{x})]^{\alpha} h(\mathbf{x})$.

Suppose that f is a signomial, that t and h are positive-valued signomials, and that α is a positive constant. If α is noninteger, z may not necessarily be a signomial. Introduce now an additional variable $u > 0$ as an upper bound on $t(\mathbf{x})$, i.e. $t(\mathbf{x}) \leq u$, so that $u^{-1}t(\mathbf{x}) \leq 1$. Then z is replaced by $z' = f(\mathbf{x}) + u^{\alpha}h(\mathbf{x})$ and the objective function z' is now a signomial which is minimized subject to the constraint $u^{-1}t(\mathbf{x}) \leq 1$ in addition to the original constraints. It is clear that the additional constraint must be tight at optimality, i.e., u will be as small as possible. Moreover, the optimal solution $\bar{\mathbf{x}}$ is the same for both problems and $\bar{z} = \bar{z}'$.

Transformation 2 $\operatorname*{Min}_{\mathbf{x}>0} z = f(\mathbf{x}) + [r(\mathbf{x}) - t(\mathbf{x})]^{\alpha} h(\mathbf{x})$.

Assume that f and t are signomials, that r is a single-term signomial with $r(\mathbf{x}) > t(\mathbf{x})$, that h is a positive-valued signomial, and that α is a positive constant. Since the objective function is not necessarily a signomial, we introduce an additional variable $u > 0$ as a lower bound on the signomial $r - t$, i.e., $u \leq r(\mathbf{x}) - t(\mathbf{x})$. This can be rewritten as $u[r(\mathbf{x})]^{-1} + t(\mathbf{x})[r(\mathbf{x})]^{-1} \leq 1$, which is now in proper signomial form. Then z is replaced by $z' = f(\mathbf{x}) + u^{-\alpha}h(\mathbf{x})$, which is signomial, and this new objective function is now minimized with the constraint $u[r(\mathbf{x})]^{-1} + t(\mathbf{x})[r(\mathbf{x})]^{-1} \leq 1$ added to the original constraints (if any).

Transformation 3 $\operatorname*{Min}_{\mathbf{x}>0} z = f(\mathbf{x}) + [\ln t(\mathbf{x})]^{\alpha} h(\mathbf{x})$.

Suppose that f is a signomial, that t and h are signomials where $t(\mathbf{x}) > 1$ and $h(\mathbf{x}) > 0$, and that α is a positive constant. Note that z is not a signomial. The idea in this transformation is to use the integral representation of the logarithm function. For any positive constant y we have $\ln y = \int_1^y s^{-1}ds$ and for any $y > 0$, one can show that $\lim_{\varepsilon \to +0} \int_1^y s^{\varepsilon-1}ds = \int_1^y s^{-1}ds$. Since $\int_1^y s^{\varepsilon-1}ds = \frac{y^{\varepsilon}}{\varepsilon} - \frac{1}{\varepsilon}$, it follows that $\lim_{\varepsilon \to +0} \left(\frac{y^{\varepsilon}}{\varepsilon} - \frac{1}{\varepsilon}\right) = \ln y$. Consequently, $\ln t(\mathbf{x})$ can be approximated by $\varepsilon^{-1}\{[t(x)]^{\varepsilon} - 1\}$ and the approximation gets better with smaller values of ε. Therefore we can write $z \approx f(\mathbf{x}) + {}^{-\alpha}\{[t(\mathbf{x})]^{\varepsilon} - 1\}^{\alpha}h(\mathbf{x})$. Transformation 1 is then applied to the approximation function if necessary.

Transformation 4 $\underset{\mathbf{x}>0}{\mathrm{Min}}\ z = f(\mathbf{x}) + \alpha^{t(\mathbf{x})}h(\mathbf{x})$.

Assuming that f and t are signomials, that α is a positive constant, and that h is a positive-valued signomial, then the objective function z is not a signomial. The idea is to use the well-known representation of e^y as a limit, i.e. $e^y = \underset{s \to +\infty}{\lim}\left(1 + \dfrac{y}{s}\right)^s$. Thus we can write

$$\alpha^{t(\mathbf{x})} = e^{t(\mathbf{x})\ln\alpha} = \lim_{s \to +\infty}\left[1 + \frac{t(x)\ln\alpha}{s}\right]^s,$$

and $\alpha^{t(\mathbf{x})}$ can be approximated by $\left[1 + \dfrac{t(x)\ln\alpha}{s}\right]^s$; the approximation gets better with larger values of s, so that

$$z \approx f(\mathbf{x}) + \left[1 + \frac{t(x)\ln\alpha}{s}\right]^s h(\mathbf{x}).$$

For sufficiently large s, the expression in brackets will be a positive-valued signomial in the vicinity of the optimal solution $\bar{\mathbf{x}}$ and Transformation 1 above can be applied to the approximating function.

We will finish this section by looking at a few simple examples, in which geometric programming can be used advantageously.

Illustration 1 Engineering design of a cylindrical tank.

Consider the problem of designing an open cylindrical tank of at least a given volume $V \geq 0$. By "open" we mean a tank that has a circular bottom, but no lid. Assuming that the material used to build the tank is the same for the bottom and the side, find the minimal amount of material required for the task. To formulate this problem, let r denote the radius of the circular bottom and let h denote the height of the tank. Then the volume of our tank will be $\pi r^2 h$, and the problem can be formulated as

$$\text{P: } \underset{r>0, h>0}{\mathrm{Min}}\ z = \pi r^2 + 2\pi r h$$

$$\text{s.t. } \pi r^2 h \geq V.$$

If we let $x_1 := r$, $x_2 := h$, $c_1 := \pi$, $c_2 := 2\pi$, and $d_1 := V/\pi$, then problem P can be written as a posynomial single-constraint geometric programming problem

$$P': \min_{x_1, x_2 > 0} z' = c_1 x_1^2 + c_2 x_1 x_2$$

$$\text{s.t. } d_1 x_1^{-2} x_2^{-1} \le 1.$$

With $T = 3$ terms and $n = 2$ variables, the degree of difficulty of this problem is $T - (n + 1) = 0$. The normality and orthogonality conditions are

$$\lambda_1 + \lambda_2 = 1$$

$$2\lambda_1 + \lambda_2 - 2\mu_1 = 0$$

$$\lambda_2 - \mu_1 = 0,$$

and the unique solution to this system is $\bar{\lambda}_1 = \frac{1}{3}, \bar{\lambda}_2 = \frac{2}{3}$, and $\bar{\mu}_1 = \frac{2}{3}$. At optimality, the bottom surface must therefore be half of the surface of the side. Therefore, the optimal values $\bar{r}$ and $\bar{h}$ must be such that $2\pi \bar{r}^2 = 2\pi \bar{r} \bar{h}$, so that $\bar{r} = \bar{h}$, i.e., the radius of the cylinder will equal its height. With $\bar{\mu}_1 = \frac{2}{3} > 0$, the constraint has to be tight at optimality, i.e., $\pi \bar{r}^2 \bar{h} = V$, and with $\bar{r} = \bar{h}$, we find that $\bar{r} = \bar{h} == \sqrt[3]{V/\pi}$. For a slight change of view, imagine that the cylinder is now a drink or food can, in which case we will have both, a bottom and a lid. We should mention that the above result changes, once we optimize for a closed can. In that case, we obtain $h = 2r$, i.e., the height of the optimal can equals its diameter. Clearly, this is not like any can that we encounter in practice. Many cans we purchase in the store have a ratio between height and diameter of roughly 4/3:1 rather than 1:1 as prescribed in our optimization. Important practical considerations include the ability to physically handle the can (which puts limits on the diameter of the can), the cost of manufacturing (other than just the raw material), the costs of packaging cans in manageable 6- or 12-packs, aesthetic considerations, and others.

Natural extensions of this design problem include having a bottom thickness proportional, rather than equal, to the thickness of the side, considering the thickness of the side panels (as the volume depends on the inside of the tank, while the material usage depends on the outside), and others. Another straightforward extension includes the waste that is generated by cutting the round top and bottom. In the simplest case, we approximate the waste by the smallest square that circumscribes the circle. In other words, rather than using the *used* area $2\pi r^2$ for top and bottom combined, we could use the *required* area $8r^2$. The result is that the cans are getting aller than in the usual case. The waste in this case does, however, depend on the way the disks are cut. A cutting as the one in Fig. 3.9b in Sect. 3.8 of this volume requires less waste than the one shown in Fig. 3.9a, which is assumed in the above argument.

The required modifications of the above formulation are obvious, and will not change the degree of difficulty of zero. If we were to add a term representing the cost of welding with the length of the weld being $2\pi r$ for welding of the bottom plus h for welding of the side, the resulting problem will have two additional terms, thus increasing the degree of difficulty to $T - (n + 1) = 5 - (2 + 1) = 2$, making a geometric programming approach less attractive.

A geometric programming approach can also be employed for the design of a closed rectangular box with bottom area, weight and volume constraints. Some manipulations and transformations are then used, leading to a problem with one degree of difficulty. For details, readers are referred to Eiselt et al. (1987).

Illustration 2   Extensions to the Economic Order Quantity (*EOQ*).

The economic order quantity was discussed in Sect. 3.9 and a geometric programming formulation was provided in Sect. 8.1. Building on the treatment of the latter, we will add the cost of obsolescence, which we assume to be proportional to Q^2, i.e., the square of the order size. Note that this differs from the approach taken in the example that illustrates the unconstrained posynomial geometric programming algorithm in Sect. 8.3.1, where obsolescence costs were assumed to be proportional to $\sqrt{Q}$. The problem can then be written as

$$P: \underset{x>0}{\text{Min}}\ z = c_1 x + c_2 x^{-1} + c_3 x^2.$$

The three terms of the objective function refer to carrying costs, ordering costs, and obsolescence costs, respectively. With $T = 3$ terms and $n = 1$ variable, the degree of difficulty is $T - (n + 1) = 1$. The normality and orthogonality conditions are

$$\lambda_1 + \lambda_2 + \lambda_3 = 1$$

$$\lambda_1 - \lambda_2 + 2\lambda_3 = 0,$$

so that a unique solution does not exist. However, from the discussion in Sect. 8.3.1 where precisely this problem was discussed, we found by bounding of the weights that at optimality we must have $\bar{\lambda}_1 \in\ ]\,0, \frac{1}{2}\,[, \bar{\lambda}_2 \in\ ]\,\frac{1}{2}, \frac{2}{3}\,[$, and $\bar{\lambda}_3 \in\ ]\,0, \frac{1}{3}\,[$. Thus we can conclude that the optimal carrying cost is less than half of the total cost, the ordering costs are more than half, but less than two thirds of the total costs, and obsolescence costs are less than one third of the minimal total cost.

Another version of the economic order quantity involves shortage costs. Consider an inventory model, in which shortages are permitted at a cost of c_s dollars per year and quantity unit, and let S denote the inventory shortage level at the time of replenishment, i.e., the maximal shortage. As usual, D symbolizes the annual demand, the variable Q is the order size, and TC are the total annual costs. From inventory theory we know that

$$TC = c_h\frac{(Q-S)^2}{2Q} + c_o\frac{D}{Q} + c_s\frac{S^2}{2Q}.$$

The three terms represent holding, ordering, and shortage costs, respectively. Setting $c_1 := \frac{1}{2}c_h$, $c_2 := c_o D$, and $c_3 := \frac{1}{2}c_s$, we obtain problem P for minimizing total annual inventory costs TC as

$$P: \underset{Q,S>0}{\text{Min}} \ z = c_1(Q-S)^2Q^{-1} + c_2Q^{-1} + c_3S^2Q^{-1},$$

or, equivalently,

$$P: \underset{Q,S>0}{\text{Min}} \ z = c_1Q - 2c_1S + c_2Q^{-1} + (c_1+c_3)S^2Q^{-1}.$$

With this formulation, the objective function is a signomial, but not a posynomial; furthermore, the degree of difficulty is $T - (n+1) = 4 - (2+1) = 1$. The problem can be solved by the procedure outlined in Sect. 8.4 for signomial geometric programming. However, using a standard trick in geometric programming, the objective function can be transformed into posynomial form. The idea is to set $x_1 := Q$, $x_2 := S$, and let x_3 be an upper bound on $Q-S$, i.e., $Q-S \le x_3$, or $x_1 - x_2 \le x_3$, or $x_1x_3^{-1} - x_2x_3^{-1} \le 1$. Now P can be written as the constrained problem

$$P': \underset{x>0}{\text{Min}} \ z' = c_1x_1^{-1}x_3^2 + c_2x_1^{-1} + c_3x_1^{-1}x_2^2$$

$$\text{s.t. } x_1x_3^{-1} - x_2x_3^{-1} \le 1.$$

Comparing problems P and P′, we find that $z' \ge z$ and that equality holds whenever the signomial constraint in P′ is tight. It is now clear that the constraint must be tight at optimality, since x_3 must be as small as possible, so that problems P and P′ will have the same optimal solution $\bar{x}$ with $\bar{z} = \bar{z}'$. We realize that the conversion of problem P into problem P′ uses the logic of transformation 1 above. The degree of difficulty is now $T - (n+1) = 5 - (3+1) = 1$, which is the same as before. However, we are better off, since the objective function is now posynomial. Recall from the previous section that signomials are easier to handle when they appear in the constraints rather than in the objective function. The normality and orthogonality conditions for P′ are

$$\lambda_1 + \lambda_2 + \lambda_3 = 1$$

$$-\lambda_1 - \lambda_2 - \lambda_3 + \mu_1 = 0$$

$$2\lambda_3 - \mu_2 = 0$$

$$2\lambda_1 - \mu_1 + \mu_2 = 0.$$

We can see that at optimality $\bar{\lambda}_2 = \frac{1}{2}$, $\bar{\mu}_1 = 1$, from which we can conclude that $\lambda_1 \in \,]0,\frac{1}{2}[$ and $\lambda_3 \in \,]0,\frac{1}{2}[$. For our inventory problem, this implies that at optimality, the ordering costs must be exactly half of the total costs, and each of the ordering and shortage costs must be less than that. For further details, see Eiselt et al. (1987).

Illustration 3 Economic optimization using the Cobb-Douglas production function. In economic theory, the famous *Cobb-Douglas production function* (see Cobb and Douglas 1928) expresses the manufacturing output Y as a function of labor L and capital K in the monomial $Y = \gamma L^{\alpha} K^{\beta}$, where γ, α, and β are constants. From extensive empirical data it appears that $\alpha = \frac{2}{3}$ and $\beta = \frac{1}{3}$ are typical values over a large range of actual cases. Since then $\alpha + \beta = 1$, the Cobb-Douglas production function will exhibit constant returns to scale, meaning that if the inputs change in the same proportion, the output will also change in that same proportion. Specifically, if given inputs $\hat{L}$ and $\hat{K}$ generate an output $\hat{Y}$, then for any proportionality factor $t > 0$, the inputs $t\hat{L}$ and $t\hat{K}$ will generate the output $t\hat{Y}$, since $\gamma(t\hat{L})^{\frac{2}{3}}(t\hat{K})^{\frac{1}{3}} = t\gamma\hat{L}^{\frac{2}{3}}\hat{K}^{\frac{1}{3}} = t\hat{Y}$.

To further the discussion, we assume that p is the unit price of output, w is the wage rate, and q is the price of capital. With a prespecified budget level M, the problem of maximizing the revenue $pY = p\gamma L^{\frac{2}{3}} K^{\frac{1}{3}}$ subject to a budget constraint becomes

$$\text{P:} \underset{L,K>0}{\text{Max}}\ z = p\gamma L^{\frac{2}{3}} K^{\frac{1}{3}}$$

$$\text{s.t. } wL + qK \leq M.$$

Since $z > 0$, an equivalent problem is to minimize $z' = z^{-1} = p^{-1}\gamma^{-1}L^{-\frac{2}{3}}K^{-\frac{1}{3}}$, so that we obtain

$$\text{P':} \underset{L,K>0}{\text{Min}}\ z' = p^{-1}\gamma^{-1}L^{-\frac{2}{3}}K^{-\frac{1}{3}},$$

$$\text{s.t. } \left(\frac{w}{M}\right)L + \left(\frac{q}{M}\right)K \leq 1.$$

By setting $x_1 := L$, $x_2 := K$, $c_1 := p^{-1}\gamma^{-1}$, $d_1 := \left(\frac{w}{M}\right)$ and $d_2 := \left(\frac{q}{M}\right)$, we can write P' as a single-constraint posynomial geometric programming problem

$$\text{P'':} \underset{x>0}{\text{Min}}\ z' = c_1 x_1^{-\frac{2}{3}} x_2^{-\frac{1}{3}}$$

$$\text{s.t. } d_1 x_1 + d_2 x_2 \leq 1.$$

With $T = 3$ terms and $n = 2$ variables, the degree of difficulty is $T - (n+1) = 3 - (2+1) = 0$ and the normality and orthogonality conditions are

$$\lambda_1 = 1$$

$$-\tfrac{2}{3}\lambda_1 + \mu_1 = 0$$

$$-\tfrac{1}{3}\lambda_1 + \mu_2 = 0.$$

The solution to this system is $\bar{\lambda}_1 = 1$, $\bar{\mu}_1 = \tfrac{2}{3}$, $\bar{\mu}_2 = \tfrac{1}{3}$, and since $\bar{\mu}_1$ and $\bar{\mu}_2$ are strictly positive, the constraint must be tight at optimality, according to Theorem 8.13. The fact that the constraint must be tight also follows from the fact that the objective function is decreasing with respect to both variables x_1 and x_2 in the minimizing objective. Since $\bar{\lambda}_1 \bar{z}' = c_1\bar{p}_1$, $\bar{\mu}_1 = \bar{\mu}d_1\bar{q}_1$ and $\bar{\mu}_2 = \bar{\mu}d_2\bar{q}_2$, where $\bar{\mu} = \bar{\mu}_1 + \bar{\mu}_2$, we find that $\bar{z}' = c_1\bar{x}_1^{-\frac{2}{3}} + c_2\bar{x}_2^{-\frac{1}{3}}$, $\tfrac{2}{3} = d_1\bar{x}_1$ and $\tfrac{1}{3} = d_2\bar{x}_2$. Therefore, $\bar{L} = \bar{x}_1 = \tfrac{2}{3}d_1^{-1} = 2M/3w$, $\bar{K} = \bar{x}_2 = \tfrac{1}{3}d_2^{-1} = M/3q$, and $\bar{z} = p\bar{Y} = p\gamma\left(\dfrac{2M}{3w}\right)^{\frac{2}{3}}\left(\dfrac{M}{3q}\right)^{\frac{1}{3}} = \tfrac{1}{3}p\gamma M\left(\dfrac{2}{w}\right)^{\frac{2}{3}}\left(\dfrac{1}{q}\right)^{\frac{1}{3}}$, a result well known in macroeconomic theory. By the way, the result could also have been obtained by way of the Karush-Kuhn-Tucker optimality conditions using Theorem 4.18 in Chap. 4.

Suppose now that instead of maximizing revenue, we wish to maximize the profits that derive from the production process, a potentially more realistic objective. We then obtain the unconstrained problem formulation

$$\text{P: } \underset{L,K>0}{\text{Max}} \; z = pY - (wL + qK) = p\gamma L^\alpha K^\beta - (wK + qK),$$

assuming implicitly that we only consider solutions with positive profit. We can then convert problem P into the equivalent minimization problem

$$\text{P}': \underset{L,K>0}{\text{Min}} \; z' = \left(p\gamma L^\alpha K^\beta - (wK + qK)\right)^{-1}.$$

Unfortunately, the objective function is now not a posynomial, it is not even a signomial. We can still use geometric programming techniques, if we apply Transformation 2 from the previous section. In order to do so, we first set $x_1 := L$, $x_2 := K$, and define the two posynomials $g(x_1, x_2) = p\gamma L^\alpha K^\beta$ and $h(x_1, x_2) = wL + qK$. Now problem P' can be written as

$$\text{P}'': \underset{\mathbf{x}>0}{\text{Min}} \; z'' = \dfrac{1}{g(x_1,x_2) - h(x_1,x_2)}.$$

We can now employ the idea of Transformation 2 by introducing the additional variable x_3 as a lower bound on the signomial $g(x_1, x_2) - h(x_1, x_2)$, from which we obtain problem P''', which can be written as

$$\text{P}''': \underset{\mathbf{x}>0}{\text{Min}} \; z''' = x_3^{-1}$$

$$\text{s.t. } x_3 g^{-1}(x_1, x_2) + h(x_1, x_2) g^{-1}(x_1, x_2) \leq 1.$$

Problem P''' is now in proper posynomial form, and since the constraint must be tight at optimality in P''', it follows that with $(\bar{x}_1, \bar{x}_2, \bar{x}_3)$ denoting an optimal solution to P''', the point $(\bar{x}_1, \bar{x}_2)$ must also be optimal for P'', P', and P.

Proceeding with the normality and orthogonality conditions for problem P''', we obtain the system

$$\lambda_1 = 1$$

$$-\alpha\mu_1 + (1-\alpha)\mu_2 - \alpha\mu_3 = 0$$

$$-\beta\mu_1 - \beta\mu_2 + (1-\beta)\mu_3 = 0$$

$$-\lambda_1 + \mu_1 = 0,$$

which for $\alpha + \beta < 1$ has the unique solution

$$\bar{\lambda}_1 = \bar{\mu}_1 = 1, \bar{\mu}_2 = \frac{\alpha}{1-\alpha-\beta}, \bar{\mu}_3 = \frac{\beta}{1-\alpha-\beta}.$$

From this result the optimal values $\bar{L}$, $\bar{K}$, and $\bar{Y}$ can be computed and expressed in closed, although fairly complex, form. The interesting conclusion to be drawn from this model is that for a unique optimal solution to exist, $\alpha + \beta < 1$ is required. This becomes clear when considering the fact that for $\alpha + \beta = 1$—considered the standard case for the Cobb-Douglas production function—the model will exhibit constant returns to scale. This also applies to the profit, which therefore will not have a finite maximal value.

Illustration 4 Queuing.

As a final illustration of geometric programming, we will consider a single-queue single-server $M/M/1$ waiting line model; see, e.g., Eiselt and Sandblom (2012). Here we assume that customer arrivals occur at a service facility randomly according to a Poisson process with an arrival rate of λ customers per time unit. Service completion also occurs randomly in Poisson fashion, with a service rate (the number of customers who can be served) of $\mu > \lambda$ customers per time unit. It can then be shown that L_s, the expected number of customers in the system (i.e. waiting in line or being served) is given by the expression $L_s = \frac{\lambda}{\mu - \lambda}$. Assign now a cost of c dollars per time unit each customer spends in the system, and a cost $r\mu$ for operating the service facility, where r is a given positive constant. The total cost TC of expected cost is therefore given by $TC = r\mu + \frac{c\lambda}{\mu - \lambda}$, and we wish to determine the service rate so as to minimize the total expected cost. In order to formulate this as a geometric programming problem, we will make use of Transformation 1 in the following way. We first set $x_1 := \mu$ and introduce a new variable x_2 as a lower positive bound on $\mu - \lambda$, i.e., $0 < x_2 \leq \mu - \lambda = x_1 - \lambda$. We then write the constraint $x_2 \leq x_1 - \lambda$ as $x_2 + \lambda \leq x_1$ or $x_1^{-1}x_2 + x_1^{-1}\lambda \leq 1$, which is a proper posynomial constraint. We can therefore formulate the minimization problem P, where $c_1 := r$, $c_2 := c\lambda$, $d_1 := \lambda$, and $d_2 := 1$ as

$$\text{P: } \underset{x>0}{\text{Min }} z = c_1 x_1 + c_2 x_2^{-1}$$

$$\text{s.t. } d_1 x_1^{-1} + d_2 x_1^{-1} x_2 \leq 1.$$

With $T = 4$ terms and $n = 2$ variables, this is a single-constraint posynomial programming problem with $T - (n + 1) = 4 - (2 + 1) = 1$ degree of freedom, and the normality and orthogonality conditions are

$$\lambda_1 + \lambda_2 = 1$$

$$\lambda_1 - \mu_1 - \mu_2 = 0$$

$$-\lambda_2 + \mu_2 = 0.$$

From this system, we conclude that at optimality, $\bar{\lambda}_1 - \bar{\lambda}_2 = \bar{\mu}_1 \geq 0$, so that $\bar{\lambda}_1 \in [½, 1]$, and $\bar{\lambda}_2 \in [0, ½]$, implying that the service facility cost must be at least half of the total cost, whereas the cost of customer time can be at most half of the total.

Returning to the original problem, we can actually find a better formulation. By setting $x := \mu - \lambda$, we can write the total cost as $TC = r(x + \lambda) + c\lambda x^{-1} = rx + c\lambda x^{-1} + r\lambda$. Omitting the constant term $r\lambda$, which does not influence the optimization, we set $c_1 := r$, $c_2 := c\lambda$, and obtain the unconstrained geometric programming problem

$$\text{P': } \underset{x>0}{\text{Min }} z' = c_1 x + c_2 x^{-1},$$

for which the degree of difficulty is zero. With the normality and orthogonality conditions

$$\lambda_1 + \lambda_2 = 1$$

$$\lambda_1 - \lambda_2 = 0,$$

we find $\bar{\lambda}_1 = \bar{\lambda}_2 = ½$, so that at optimality the service facility cost must equal the cost of customer time.

Other applications of geometric programming have been reported in the literature, e.g., in marketing (Balachandran and Gensch 1974), optimal gas compression in networks (Misra et al. 2015), and designs of networks with randomly switching structure (Ogura and Preciado 2017). See also Chandrasekaran and Shah (2014), Liu (2008), Lin and Tsai (2012), Ojha and Biswal (2010), and Toussaint (2015).

References

Avriel M (ed) (2013) *Advances in geometric programming*. Springer, Cham

Balachandran Y, Gensch DH (1974) Solving the marketing mix problem using geometric programming. *Management Science* **21**:160–171

Bazaraa MS, Sherali HD, Shetty CM (2013) *Nonlinear programming: theory and algorithms*. 3rd
edn. Wiley, New York, NY

Beightler CS, Phillips DT (1976) *Applied geometric programming*. Wiley, New York

Chandrasekaran V, Shah P (2014) Conic geometric programming. Paper presented at the 48th
Annual Conference on Information Sciences and Systems (CISS), Princeton, NJ. https://arxiv
org/pdf/1310.0899.pdf. Accessed 2 Oct 2019

Cobb CW, Douglas PH (1928) A theory of production. *Am Econ Rev* **18** (Suppl):139–165

Duffin RJ (1962) Cost minimization problems treated by geometric means. *Oper Res* **10**: 668–675

Eiselt HA, Pederzoli G, Sandblom, C-L (1987) *Continuous optimization models*. W. de Gruyter
Berlin – New York

Eiselt HA, Sandblom C-L (2012) *Operations research: a model-based approach*. 2nd edn
Springer, Berlin

Lin M-H, Tsai J-F (2012) Range reduction techniques for improving computational efficiency in
global optimization of signomial geometric programming problems. *Eur J Oper Res* **216(1)**
17–25

Liu S-T (2008) Posynomial geometric programming with interval exponents and coefficients
European Journal of Operational Research **186/1**: 17–27

Luptáčik M (1977) *Geometrische Programmierung und ökonomische Analyse*. Mathematical
Systems in Economics, vol 32. Anton Hain, Meisenheim am Glan

Luptáčik M (1981a) Geometric programming, methods and applications. *OR Spektrum* **2**: 129–14

Luptáčik M (1981b) *Nichtlineare Programmierung mit ökonomischen Anwendungen*. Athenäum
Königstein

Misra S, Fisher MW, Backhaus S, Bent R, Chertkov M, Pan F (2015) Optimal compression in
natural gas networks: a geometric programming approach. *IEEE Trans Control Netw Syst* **2(1)**
47–56

Ogura M, Preciado VM (2017) Optimal design of switched networks of positive linear systems vi
geometric programming. *IEEE Trans Control Netw Syst* **4(2)**: 213–222

Ojha AK, Biswal KK (2010) Multi-objective geometric programming problem with weighted mean
method. *Int J Comput Sci Inf Secur* **7(2)**. https://arxiv.org/ftp/arxiv/papers/1003/1003.1477.pdf
Accessed 1 Sept 2019

Toussaint M (2015) Logic-geometric programming: an optimization-based approach to combined
task and motion planning. *IJCAI'15 Proceedings of the 24th International Conference on
Artificial Intelligence* 1930–1936. http://ipvs.informatik.uni-stuttgart.de/mlr/papers/15
toussaint-IJCAI.pdf. Accessed 1 Sept 2019

Woolsey RED, Swanson HS (1975) *Operations research for immediate application-a quick an
dirty manual*. Harper & Row, New York

Zener C (1961) A mathematical aid in optimizing engineering designs. *Proc. Natl. Acad. Sci. U.S.A*
47/4: 537–539

Appendices

Appendix A Matrices

Definition A.1 A *matrix* is a two-dimensional array with m rows and n columns. If $m = n$, it is said to be *square*; if $m = 1$, it is called a *row vector*, if $n = 1$, it is a *column vector*, and if $m = n = 1$, it is a *scalar*.

Typically, matrices are denoted by boldface capital letters, vectors are shown as boldface small letters, and scalars are shown as italicized small letters. The i-th row of a matrix $\mathbf{A} = (a_{ij})$ is $\mathbf{a}_{i\bullet}$ and the j-th column of the matrix $\mathbf{A}$ is $\mathbf{a}_{\bullet j}$.

Definition A.2 An $[n \times n]$-dimensional matrix $\mathbf{A} = (a_{ij})$ is called an *identity matrix*, if $a_{ij} = 1$ if $i = j$, and zero otherwise. It is called a *diagonal matrix*, if $a_{ij} = 0 \ \forall \ i \neq j$.

An identity matrix is usually denoted by $\mathbf{I}$. The i-th row of an identity matrix is called a *unit vector* $\mathbf{e}_i = [0, 0, \ldots 0, 1, 0, 0, \ldots 0]$ which has the "1" in the i-th position, and zeroes otherwise. The vector $[1, 1, \ldots, 1]$ is called a *summation vector* and is denoted by $\mathbf{e}$.

Definition A.3 The *sum* of two $[m \times n]$-dimensional matrices $\mathbf{A}$ and $\mathbf{B}$ is an $[m \times n]$-dimensional matrix $\mathbf{C}$, such that $c_{ij} = a_{ij} + b_{ij} \ \forall \ i = 1, \ldots, m$; $j = 1, \ldots, n$. The *difference* of two matrices is defined similarly.

Definition A.4 The *product* of an $[m \times n]$-dimensional matrix $\mathbf{A} = (a_{ij})$ and an $[n \times p]$-dimensional matrix $\mathbf{B} = (b_{jk})$ is an $[m \times p]$-dimensional matrix $\mathbf{C} = (c_{ik})$, such that

$$c_{ik} = \sum_{j=1}^{n} a_{ij} b_{jk} \ \forall \ i, k.$$

© Springer Nature Switzerland AG 2019
H. A. Eiselt, C.-L. Sandblom, *Nonlinear Optimization*, International Series in
Operations Research & Management Science 282,
https://doi.org/10.1007/978-3-030-19462-8

Definition A.5 The *transpose* of an $[m \times n]$-dimensional matrix $\mathbf{A} = (a_{ij})$ is an $[n \times m]$-dimensional matrix $\mathbf{A}^T = (a_{ij}^T)$, such that $a_{ij}^T = a_{ji} \ \forall \ i,j$. If $m = n$ and $\mathbf{A} = \mathbf{A}^T$, then $\mathbf{A}$ is said to be *symmetric*; if $\mathbf{A} = -\mathbf{A}^T$, then $\mathbf{A}$ is *skew-symmetric*. The *trace* of an $[n \times n]$-dimensional matrix $\mathbf{A}$ is defined as

$$tr(\mathbf{A}) = \sum_{j=1}^{n} a_{jj},$$

and the *inverse* of an $[n \times n]$-dimensional matrix $\mathbf{B} = (b_{ij})$ is a matrix $\mathbf{B}^{-1}$, such that $\mathbf{B}\mathbf{B}^{-1} = \mathbf{B}^{-1}\mathbf{B} = \mathbf{I}$.

Example Let $\mathbf{A} = \begin{bmatrix} 0 & 1 & 4 \\ 2 & 1 & 1 \\ 4 & 0 & 1 \end{bmatrix}$. Then $\mathbf{A}^{-1} = \begin{bmatrix} -\dfrac{1}{14} & \dfrac{1}{14} & \dfrac{3}{14} \\ \dfrac{2}{14} & \dfrac{16}{14} & -\dfrac{8}{14} \\ \dfrac{4}{14} & -\dfrac{4}{14} & \dfrac{2}{14} \end{bmatrix}$. Multiplying these two matrices results again in the identity matrix $\mathbf{I}$.

Proposition A.6 The following results hold when multiplying, transposing, and inverting matrices $\mathbf{A}$, $\mathbf{B}$, and $\mathbf{C}$:

- $\mathbf{AI} = \mathbf{IA} = \mathbf{A}$
- $\mathbf{A}(\mathbf{BC}) = (\mathbf{AB})\mathbf{C}$
- $(\mathbf{A}^T)^T = \mathbf{A}$
- $(\mathbf{AB})^T = \mathbf{B}^T\mathbf{A}^T$
- $(\mathbf{A}^{-1})^{-1} = \mathbf{A}$
- $(\mathbf{AB})^{-1} = \mathbf{B}^{-1}\mathbf{A}^{-1}$
- $(\mathbf{A}^T)^{-1} = (\mathbf{A}^{-1})^T$.

Definition A.7 The *determinant* of an $[n \times n]$-dimensional matrix $\mathbf{A}$ is defined recursively as $\det(\mathbf{A}) = \sum_{j=1}^{n} (-1)^{i+j} a_{ij} \det(\mathbf{A}_{ij})$, where $\mathbf{A}_{ij}$ denotes the matrix that results from the given matrix $\mathbf{A}$ by deleting the i-th row and the j-th column.

This development of the determinant via the i-th row is due to Laplace. As a starting condition, the determinant of a $[1 \times 1]$-dimensional matrix $\mathbf{A}$ is de $(\mathbf{A}) = a_{11}$. Applying Laplace's formula to a $[2 \times 2]$-dimensional matrix $\mathbf{A} = \begin{bmatrix} a_{11} & a_{12} \\ a_{21} & a_{22} \end{bmatrix}$, we obtain $\det(\mathbf{A}) = a_{11}a_{22} - a_{12}a_{21}$. The inverse $\mathbf{A}^{-1}$ of a square matrix $\mathbf{A}$ exists if and only if $\det(\mathbf{A}) \neq 0$.

Example Let

$$\mathbf{A} = \begin{bmatrix} 0 & 5 & 2 \\ 3 & 1 & 0 \\ 2 & 2 & 1 \end{bmatrix}.$$

Evaluating the determinant with respect to the first row, we obtain

$$\mathbf{A}_{11} = \begin{bmatrix} 1 & 0 \\ 2 & 1 \end{bmatrix}, \quad \mathbf{A}_{12} = \begin{bmatrix} 3 & 0 \\ 2 & 1 \end{bmatrix}, \quad \text{and} \quad \mathbf{A}_{13} = \begin{bmatrix} 3 & 1 \\ 2 & 2 \end{bmatrix}$$

so that $\det(\mathbf{A}) = (-1)^{1+1}(0)\det(\mathbf{A}_{11}) + (-1)^{1+2}(5)\det(\mathbf{A}_{12}) + (-1)^{1+3}(2)\det(\mathbf{A}_{13}) = (1)(0)(1) + (-1)(5)(3) + (1)(2)(4) = -7$. The determinant can be interpreted as the volume of the n-dimensional parallelepiped spanned by the column vectors of the matrix.

Definition A.8 An *eigenvalue* or *characteristic value* λ of a square matrix $\mathbf{A}$ is a number, possibly complex, which solves the equation $\det(\mathbf{A} - \lambda\mathbf{I}) = 0$.

One can show that if λ is a real-valued eigenvalue of the square matrix $\mathbf{A}$, then there exists a corresponding vector $\mathbf{x} \neq \mathbf{0}$, such that $\mathbf{A}\mathbf{x} = \lambda\mathbf{x}$. This vector is called an *eigenvector* of $\mathbf{A}$ corresponding to the eigenvalue λ. It may happen that no real-valued eigenvalue exists. Geometrically, an eigenvalue can be interpreted as some form of "scaling" resulting from the linear transformation $\mathbf{x} \to \mathbf{A}\mathbf{x}$. For instance, if 1, 0] and [0, 1] are the eigenvectors for a [2 × 2]-dimensional matrix $\mathbf{A}$, with eigenvalues ½ and 3, respectively, then every transformed vector $\mathbf{x}$ will have its first component halved and second component tripled.

Example Using again the matrix $\mathbf{A}$ in the example following Definition A.7, the eigenvalues of $\mathbf{A}$ can be calculated as 5.2017, -3.5778, and 0.3761, and the corresponding eigenvectors are $[-0.6780, -0.4841, -0.5532]$, $[-0.8298, 0.5438, 0.1250]$, and $[0.0760, -0.3654, 0.9278]$, respectively.

If a matrix is symmetric, all its eigenvalues are real-valued; if a matrix $\mathbf{A}$ is symmetric and positive definite, all its eigenvalues are positive. Denote its smallest eigenvalue by $\lambda_{\min}$ and its largest eigenvalue by $\lambda_{\max}$, so that we have $0 < \lambda_{\min} < \lambda_{\max}$. We can then write

Proposition A.9 The *Kantorovich matrix inequality* states that for any $\mathbf{0} \neq \mathbf{x} \in \mathbb{R}^n$ we have

$$\frac{4\lambda_{\min}\lambda_{\max}}{(\lambda_{\min} + \lambda_{\max})^2} \leq \frac{(\mathbf{x}^T\mathbf{x})^2}{(\mathbf{x}^T\mathbf{A}\mathbf{x})(\mathbf{x}^T\mathbf{A}^{-1}\mathbf{x})} \leq 1.$$

For a proof, see, e.g., Bertsekas (2016).

Proposition A.10 The following rules apply to determinants of $[n \times n]$-dimensional matrices $\mathbf{A}$ and $\mathbf{B}$:

det $(\mathbf{I}) = 1$

det $(\mathbf{AB}) = \det(\mathbf{A})\det(\mathbf{B})$

det $(\mathbf{A}) = \det(\mathbf{A}^T)$

det $(\alpha\mathbf{A}) = \alpha^n \det(\mathbf{A})$

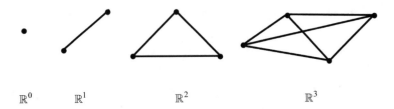

Fig. A.1 Simplices in different dimensions

$$\det (\mathbf{A}^{-1}) = \frac{1}{\det (\mathbf{A})}.$$

Definition A.11 A *simplex S* in $\mathbb{R}^n$ is defined as a set of $(n + 1)$ points $\mathbf{x}^k$, $k = 1$, $2, \ldots, n{+}1$, such that there exists no hyperplane H: $\mathbf{a}\mathbf{x} = b$ with the property that $\mathbf{a}\mathbf{x}^k = b \ \forall \ k = 1, 2, \ldots, n{+}1$. In other words, a simplex is a minimal independence structure in the sense that all but one of its extreme points are located on a hyperplane. In order to determine the volume of a n-dimensional polyhedron P, it is first necessary to subdivide P into simplices. The *volume of a simplex S* with extreme points $\mathbf{x}^k$, $k = 1, 2, \ldots, n{+}1$ (each written as a row vector), can then be expressed as

$$v(S) = \tfrac{1}{n}\det \begin{bmatrix} (\mathbf{x}^1)^T & 1 \\ (\mathbf{x}^2)^T & 1 \\ \vdots & \vdots \\ (\mathbf{x}^{n+1})^T & 1 \end{bmatrix}$$

As an example, a simplex in $\mathbb{R}^2$ is a triangle, and in $\mathbb{R}^3$ it is a tetrahedron. Figure A.1 shows simplices in 0-, 1-, 2-, and 3-dimensional real space.

Appendix B Systems of Simultaneous Linear Relations

The purpose of this section is to provide some of the tools which are necessary for the topics in this book and thus establish a common background of knowledge. In order to do so, we will introduce those elements of linear algebra that are mandatory for the understanding of linear programming. For further details, readers are referred to texts on linear algebra, e.g., Nicholson (2013) or Bretscher (2013).

Throughout this section we assume that $x_1, x_2, \ldots, x_n$ are variables or unknowns of some given system, and $a_1, a_2, \ldots, a_n$ and b are given real numbers. The symbol R defines a relation of the type $=, \leq, \geq, <$, or $>$, and $f(\mathbf{x})$ is a function of the variables $x_1, x_2, \ldots, x_n$.

Definition B.1 A *relation* $f(\mathbf{x}) \ R \ b$ is said to be *linear* if $f(\mathbf{x}) = \sum\limits_{j=1}^{n} a_j x_j$. We will refer to $f(\mathbf{x})$ as the left-hand side and b as the right-hand side of this relation.

Example The relation

$$\sqrt{5}x_1 + 4x_2 - 27x_3 \leq \frac{28}{13} \tag{B.1}$$

is linear, while the relations

$$\sqrt{5}x_1 + 4\sqrt{x_2} - 27x_3 \leq \frac{28}{13}, \tag{B.2}$$

$$\sqrt{5}x_1 + 4x_2^2 - 27x_3 \leq \frac{28}{13}, \quad \text{and} \tag{B.3}$$

$$\sqrt{5}x_1 + 4x_2x_4 - 27x_3 \leq \frac{28}{13} \tag{B.4}$$

are not, as relation (B.2) includes the square root of a variable, (B.3) has the variable x_2 raised to the second power, and (B.4) includes the product of two variables. Loosely speaking, a function is linear if all variables are not multiplied by other variables and appear only raised to the 0-th or first power.

In order to formalize, consider the system of simultaneous linear equations $\mathbf{Ax} = \mathbf{b}$, where $\mathbf{A}$ is an $[m \times n]$-dimensional matrix, $\mathbf{x}$ is an n-dimensional column vector, and $\mathbf{b}$ is an m-dimensional column vector.

Definition B.2 A set of vectors $\mathbf{a}_{1\cdot}, \mathbf{a}_{2\cdot}, \ldots, \mathbf{a}_{m\cdot}$ is said to be *linearly dependent*, if there exists a vector $\boldsymbol{\lambda} = [\lambda_1, \lambda_2, \ldots, \lambda_m] \neq \mathbf{0}$ of real numbers, such that $\sum_{i=1}^{m} \lambda_i \mathbf{a}_{i\cdot} = \mathbf{0}$. If the vectors $\mathbf{a}_{1\cdot}, \mathbf{a}_{2\cdot}, \ldots, \mathbf{a}_{m\cdot}$ are not linearly dependent, they are said to be *linearly independent*.

It is not difficult to demonstrate that if the vectors $\mathbf{a}_{1\cdot}, \mathbf{a}_{2\cdot}, \ldots, \mathbf{a}_{m\cdot}$ are linearly dependent, then there exists some number k, $1 \leq k \leq m$ and some real numbers λ_i, such that $\mathbf{a}_{k\cdot} = \sum_{\substack{i=1 \\ i \neq k}}^{m} \lambda_i \mathbf{a}_{i\cdot}$. In other words, at least one of the vectors in the system can be generated as a weighted sum of the others. It also follows that if the vectors are linearly independent, then $\sum_{i=1}^{m} \lambda_i \mathbf{a}_{i\cdot} = \mathbf{0}$ implies that $\lambda_i = 0 \; \forall \; i = 1, 2, \ldots, m$. A similar definition and results can be developed for column vectors. As an example, consider the matrix

$$\mathbf{A} = \begin{bmatrix} 2 & -3 \\ 5 & 1 \\ -4 & -11 \\ 9 & -5 \end{bmatrix}.$$

Here, the first row can be expressed as $a_1. = [2, -3] = \lambda_2 a_2. + \lambda_3 a_3. + \lambda_4 a_4. = \lambda_2 [5, 1] + \lambda_3 [-4, -11] + \lambda_4 [9, -5]$ with $\lambda_2 = \frac{2}{3}$, $\lambda_3 = \frac{1}{3}$, and $\lambda_4 = 0$. In other words, a weighted combination of the second and third rows can generate the first row, so that row 1 does not include any information beyond what is provided by rows 2, 3, and 4. Using the same argument, the fourth row can be written as $a_4. = [9, -5] = \lambda_2 a_2. + \lambda_3 a_3. = \lambda_2 [5, 1] + \lambda_3 [-4, -11]$ with $\lambda_2 = \frac{7}{3}$ and $\lambda_3 = \frac{2}{3}$, so that the second and third rows can be used to generate the fourth row, which, again, does not provide any information beyond what is already included in rows 2 and 3. While we have just shown that rows 2 and 3 contain as much information as the four rows combined, it is also true that rows 1 and 2 can be used to generate rows 3 and 4, respectively, so they also contain all the information included in the system. The message here is that different combinations of rows may include all the information contained in the system. We will now formalize the discussion. For that purpose, we can write

Definition B.3 The *rank of a matrix* A, written rkA, is the maximal number of linearly independent rows and columns of **A**.

For instance, in the above example $rkA = 2$. It should be noted that in an $[m \times n]$-dimensional matrix the rank is not necessarily equal to min $\{m, n\}$, but we certainly have $rkA \leq$ min $\{m, n\}$. If, however, $rkA =$ min $\{m, n\}$, we say that the matrix **A** has *full rank*.

Let now **A** be an $[m \times n]$-dimensional matrix, so that m equations and n variables are included in the system $Ax = b$, and let $[A, b]$ denote the matrix that includes all columns of **A** as well as the vector **b**. Clearly, $rk [A, b] \geq rkA$. We can now state the following

Theorem B.4 A system of simultaneous linear equations $Ax = b$ has

(i)   no solution if $rkA < rk [A, b]$,
(ii) exactly one solution, if $rkA = rk [A, b] = n$, and
(iii) an infinite number of solutions, if $rkA = rk [A, b] < n$.

A proof of Theorem B.4 can be found in many books on linear algebra; see, e.g. Nicholson (2013) or Bretscher (2013). The following examples may explain the three cases outlined in the theorem.

Example 1 Let $A = \begin{bmatrix} 2 & 4 & -2 \\ 5 & -3 & 7 \\ -4 & 18 & -20 \end{bmatrix}$ and $b = \begin{bmatrix} 6 \\ 0 \\ 12 \end{bmatrix}$.

Then $rkA = 2$, since $a_3. = 3a_1. - 2a_2.$. On the other hand,

$$rk[\mathbf{A}, \mathbf{b}] = \begin{bmatrix} 2 & 4 & -2 & 6 \\ 5 & -3 & 7 & 0 \\ -4 & 18 & -20 & 12 \end{bmatrix} = 3,$$

since none of the three rows can be generated by the two other rows of the matrix. Hence, the above system has no solution.

Example 2 Let $\mathbf{A} = \begin{bmatrix} 1 & 2 & -1 \\ 3 & 1 & -2 \\ -2 & 3 & 0 \end{bmatrix}$ and $\mathbf{b} = \begin{bmatrix} 8 \\ 10 \\ 6 \end{bmatrix}$. Here, $rk\,\mathbf{A} = 3$ and

$$rk[\mathbf{A}, \mathbf{b}] = \begin{bmatrix} 1 & 2 & -1 & 8 \\ 3 & 1 & -2 & 10 \\ -2 & 3 & 0 & 6 \end{bmatrix} = 3,$$

so that the system will have exactly one solution, which, incidentally, is $x_1 = 0$, $x_2 = 2$, and $x_3 = -4$.

Example 3 Let $\mathbf{A} = \begin{bmatrix} 1 & 2 & 3 \\ 2 & -3 & -2 \\ -1 & 5 & 5 \end{bmatrix}$ and $\mathbf{b} = \begin{bmatrix} 20 \\ 12 \\ 8 \end{bmatrix}$. Now $rk\,\mathbf{A} = 2$ and $rk\,[\mathbf{A}, \mathbf{b}] = 2 < 3 = n$.

Thus, there are infinitely many solutions to this system, e.g., $\mathbf{x} = [12, 4, 0]^T$, $\mathbf{x} = [9.5, 0, 3.5]^T$, $\mathbf{x} = [10.75, 2, 1.75]^T$, etc.

Definition B.5 Let $\mathbf{A}$ be an $[m \times n]$-dimensional matrix with $n \geq m$. Any collection of m linearly independent columns of $\mathbf{A}$ is said to be a *basis* of $\mathbf{A}$.

It follows that if a matrix $\mathbf{A}$ has full rank, i.e., if $rk\,\mathbf{A} = m$, then $\mathbf{A}$ has at least one basis. Consider now an $[m \times n]$-dimensional matrix $\mathbf{A}$ with $n \geq m$ and full rank and one of its bases. The columns forming this basis constitute an $[m \times m]$-dimensional submatrix $\mathbf{B}$, which we will call a *basis matrix*. We can then partition the original matrix $\mathbf{A} = [\mathbf{B}, \mathbf{N}]$, maybe after reordering the columns, where the $[m \times (n-m)]$-dimensional matrix $\mathbf{N}$ consists of all columns of $\mathbf{A}$ that are not in the basis. The vector of variables $\mathbf{x}$ can be partitioned accordingly into $\mathbf{x}^T = \left[\mathbf{x}_B^T, \mathbf{x}_N^T\right]$.

The components of the m-dimensional vector $\mathbf{x}_B$ are called *basic variables*, whereas the components of the $(n-m)$-dimensional vector $\mathbf{x}_N$ are called *nonbasic variables*. Then the system $\mathbf{A}\mathbf{x} = \mathbf{b}$ can be written $\mathbf{B}\mathbf{x}_B + \mathbf{N}\mathbf{x}_N = \mathbf{b}$. Setting $\mathbf{x}_N := \mathbf{0}$, we see that this system has a solution $\mathbf{x}_B = \mathbf{B}^{-1}\mathbf{b}$, $\mathbf{x}_N = \mathbf{0}$. The concept of the basis of a matrix is of fundamental importance in linear programming.

Looking at the graphical representation in two dimensions, it is well known that each equation can be represented by a straight line. The combinations of all coordinates (x_1, x_2) on the line satisfy this equation. Given two equations, the two lines are either parallel (in which case there is no pair of coordinates that satisfies both of them, i.e., there is no solution), they intersect at a single point (in which case there is a unique solution), or they coincide (in which case any point on the two identical lines is a solution, i.e., there is an infinite number of solutions). These are exactly the three cases (i), (ii), and (iii) outlined in Theorem B.4 above.

Definition B.6 The set of points $\{x: a_i.x \; R \; b_i\}$ defines

(i) a *hyperplane*, if $R = \{=\}$,
(ii) a *closed halfspace*, if $R \in \{\le, \ge\}$, or
(iii) an *open halfspace*, if $R \in \{<, >\}$.

In the case of a single dimension, i.e., if a single variable is given, a hyperplane is a point (e.g., $x = 2$), while a halfspace is a *halfline* (e.g., $x \le 2$ or $x \ge 2$). In two dimensions a hyperplane defines a straight line and a halfspace half of a plane. Finally, in three dimensions a hyperplane defines a plane and a halfspace half of the three-dimensional space bordered by a hyperplane.

Generally speaking, a hyperplane in n-dimensional real space $\mathbb{R}^n$ is $(n-1)$-dimensional, while a halfspace in $\mathbb{R}^n$ is n-dimensional. Note that every halfspace $a_i.x \; R \; b_i$ with $R \in \{\le, \ge, <, >\}$ is bordered by the hyperplane $a_i.x = b_i$, whereas a closed halfspace includes all points of its bordering hyperplane, while an open halfspace includes none of them. Formally we can write

Definition B.7 A *set S* is called *closed*, if all points on the boundary of S belong to the set as well. On the other hand, a *set S* is called *open*, if no point on the boundary of S belongs to S.

As an example, the set $\{x \in \mathbb{R}: x \in [1, 5]\}$ is closed, while the set $\{x \in \mathbb{R}: x \in]1, 5[\}$ is open. We should note here that linear programming exclusively deals with closed sets. Hence, if we simply refer to halfspaces in linear programming, this means closed halfspaces.

Consider now the case of more than one single inequality. Equivalent to the above discussion regarding equations, a point that satisfies all inequalities will graphically be located in the intersection of the halfspaces that are defined by the given inequalities. As an illustration, consider the system of linear inequalities

$$2x_1 + 1x_2 \le 4 \qquad (I)$$
$$1x_1 - 1x_2 \le 1. \qquad (II)$$

Figure B.1 shows the straight lines I and II representing the equations $2x_1 + 1x_2 = 4$ and $1x_1 - 1x_2 = 1$, while the flags at the ends of the lines indicate the halfspace defined by the relations. These halfspaces subdivide the two-dimensional plane into the four sets A, B, C and D. All interior points in the set A violate constraint I and satisfy constraint II, all interior points in B violate both constraints, all interior points in the set C satisfy constraint I and violate constraint II, and all points in the set D (including the boundaries) satisfy both constraints. This is why D would be called the *feasible set*, or, equivalently, the *set of feasible solutions*.

Fig. B.1 A feasible set

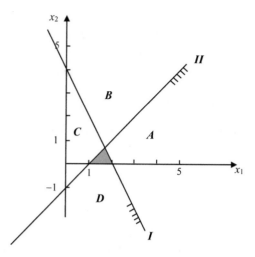

Definition B.8 The *nonnegative orthant* $\mathbb{R}_+^n$ is the set of all nonnegative points in $\mathbb{R}^n$, i.e., $\bigcap_{j=1}^{n} \{x_j : x_j \geq 0\}$ or simply $\{\mathbf{x} \in \mathbb{R}^n : \mathbf{x} \geq \mathbf{0}\}$.

In $\mathbb{R}^2$, the nonnegative orthant is the first quadrant including the positive axes and the origin. If we were to add the relations *III*: $x_1 \geq 0$ and *IV*: $x_2 \geq 0$ to the above system of simultaneous linear inequalities, then the set of solutions is the intersection of the set D in Fig. B.1 and the first quadrant, i.e. the shaded triangle with the vertices (1, 0), (2, 0), and (5/3, 2/3).

Definition B.9 A set S is said to be *bounded*, if there exists a finite number $c \in \mathbb{R}$, so that $\|\mathbf{x}\| < c$ for every point $\mathbf{x} \in S$. A set that is not bounded is said to be *unbounded*. A set S is called *compact*, if it is closed and bounded.

As an example, the sets A, B, C, and D in Fig. B.1. are unbounded, while the shaded set of points that satisfies the constraints *I*, *II*, *III*, and *IV* is bounded and closed and hence compact.

Definition B.10 The intersection of a finite number of hyperplanes and/or closed halfspaces in $\mathbb{R}^n$ is called a *polytope*. A bounded polytope is called a *polyhedron*.

In general, we will refer to the set that is defined by a system of simultaneous linear inequalities of type $\leq$ and/or $\geq$ as a polytope since it is not obvious whether or not the set is bounded. Also, a polytope and a polyhedron may degenerate to the empty set or a single point in $\mathbb{R}^n$. Again, the set D in Fig. B.1 is a polytope since it is generated by the halfspaces of constraints *I* and *II*, but it is not a polyhedron, since it is unbounded from below. On the other hand, the shaded set of points in the triangle with vertices (1, 0), (2, 0), and (5/3, 2/3) is a polytope since it is generated by the halfspaces of constraints *I*, *II*, *III*, and *IV*, and since it is bounded, it is also a polyhedron. We wish to point out that the terms polytope and polyhedron are sometimes defined differently by other authors.

Let now H_i denote the set of points that satisfy the i-th linear relation $\sum_{j=1}^{n} a_{ij} x_j \, R_i \, b_i$, $i = 1, 2, \ldots, m$; then $S = \bigcap_{j=1}^{m} H_i$ is the set of points that satisfy all m relations simultaneously.

Definition B.11 The k-th relation is said to be *redundant*, if $\bigcap_{\substack{i=1 \\ i \neq k}}^{m} H_i = S$. If the k-th relation is not redundant, i.e., if $\bigcap_{\substack{i=1 \\ i \neq k}}^{m} H_i \supset \bigcap_{\substack{j=1}}^{m} H_i = S$, it is called *essential*.

Note that both properties, redundant and essential, refer to the relation between a constraint and the feasible set. Moreover, we can state

Definition B.12 If for some point $\tilde{\mathbf{x}} \in S$ the i-th linear relation is satisfied as an equation, i.e., if $\mathbf{a}_i . \tilde{\mathbf{x}} = b_i$, then the i-th relation is said to be *binding* or *tight* at $\tilde{\mathbf{x}}$. If the i-th relation is satisfied as a strict inequality, i.e., $\mathbf{a}_i . \tilde{\mathbf{x}} < b_i$, the constraint is said to be *loose* or *nonbinding*.

Both properties, binding and loose, refer to the relation between a constraint and a point. Figure B.2 may explain the concept.

Fig. B.2 A feasible set and redundancy

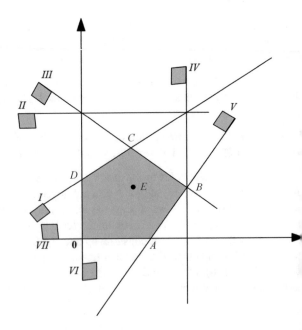

Fig. B.3 Redundancy

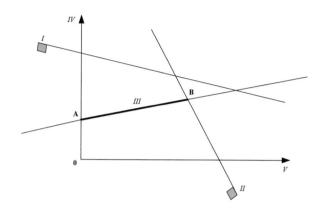

In Fig. B.2, the polytope generated by the halfspaces *I*, *II*, ..., *VII* is the shaded area with corner points *0*, *A*, *B*, *C*, and *D*. Clearly, inequality *II* is redundant since its inclusion (or removal) does not change the shape of the polytope. The same argument can be applied to inequality *IV*, which therefore is also redundant. Note, however, that inequality *IV* is binding at point *B*, whereas inequality *II* is not binding at any point of the polytope, which makes constraint *II strongly redundant*, whereas constraint *IV* is *weakly redundant*. At point *C* in Fig. B.2, relations *I* and *III* are binding, at point *0* it is relations *VI* and *VII* that are binding, whereas at the interior point *E* none of the relations is binding.

In Fig. B.3, relations *I* and *II* are inequalities while relation *III* is an equation, thus the polytope defined by the relations is the set of points on the straight line between the points *A* and *B*. This implies that relations *I* and *V* are redundant and could be deleted without changing the polytope.

Definition B.13 A point $\mathbf{y} \in \mathbb{R}^n$ is said to be a *linear combination* of a given set of points $\mathbf{x}^1, \mathbf{x}^2, \ldots, \mathbf{x}^r$ if there exist real numbers $\lambda_1, \lambda_2, \ldots, \lambda_r$, such that $\mathbf{y} = \sum_{k=1}^{r} \lambda_k \mathbf{x}^k$. The linear combination $\mathbf{y}$ is said to be a *nonnegative linear combination*, if $\lambda_k \geq 0 \ \forall \ k = 1, \ldots, r$; it is called an *affine linear combination*, if $\sum_{k=1}^{r} \lambda_k = 1$, and it is called a *linear convex combination (lcc)* if $\lambda_k \geq 0 \ \forall \ k = 1, \ldots, r$, and $\sum_{k=1}^{r} \lambda_k = 1$.

As an illustration, consider the five points $\mathbf{x}^1 = (0, 0)$, $\mathbf{x}^2 = (3, 0)$, $\mathbf{x}^3 = (0, 2)$, $\mathbf{x}^4 = (3, 2)$, and $\mathbf{y} = (1\frac{1}{2}, \frac{1}{2})$. Inspection reveals that $\lambda_1 = \lambda_2 = 0$, $\lambda_3 = -\frac{1}{4}$, and $\lambda_4 = \frac{1}{2}$ generate point $\mathbf{y}$, which is therefore a linear combination of the other points. However, $\lambda_1 = 0$, $\lambda_2 = \frac{1}{2}$, $\lambda_3 = \frac{1}{4}$, and $\lambda_4 = 0$ also generate point $\mathbf{y}$ which makes it a nonnegative linear combination as well. In order to find out whether or not $\mathbf{y}$ is an affine linear combination or even a linear convex combination of $\mathbf{x}^1$, $\mathbf{x}^2$, $\mathbf{x}^3$, and $\mathbf{x}^4$, we have to find a set of solutions to the system

$$0\lambda_1 + 3\lambda_2 + 0\lambda_3 + 3\lambda_4 = 1\frac{1}{2}$$

$$0\lambda_1 + 0\lambda_2 + 2\lambda_3 + 2\lambda_4 = \frac{1}{2}$$

$$\lambda_1 + \lambda_2 + \lambda_3 + \lambda_4 = 1.$$

One of the solutions to this system is $\lambda_1 = \frac{3}{4}$, $\lambda_2 = 0$, $\lambda_3 = \frac{1}{4}$, and $\lambda_4 = 0$, indicating that $\mathbf{y}$ is indeed an affine linear combination of $\mathbf{x}^1$, $\mathbf{x}^2$, $\mathbf{x}^3$, and $\mathbf{x}^4$. Finally, for $\mathbf{y}$ to be a linear convex combination of the given points $\mathbf{x}^1$, $\mathbf{x}^2$, $\mathbf{x}^3$, and $\mathbf{x}^4$, the following conditions must be satisfied: $\lambda_1 = \frac{3}{4} - \lambda_2 \geq 0$, $\lambda_3 = -\frac{1}{4} + \lambda_2 \geq 0$, and $\lambda_4 = \frac{1}{2} - \lambda_2 \geq 0$, implying that $\lambda_2 \leq \frac{3}{4}$, $\lambda_2 \geq \frac{1}{4}$, and $\lambda_2 \leq \frac{1}{2}$. It follows that for any $\lambda_2 \in [\frac{1}{4}, \frac{1}{2}]$, the resulting multipliers λ_1, λ_3, $\lambda_4 \geq 0$ and $\mathbf{y}$ is therefore also a linear convex combination of the given points.

As another example, consider the same given points $\mathbf{x}^1$, $\mathbf{x}^2$, $\mathbf{x}^3$, and $\mathbf{x}^4$ and let $\mathbf{z} = (1, 3)$ be an additional point. The reader will then be able to verify that $\mathbf{z}$ is both a nonnegative and an affine linear combination of the given points, but it is not a linear convex combination of these points.

In the following we will assume that $\mathbf{A}$ is an $[m \times n]$-dimensional matrix with rank $rk \, \mathbf{A} = r$, $\mathbf{x}$ is an $[n \times 1]$-vector of variables, $\mathbf{b}$ is an $[m \times 1]$-vector of parameters, and $\mathbf{u}$ is a $[1 \times m]$-vector of variables. Then we can state

Lemma B.14 (Farkas' Lemma) Either the system I: $\mathbf{Ax} = \mathbf{b}$, $\mathbf{b} \neq \mathbf{0}$ has a solution $\mathbf{x} \geq \mathbf{0}$, or the system II: $\mathbf{uA} \geq \mathbf{0}$, $\mathbf{ub} < \mathbf{0}$ has a solution $\mathbf{u} \in \mathbb{R}^m$, but never both.

This theorem is crucial for the development of duality theory in linear programming; for details, readers are referred to Eiselt and Sandblom (2007). Considering the following three systems P, P_D and P*:

P :   Max $z = \mathbf{cx}$ P_D :   Min $z_D = \mathbf{ub}$ P* : $\mathbf{Ax} \leq \mathbf{b}$

$$\begin{array}{lll}
\text{P}: & \text{Max } z = \mathbf{cx} & \text{P}_D: \text{ Min } z_D = \mathbf{ub} & \text{P*}: \mathbf{Ax} \leq \mathbf{b} \\
& \text{s.t.} \mathbf{Ax} \leq \mathbf{b} & \text{s.t.} \mathbf{uA} \geq \mathbf{c} & \mathbf{x} \geq 0 \\
& \mathbf{x} \geq 0 & \mathbf{u} \geq 0 & \mathbf{uA} \geq \mathbf{c} \\
& & & \mathbf{u} \geq 0 \\
& & & \mathbf{cx} - \mathbf{ub} = 0,
\end{array}$$

then we obtain the following result:

Theorem B.15 The following two statements are equivalent:

(i) $\bar{\mathbf{x}}$ is an optimal solution for P and $\bar{\mathbf{u}}$ is an optimal solution for P_D.

(ii) $(\bar{\mathbf{x}}, \bar{\mathbf{u}})$ is a feasible solution for P*.

Appendix C Convexity and Cones

Definition C.1 A set S is said to be *convex*, if the linear combination of any two elements of S is also an element of S, i.e., if $\mathbf{x}, \mathbf{y} \in S$ and $\lambda \in [0, 1]$, then $\lambda\mathbf{x} + (1-\lambda)\mathbf{y} \in S$. Geometrically speaking, a set is convex, if all points on a straight line segment that joins any pair of arbitrary elements of S are also elements of S.

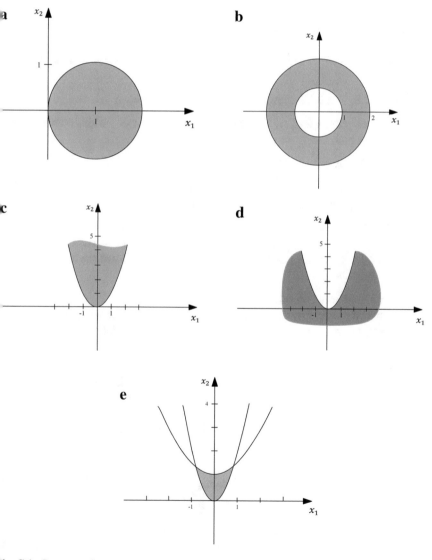

Fig. C.1 Convex and nonconvex sets

Figure C.1a–e are examples for convex and nonconvex sets, where the shaded areas denote the sets defined by the relations shown below.

Figure C.1a shows a disk that is described by the inequality $x_1^2 + x_2^2 - 2x_1 \le 0$. It is apparent that the set is convex. The shaded set in Fig. C.1b is the difference of two disks. The inequalities that describe the set are $x_1^2 + x_2^2 \le 4$ and $x_1^2 + x_2^2 \ge 1$. The set is not convex. The shaded area above the parabola in Fig. C.1c is described by the inequality $x_1^2 - x_2 < 0$. It is a convex set. The shaded set in Fig. C.1d is the complement of the set of Fig. C.1c. It is described by the nonlinear inequality $x_1^2 - x_2 \ge 0$ and it is not convex. The shaded set shown in Fig. C.1e is the difference between two parabolas. The set satisfies the conditions $x_1^2 - x_2 \le 0$ and $-\dfrac{1}{2}x_1^2 + x_2 \le 1$, and it is not convex.

Lemma C.2 Every linear relation of the type $>$, $\ge$, $=$, $\le$, and $<$ defines a convex set.

Theorem C.3 The intersection of a finite number of convex sets is a convex set.

Since each polytope is, by definition, the intersection of a finite number hyperplanes and/or halfspaces, we can conclude that

Corollary C.4 A polytope is a convex set.

By definition, the empty set as well as a single point are also convex sets. With respect to linear relations, we will also need the following

Definition C.5 A *basic point* in $\mathbb{R}^n$ is the intersection of at least n hyperplanes at one single point; a *basic feasible point* is a basic point that satisfies all given linear relations. A basic feasible point is also called an *extreme point* of the set described by the linear relations.

As an illustration, consider Fig. C.2.

The shaded area in Fig. C.2 represents a polytope with extreme points 0, A, C, and F. The points 0, A, B, C, D, E, F, G, and H are basic points defined by the problem.

Fig. C.2 Basic points and extreme points

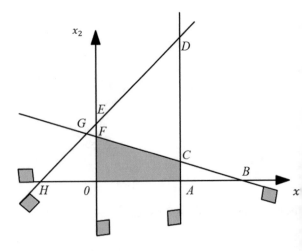

Definition C.6 The set of all points that can be expressed as linear convex combinations of extreme points is called the *convex hull* of the given extreme points.

Proposition C.7 A polyhedron is the convex hull of its extreme points.

Definition C.8 A *convex polyhedral cone* is the intersection of any number of closed halfspaces whose bordering hyperplanes intersect at a single point.

Hence, each convex polyhedral cone is an unbounded polytope with just one extreme point, called its *vertex*. One can show that each convex polyhedral cone with its vertex at the origin can be generated as the set of all nonnegative linear combinations of a finite number of given points. Algebraically, we define a cone in more general terms in

Definition C.9 A *cone* is a set C with the property that if $\mathbf{x} \in C$, then $\lambda \mathbf{x} \in C$ for all $\lambda \geq 0$.

Often, we consider cones $C \in \mathbb{R}^n$. For instance, the nonnegative orthant $\mathbb{R}^n_+$ is a cone that happens to be closed and convex. Its complement $\mathbb{R}^n \backslash \mathbb{R}^n_+$ is also a cone, but it is neither closed nor convex. In $\mathbb{R}^n$, we say that a cone C is *pointed*, if $\mathbf{x} \in C$ implies that $-\mathbf{x} \notin C$. It is apparent that $\mathbb{R}^n_+$ is a pointed cone, whereas $\mathbb{R} \backslash \mathbb{R}^n_+$ is not. One can form cones in other sets than $\mathbb{R}^n$. For instance, in the set of square matrices, the subset of positive semidefinite matrices forms a cone. This cone happens to be closed as well as convex.

Appendix D The Simplex Method for Linear Programming

This appendix will provide a very brief account of Dantzig's simplex method. While linear programming is not the subject of this book, its main solution technique (or slight deviations thereof) is used in some of the nonlinear optimization techniques described in this book. This appendix is a somewhat modified version of the material in Eiselt and Sandblom (2000).

Given an n-dimensional row vector of coefficients $\mathbf{c}$, an m-dimensional column vector $\mathbf{b}$ and an $[m \times n]$-dimensional matrix $\mathbf{A}$ as well as the variable scalar z and an n-dimensional column vector $\mathbf{x}$ of *decision variables*, then the problem

$$
\begin{array}{llll}
\text{P}: & \underset{\mathbf{x}}{\text{Max}}\ z & = & \mathbf{c}\mathbf{x} & \text{objective (function)} \\
& \text{s.t.} & \mathbf{A}\mathbf{x} \leq & \mathbf{b} & \text{(structural) constraints} \\
& & \mathbf{x} \geq & 0 & \text{nonnegativity constraints,}
\end{array}
$$

is called a *linear programming (LP) problem in canonical form*. Before we can deal with the problem, we first have to being it into *standard form*. In order to do so, we first ensure that all right hand side values $b_i \geq 0$, then add *slack variables* S_i to the left-hand sides of $\leq$ constraints, add *artificial variables* A_i (for technical reasons) to the left-hand sides of $=$ constraints, and subtract *excess* (or *surplus*) *variables* E_i

Table D.1 Transformations into standard form

Given	Transform to
$\mathbf{a}_{.i}\mathbf{x} \leq b_i$	$\mathbf{a}_{.i}\mathbf{x} + S_i = b_i,\, S_i \geq 0$
$\mathbf{a}_{.i}\mathbf{x} = b_i$	$\mathbf{a}_{.i}\mathbf{x} + A_i = b_i,\, A_i \geq 0$
$\mathbf{a}_{.i}\mathbf{x} \geq b_i$	$\mathbf{a}_{.i}\mathbf{x} - E_i + A_i = b_i,\, A_i,\, E_i \geq 0$

Table D.2 General tableau setup

x	S	E	A	1	Variables
A	I	0	0	b	original $\leq$ constraints
	0	$-$I	I		original $\geq$ constraints
		0			original equalities
$-$c	0			z_0	gof: given objective function
w	0	e	0	w_0	aof: artificial objective function

from and add artificial variables to the left-hand sides of $\geq$ constraints. Table D.1 summarizes this procedure.

The simplex method, the best-known and to this day predominant solution technique for linear programming problems, proceeds in two phases: Phase 1 starts with a so-called artificial solution and ends with either a basic feasible solution or a message that no feasible solution exists, while Phase 2 starts with some feasible basic solution (if it follows Phase 1, it starts with the solution that was obtained at the end of Phase 1) and optimizes.

Assume now that we need a Phase 1 procedure. The general setup for the tableau, in which the computations will take place, is shown in Table D.2.

In general, all variables that have a unit vector under them, are basic variables. A variable that has a unit vector under it with the "1" in the i-th position, has a present value that equals the value found on the i-the right-hand side. The values of all nonbasic variables are zero. Initially, all slack variables and all artificial variables are in the basis. In each step of the method, one basic variable leaves the basis and becomes nonbasic, while one nonbasic variable enters the basis and becomes a basic variable. Essentially, this is done by a so-called pivoting step. After a *pivot element*, say a_{rs}, is chosen (for details, see the descriptions of the Phase 1 and Phase 2 methods), the tableau transformation applies the rules shown in the *Gauss-Jordan pivoting method*. The description provides the rules for all elements, even though it states only the transformation for elements in the body of the tableau **A**, the right hand side values **b**, and the objective function **c**.

The Gauss-Jordan Pivoting Method

With a_{rs} as the pivot element, do a tableau transformation as follows:

$$a_{ij} := \begin{cases} \left. \begin{array}{l} 1 \quad \text{if } i = r \text{ and } j = s \\ 0 \quad \text{if } i \neq r \text{ and } j = s \end{array} \right\} \text{ (pivot column)} \\ \dfrac{a_{ij}}{a_{rs}} \quad \text{if } i = r \text{ and } j \neq s \quad \text{(pivot row)} \\ a_{ij} - \dfrac{a_{rj}\, a_{is}}{a_{rs}} \quad \text{if } i \neq r \text{ and } j \neq s \quad \text{(all other elements)}, \end{cases}$$

$$b_i := \begin{cases} \dfrac{b_r}{a_{rs}} \quad \text{if } i = r \quad \text{(pivot row)} \\ b_i - \dfrac{b_r\, a_{is}}{a_{rs}} \quad \text{if } i \neq r \quad \text{(all other rows)}, \end{cases}$$

$$c_j := \begin{cases} 0 \quad \text{if } j = s \quad \text{(pivot column)} \\ c_j - \dfrac{a_{rj}\, c_s}{a_{rs}} \quad \text{(all other columns)}, \text{and} \end{cases}$$

$$z_0 := z_0 - \frac{b_r\, c_s}{a_{rs}}.$$

Given these rules, we can now formally state

The Primal Simplex Algorithm: Phase 1

Step 1: Is $w_j \geq 0 \; \forall \, j$?

If yes: Go to Step 3.
If no: Go to Step 2.

Step 2: Select some $w_s < 0$; the s-th column is then the pivot column.

Select the pivot row r, so that $\dfrac{b_r}{a_{rs}} = \min\limits_{i=1,\dots,m} \left\{ \dfrac{b_i}{a_{is} : a_{is} > 0} \right\}$. The element $a_{rs} >$ 0 is the pivot. Perform a tableau transformation, using the Gauss-Jordan Pivoting Method. Go to Step 1.

Step 3: Are all artificial variables nonbasic?

If yes: Drop the artificial objective function as well as all artificial variables and their columns from the tableau and go to Phase 2 of the primal simplex algorithm.
If no: Go to Step 4.

Step 4: Is $w_0 = 0$?

If yes: The current solution is feasible. Select any pivot $a_{rs} \neq 0$, so that $b_r = 0$ and some artificial variable A_i is basic in row r. Perform a tableau transformation with the Gauss-Jordan Pivoting Method. Repeat this procedure until all artificial variables are nonbasic. (If at some point $b_r = 0$, A_s is basic in row r, and all elements in row r except $a_{rs} = 1$ are zero, drop row r and column s from the tableau.) Delete all artificial variables and their columns and the artificial objective function from the tableau and go to Phase 2 of the primal simplex algorithm.

If no: Stop, the problem has no feasible solution.

Phase 2 of the regular primal simplex algorithm starts from any basic feasible solution. We now summarize the Phase 2 method in algorithmic form.

The Primal Simplex Algorithm: Phase 2

Step 0:
Start with a tableau that represents a basic feasible solution.

Step 1:
Is $c_j \geq 0 \ \forall j = 1, \ldots, n$?

If yes: Stop; the current solution is optimal.
If no: Go to Step 2.

Step 2:
Select any nonbasic variable x_s as entering variable, so that $c_s < 0$. The s-th column is called the *pivot column*.

Step 3:
Is there any positive element in the pivot column $a_{is} > 0$, $i = 1, \ldots, m$?

If yes: Go to Step 4.
If no: Stop, unbounded optimal solutions exist.

Step 4:
Determine the r-th row as *pivot row*, i.e., find r, such that
$$\frac{b_r}{a_{rs}} = \min_{i=1,\ldots,m} \left\{ \frac{b_i}{a_{is}} : a_{is} > 0 \right\}.$$
The variable which is in the basis in the r-th row leaves the basis. The element $a_{rs} > 0$ is called the *pivot element* or simply *pivot*.

Step 5:
With a_{rs} as the pivot element, perform a tableau transformation with the Gauss Jordan Pivoting Method. Go to Step 1.

If the algorithm stops with an optimal solution, z_0 will be the optimal value of the objective function. We will now illustrate the two-phase method by means of the following

Example (The Two-Phase Simplex Method) Consider the following linear programming problem

$$P : \text{Max } z = 3x_1 + x_2$$
$$\text{s.t.} \quad 3x_1 + 2x_2 \leq 24 \quad (I)$$
$$4x_1 - x_2 \geq 8 \quad (II)$$
$$x_1 - 2x_2 = 0 \quad (III)$$
$$x_1, \quad x_2 \geq 0.$$

Adding a slack variable S_1 to the first constraint, subtracting an excess variable E_2 and adding an artificial variable A_2 to the second constraint, and adding an artificial variable A_3 to the third constraint, we obtain the initial tableau T^1.

Again, the artificial objective function coefficients are obtained by adding the coefficients in the columns in all rows, in which artificial variables are in the basis. In this example, these are the second and third rows. The meaning of this procedure becomes apparent when we rewrite the artificial variables in terms of nonbasic variables as $A_2 = 8 - 4x_1 + x_2 + E_2$ and $A_3 = -x_1 + 2x_2$, so that the artificial objective function Min $w = A_2 + A_3$ or Min $w - 8 = -5x_1 + 3x_2 + E_2$ which is exactly the artificial objective function in tableau T^1.

T^1:

	x_1	x_2	S_1	E_2	A_2	A_3	1
	3	2	1	0	0	0	24
	4	-1	0	-1	1	0	8
	(1)	-2	0	0	0	1	0
gof	-3	-1	0	0	0	0	0
aof	-5	3	0	1	0	0	-8
	↑						

Since w_1 is the only negative entry in the artificial objective function, the x_1 column must be selected as pivot column and is thus the entering variable (the fact that c_1 and $c_2 < 0$ does not matter at this stage). The minimum ratio rule then determines a_{31} as the pivot. Since x_1 enters the basis in the third row, A_3 leaves the basis and is immediately deleted. After one iteration, the new tableau is

T^2:

	x_1	x_2	S_1	E_2	A_2	1
	0	8	1	0	0	24
	0	(7)	0	-1	1	8
	1	-2	0	0	0	0
gof	0	-7	0	0	0	0
aof	0	-7	0	1	0	-8
		↑				

Now w_2 is the only negative entry in the artificial objective function, and thus x_2 is introduced into the basis with a_{22} as pivot. This means that A_2 leaves the basis and at this point, no more artificial variables are in the basis; Phase 1 has now been terminated successfully (i.e., with a feasible solution), and the artificial objective function can be dropped from the tableau. (If we had kept the A_2 and A_3 columns as well as the artificial objective function, all coefficients in the artificial objective function would be zero except those under A_2 and A_3 which would be one.)

T^3:

x_1	x_2	S_1	E_2	1
0	0	1	$\frac{8}{7}$	$\frac{104}{7}$
0	1	0	$-\frac{1}{7}$	$\frac{8}{7}$
1	0	0	$-\frac{2}{7}$	$\frac{16}{7}$
gof	0	0	0	-1	8

T^3 is also the first tableau in Phase 2 and the optimization proceeds with the usual optimality test that investigates the signs of the indicators in the given objective function row. Since $c_4 = -1 < 0$, the solution in T^3 is not yet optimal and the variable E_2 has to enter the basis. The variable enters the basis in the first row, so that the slack variable S_1 will leave the basis. After one more iteration we obtain tableau T^4 which is optimal with $\bar{\mathbf{x}} = (6, 3)$, $\bar{S}_1 = 0$, $\bar{E}_2 = 13$, and $\bar{z} = 21$.

T^4:

x_1	x_2	S_1	E_2	1
0	0	$\frac{8}{7}$	1	13
0	1	$\frac{1}{8}$	0	3
1	0	$\frac{1}{4}$	0	6
gof	0	0	$\frac{7}{8}$	0	21

References

Abadie J (1970) *Integer and nonlinear programming*. North Holland, Amsterdam

Abadie J (1978) The GRG method for nonlinear programming. In: Greenberg HJ (ed.) *Design and implementation of optimization software*. Sijthoff and Noordhoff, Netherlands, pp. 335-362

Abadie J, Carpentier J (1969) Generalization of the Wolfe reduced gradient method to the case of nonlinear constraints. In: Fletcher R (ed.) *Optimization*. Academic Press, New York, pp. 37-47

Abadie J, Guigou J (1970) Numerical experiments with the GRG method, pp. 529-536 in Abadie J (ed.) *Integer and nonlinear programming*. North Holland, Amsterdam,

Aganagić M (1984) Newton's method for linear complementarity problems. *Mathematical Programming* **28**: 349-362

Aho AY, Hopcroft JE, Ullman JD (1974) *The design and analysis of computer algorithms*. Addison-Wesley, Reading, MA

Alizadeh F, Goldfarb D (2003) Second-order cone programming. *Mathematical Programming* **95**: 3-51

Andréasson N, Evgrafov A, Patriksson M (2005) *An introduction to continuous optimization*. Studentlitteratur, Lund, Sweden

Anjos MF, Lasserre JB (2012) Introduction to semidefinite, conic, and polynomial optimization, pp 1-22 in Anjos MF, Lasserre JB (eds.) *Handbook on semidefinite, conic, and polynomial optimization*. Springer, New York

Armijo L (1966) Minimization of functions having Lipschitz continuous first partial derivatives. *Pacific Journal of Mathematics* **16**: 1-3

Arrow KJ, Hurwicz L, Uzawa H (1958) *Studies in linear and nonlinear programming*. Stanford University Press, Stanford, California

Avriel M (1976) *Nonlinear programming: analysis and methods*. Prentice-Hall, New Jersey

Avriel M (ed.) (1978a) Geometric programming issue, part I. *Journal of Optimization Theory and Applications* **26/1**

Avriel M (ed.) (1978b) Geometric programming issue, part 2. *Journal of Optimization Theory and Applications* **26/2**

Avriel M (ed.) (2013) *Advances in geometric programming*. Springer Publishers, Cham, Switzerland

Avriel M, Wilde DJ (1966) Optimality proof for the symmetric Fibonacci search technique. *Fibonacci Quarterly Journal* **4**: 265-269

Balachandran Y, Gensch DH (1974) Solving the Marketing Mix Problem Using Geometric Programming. Management Science 21, 160-171

© Springer Nature Switzerland AG 2019

H. A. Eiselt, C.-L. Sandblom, *Nonlinear Optimization*, International Series in Operations Research & Management Science 282,

https://doi.org/10.1007/978-3-030-19462-8

Balinski ML, Cottle RW (eds.) (1978) Complementarity and fixed point problems. *Mathematical Programming Study* **7**, North Holland, Amsterdam

Bard Y (1968) On a numerical instability of Davidon-like methods. *Mathematical Computation* **22**: 665-666

Bazaraa MS, Sherali HD, Shetty CM (2013) *Nonlinear programming: theory and algorithms* (3rd ed.) Wiley, New York

Beale EML (1955) On minimizing a convex function subject to linear inequalities. *Journal of the Royal Statistical Society* **17B**: 173-184

Beale EML (1959) On quadratic programming. *Naval Research Logistics Quarterly* **6**: 227-243

Beightler CS, Phillips DT (1976) *Applied geometric programming.* J. Wiley & Sons, New York

Ben-Israel A, Ben-Tal A, Zlobec S (1981) *Optimality in nonlinear programming: a feasible directions approach.* J. Wiley & Sons, New York

Ben-Israel A, Ben-Tal A, Zlobec S (1982) Optimality in convex programming: a feasible direction approach. *Mathematical Programming Study* **19**: 16-38

Ben-Israel A, Greville T (2003) *Generalized inverses.* Springer-Verlag, Berlin Heidelberg-New York

Benveniste R (1979) A quadratic programming algorithm using conjugate search directions *Mathematical Programming* **16**: 63-80

Benveniste R (1981) One way to solve the parametric quadratic programming problem. *Mathematical Programming* **21**: 224-228

Benveniste R (1982) *Constrained optimization and Lagrange multiplier methods.* Academic Press New York

Bertsekas DP (2015) *Convex optimization algorithms.* Athena Scientific, Belmont, MA

Bertsekas DP (2016) *Nonlinear programming.* (3rd ed.) Athena Scientific, Belmont, MA

Boyd S, Vandenberghe L, (2004) *Convex optimization.* Cambridge University Press, Cambridge UK

Bretscher O (2013) *Linear algebra with applications.* (5th ed.) Pearson, New York

Broise P, Huard P, Sentenac J (1968) *Décomposition des programmes mathématiques.* Dunod Paris

Broyden CG (1970) The convergence of a class of double rank minimization algorithms. Parts I and II of the *Journal for the Institute of Mathematics and its Applications* **6**: 76-90, 222-231

Chandrasekaran V, Shah P (2014) Conic geometric programming. Paper presented at the 48th Annual Conference on Information Sciences and Systems (CISS) in Princeton, NJ. Paper available via https://arxiv.org/pdf/1310.0899.pdf, last accessed on 2/10/2019

Chapman DG (1961) Statistical problems in population dynamics. *Proceedings of the Fourth Berkeley Symposium on Mathematical Statistics and Probability.* University of California Press, Berkeley and Los Angeles, pp. 153-168

Cobb CW, Douglas PH (1928) A theory of production. *American Economic Review* **18** (suppl.) 139-165

Collette Y, Siarry P (2003), *Multiobjective optimization: principles and case studies.* (2nd ed Springer, Berlin-Heidelberg-New York

Cornuejols G, Tütüncü R (2007) *Optimization methods in finance.* Cambridge University Press Cambridge, UK

Cottle RW (1963) Symmetric dual quadratic programs. *Quarterly of Applied Mathematics* **21** 237-243

Cottle RW, Dantzig GB (1968) Complementary pivot theory of mathematical programming. *Linear Algebra and its Applications* **1**: 103-125

Cottle RW, Pang JS (1978) On solving linear complementarity problems as linear programs *Mathematical Programming Study* **7**: 88-107

Cottle RW, Pang JS, Stone RE (2009) *The linear complementarity problem.* Classics in Applied Mathematics, SIAM, Philadelphia

Cottle RW, Thapa MN (2017) *Linear and nonlinear optimization.* Springer-Verlag, Berlin Heidelberg-New York

Dantzig GB, Thapa MN (2003) *Linear programming 2: theory and extensions.* Springer, New York, Berlin, Heidelberg

Daskin MS (2013) *Network and discrete location.* Wiley-Interscience, New York, NY

Davidon WC (1959) Variable metric method for minimization. *Research and Development Report* ANL-5990 (Rev.) Argonne National Laboratory, U. S. Atomic Energy Commission

Davis LS, Johnson KN, Bettinger PS, Howard TE (2001) *Forest management: to sustain ecological, economic, and social values.* 4th ed., Mc Graw-Hill

de Boor C (2001) *A practical guide to splines.* (rev. ed.) Springer-Verlag, New York

Dixon LCW (1972) *Nonlinear optimisation.* English Universities Press, London

Djang A (1980) *Algorithmic equivalence in quadratic programming.* Ph.D. Thesis, Stanford University, California

Dorn WS (1960a) Duality in quadratic programming. *Quarterly of Applied Mathematics* **18**: 155-162

Dorn WS (1960b) A duality theorem for convex programs. *IBM Journal of Research and Development* **4**: 407-413

Duffin R, Peterson E, Zener C (1967) *Geometric programming: theory and application.* J. Wiley & Sons, New York

Duffin RJ (1962) Cost minimization problems treated by geometric means. *Operations Research* **10**: 668-675

Eaves BC (1971) The linear complementarity problem. *Management Science* **17**: 612-634

Eaves BC (1978) Computing stationary points. *Mathematical Programming Study* **7**: 1-14

Eiselt HA, Sandblom C-L (2000) *Integer programming and network models.* Springer-Verlag, Berlin-Heidelberg-New York

Eiselt HA, Sandblom C-L (2004) *Decision analysis, location models, and scheduling problems.* Springer-Verlag, Berlin-Heidelberg-New York

Eiselt HA, Sandblom C-L (2007) *Linear programming and its applications.* Springer-Verlag, Berlin-Heidelberg

Eiselt HA, Sandblom C-L (2012) *Operations research: a model-based approach*, 2nd ed. Springer-Verlag, Berlin-Heidelberg-New York

Elton EJ, Gruber MJ (1981) *Modern portfolio theory and investment analysis.* J. Wiley & Sons, New York

Erlander S (1981) Entropy in linear programs. *Mathematical Programming* **21**: 137-151

Estrada J (2008) Mean-semivariance optimization: a heuristic approach. *Journal of Applied Finance*, Spring/summer: 57-72

Everett H (1963) Generalized Lagrange multiplier method for solving problems of optimum allocation of resources. *Operations Research* **11**: 399-417

Feist AM, Palsson BO (2010) The biomass objective function. *Current Opinion in Microbiology* **13**: 344-349

Fenchel W (1951) *Convex cones, sets and functions.* Lecture Notes, Princeton University, Princeton, NJ

Fiacco AV, McCormick GP (1964a) Computational algorithm for the sequential unconstrained minimization technique for nonlinear programming. *Management Science* **10**: 601-617

Fiacco AV, McCormick GP (1964b) The sequential unconstrained minimization technique for nonlinear programming-a primal-dual method. *Management Science* **10**: 360-366

Fiacco AV, McCormick GP (1966) Extensions of SUMT for nonlinear programming equality constraints and extrapolation. *Management Science* **12**: 816-828

Fiacco AV, McCormick GP (1967a) The sequential unconstrained minimization technique (SUMT) without parameters. *Operations Research* **15**: 820-827

Fiacco AV, McCormick GP (1967b) The slacked unconstrained minimization technique for convex programming. *The SIAM Journal of Applied Mathematics* **15**: 505-515

Fiacco AV, McCormick GP (1968) *Nonlinear programming.* J. Wiley & Sons, New York

Fiacco AV (1979): Barrier methods for nonlinear programming, pp. 377-400 in: Holzman, AG (ed.) *Operations research support methodology.* Marcel Dekker, New York

Fletcher R (1970) A new approach to variable metric algorithms. *The Computer Journal* **13** 317-322

Fletcher R (1971) A general quadratic programming algorithm. *Journal for the Institute of Mathematics and its Applications* **7**: 76-91

Fletcher R (1983) Penalty Functions, pp. 87-114 in Bachem A, Grötschel M, Korte B (eds.) *Mathematical programming-the state of the art.* Springer, Berlin-Heidelberg-New York

Fletcher R (2000) *Practical methods of optimization* (2nd ed.). J. Wiley & Sons, New York

Fletcher R (ed.) (1969) *Optimization.* Academic Press, New York

Fletcher R, Powell MJD (1963) A rapidly convergent descent method for minimization. *The Computer Journal* **6**: 163-168

Fletcher R, Reeves CM (1964) Function minimization by conjugate gradients. *The Computer Journal* **7**: 149-154

Floudas CA (2000) Deterministic global optimization. Theory, methods, and applications. Springer Science + Business Media, Dordrecht

Floudas CA, Pardalos PM (eds.) (2003) *Frontiers in global optimization.* Kluwer Academic Publishers, Dordrecht

Floudas CA, Visweswaran V (1995) Quadratic optimization. In: Horst R, Pardalos PM (eds.) *Handbook of global optimization.* Kluwer, Boston, MA

Foulds LR (1981) *Optimization techniques-an introduction.* Springer, Berlin Heidelberg-New York

Francis JC, Archer SH (1979) *Portfolio analysis* (2nd ed.), Prentice-Hall, Englewood Cliffs, NJ

Frank M, Wolfe P (1956) An algorithm for quadratic programming. *Naval Research Logistics Quarterly* **3**: 95-110

Frisch KR (1956) La résolution des problèmes de programme lineaire pour la méthode du potentia logarithmique. *Cahiers du seminaire d'Économetrie* **4**: 7-20

Fröberg C-E (1969) *Introduction to numerical analysis* (2nd ed.) Addison-Wesley, Reading, MA

Fukushima M (1984) A descent algorithm for nonsmooth convex optimization. *Mathematical Programming* **30**: 163-175

Fuller R, Carlsson C (1996) Fuzzy multiple criteria decision making: recent developments. *Fuzzy Sets and Systems* **78**: 139-153

Gale D, Kuhn HW, Tucker AW (1951) Linear programming and the theory of games, pp 317-329 ii Koopmans TC (ed.) Activity analysis of production and allocation. *Cowles Commission mono graph* **13**: 317-329

Garey MR, Johnson DS (1979) *Computers and intractability: a guide to the theory o NP-completeness.* W. H. Freeman and Co., San Francisco

Geoffrion AM (1968) Proper efficiency and the theory of vector optimization. *Journal of Mathe matical Analysis and Applications* **22**: 618-630

Geoffrion AM (1970a) Elements of large-scale mathematical programming. *Management Science* **16**: 652-691

Geoffrion AM (1970b) Primal resource-directive approaches for optimizing nonlinear decompos able systems. *Operations Research* **18**: 375-403

Geoffrion AM (1971) Duality in nonlinear programming: a simplified applications-oriented devel opment. *SIAM Review* **13** 1-37

Geoffrion AM (1972) Generalized Benders decomposition. *Journal of Optimization Theory an Applications* **10**: 237-260

Ghosh A, Buchanan B (1988) Multiple outlets in a duopoly: a first entry paradox. *Geographica Analysis* **20**: 111–121

Giese RF, Jones PC (1984) An economic model of short-rotation forestry. *Mathematical Program ming* **28**: 206-217

Gill PE, Murray W (1978) Numerically stable methods for quadratic Programming. *Mathematica Programming* **14**: 349-372

Gill PE, Murray W, Wright MH (1981) *Practical optimization.* Academic Press, New York

Goffin JL (1977) On the convergence rates of subgradient optimization methods. *Mathematical Programming* **13**: 329-347

Goffin JL (1984) Variable metric relaxation methods, part II: the ellipsoid method. *Mathematical Programming* **30**: 147-162

Goldfarb D (1969a) Extension of Davidon's variable metric method to maximization under linear inequality constraints. *SIAM Journal on Applied Mathematics* **17**: 739-764

Goldfarb D (1969b) Sufficient conditions for the convergence of a variable metric algorithm, pp. 273-281 in: Fletcher R (ed.) *Optimization*. Academic Press, New York

Goldfarb D (1970) A family of variable metric methods derived by variational means. *Mathematics of Computation* **24**: 23-26

Goldfarb D (1972) Extensions of Newton's method and simplex methods for solving quadratic programs, pp. 239-254 in Lootsma FA (ed.) *Numerical methods for non-linear optimization*. Academic Press, New York

Goldfarb D, Idnani A (1983) A numerically stable dual method for solving strictly convex quadratic programs. *Mathematical Programming* **27**: 1-33

Goldman AJ, Dearing PM (1975) Concepts of optimal location for partially noxious facilities. *Bulletin of the Operations Research Society of America* **23**, Supplement 1: 331

Goldstein AA (1965) On steepest descent. *SIAM Journal on Control* **3**: 147-151.

Graupe D (2013, 3rd ed.) *Principles of artificial neural networks*. World Scientific, NJ

Graves RL, Wolfe P (eds.) (1963) *Recent advances in mathematical programming*. McGraw-Hill, New York

Griffith RE, Stewart RA (1961) A nonlinear programming technique for the optimization of continuous processing systems. *Management Science* **7**: 379-392

Grünbaum B (1967) *Convex polytopes*. J. Wiley & Sons, New York

Hadley G (1964) *Nonlinear and dynamic programming*. Addison-Wesley, Reading, MA

Hancock H (1917) *Theory of maxima and minima*. Ginn, Boston, MA

Hansen P, Jaumard B (1995) Lipschitz optimization, pp 407-493 in Horst R, Pardalos PM (eds.) *Handbook of global optimization*. Kluwer, Boston, MA

Held M, Wolfe P, Crowder HP (1974) Validation of subgradient optimization. *Mathematical Programming* **6**: 62-88

Hellinckx LJ, Rijckaert MJ (1972) Optimal capacities of production facilities: an application of geometric programming. *Canadian Journal of Chemical Engineering* **50**: 148-150

Hendrix EMT, G.-Tóth B (2010) *Introduction to nonlinear and global optimization*. Springer-Verlag, Berlin-Heidelberg-New York

Hillier FS, Lieberman GJ (1972) A uniform evaluation of unconstrained optimization techniques, pp. 69-97 in Lootsma FA (ed.) *Numerical methods for non-linear optimization*. Academic Press, New York

Hillier FS, Lieberman GJ, Nag B, Basu P (2017) *Introduction to operations research*. (10th ed.), McGraw Hill India

Himmelblau DM (1972) *Applied Nonlinear Programming*. McGraw-Hill, New York

Himmelblau DM (ed.) (1973) *Decomposition of large-scale problems*. North-Holland, Amsterdam

Hitchcock FL (1941) The distribution of a product from several sources to numerous localities. *Journal of Mathematical Physics* **20**: 224-230

Holloway CA (1974) An extension of the Frank and Wolfe method of feasible directions. *Mathematical Programming* **6**: 14-27

Horst R, Pardalos PM (eds.) (1995) *Handbook of global optimization*. Kluwer, Boston, MA

Horst R, Pardalos PM, Thoai NV (2000) *Introduction to global optimization*, vol. 2. Kluwer, Dordrecht, The Netherlands

Hwang CL, Yoon K (1981) *Multiple attribute decision making: methods and applications*. Springer-Verlag, New York

Intriligator MD (1971) *Mathematical optimization and economics theory*. Prentice-Hall, Englewood Cliffs, NJ

Jahn J (2007) *Introduction to the theory of nonlinear optimization* (3rd ed.), Springer-Verlag, Berlin-Heidelberg-New York

Janssen JML, Pau LF, Straszak A (eds.) (1981) *Dynamic modelling and control of national economies*. Pergamon Press, Oxford

John F (1948) Extremum problems with inequalities as side constraints, pp. 187-204 in: Friedrichs KO, Neugebauer OE, Stoker JJ (eds.) *Studies and Essays, Courant Anniversary Volume*. Wiley Interscience, New York

Jones PC (1983) A note on the Talman, van der Heyden linear complementarity algorithm. *Mathematical Programming* **25**: 122-124

Jörnsten KO, Sandblom C-L (1985) Optimization of an economic system using nonlinear decomposition. *Journal of Information and Optimization Sciences* **6**: 17-40

Jörnsten KO, Sandblom C-L (1984) A nonlinear economic model with bounded controls and an entropy objective, pp. 123-134 in Thoft-Christensen P (ed) *System Modelling and Optimization* Vol. 59 in System modelling and optimization. Lecture Notes in Control and Information Sciences. Springer, Berlin, Heidelberg

Karamardian S (1969a) The nonlinear complementarity problem with applications, part I. *Journal of Optimization Theory and Applications* **4**: 87-98

Karamardian S (1969b) The nonlinear complementarity Problem with Applications, part II. *Journal of Optimization Theory and Applications* **4**: 167-181

Karamardian S (1972) The complementarity problem. *Mathematical Programming* **2**: 107-129

Karlin S (1959) *Mathematical methods and theory in games, programming and economics*. Vols. and II. Addison-Wesley, Reading, MA

Karush W (1939): *Minima of functions of several variables with inequalities as side conditions*. MS Thesis, Department of Mathematics, University of Chicago, IL

Kelley JE (1960) The cutting-plane method for solving convex programs. *SIAM Journal* **8**: 703-712

Kiefer J (1953) Sequential minimax search for a maximum. *Proceedings of the American Mathematical Society* **4**: 502-506

Kiwiel, KC (1983) An aggregate subgradient method for nonsmooth convex minimization. *Mathematical Programming* **27**: 320-341

Klein R (1989) *Concrete and abstract Voronoi diagrams*. Lecture Notes in Computer Science 400 Springer-Verlag, Berlin-New York

Kuhn HW (1976) Nonlinear Programming: A Historical View. In: Cottle RW, Lemke CE (eds. *Nonlinear Programming*. American Mathematical Society, Providence, RI

Kuhn HW, Tucker AW (1951) Nonlinear programming, pp. 481-492 in Neyman J (ed.) *Proceedings of the Second Berkeley Symposium on Mathematical Statistics and Probability* University of California Press, Berkeley, CA

Künzi HP, Krelle W, Oettli W (1966) *Nonlinear programming*. Blaisdell Publ. Co., MA

Kutner M, Nachtsheim C, Neter J, Li W (2004) *Applied linear statistical models*. (5th ed.) McGraw Hill/Irwin, New York

Laffer A (2004) The Laffer curve: past, present, and future. Available online at https://www.heritage.org/taxes/report/the-laffer-curve-past-present-and-future. Last accessed 2/10/2019

Lakhera S, Shanbhag UV, McInerney MK (2011) Approximating electrical distribution network via mixed-integer nonlinear programming. *Electrical Power and Energy Systems* **33**: 245-25

Larsson T, Smeds PA (eds.) (1983) *Proceedings of a Nordic symposium on linear complementarity problems and related areas*. Linköping University, Sweden

Lasdon LS (1970) *Optimization theory for large systems*. MacMillan, New York

Lasdon LS, Waren AD (1978) Generalized reduced gradient software for linearly and nonlinearly constrained problems. pp 363-396 in Greenberg H (ed.) *Design and implementation of optimization software*. Sijthoff and Noordhoff, The Netherlands

Lasdon LS, Waren AD (1980) Survey of nonlinear programming applications. *Operations Research* **28**: 1029-1073

Lemarechal C (1975) An extension of Davidon methods to nondifferentiable problems. *Mathematical Programming Study* **3**: 95-109

Lemarechal C, Mifflin R (eds.) (1978) *Nonsmooth optimization*. Pergamon Press, Oxford

Lemke CE (1954) The dual method of solving the linear programming problem. *Naval Research Logistics Quarterly* **1**: 36-47

Lemke CE (1968) On complementary pivot theory, pp. 95-114 in Dantzig GB, Veinott AF (eds.) *Mathematics of the Decision Sciences, Part 1*. American Mathematical Society, Providence, RI

Lenstra JK, Rinnooy Kan AHG, Schrijver A (1991) *History of Mathematical Programming: A Collection of Personal Reminiscences*. CWI–North Holland, Amsterdam

Levin AY (1965) On an algorithm for the minimization of convex functions. *Soviet Mathematics Doklady* **6**, 286-290

Lin M-H, Tsai J-F (2012) Range reduction techniques for improving computational efficiency in global optimization of signomial geometric programming problems. *European Journal of Operational Research* **216/1**: 17-25

Lin M-H, Tsai J-F, Yul C-S (2012) A review of deterministic optimization methods in engineering and management. *Mathematical Problems in Engineering*, Article ID 756023, available online via https://www.hindawi.com/journals/mpe/2012/756023/, last accessed 2/10/2019

Lootsma FA (ed.) (1972) *Numerical methods for non-linear optimization*. Academic Press, New York

Luenberger DL (1979) *Introduction to dynamic systems: theory, models and applications*. Wiley & Sons, New York

Luenberger DL, Ye Y (2008) *Linear and nonlinear programming*. (3rd ed.) Springer-Verlag, Berlin-Heidelberg-New York

Luptáčik M (1977) *Geometrische Programmierung und ökonomische Analyse*. Vol. 32 in Mathematical Systems in Economics. Anton Hain, Meisenheim am Glan

Luptáčik M (1981a) Geometric programming, methods and applications. *OR Spektrum* **2**:129-143

Luptáčik M (1981b) *Nichtlineare Programmierung mit ökonomischen Anwendungen*. Athenäum, Königstein

Magnanti TL (1974) Fenchel und Lagrange duality are equivalent. *Mathematical Programming* **7**: 253-258

Mangasarian OL (1969) *Nonlinear programming*. McGraw-Hill, New York

Mangasarian OL (1976) Linear complementarity problems solvable by a single linear program. *Mathematical Programming* **10**: 263-270

Markowitz HM (1952) Portfolio selection. *Journal of Finance* **7**: 77-91

Markowitz HM (1959) *Portfolio selection: efficient diversification of investments*. Cowles Foundation Monographs #16, Wiley, New York

McCormick GP (1983) *Nonlinear programming: theory, algorithms, and applications*. J. Wiley & Sons, New York

Michalewicz Z, Foge DB (2004) *How to solve it: modern heuristics*. (2nd ed.) Springer-Verlag, Berlin-Heidelberg-New York

Mifflin R (1982) A modification and an extension of Lemarechal's algorithm for nonsmooth minimization. *Mathematical Programming Study* **17**: 77-90

Minkowski H (1904) Dichteste gitterförmige Lagerung kongruenter Körper. *Nachrichten von der Gesellschaft der Wissenschaften zu Göttingen, Mathematisch-Physikalische Klasse II* 311–355

Minoux M (1983) *Programmation mathématique: théorie et algorithmes*. Volumes 1 and 2, Dunod, Paris

Misra S, Fisher MW, Backhaus S, Bent R, Chertkov M, Pan F (2015) Optimal compression in natural gas networks: A geometric programming approach. *IEEE Transactions on Control of Network Systems* **2/1**: 47-56

Molzahn DK, Dörfler F, Sandberg H, Low SH, Chakrabarti S, Baldick R, Lavaei J (2017) A survey of distributed optimization and control algorithms for electric power systems. *IEEE Transactions on Smart Grid* **8/6**: 2941-2962

Momoh JA (2017, 2nd ed.) *Electric power system applications of optimization*. CRC Press, Boc Raton, FL

Moore EH (1935) *General analysis, part I*. Memoranda of the American Philosophical Society Vol. I

Nelder JA, Mead R (1965) A simplex method for function minimization. *The Computer Journal* 7 308-313

Nemirovski A (2007) Advances in convex optimization: conic programming. *Proceedings of the International Congress of Mathematics*, Madrid 2006. European Mathematical Society 413-444

Neyman J (ed.) (1951) *Proceedings of the Second Berkeley Symposium on Mathematical Statistic and Probability*. University of California Press, Berkeley, CA

Nicholson WK (2013) Linear algebra with applications. (7th ed.) McGraw-Hill Ryerson, New Yor

Nocedal J, Wright SJ (2006) *Numerical Optimization*. (2nd ed.) Springer-Verlag, Berlir Heidelberg-New York

Ogura M, Preciado VM (2017) Optimal design of switched networks of positive linear systems vi geometric programming. *IEEE Transactions on Control of Network Systems* **4/2**: 213-222

Ojha AK, Biswal KK (2010) Multi-objective geometric programming problem with weighted mea method. *International Journal of Computer Science and Information Security* **7/2**. Available online at https://arxiv.org/ftp/arxiv/papers/1003/1003.1477.pdf, last accessed on 2/10/2019

Okabe A, Boots B, Sugihara K, Chiu S-N (2000) *Spatial tessellations: concepts and applications of Voronoi diagrams*. (2nd ed.) Wiley, Chichester

Pardalos RM, Romeijn HE (eds.) (2002) *Handbook of global optimization*, vol. 2. Springer Scienc + Business Media Dordrecht

Parkinson JM, Hutchinson D (1972) A consideration of non-gradient algorithms for th unconstrained optimization of functions of high dimensionality, pp. 99-113 in Lootsma F. (ed.) *Numerical methods for non-linear optimization*. Academic Press, New York

Pearson JD (1969) Variable metric methods of minimization. *The Computer Journal* **12**: 171-17

Penot J-P (1982) On regularity conditions in mathematical programming. *Mathematical Program ming Study* **19**: 167-199

Penrose R (1955) A generalized inverse for matrices. *Proceedings of the Cambridge Philosophic Society* **51**: 406-413

Peterson EL (1978) Geometric programming, pp. 207-244 in Moder JP, Elmaghraby SE (eds Handbook of operations research, vol. 2. Van Nostrand Reinhold, New York

Pintér JD (1996) *Global optimization in action*. Kluwer, Boston, MA

Polyak BT (1967) A general method of solving extremum problems. *Soviet Mathematics Doklad* **8**: 593-597 (translated from the Russian)

Polyak BT (1969) Minimization of unsmooth functionals. *USSR Computational Mathematics an Mathematical Physics* **9**: 509-521 (translated from the Russian)

Ponstein J (1983) Comments on the general duality survey by J Tind and LA Wolsey. *Mathematic Programming* **25**: 240-244

Powell MJD (1964) An efficient method for finding the minimum of function of several variable without calculating derivatives. *The Computer Journal* **7**: 155-162

Powell MJD (1981) An example of cycling in a feasible point algorithm. *Mathematical Program ming* **20**: 353-357

Preciado VM, Zargham M, Enyioha C, Jadbabaie A, Pappas G (2013) Optimal resource allocatio for network protection: a geometric programming approach. Available online at https georgejpappas.org/papers/1309.6270.pdf, last accessed on 2/10/2019

Rao SS (1979) *Optimization theory and applications*. Wiley Eastern, New Delhi, India

Ratner M, Lasdon LS, Jain A (1978) Solving geometric programs using GRG: results an comparisons. *Journal of Optimization Theory and Applications* **26**: 253-264

Ravindran A (1973) A comparison of primal simplex and complementary pivot methods for linea programming. *Naval Research Logistics Quarterly* **20**: 95-100

Richards FJ (1959) A flexible growth function for empirical use. *Journal of Experimental Botany* **10/2**: 290–300

Rijckaert MJ, Martens XM (1978) Bibliographical note on geometric programming. *Journal of Optimization Theory and Applications* **26**: 325-337

Robinson SM (1974) Perturbed Kuhn-Tucker points and rates of convergence for a class of nonlinear programming algorithms. *Mathematical Programming* **7**: 1-16

Rockafellar RT (1970) *Convex analysis.* University Press, Princeton, NJ

Rockafellar RT (1981) *The theory of subgradients and its applications to problems in optimization: convex and nonconvex functions.* Heldermann Verlag, Berlin

Rockafellar RT (1983) Generalized subgradients in mathematical programming, pp. 368-390 in Bachem A, Grötschel M, Korte B (eds.) *Mathematical programming: the state of the art.* Springer, Berlin-Heidelberg-New York

Rönnquist M (2003) Optimization in forestry. *Mathematical Programming* **97**: 267-284

Rosen JB (1960) The gradient projection method for non-linear programming: part I, linear constraints. *Journal of SIAM* **8**: 181-217

Rosen JB (1961) The gradient projection method for non-linear programming: part II. *Journal of SIAM* **9**: 514-532

Sandblom C-L (1974) On the convergence of SUMT. *Mathematical Programming* **6**: 360-364

Sandblom C-L (1984) Optimizing economic policy with sliding windows. *Applied Economics* **6**: 45-56

Sandblom C-L (1985) Economic policy with bounded controls. *Economic Modelling* **2**: 135-148

Sandi C (1979) Subgradient optimization, pp. 73-91 in Christofides N, Mingozzi A, Toth P (eds.) *Combinatorial optimization.* J. Wiley & Sons, New York

Sargent RWH, Sebastian DJ (1972) Numerical experience with algorithms for unconstrained minimization, pp. 45-68 in Lootsma FA (ed.) *Numerical methods for non-linear optimization.* Academic Press, New York

Sarma PVLN, Martens XM, Reklaitis G, Rijckaert MJ (1978) Comparison of computational strategies for geometric programs. *Journal of Optimization Theory and Applications* **26**: 185-203

Schaible S, Ziemba NT (eds.) (1981) *Generalized concavity in optimization and economics.* Academic Press, New York

Schnute J (1981) A versatile growth model with statistically stable parameters. *Canadian Journal of Fisheries Aquatic Sciences* **38**: 1128-1140

Scott C, Fang S-C (eds.) (2002) Geometric programming. Vol 105 of the *Annals of Operations Research*

Shamos MI, Hoey D (1975) Closest-point problems. *Proceedings of the 16th Annual IEEE Symposium on Foundations of Computer Science*: 151-162

Shanno DF (1970) Conditioning of quasi-Newton methods for function minimization. *Mathematics of Computation* **24**: 647-656

Shapiro JF (1979) *Mathematical programming: structures and algorithms.* J. Wiley & Sons, New York

Shetty CM (1963) A simplified procedure for quadratic programming. *Operations Research* **11**: 248-260

Shor NZ (1983) Generalized gradient methods of nondifferentiable optimization employing space dilatation operations, pp. 501-529 in Bachem A, Grötschel M, Korte B (eds.) *Mathematical programming, the state of the art.* Springer, Berlin-Heidelberg-New York

Shor Z (1964) *On the structure of algorithms for the numerical solution of optimal planning and design problems.* Dissertation, Cybernetics Institute, Kiev, Ukraine

Slater M (1950) *Lagrange multipliers revisited: a contribution to nonlinear programming.* Cowles Commission Discussion Paper, Mathematics 403, Yale University, New Haven, CT

Sorenson DC, Wets RJ-B (eds.) (1982) Nondifferential and variational techniques in optimization. *Mathematical Programming Study* **17**

Spendley W, Hext GR, Himsworth FR (1962) Sequential application of simplex designs i optimization and evolutionary operation. *Technometrics* **4**: 441-461

Stackelberg H von (1943) *Grundlagen der theoretischen Volkswirtschaftslehre* (translated as: *Th Theory of the Market Economy*), W. Hodge & Co. Ltd., London, 1952

Sun W, Yuan Y (2006) *Optimization theory and methods. Nonlinear programming.* Springer Verlag, Berlin-Heidelberg-New York

Svanberg K (1982) An algorithm for optimum structural design using duality. *Mathematic Programming Study* **20**: 161-177

Talman D, Van der Heyden L (1981) *Algorithms for the linear complementarity problem whic allow an arbitrary starting point.* Cowles Foundation Discussion Paper No. 600. Yale Univer sity, New Haven, CT

Tind J, Wolsey LA (1979) A unifying framework for duality theory in mathematical programming *CORE Discussion Paper* **7834** (Revised)

Tind J, Wolsey LA (1981) An elementary survey of general duality theory in mathematic programming. *Mathematical Programming* **21**: 241-261

Topkis DM, Veinott AF (1967) On the convergence of some feasible direction algorithms fc nonlinear programming. *SIAM Journal on Control* **5/2**: 268-279

Toussaint GT (1983) Computing largest empty circles with location constraints. *Internationc Journal of Computer and Information Sciences* **12/5**: 347-358

Toussaint M (2015) Logic-geometric programming: an optimization-based approach to combine task and motion planning. IJCAI'15 *Proceedings of the 24th International Conference o Artificial Intelligence 1930-1936.* Available online at http://ipvs.informatik.uni-stuttgart.d mlr/papers/15-toussaint-IJCAI.pdf, last accessed on 2/10/2019

Tuy H (1995) *D.C. optimization theory, methods and algorithms.* In Horst R, Pardalos P (eds *Handbook of global optimization.* Kluwer, Boston, MA

Tuy H (1998) *Convex analysis and global optimization.* Kluwer, Boston, MA

Tuy H (2005) Polynomial optimization: a robust approach. *Pacific Journal of Optimization* **1** 257-274

Tuy H, Al-Khayyal F, Thach PT (2005) Monotonic optimization: branch-and-cut methods. I Audet C, Hansen P, Savard G (eds.) *Surveys in global optimization.* Springer, Berli Heidelberg-New York

Vajda S (1961) *Mathematical programming.* Addison-Wesley, Reading, MA

Vajda S (1974) Tests of optimality in constrained optimization. *Journal of the Institute f Mathematics and its Applications* **13**: 187-200

Van de Panne C (1975) *Methods for linear and quadratic programming.* North Hollan Amsterdam

Vandenberghe L, Boyd S (1996) Semidefinite programming. *SIAM Review* **38**: 49-95

von Bertalanffy L (1951) *Theoretische Biologie.* (Band II). Franke, Bern

Wanniski J (1978) Taxes, revenues, and the "Laffer curve." *National Affairs* **50**: 3-16, Winter 197 Available online at https://nationalaffairs.com/public_interest/detail/taxes-revenues-and-th laffer-curve, last accessed on 2/10/2019

Wierzbicki AP (1982) A mathematical basis for satisficing decision making. *Mathematical Mode ling* **3**: 391–405

Wilde DJ (1967) *Foundations of optimization.* Prentice-Hall, Englewood Cliffs, NJ

Williams HP.(1978) *Model building and mathematical programming.* J. Wiley & Sons, New Yor

Willis HL, Tram H, Engel MV, Finley L (1995) Optimization applications to power distributio *IEEE Computer Applications in Power* **10**: 12-17

Wismer DA (ed.) (1971) *Optimization methods for large-scale systems.* McGraw-Hill, New Yor

Wolfe P (1959) The simplex method for quadratic programming. *Econometrica* **27**: 382-398

Wolfe P (1961) A duality theorem for non-linear programming. *Quarterly of Applied Mathemati* **19**: 239-244

Wolfe P (1962) Recent developments in nonlinear programming. *Advances in Computers* **3**: 155-187, Academic Press, New York

Wolfe P (1963) Methods of nonlinear programming, pp. 67-86 in Graves RL, Wolfe P (eds.) *Recent advances in mathematical programming*. McGraw Hill, New York

Wolfe P (1969) Convergence conditions for ascent methods. *SIAM Review* **11**, 226-235

Wolfe P (1970) Convergence theory in nonlinear programming, pp. 1-36 in Abadie J (ed.) *Integer and nonlinear programming*. North Holland, Amsterdam

Wolfe P (1972) On the convergence of gradient methods under constraints. *IBM Journal of Research and Development* **16**: 407-411

Wolfe P (1975) A method of conjugate subgradients for minimizing nondifferentiable functions. *Mathematical Programming Study* **3**: 145-173

Wolkowicz H, Saigal R, Vanderberghe L (eds.) (2000) *Handbook of semidefinite programming*. Kluwer, Boston, MA

Woolsey RED, Swanson HS (1975) *Operations research for immediate application-a quick and dirty manual*. Harper & Row, New York

Zhu JZ (2009) *Optimization of power systems operations*. Wiley-IEEE Press

Index

© Springer Nature Switzerland AG 2019
H. A. Eiselt, C.-L. Sandblom, *Nonlinear Optimization*, International Series in Operations Research & Management Science 282,
https://doi.org/10.1007/978-3-030-19462-8